endorsed for edexcel

Active Book included

Edexcel AS/A level Geography

Series editor:
Lindsay Fros[t]
Paul Wraight

Book 1

ALWAYS LEARNING

PEARSON

Published by Pearson Education Limited, 80 Strand, London, WC2R 0RL.

www.pearsonschoolsandfecolleges.co.uk

Copies of official specifications for all Edexcel qualifications may be found on the website: www.edexcel.com

Designed by Elizabeth Arnoux for Pearson

Typeset, illustrated and produced by Phoenix Photosetting, Chatham, Kent

Cover design by Elizabeth Arnoux for Pearson

Picture research by Rebecca Sodergren

Cover photo/illustration © Poorfish / Getty Images

First published 2016

19 18 17 16

10 9 8 7 6 5 4 3 2

British Library Cataloguing in Publication Data

A catalogue record for this book is available from the British Library

ISBN 978 1 292 13960 9

Printed in Slovakia by Neografia

Websites

Pearson Education Limited is not responsible for the content of any external internet sites. It is essential for tutors to preview each website before using it in class so as to ensure that the URL is still accurate, relevant and appropriate. We suggest that tutors bookmark useful websites and consider enabling students to access them through the school/college intranet.

A note from the publisher

In order to ensure that this resource offers high-quality support for the associated Pearson qualification, it has been through a review process by the awarding body. This process confirms that this resource fully covers the teaching and learning content of the specification or part of a specification at which it is aimed. It also confirms that it demonstrates an appropriate balance between the development of subject skills, knowledge and understanding, in addition to preparation for assessment.

Endorsement does not cover any guidance on assessment activities or processes (e.g. practice questions or advice on how to answer assessment questions), included in the resource nor does it prescribe any particular approach to the teaching or delivery of a related course.

While the publishers have made every attempt to ensure that advice on the qualification and its assessment is accurate, the official specification and associated assessment guidance materials are the only authoritative source of information and should always be referred to for definitive guidance.

Pearson examiners have not contributed to any sections in this resource relevant to examination papers for which they have responsibility.

Examiners will not use endorsed resources as a source of material for any assessment set by Pearson.

Endorsement of a resource does not mean that the resource is required to achieve this Pearson qualification, nor does it mean that it is the only suitable material available to support the qualification, and any resource lists produced by the awarding body shall include this and other appropriate resources.

Contents

How to use this book

Introductory pages

The introductory pages outline the objectives of the topic and the skills that are covered within it. There is also a paragraph about synoptic links, which is a theme running throughout the specification. If you have studied geography at KS3 or GCSE level there is a reminder of the content and concepts that you should know linked to the AS/A level topic.

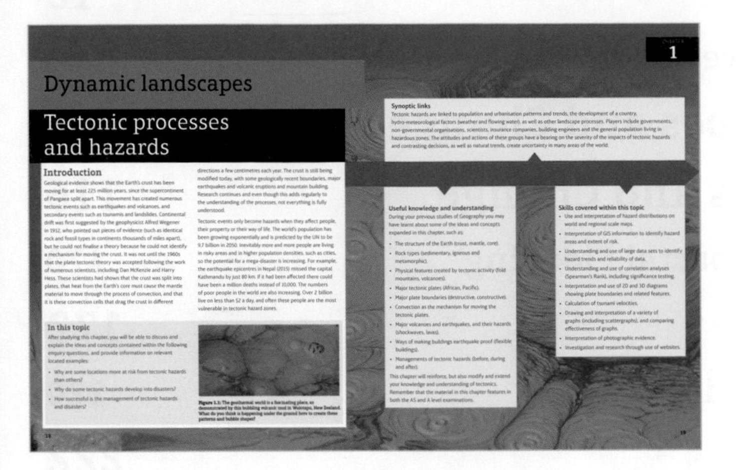

1

Dynamic landscapes

Tectonic processes and hazards

Introduction

Synoptic links

Useful knowledge and understanding

Skills covered within this topic

In this topic

Key terms and glossary

The key terms you need to know are highlighted for you in bold and there is an extensive glossary at the back of the book to help with your revision.

Activities and extension

Frequent activities contain academic questions to reinforce your learning and help you to develop and apply numerical, statistical, cartographic, graphical, ICT and literacy skills in geographical contexts. Extension activities will also stretch and challenge your knowledge.

ACTIVITY

TECHNOLOGY/ICT SKILLS

Investigate the aims and actions of a UK coastal ICZM, perhaps an area in which you have undertaken fieldwork. Assess how successful the ICZM is for managing the coastal zone sustainably.

Extension 

The Shoreline Management Plans for England and Wales are on the Environment Agency website, and often on local council websites. Investigate the SMP for your fieldwork area and examine the local coastal processes, land use, defences and plans, and the reasons behind the SMP.

Exam support

There are AS and A Level style exam style questions with guidance on every enquiry question, providing practice for your exams.

AS level exam-style question

With reference to earthquake (seismic) waves, explain two reasons why it is difficult for buildings to remain intact during an earthquake event. (4 marks)

Guidance

Each type of earthquake (seismic) wave moves through the ground differently, and all the waves arrive within seconds of each other. Anything attached to the ground would also move.

A level exam-style question

Explain the link between plate boundary type and the strength of earthquake (seismic) waves. (4 marks)

Guidance

Think about which types of plate boundary create the most strain and release the greatest amount of seismic energy and why, and which release very little and why. Make reference to body and surface waves.

There are focussed exam pages at the end of every topic with sample questions, sample answers and guidance for both AS and A level.

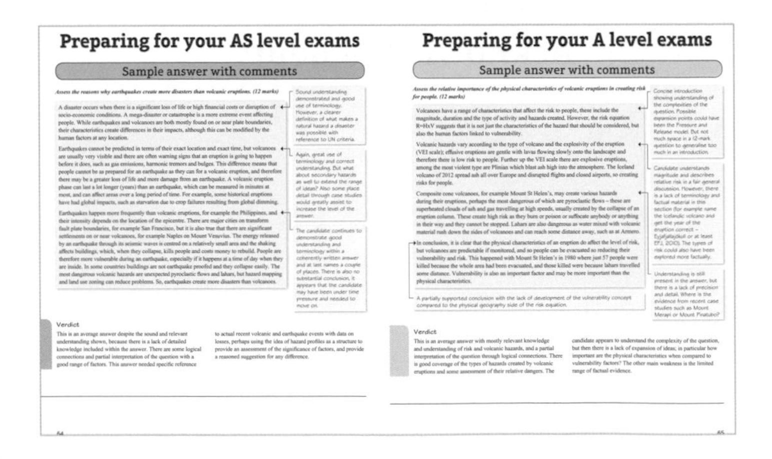

Preparing for your AS level exams

Sample answer with comments

Verdict

Preparing for your A level exams

Sample answer with comments

Verdict

Synoptic links

The synoptic links from the specification are highlighted in the text and there are developed explanations of these synoptic links throughout to help you make links of your own.

Synoptic link 

One effect of globalisation is increased migration, especially of refugees. While environmental refugees may only be a small part of this process at the moment, the global link between countries is likely to become much more important in the future, as more people are displaced by sea-level rise and coastal storms. This will also affect sovereignty, for example, if whole island communities move to another country; what happens to territorial waters if the land is flooded? (See page 207.)

The key and an example of the text highlighted

Such criteria-based ethical and environmental ratings have subsequently become commonplace for business-to-business corporate social responsibility and sustainability ratings, such as those provided by Innovest, Calvert and Domini. Businesses have become aware of the importance of ethical considerations, and increasingly present themselves to their consumers as morally and environmentally aware. For example, Marks & Spencer announced their 'Plan A' in 2007. This set out 100 commitments to source responsibly, reduce waste and help communities over five years. To support their goal of becoming the world's most sustainable retailer Marks & Spencer have launched their Plan A 2020. The plan combines 100 existing commitments with revised and new ones.

PAF themes

Yellow = Players (P), Orange = Attitudes and actions (A), Purple = Futures and uncertainties (F)

Maths and literacy tips

We have included maths and literacy tips in the topics to support your learning. There are also answers to the statistical questions at the back of the book so that you can check your learning.

Thinking synoptically pages

Our thinking synoptically pages provide interesting material to take you beyond your studies, linking topics together and giving you an opportunity to apply your skills and knowledge and prepare for synoptic assessment.

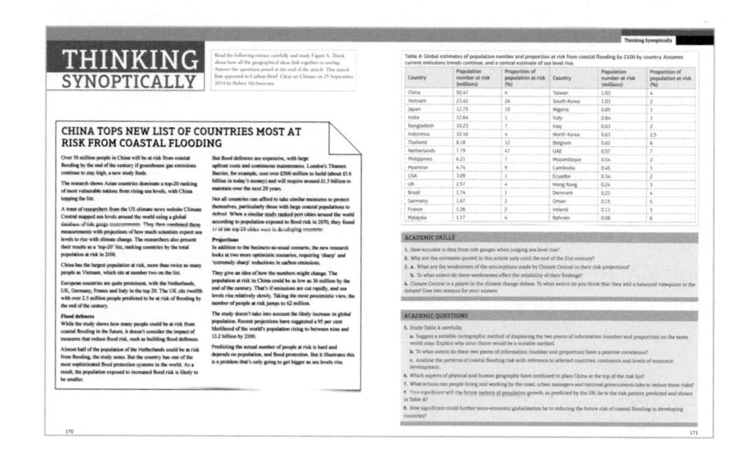

THINKING SYNOPTICALLY

CHINA TOPS NEW LIST OF COUNTRIES MOST AT RISK FROM COASTAL FLOODING

ACADEMIC SKILLS

ACADEMIC QUESTIONS

Summary knowledge check

At the end of each topic is a summary knowledge check with questions which test your learning and a checklist of key links and ideas you will need when preparing for your exams.

This Book 1 covers all of the Edexcel AS level Geography topics, so if you are just studying for an AS Geography qualification you just need to use this book. The same topics also form part of the Edexcel A level Geography specification, and if you are following this course you will need to study the topics in this Book 1 and also the topics featured in Book 2 of this series.

Assessment outline for AS and A Level

Table 1: Specification content

Year 1: AS	Year 1 and Year 2: A
Compulsory topics: Topic 1: Tectonic Processes and Hazards *and* Topic 3: Globalisation	Compulsory topics: Topic 1: Tectonic Processes and Hazards Topic 3: Globalisation Topic 5: The Water Cycle and Water Insecurity Topic 6: The Carbon Cycle and Energy Security Topic 7: Superpowers
Optional topics: Topic 2A: Glaciated Landscapes and Change *or* Topic 2B: Coastal Landscapes and Change *and* Topic 4A: Regenerating Places *or* Topic 4B: Diverse Places	Optional topics: Topic 2A: Glaciated Landscapes and Change *or* Topic 2B: Coastal Landscapes and Change *and* Topic 4A: Regenerating Places *or* Topic 4B: Diverse Places *and* Topic 8A: Health, Human Rights and Intervention *or* Topic 8B: Migration, Identity and Sovereignty

Table 2: Specification assessment

Year 1: AS	Year 2: A
Paper 1: 1 hour and 45 minutes (90 marks): 50% of qualification Section A: Topic 1 *and* Section B: Topic 2A *or* Section C: Topic 2B. Including questions on glaciation *or* coastal fieldwork, *and* a synoptic question linking Topic 1 to Topic 2A *or* Topic 2B (and P, A, F themes).	Paper 1: 2 hours and 15 minutes (105 marks): 30% of qualification Section A: Topic 1 Section B: Topic 2A *or* Topic 2B Section C: Topic 5 *and* Topic 6 Includes 20 mark extended writing questions.
Paper 2: 1 hour and 45 minutes (90 marks): 50% of qualification Section A: Topic 3 *and* Section B: Topic 4A *or* Section C: Topic 4B. Including questions on regeneration *or* diversity fieldwork, *and* a synoptic question linking Topic 3 to Topic 4A *or* Topic 4B (and P, A, F themes).	Paper 2: 2 hours and 15 minutes (105 marks): 30% of qualification Section A: Topic 3 *and* Topic 7 Section B: Topic 4A *or* Topic 4B Section C: Topic 8A *or* Topic 8B Includes 20 mark extended writing questions.
	Paper 3: 2 hours and 15 minutes (70 marks): 20% of the qualification Sections A, B and C are all synoptic: Based on linked topics and Players, Attitudes and actions, and Futures and uncertainties, within a resource booklet about a geographical issue. There are 18 and 24 mark extended writing questions.
	Paper 4: Independent Investigation (70 marks): 20% of the qualification.

Exam command words

It is important for you to know the meaning of command words in exams so that your answer is phrased in the correct way so that it answers the questions. Marks can easily be lost if you do not answer the question, even if you write lots of correct geography. Use the exam question analysis at the end of each chapter to help you.

Table 3: AS and A examination command words

Command word and use	Explanation
Analyse (A)	You must use all of your academic geographical skills to investigate an issue. You must divide the issue into its main parts and make logical connections on the causes and effects of the links between these parts. You must use evidence (e.g. examples and case studies) wherever possible to support the points you make. This command word will be used with longer answer questions worth more marks.
Assess (AS & A)	You must use evidence to show the significance of something relative to other factors. Spend equal time on each factor in the main part of your answer, and in a conclusion identify the most important.
Calculate (AS)	A mathematical or statistical calculation, but always show your full working.
Compare (AS)	You must explore the similarities and differences between two elements in a question. You must try to balance the similarities and differences but recognise if one is stronger.
Complete (AS & A)	You must finish a diagram, map or graph using the data provided.
Define (AS)	You must give the meaning of a geographical term.
Describe (AS)	You must give a geographical account of the main characteristics of something or the stages of a process (e.g. where and what). Your points or concepts should be well developed, with examples, but do not need reasons or justification.
Draw or Plot (AS & A)	You must create a graphical representation of data provided.
Evaluate (AS & A)	You must write about the success or importance of something, and provide a balanced judgement based on evidence (e.g. data) that you have included in your answer. You should have a powerful conclusion that brings together all of the ideas, including strengths and weaknesses, and explores alternatives. This command word will be used with longer answer questions worth more marks.
Explain (AS & A)	You must provide reasons (e.g. how and why) for a geographical pattern or process. You must show your understanding within your answer, supported by justification and evidence – often in the form of examples.
Identify or Give or Name or State (AS)	You must recall specific pieces of information.
Suggest (AS & A)	You must provide a reasoned explanation of how or why a geographical pattern or process may have occurred in an unfamiliar situation. You should provide justification and evidence for your suggestions based on your geographical understanding and knowledge of similar situations.

Fieldwork investigations

An essential part of geographical study is the collection of data and information. This requires you to develop and enhance your

- Practical fieldwork and investigative skills (Table 3)
- Data analysis and presentation skills (Tables 3 and 4)
- Academic skills needed for analysing and concluding, and
- Ability to evaluate data, information and methodologies (including risk assessments) (Table 2)

The Edexcel specification requires a minimum of 2 days of fieldwork to be completed during the AS year as part of the Geography course, including one physical geography (glaciation or coasts) investigation, and one human geography (regenerating places or diverse places) investigation. This will be organised and structured by a school's geography department. In the second year for those studying the A Level you are required to complete an independent investigation based on further fieldwork, this will be organised and structured by you. A minimum of 2 further days of fieldwork are required to meet the specification requirement of 4 days in total for the two-year A level course. This could be completely new fieldwork or extensions arising from the AS fieldwork days. Guidance and advice for the independent investigation will be available from the school's geography department and teachers, and group work may be used to collect data. The fieldwork must include both a physical and a human geography element based on the specification. The Independent Investigation must be based on a physical or human geography theme, or both, taken from the AS or A level specification. In many cases it may be that the fieldwork for the individual investigation will be collected during the summer between the AS and A parts of the two year course (i.e. June to September).

It is essential that you learn a lot during the fieldwork organised for you during the AS year of study so that you are prepared to complete an independent investigation yourself; there are limitations imposed by the specification regulations about how much assistance teachers may provide on the independent investigation.

ACTIVITY

Find out when your AS fieldwork will take place, and when you will be expected to collect data for your independent investigation.

ACTIVITY

Browse the four examples of fieldwork provided in this book. Note how these compare (similarities and differences) with fieldwork and investigations that you completed at GCSE level. These examples are suggested projects, there are many others you may wish to choose, just as you may choose to use different approaches to analysing and presenting the results.

Exam tip

To help students with AS fieldwork and the A level independent investigation, this text book provides four detailed exemplars of fieldwork investigations completed at AS level, but these could also act as starters for the A level investigation. There is one at the end of each of the chapters on the topics for which fieldwork could be chosen for the AS specification. These exemplars are not meant to be complete answers but are designed to show the parts of an investigation, both in terms of organisation and main stages. They also show the writing style, level of detail, and the quality of data presentation required. Further detailed advice is also provided in this section of the text book, which includes possible AS exam questions and advice on how to answer them.

Assessment of fieldwork

AS level: Two examination questions, divided into several parts, one in Paper 1 and one in Paper 2 will assess fieldwork skills linked to Glaciation *or* Coasts, and Regenerating Places *or* Diverse Places. These questions will test (i) knowledge and understanding of methodologies, (ii) interpretation, analysis and evaluation of fieldwork information, and (iii) the ability to synthesise conclusions from the your own fieldwork experience. (See example questions and the end of this section.) A total of 18 marks are available in each examination paper for fieldwork questions (20% of the total AS marks).

A level: An Independent Investigation will be completed for 20% of the total A level. This will be a 3,000 to 4,000 word structured report based on fieldwork and secondary information. Marks are divided as follows: Purpose (introduction) – 12 marks; Planning and data collection – 10 marks; Data presentation, methodology, and analysis – 24 marks; Conclusions and evaluation – 24 marks. (See the specification mark scheme for the detail.)

ACTIVITY

After you have completed your first AS fieldwork day and follow up work, answer the fieldwork questions posed at the end of this section.

Stages in a fieldwork independent investigation

1. Study the specification requirements carefully

Follow the guidance given in this section, and study mark schemes provided by Edexcel to know what you will be judged on. Read and understand the relevant geographical topic and synoptic links to other topics.

2. Decide on an enquiry question for study

The specification states three enquiry themes for each AS topic; the ones in italics in Table 1 are featured at the end of the relevant chapter in this book. An enquiry question is a **broad question** based on the enquiry theme (see exemplars at end of each chapter).

Table 1: AS enquiry themes

Dynamic Landscapes (physical geography)	Dynamic Places (human geography)
Glaciated Landscapes and Change: • Changing glacial and/or fluvioglacial sediments • *Glacial and/or fluvioglacial landform morphology and orientation* • The impact of human activity on fragile glaciated landscapes	Regenerating Places: • *Evidence of regeneration strategies* • Public opinion on local regeneration strategies • Historical change in an area.
Coastal Landscapes and Change: • Changing coastal sediments • *Changing coastal profiles* • Success of coastal management approaches	Diverse Places: • Evaluation of areas that have potential for improvement • Attitudes towards geo-demographic change • *Extent of deprivation in an area*

A level students may choose to investigate one of the same enquiry themes as AS, or one of those featured in this book, or can investigate another covering any topic within the specification. However, students must ensure that meaningful fieldwork and primary data collection (qualitative and quantitative) can take place in a local (small scale) place.

3. Choose a hypothesis

You should consider several hypotheses that link to an enquiry question, and then select the one that you think you can answer the best. A hypothesis consists of a **specific statement** that can be proved true or false, this provides the whole focus of a fieldwork investigation and allows a concise but well balanced report to be produced. This is sometimes called the 'working hypothesis'. It will be based on geographical understanding and consist of a simple statement that is located and/or time specific, and provable through fieldwork measurement and wider investigation. Paired null and alternative hypotheses may be used (see statistical exercises pages 98–99): A null hypothesis (H_0) is the one which, using geographical understanding, is expected to be false, while the alternative hypothesis (H_1) is the one expected to be true.

ACTIVITY

Study the four suggested hypotheses. (a) Which one would you choose and why? (b) For each what would the null hypothesis be?

FOUR POSSIBLE HYPOTHESES (H_1)

- Centuries of farming activity has significantly altered the original shape of drumlins in Ribblesdale, North Yorkshire.
- Hard engineering defences at Overstrand, North Norfolk, have effectively prevented coastal recession since they were built.
- There have been a greater number of regeneration strategies in the inner city of Liverpool than in the suburbs since 2000.
- Local people in the age group 40 to 50 years of age are more concerned about change in the population characteristics of Stratford, London than the local 20 to 30 age group.

ACTIVITY

Where would you go to study the following (a) a coast, (b) a glacial area, (c) an urban area, (d) a rural area. Remember to specify a small area.

4. Decide on a location that is suitable for the chosen enquiry question and hypothesis

Clearly the geographical location selected must be able to provide accurate and relevant information so that the hypothesis and enquiry question can be answered. The small local area must also be accessible. For the Independent Investigation the location should be accessible for at least 2 days, and sometimes revisits are desirable to check or expand on the first set of results. For example, it is no good setting out to measure drumlins in an area where they do not exist, or measure population change where there has been none. A small local area should be studied so that it is manageable, for example, it will not be possible to study the whole of London or the whole of the Lake District. However, sometimes it may be valid to compare two small areas.

ACTIVITY

Study Table 2. (a) In a methodology, what is the difference between explaining how and why? (b) Compare the following two methodologies. Assess why one is much better than the other.

5. Plan the methodology, including collection of primary and secondary data, and the time allocation

Primary data consists of original measurements and information which are collected by you independently or in a group through your fieldwork. Individual tasks may need to be allocated time slots, or combined if a point analysis is being completed. Secondary data consists of measurements or information that has already been collected, such as analysed census and index of multiple deprivation data. Searching for secondary data can take longer than you think. Both will be required in the AS fieldwork exercises and the Independent Investigation. Two important aspects of planning are developing methodologies that will reduce errors and increase the reliability of the data, such as interviewing 50 people instead of just 5, or taking 30 point readings rather than just 3. It is also important to think about natural and human patterns, for example, measuring wave action in different seasons or on days with different wind and wave directions, or measuring patterns in urban areas at different times of day or different days of the week. Consider how much time is needed and when.

Table 2: Extracts from two pupil methodologies

Primary or secondary data required	Why was this data required?	How was this data collected?	Why was the data collected in this way?
Quality of urban environment (QUE) (primary)	The main focus/aim of the study was to investigate the pattern deprivation through the quality of urban environment and this primary data is central to the study as it will show changes across the city.	A 16 km transect from east to west across Norwich was selected (stratified sample). 26 systematic points were then pre-selected along the transect. At each point a pre-prepared QUE table covering a range of externality sources (e.g. buildings) and effects (e.g. visual) was completed (+3 to −3 scoring). (Noise and NOx readings were also completed along with qualitative notes and sketches).	The stratified transect bisected all the urban zones (predicted by the Burgess model) so revealing the QUE levels in each urban area. The systematic points ensured that bias was reduced by having evenly spaced points. There were sufficient points to provide a large data base. The recording table ensured that everyone evaluated the points in the same way, and gave observers a structure to use that enabled a judgement to be made on many factors. The additional information provided evidence to support and correlate with the QUE scores.
Quality of urban environment (QUE) (primary)	This gave an idea of what we thought each transect point was like, and an outer visual idea of what the area came across to be like.	At every transect point a table was filled in with −3 – +3 scoring for many aspects that may affect the quality of the transect.	An average score can be taken of each transect point, and it covered most aspects of the transect points.

Primary or secondary data required	Weaknesses with the data collection method	How problems were (or could have been) reduced	What risks were present with this method	How were the risks reduced
Quality of urban environment (QUE) (primary)	Only two directions were covered, so the whole of north and south of city were not covered. All 26 points were covered at different times of the day and urban processes change (e.g. traffic flows). The original data was scored individually (subjective or qualitative). Data was collected on a weekday which may have different patterns to a weekend.	More transects could have been completed from different directions. Could have recorded at each point at the same time but subjectivity would increase as 26 individual judgements would have existed. The qualitative data was made quantitative (objective) by calculating an arithmetic mean from all scores. Surveys could have been completed on different days of the week in the same places or in different places.	The major risks of working in an urban area include the risk of students being hit by a moving vehicle. The work was completed outside in February so it was cold and there was a risk of hypothermia. Some urban areas may be less safe than others and so students had to be aware of approaches from strangers.	Students took care when by roads and used designated crossing places and paid full attention to traffic movements. Students wore clothing appropriate to the weather conditions and had hot drinks occasionally. Students stayed together in a small group at all times and behaved in a manner that did not draw attention to the group.
Quality of urban environment (QUE) (primary)	They were based on opinion and may be biased.	They were not solved, but an average was taken across the group to give a more general idea.	Cars may hit people and it was very cold and wet.	We watched for cars especially when crossing the road and we went into shops as much as we could to get warm.

ACTIVITY

Before visiting a coastal location to examine glacial deposits in a cliff or beach sediment distribution, what risks do you need to be aware of?

6. Collect primary and secondary data

It will need to be decided if the data is going to be collected individually or through group work, and, if the latter, responsibility for each task needs to be allocated, along with the production of record sheets. A risk assessment for the primary data collection must be devised and considered; you need to be fully aware of health and safety and how to reduce risks effectively.

7. Present the data in a wide variety of ways

You should use a wide range of presentation techniques for your AS fieldwork or Independent Investigation data. These should include cartographic, graphical, statistical and visual techniques. In geography it may still be desirable to produce data presentation by hand as it is often more effective, for example choropleth maps or flow line maps. It can also often be quicker by hand and be difficult to construct more complex graphs using ICT, unless more sophisticated software is available. Students must ensure that they are fully aware of the best ways of presenting data before starting an Independent Investigation.

8. Interim deadlines and final deadline

The school geography department will set deadlines by which to complete certain fieldwork tasks during the AS course. Deadlines will also be set for the Independent Investigation; you must use these to plan your work. The Independent Investigation is internally assessed so teachers will have a fixed deadline from Edexcel to work to and they will have to mark the reports and complete internal moderation, so your deadline will be well before the Edexcel deadline. You must stick to interim deadlines to ensure that tasks are completed in good time, in this way you avoid rushing and producing inferior work.

9. Writing analysis, conclusions, evaluation

The specification makes it very clear that it is the quality of the thinking (reasons for links, reasoned arguments) that is important in any fieldwork report or Independent Investigation, and these must concentrate on quality and not quantity. A mature academic writing style is needed and you should look at the examples at the end of Chapters 2, 3, 5 and 6 for help.

10. Evaluation and reflection

In several places within a fieldwork enquiry or investigation it will be necessary to consider strengths and weaknesses, and show an understanding of how the results fit into wider geography. The reliability, accuracy and validity of results must be considered, especially in the methodology and the evaluation sections.

Table 3: Fieldwork and Investigation skills

Skill	Detail
Collecting information	Primary data – original data collected by fieldwork Secondary data – information already published and available for use Printed material – factual, descriptive or creative Digital data – Geographical information systems, data bases, satellite or photographic images, blogs Diagrammatic data – Graphical, cartographic, diagrams (including 2D and 3D).
Quantitative techniques	Using statistics for establishing central tendency, dispersion or correlation. May include Student t-test, Spearman's rank correlation analysis, Chi-squared, Gini co-efficient, and Lorenz curve. Including significance testing.
Qualitative techniques	Using judgements through coding and sampling techniques (random, stratified, systematic and sampling error), questionnaires and interviews, oral accounts, newspaper reports, social media sites, photographic representations, and field sketches (with annotations).

Skill	Detail
Cartographic	Construct or interpret geospatial technologies (e.g. GIS), dot maps, flow maps, OS maps.
Graphical	Linear and logarithmic graphs, kite graphs, scattergraphs and lines of best fit.
Analysing information	Application of suitable analytical techniques (see above), use of relevant theory, accurate interpretation of data.
Evaluating information	Informed and critical questioning of data sources, identification of sources of error in data or the misuse of data.

Extended Project Qualification

You may have the opportunity to complete the Extended Project (EP) qualification within your school. The EP can support geographical fieldwork investigations and the Independent Investigation through developing organisation and research skills, encouraging independent wider reading, developing planning and management skills, helping in the devising of a hypothesis, developing data collection and interpretation skills, and developing the powers of evaluation and critical thinking.

AS examination fieldwork questions: Diverse Places

(a) Table A below shows a summary of Index of Multiple Deprivation (IMD) scores (secondary data) collected to support a field trip to a UK city. The higher the score the more deprived the urban zone. The scores are the means for census wards in each urban zone.

Table A: IMD data for a UK city

CBD (1 ward)	Inner urban (4 wards)	Inner Suburbs (5 wards)	Outer Suburbs (5 wards)
32.51	27.26	30.46	

(i) Complete Table A by using the following data for five wards to calculate the mean IMD score for the Outer Suburbs. (1 mark)

35.07	22.84	9.60	9.08	11.37

(ii) State one strength and one weakness of using the mean to analyse this data. (2 marks)

Strength

Weakness

(iii) Give two possible reasons for the different IMD scores for the inner urban area and the outer suburbs. (2 marks)

1

2

(b) Explain one strength and one weaknesses of using Geographical Information Systems (GIS) to represent the IMD data from this fieldwork. (4 marks)

Strength

Weakness

(c) You have collected and analysed data during fieldwork linked to Diverse Places.

Assess the validity of the conclusions made from your fieldwork investigation. (9 marks)

Research question or hypothesis ....

AS examination fieldwork questions: Coasts

(a) Table B below shows a summary of pebble size analysis (primary data) for three coastal locations along a 40 km length of UK coastline. The longshore drift is from site A towards sites B and C. The figures are mean pebble lengths for parts of the beach profile and by site.

Table B: Mean pebble sizes (mm) for three coastal UK sites

Mean pebble lengths (mm)	Upper beach	Middle beach	Lower beach	Mean for site
Site A: 0 km	19.7	26.4	23.8	23.30
Site B: 20 km	27.1	21.2	19.5	22.60
Site C: 40 km	22.3	15.5	14.7	17.50
Mean for beach section		21.03	19.33	

(i) Complete Table B by calculating the mean pebble size for the upper beach. (1 mark)

(ii) Suggest why calculating the median for the pebble lengths at these fieldwork sites may have been better than calculating the mean. (2 marks)

(iii) Suggest one reason for the change in mean pebble length between the three fieldwork sites and suggest one reason for the change in mean pebble length on different parts of the beach. (2 marks)

Fieldwork sites

Parts of the beach

(b) Explain two reasons why risk assessments are important to the planning of geography fieldwork. (4 marks)

(c) You have collected and analysed data during fieldwork in a coastal environment.

Assess how important prior knowledge and understanding of coastal processes was to developing an understanding of the results of your fieldwork investigation. (9 marks)

Research question or hypothesis ....

Synopticity

What is synopticity?

Synopticity means bringing everything together and recognising that there are many links between features, processes and influences. This is an important part of your geographical studies. Topics may be separated for the purposes of detailed study but a geographer must always recognise the links between these topics and the processes within them when developing an understanding of the real world, which is a complicated place. There are many players, or stakeholders, within all aspects of geography; these people have opinions and attitudes which determine their actions in a wide variety of situations. Their actions have an important influence on the future, and geographers are most interested in how physical and human patterns and processes may change over time and spatially. (Highlighting like this in the text will help you to pick these out.) Making geographical decisions involves synopticity; in a real world situation all geographical topics are linked together and a full range of geographical understanding is required.

Synopticity in this book

The Edexcel specification has an emphasis on synopticity and the examinations have synoptic questions. To help you prepare for this aspect of assessment this book offers several synoptic features. First, there are margin boxes that explore synoptic aspects that are linked to points made in each chapter; second, there are occasional Activities which have synoptic aspects to them, such as decision-making; and third, there are two-page spreads called Thinking Synoptically, where extracts from journal or magazine articles are presented with a series of questions designed to help you think synoptically.

Synopticity in the examinations

AS Paper 1	AS Paper 2
Answer one 16-mark synoptic question based on Glaciation and Tectonics or Coasts and Tectonics	Answer one 16-mark synoptic question based on Regenerating places and Globalisation or Diverse places and Globalisation
Mark scheme: Level 4: 'Argument comprehensively and meaningfully synthesises geographical ideas from across the course of study throughout the response.'	

A Paper 1 and Paper 2	A Paper 3
Answer two 20-mark questions in each examination which require a wide range of knowledge and understanding of a topic.	Answer all questions based on a resource booklet of information from a real world location. Two long answers require specifically synoptic answers.
Mark scheme: Level 4: 'Applies knowledge and understanding, provides coherent conclusion, interprets and substantiates ideas and evidence.' '...fully supported by a balanced argument that is drawn together coherently.'	Mark scheme: Level 3: 'Applies knowledge and understanding of geographical information/ideas to find fully logical and relevant connections/relationships.' Level 4: 'Critically investigates the questions/issue to produce a coherent interpretation of quantitative and qualitative data/evidence, comprehensively making meaningful connections to relevant geographical ideas from across the course of study throughout the response.'

Example AS synoptic questions

Glaciated landscapes and change (linked to Tectonic processes and hazards)

Study Figure 2: Evaluate the role of tectonic processes in the creation of past and present glacial environments. (16 marks)

Figure 2: Mount Pinatubo (1991) and Himalayan mountains

Coastal landscapes and change (linked to Tectonic processes and hazards)

Study Figure 3: Evaluate the role of tectonic processes in the creation of emerging and submerging coastal features. (16 marks)

Figure 3: Erupting undersea volcano (Tonga) and fjord of Milford Sound, New Zealand

Regenerating places (linked to Globalisation topic)

Study Figure 5: Evaluate the extent to which foreign direct investment (FDI) may contribute to the regeneration of an area that has experienced deindustrialisation. (16 marks)

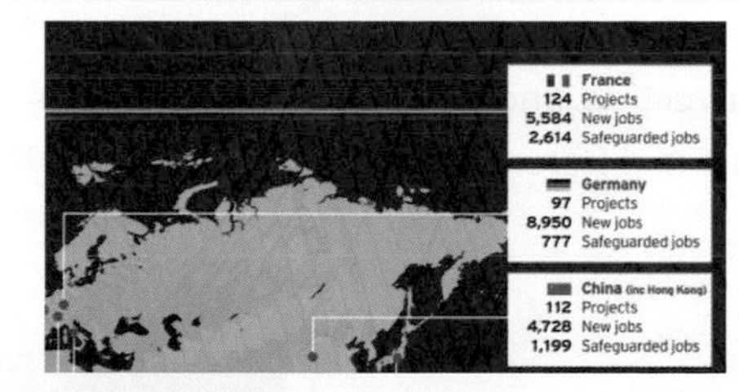

Figure 5: UK FDI resources 2014–15 and Nissan factory, Sunderland.

Diverse places (linked to Globalisation topic)

Study Figure 6: Evaluate the extent to which globalisation processes cause increased diversity of places within a country. (16 marks)

Figure 6: Shops in Romford Road, Newham and Canary Wharf financial centre

ACTIVITY

Produce a plan for each of these AS synoptic questions. Make sure that each plan identifies synoptic links. Compare your plan with that of other students and then make improvements.

Dynamic landscapes

Tectonic processes and hazards

Introduction

Geological evidence shows that the Earth's crust has been moving for at least 225 million years, since the supercontinent of Pangaea split apart. This movement has created numerous tectonic events such as earthquakes and volcanoes, and secondary events such as tsunamis and landslides. Continental drift was first suggested by the geophysicist Alfred Wegener in 1912, who pointed out pieces of evidence (such as identical rock and fossil types in continents thousands of miles apart), but he could not finalise a theory because he could not identify a mechanism for moving the crust. It was not until the 1960s that the plate tectonic theory was accepted following the work of numerous scientists, including Dan McKenzie and Harry Hess. These scientists had shown that the crust was split into plates, that heat from the Earth's core must cause the mantle material to move through the process of convection, and that it is these convection cells that drag the crust in different directions a few centimetres each year. The crust is still being modified today, with some geologically recent boundaries, major earthquakes and volcanic eruptions and mountain building. Research continues and even though this adds regularly to the understanding of the processes, not everything is fully understood.

Tectonic events only become hazards when they affect people, their property or their way of life. The world's population has been growing exponentially and is predicted by the UN to be 9.7 billion in 2050. Inevitably more and more people are living in risky areas and in higher population densities, such as cities, so the potential for a mega-disaster is increasing. For example, the earthquake epicentres in Nepal (2015) missed the capital Kathmandu by just 80 km. If it had been affected there could have been a million deaths instead of 10,000. The numbers of poor people in the world are also increasing. Over 2 billion live on less than $2 a day, and often these people are the most vulnerable in tectonic hazard zones.

In this topic

After studying this chapter, you will be able to discuss and explain the ideas and concepts contained within the following enquiry questions, and provide information on relevant located examples:

- Why are some locations more at risk from tectonic hazards than others?
- Why do some tectonic hazards develop into disasters?
- How successful is the management of tectonic hazards and disasters?

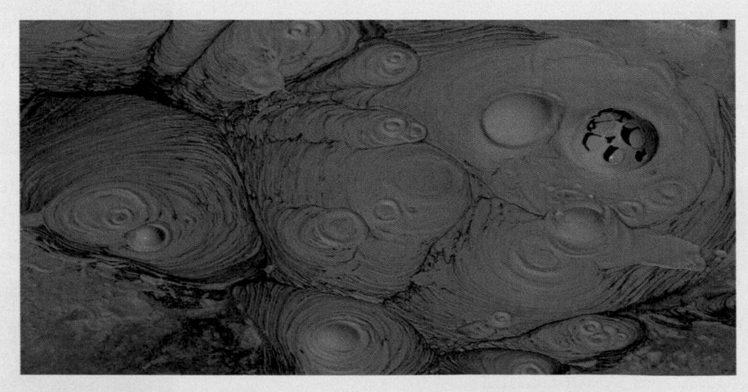

Figure 1.1: The geothermal world is a fascinating place, as demonstrated by this bubbling volcanic mud in Waiotapu, New Zealand. What do you think is happening under the ground here to create these patterns and bubble shapes?

CHAPTER 1

Synoptic links

Tectonic hazards are linked to population and urbanisation patterns and trends, the development of a country, hydro-meteorological factors (weather and flowing water), as well as other landscape processes. Players include governments, non-governmental organisations, scientists, insurance companies, building engineers and the general population living in hazardous zones. The attitudes and actions of these groups have a bearing on the severity of the impacts of tectonic hazards and contrasting decisions, as well as natural trends, create uncertainty in many areas of the world.

Useful knowledge and understanding

During your previous studies of Geography you may have learnt about some of the ideas and concepts expanded in this chapter, such as:

- The structure of the Earth (crust, mantle, core).
- Rock types (sedimentary, igneous and metamorphic).
- Physical features created by tectonic activity (fold mountains, volcanoes).
- Major tectonic plates (African, Pacific).
- Major plate boundaries (destructive, constructive).
- Convection as the mechanism for moving the tectonic plates.
- Major volcanoes and earthquakes, and their hazards (shockwaves, lavas).
- Ways of making buildings earthquake proof (flexible buildings).
- Managements of tectonic hazards (before, during and after).

This chapter will reinforce, but also modify and extend your knowledge and understanding of tectonics. Remember that the material in this chapter features in both the AS and A level examinations.

Skills covered within this topic

- Use and interpretation of hazard distributions on world and regional scale maps.
- Interpretation of GIS information to identify hazard areas and extent of risk.
- Understanding and use of large data sets to identify hazard trends and reliability of data.
- Understanding and use of correlation analyses (Spearman's Rank), including significance testing.
- Interpretation and use of 2D and 3D diagrams showing plate boundaries and related features.
- Calculation of tsunami velocities.
- Drawing and interpretation of a variety of graphs (including scattergraphs), and comparing effectiveness of graphs.
- Interpretation of photographic evidence.
- Investigation and research through use of websites.

CHAPTER 1

Why are some locations more at risk from tectonic hazards?

Learning objectives

1.1 To understand that the global distribution of tectonic hazards can be explained by the processes operating at plate boundaries and intra-plate locations.

1.2 To understand that the plate tectonic theory was developed to explain plate movements and the magnitude of tectonic hazards.

1.3 To understand that the physical processes, linked to earthquakes, volcanoes and tsunamis, explain the causes of tectonic hazards.

Distribution and causes of tectonic hazards

There are several major tectonic plates (for example Africa), minor plates (for example Cocos), and micro-plates (for example Sunda) (Figure 1.2). These plates often move in different directions and at different speeds, creating different types of plate boundary and therefore variations in earthquake and volcanic activity. Most earthquake and volcanic activity is found in zones along the plate boundaries. The boundary which creates earthquakes with a **magnitude** of 9 M_w or higher (**Moment Magnitude Scale**) is associated with convergence between an oceanic plate and a continental plate (a **destructive plate boundary** with **subduction**).

Extension

Investigate the ancient crustal plate called Zealandia. GNS Science (Te Pu Ao) may be a useful source.

ACTIVITY

1. Study Figure 1.2 and consider the plate boundaries and movements.
   a. Identify three locations with convergence of an oceanic and a continental plate.
   b. Describe what is happening in the Indian Ocean between Australia and India. Why is this significant for the people living around the coasts of this ocean?
   c. Describe what is happening in and around New Zealand.
   d. Describe what is happening in the Middle East.

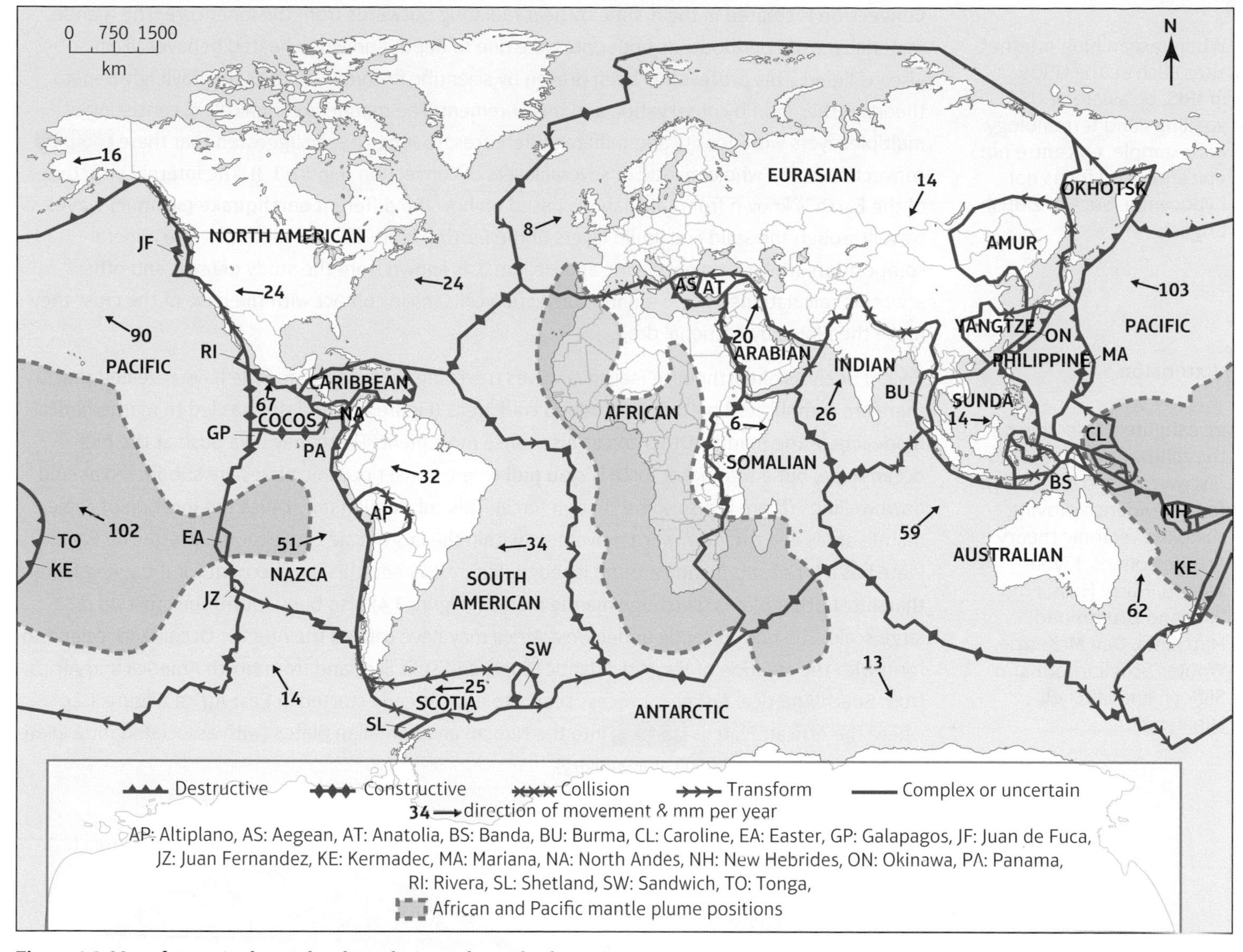

Figure 1.2: Map of tectonic plates, plate boundaries and mantle plumes

Intra-plate processes

There are also earthquakes near the middle of plates (**intra-plate**) associated with ancient faults, such as the Rhine Rift Valley, which resulted from the solid crust cracking during its long journey over millions of years. Newer large faults can also be found, such as the East African Rift valley, which may turn into new plate boundaries. These situations are associated with smaller magnitude earthquakes. Collisions of tectonic plates may also fracture the crust well away from the plate boundary, such as in the large zone of the Himalayas and Tibetan Plateau (Figure 1.8), and these may yield large magnitude earthquakes over a long period of time as the collision slowly takes place and movement is caused along faults.

There is also volcanic activity near the centre of some plates (**hot spots**). This is either the result of the upwelling of hot molten material from the core/mantle boundary, such as the Hawaiian (Figure 1.18) and Tristan da Cunha hot spots (Figure 1.4), or from the top of a large **mantle plume** just under the crust, such as Nyiragongo and Cape Verde Islands (African Plate). It is also possible that large meteorite impacts create symmetrical hot spots on opposite sides of the planet when they collide with the Earth.

ACTIVITY

CARTOGRAPHIC SKILLS

1. On a copy of Figure 1.2 plot the location of the recent 9.0+ M_w magnitude earthquakes, and the largest recent volcanic eruptions.

2. Make notes on what you notice about these locations.

Literacy tip

When researching internet sites such as the USGS or IRIS, be aware of USA spellings and terminology. For example, epicentre not epicenter and focus not hypocenter. Stick to British English.

Extension

Investigate a selection of the following scientists to discover their contribution to the evidence proving the plate tectonic theory: Arthur Holmes, Kiyoo Wadati, Harry Hess, Fred Vine and Drummond Matthews, Dan McKenzie, Wouter Schellart or Karin Sigloch and Mitchell Mihalynuk.

Processes moving the tectonic plates

Convection is created in the mantle by heat radiating outwards from the inner core. The mantle material, mostly peridotite, is under pressure due to depth and when heated behaves like a viscous liquid. This process has been proven by scientific formulae, such as the Rayleigh-Benard theory, rather than by observation and measurement. The mantle is complicated, consisting of multiple layers with density and mineral differences, so it has been suggested that there could be convection in the whole mantle or several layers of convection (Figure 1.3). The internal structure of the Earth is known from calculations based on how the different **earthquake (seismic) waves** travel through the solid and liquid layers of the Earth (Figure 1.4 and Table 1.1). The mineral composition, which includes olivine and Helium-3, is known from the study of lavas and other scientific calculations. Where mantle convection cells make contact with the base of the crust they move the plates by frictional drag.

Recent analysis of earthquake (seismic) waves travelling through the mantle have revealed that there are complicated patterns of hot and cold areas (Figure 1.4) and this has led to mathematical modelling of the mantle. Other forces also cause movement. There may be a push at the mid-ocean ridge, but a significant force is **slab pull** where denser oceanic plates are subducted at cold downwellings (there is also some suction force). This subduction may cause the location of cooler mantle areas and the downward movement within the large scale convection pattern. The Pacific Plate has a lot of subduction around its edge (Figure 1.2) and this may account for it moving faster than most other plates. Two huge mantle plumes (Figure 1.4) also buoy (push) the crust up on a large scale. The one currently under West Africa may have caused the Atlantic Ocean and Iceland to form with the creation of the Mid-Atlantic Ridge, and split Scotland from North America and Africa from South America. A similar process seems to have already started in East Africa (Figure 1.2), where the African Plate is splitting into the Nubian and Somalian plates (with associated intra-plate minor earthquakes and volcanic activity).

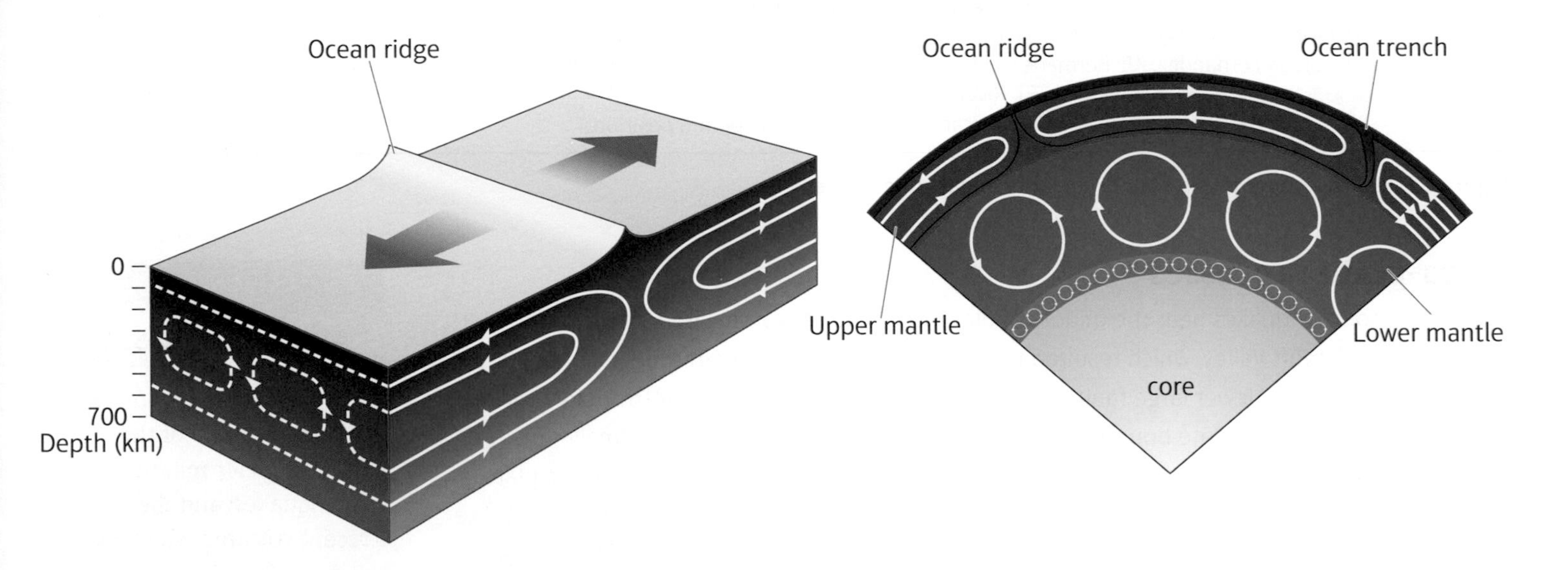

Figure 1.3: Convection cell possibilities in the Earth's mantle

Yellow = Players, Orange = Attitudes and actions, Purple = Futures and uncertainties

Synoptic link

Convection in the iron rich outer core creates the Earth's magnetic field which is not only responsible for the palaeomagnetism of the oceanic crust, providing proof of sea-floor spreading, but also for protecting the Earth from solar radiation which has enabled the biodiversity of ecosystems to develop on the planet. (See pages 124, 147 and 153.)

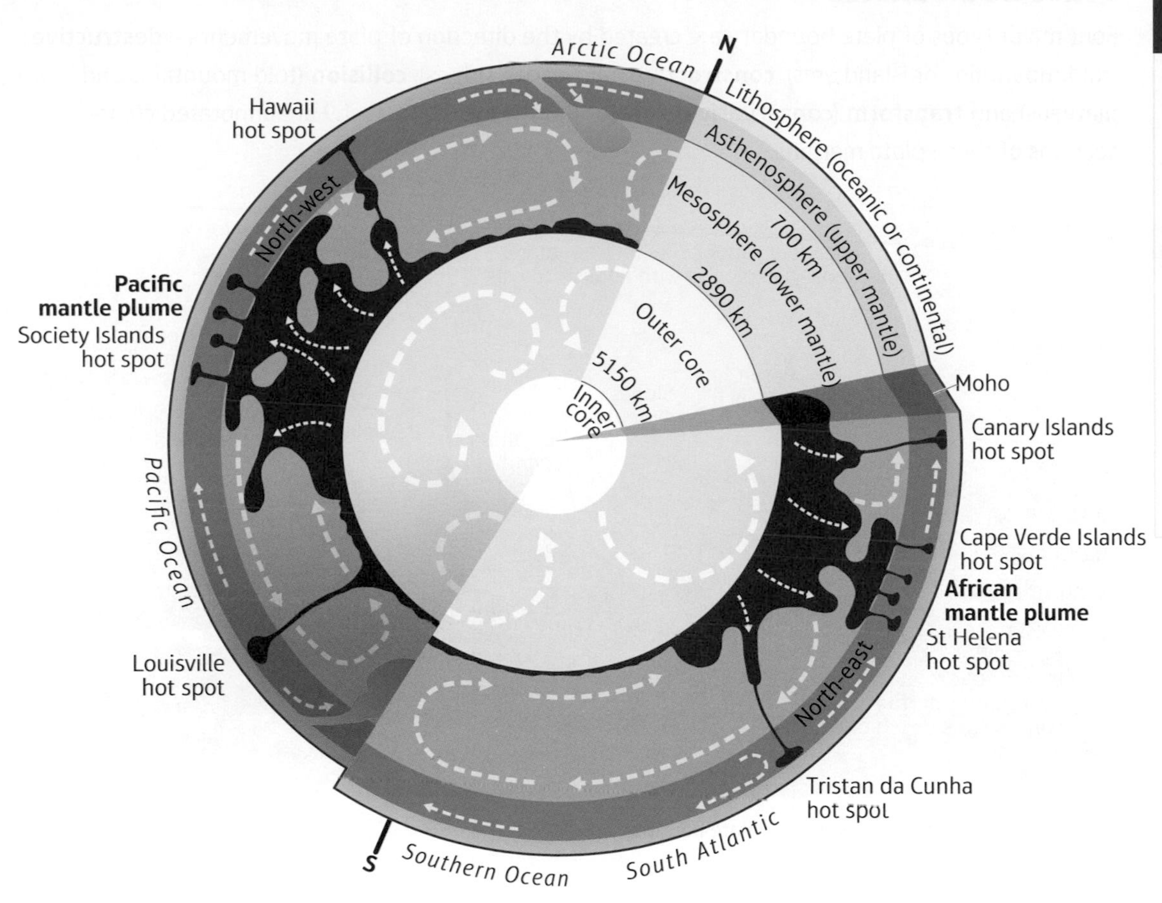

Figure 1.4: Structure of the Earth with mantle plumes and hot spots

Table 1.1: Characteristics of the Earth's structural layers

Structure part	Temperature	Density	Composition	Physical state	Earthquake (seismic) waves
Crust (oceanic 7 km thick, continental up to 70 km thick)	Surface temperature to about 400°C	Less dense (oceanic = 2.7 g/cm^3 continental = 3.3 g/cm^3)	Granite (continental) and basalt (oceanic)	Solid	Surface and body waves able to pass through
Mantle (700 km to 2890 km deep)	870°C	Less dense to medium density (3.3 to 5.4 g/cm^3)	Peridotite Upper = olivine Lower = magnesium silicate	Phases of liquid and solid in layers	Body waves pass through at variable rates due to density changes
Outer core (2890 km to 5150 km deep)	4400 to 6100°C	Dense (9.9 to 12.3 g/cm^3)	12% sulphur 88% iron	Liquid (+ generates magnetic field)	Only P waves able to pass through, an S wave 'shadow zone' is created from about 105° from the focal point
Inner core (5150 km deep to centre)	7000°C (radioactive decay)	Very dense (13.5 g/cm^3)	20% nickel 80% iron	Solid (+ radiates heat), maybe two parts with huge crystals aligned in opposite directions	Only P waves reach the inner core and pass through, but their refraction at the core-mantle boundary creates a ring 'shadow zone' between 105° and 140° from the focal point

Plate boundaries

Four major types of plate boundary are created by the direction of plate movements – **destructive** (fold mountains or island arcs), **constructive** (mid-ocean ridges), **collision** (fold mountains and plateaus) and **transform** (**conservative** or major fault). Figures 1.5 to 1.9 are annotated cross-sections of these plate margins.

ACTIVITY

1. Study Figures 1.5 to 1.9 (on page 26).
   a. For each type of plate boundary, use the labels and annotations to write an explanation of the features and processes found.
   b. Add to your explanations by finding additional points from Table 1.2 on page 26.

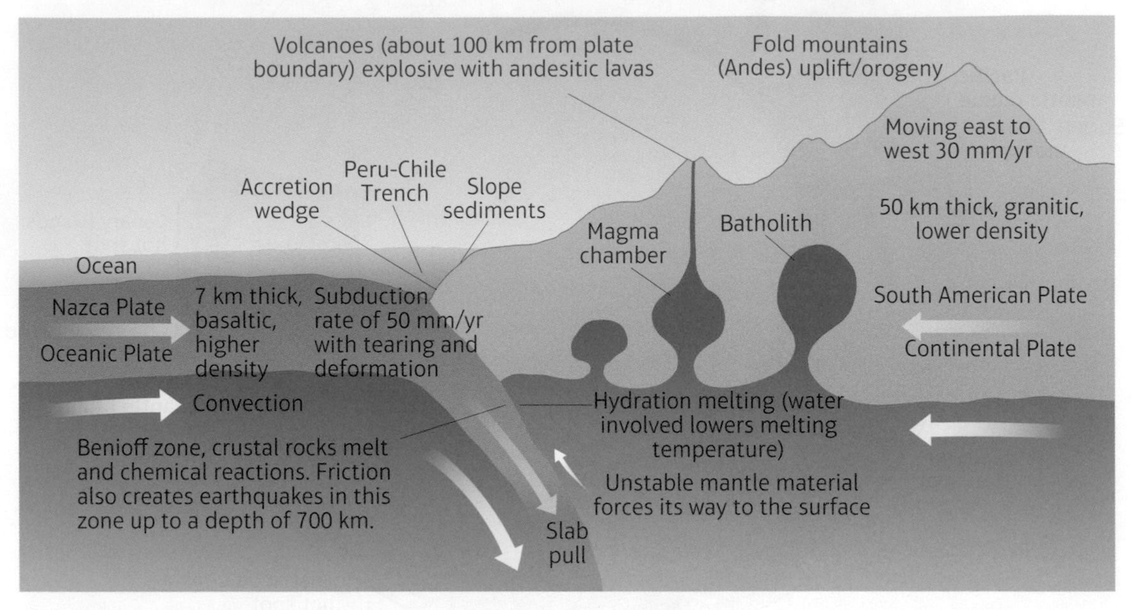

Figure 1.5: Destructive (convergent) plate boundary

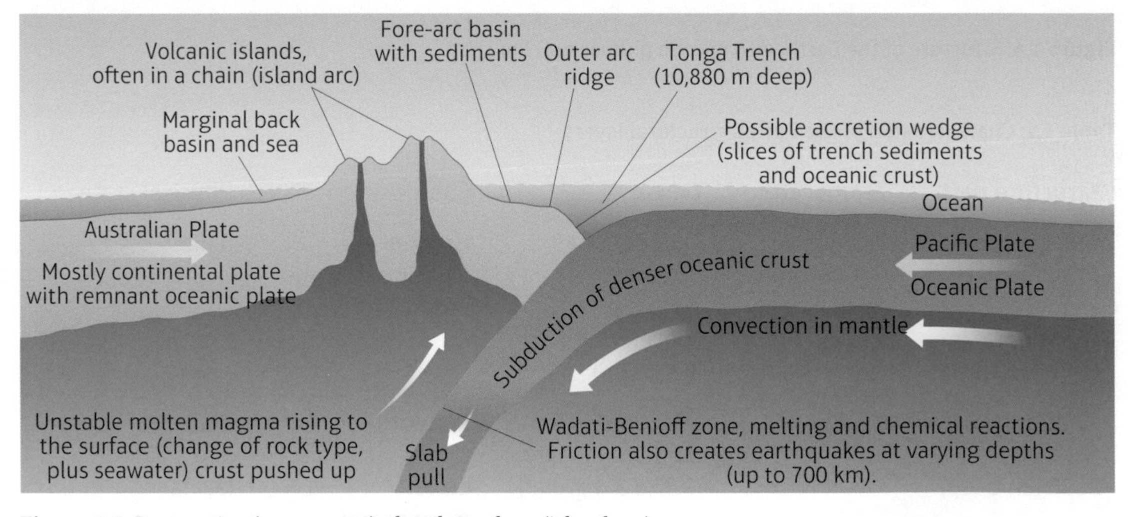

Figure 1.6: Destructive (convergent) plate boundary (island arc)

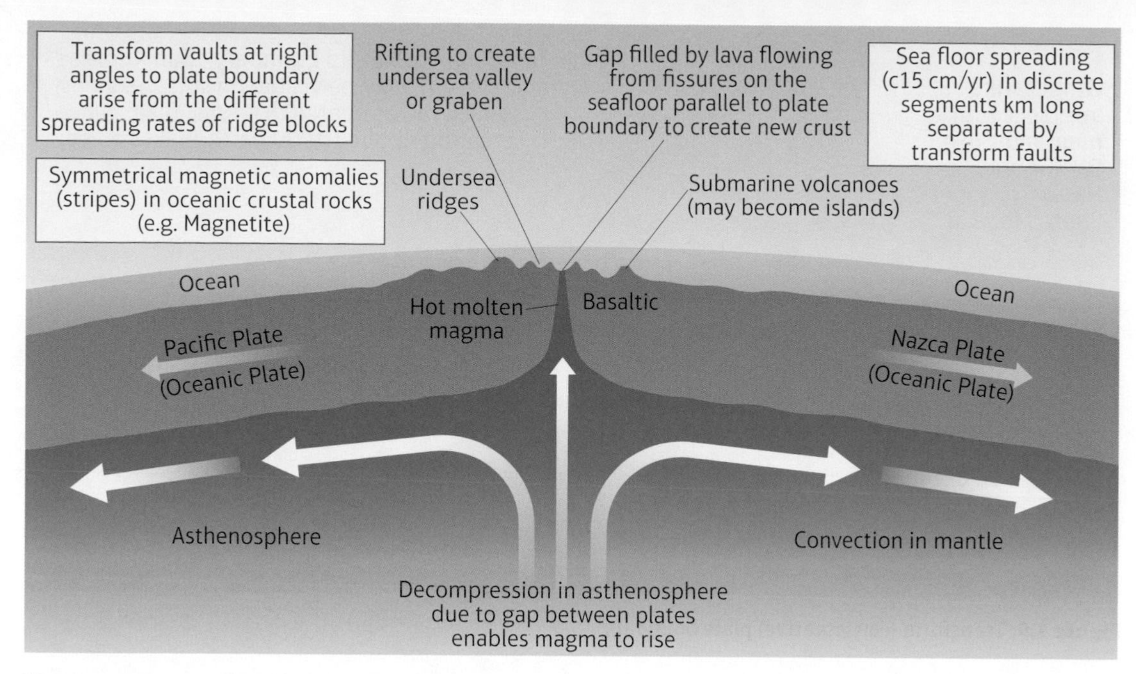

Figure 1.7: Constructive (divergent) plate boundary

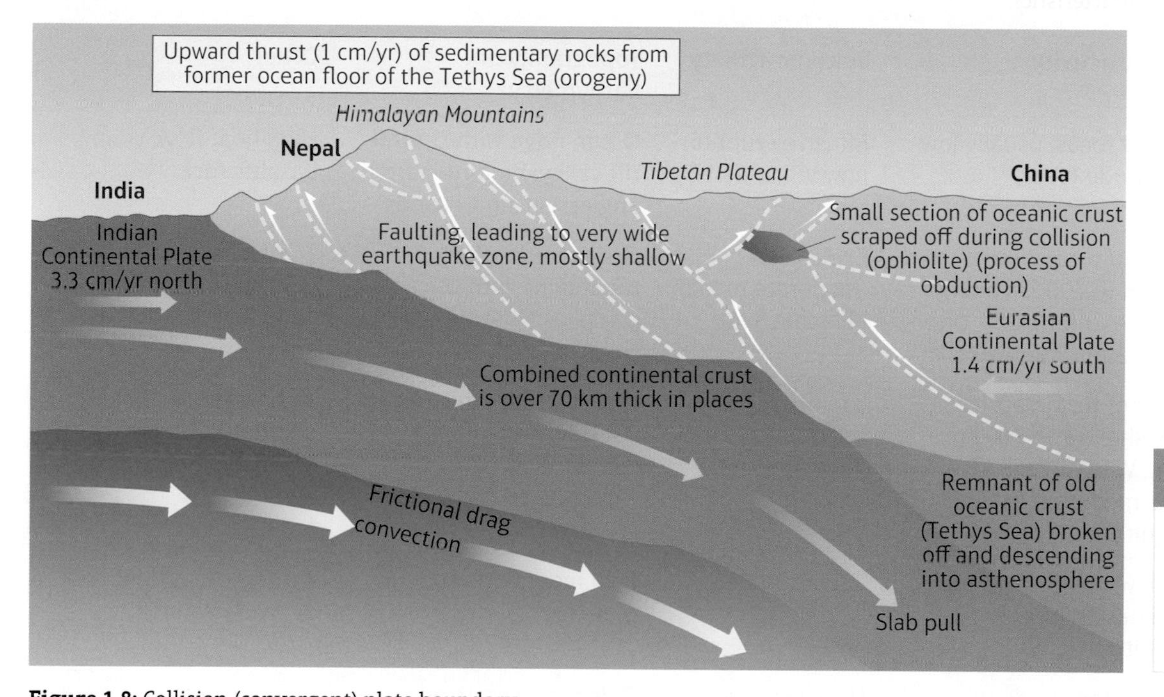

Figure 1.8: Collision (convergent) plate boundary

Extension

Investigate the Tethys Sea. What evidence is left today to show that this sea existed?

Literacy tip

Always use correct geographical terms in your written answers when there is one to use. For example, asthenosphere for upper mantle; Moho for the boundary between the crust and upper mantle.

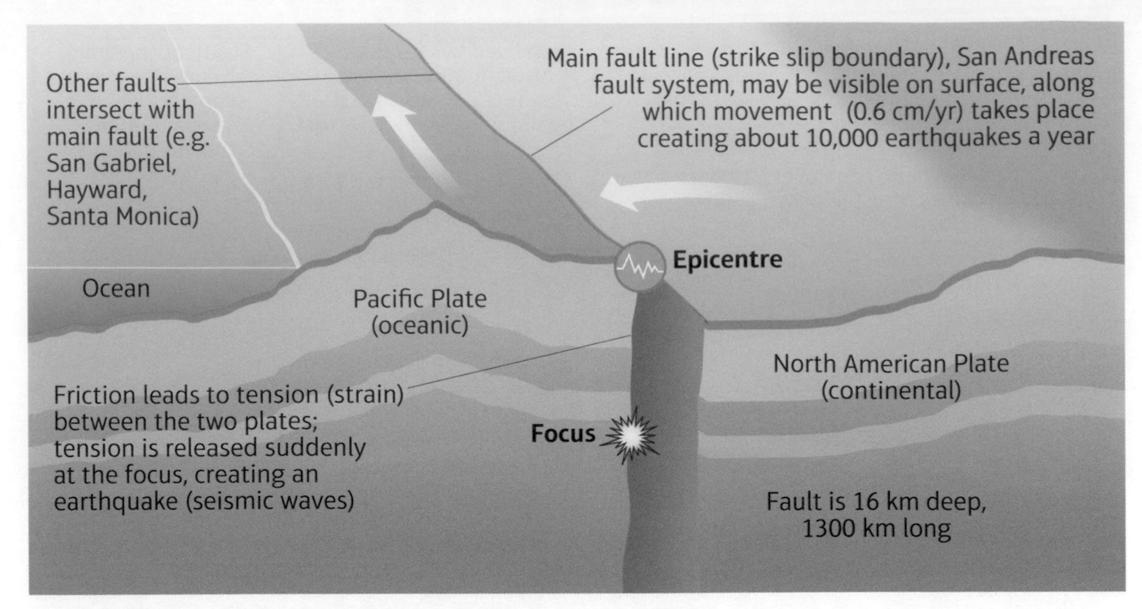

Figure 1.9: Transform (conservative) plate boundary

Table 1.2: Plate boundary tectonic characteristics.

Plate boundary type	Seismic activity	Volcanic activity	Topography or bathymetry	Other
Constructive (divergent)	Shallow focus; usually low magnitude (5–6)	Effusive eruptions (low VEI scale 1-3)	Ocean ridge with central rift valley; faulting at right angles; volcanic islands	High heat flow; young basaltic rock
Destructive (convergent): ocean and continent	Range of focal depths from shallow to 700 km along Wadati-Benioff zone; often high magnitude (8–9)	Explosive (moderate to high VEI scale 5–6)	Ocean trench; fold mountains with volcanic peaks	Trench gas low heat flow and negative gravity anomaly; range of rock age
Destructive (convergent): ocean and ocean	Range of focal depths from shallow to 700 km along Wadati-Benioff zone; moderate to high magnitude (7–9)	Explosive (moderate to high VEI scale 5–6)	Island arc; oceanic trench; back arc and fore arc zones	Trench gas low heat flow and negative gravity anomaly; range of rock age
Collision (convergent): continent and continent	Shallow to intermediate focal depth; moderate magnitude (6–8)	Usually none	Fold mountains and plateaus	Average heat flow; rock age very variable
Transform (conservative)	Shallow focus; usually moderate magnitude (6–8)	Usually none; occasional fissure eruptions	Ridges and scars on surface	Average heat flow; rock age depends on oceanic (young) or continental (old) location

Plate tectonic theory

All pieces of evidence are important for the theory. There are some relatively small pieces – such as the continental shelf shapes fitting together (for example Africa with South America), matching rock types (in ancient cratons 2.6 billion years old), matching fossils (such as the Cynognathus reptile in South America and Africa; Glossopteris fern in all southern continents), coal that forms under tropical conditions found in temperate areas, evidence of glaciation in tropical Africa, and of course the fact that the majority of earthquakes and volcanoes occur in zones near plate boundaries. More significant pieces of evidence include studying the way that earthquake (seismic) waves travel through the Earth, and the location of earthquake focal points. Along the **Wadati-Benioff** zone foci depths can reach 700 km, which shows subduction of the denser basaltic oceanic plates into the asthenosphere (upper mantle). It was not until the 1950s that the ocean floor was surveyed; this showed the presence of mid-ocean ridges where new oceanic crust was being created, and symmetrical 'magnetic stripes' (**palaeomagnetism**) in the new iron rich rocks either side of the constructive plate boundaries. The magnetic direction is locked in when rocks cool below about 250°C and shows the reversal of the Earth's magnetic field and age of the oceanic crust. Put together, these two points demonstrate the 'conveyor belt' movement of **sea floor spreading**, with youngest rocks at the ridges (Figure 1.7). Studies of palaeomagnetism also show continental rotation (polar wandering), which, when reversed, demonstrate that the continents did once fit together as Pangaea (for example Greenland and UK joined). The study of lavas reveals the nature of the Earth's interior, and hot spot volcanoes show plate movement, for example the Pacific Plate is moving over the Hawaiian fixed hot spot at 7 cm per year.

Synoptic link

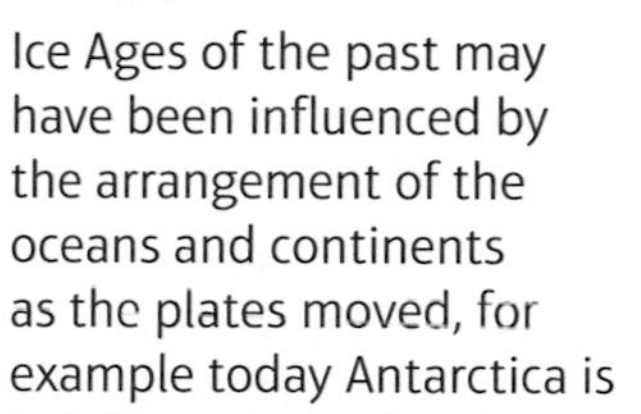

Ice Ages of the past may have been influenced by the arrangement of the oceans and continents as the plates moved, for example today Antarctica is helping to keep the planet cooler. (See page 71.)

Physical tectonic processes

Earthquakes and earthquake waves

At any plate boundary stress and strain can build up along the join, due to friction created by masses of rock trying to move past each other. This is greatest along the Wadati-Benioff zone within destructive plate boundaries (Figures 1.5 and 1.6), with the 9.5 M_w earthquake in Chile (1960) being the largest. Stress is also great along transform (conservative) boundaries (Figure 1.9), with the 7.6 M_w earthquake of 1992 on the San Andreas Fault being the largest. There is less friction at constructive plate boundaries (Figure 1.7) and so magnitudes are lower, for example the largest recent earthquake in Iceland was 6.5 M_w in 2000, but they are usually smaller. The focus of an earthquake is the point at which the strain is released. This sends earthquake (seismic) waves in all directions, being strongest at the point on the Earth's surface directly above the focus, known as the epicentre. There are several types of earthquake (seismic) wave, and all move the ground in a different way. The P waves (primary) arrive first, have a short wavelength and travel quickly through the crust. S waves (secondary) arrive a few seconds later and have a longer wavelength and a velocity of 4 km/s, causing more destruction. In the November 2011 Tohoku earthquake in Japan the S waves reached Tokyo 90 seconds after the release of stress at the focal point.

ACTIVITY

TECHNOLOGY AND ICT SKILLS

Research websites for animations of earthquake (seismic) waves and shadow zones. Useful websites include Incorporated Research Institutions for Seismology (IRIS), United States Geological Survey (USGS) and Purdue University. Study these to add to your understanding.

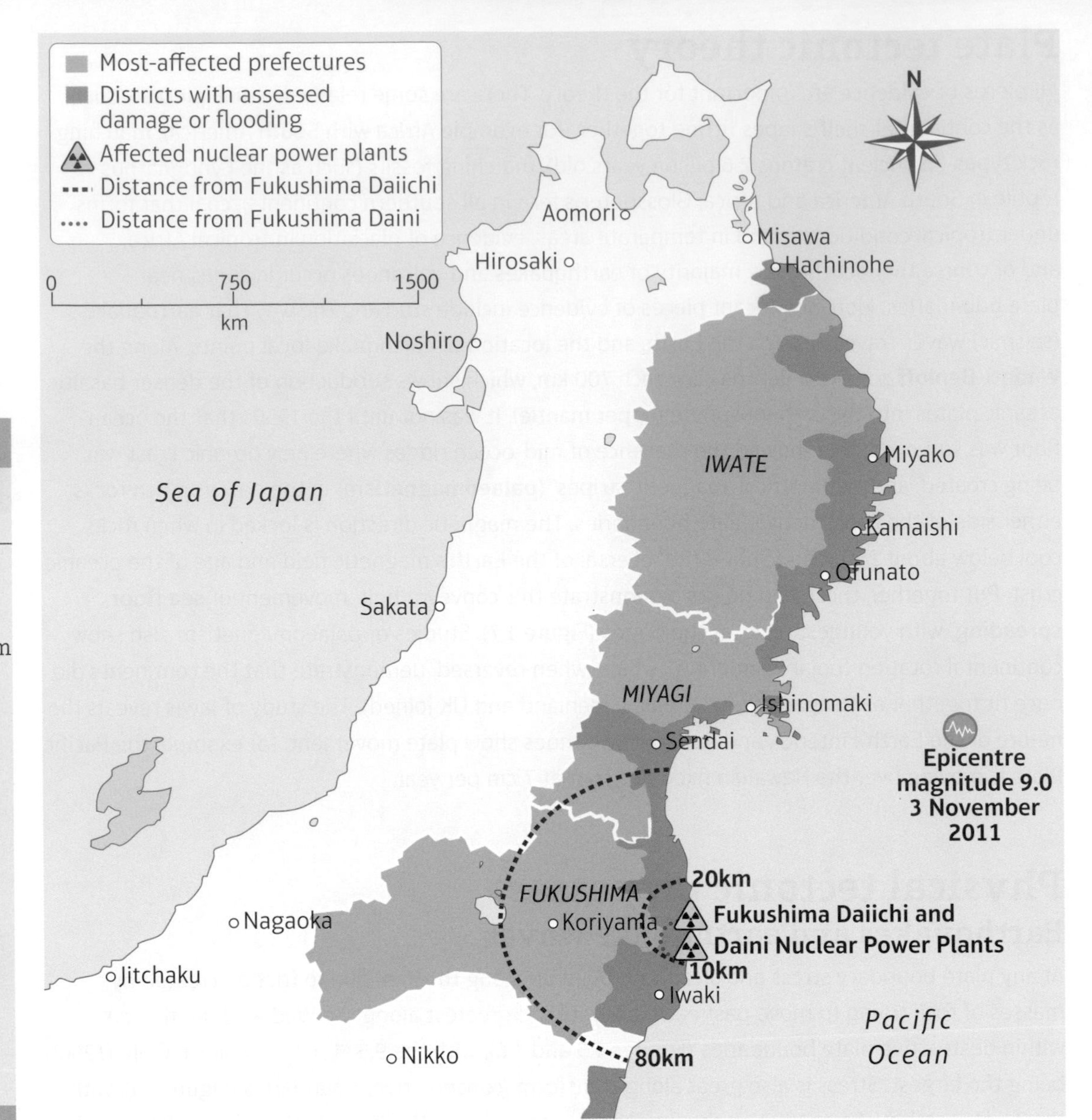

Figure 1.10: Areas of Japan affected by the Tohoku tsunami in November 2011

ACTIVITY

NUMERICAL AND STATISTICAL SKILLS

1. Study Figure 1.10 and locate Sendai.

a. Sendai is located 130 km west of the November 2011 Tohoku earthquake epicentre. How many seconds should it have taken for the S waves to reach Sendai?

b. The earthquake took place at 2:46:45 local time, and a warning was automatically issued at 2:46:48. Did local people have enough time to prepare for the event?

There are additional earthquake (seismic) waves (Figure 1.11). When all shockwaves are combined over a short period of time, maybe 30 seconds, it is no surprise that buildings without an aseismic design get severely damaged and collapse. This is because the ground moves in many different ways (see page 40 for information on earthquake and tsunami scales).

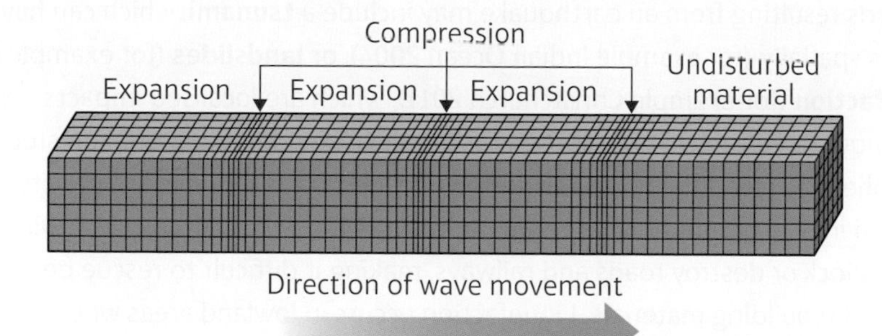

Primary wave: Arrives first, fast, moves through solid rock and fluids, pushes and pulls (compresses) in the direction of travel.

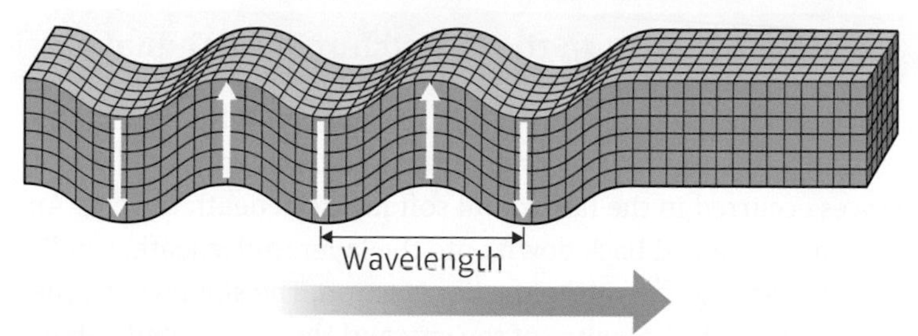

Secondary wave: Slower than P wave, only moves through solid rock, up and down movement.

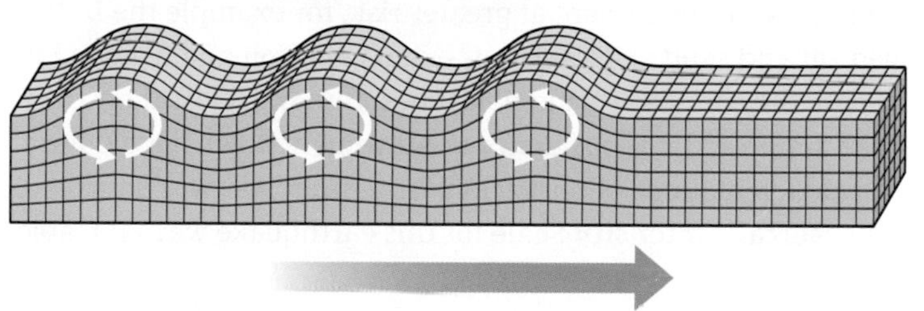

Rayleigh wave: Only travels through the surface of the crust, in a rolling motion, the ground is moved up and down and side to side. Responsible for most of the shaking felt by people.

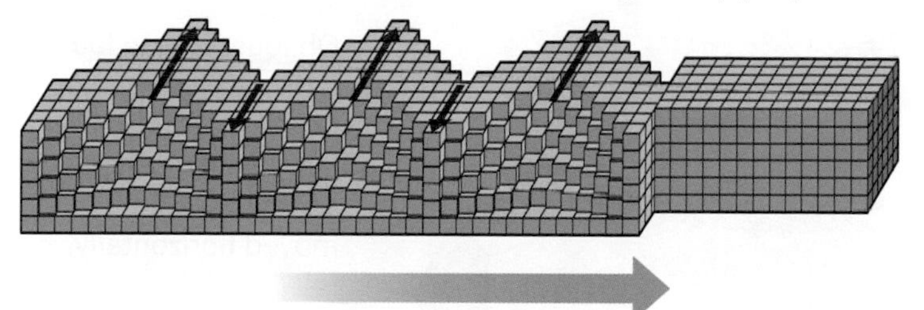

Love wave: Only travels through the surface of the crust, fastest of the surface waves and moves from side to side (horizontal) as it moves forward.

Figure 1.11: Types of earthquake (seismic) wave.

ACTIVITY

1. What is an S wave shadow zone? (See Table 1.1 on page 23)
2. How has the S wave shadow zone helped prove the plate tectonic theory?

AS level exam-style question

With reference to earthquake (seismic) waves, explain two reasons why it is difficult for buildings to remain intact during an earthquake event. (4 marks)

Guidance

Each type of earthquake (seismic) wave moves through the ground differently, and all the waves arrive within seconds of each other. Anything attached to the ground would also move.

A level exam-style question

Explain the link between plate boundary type and the strength of earthquake (seismic) waves. (4 marks)

Guidance

Think about which types of plate boundary create the most strain and release the greatest amount of seismic energy and why, and which release very little and why. Make reference to body and surface waves.

Secondary hazards

Secondary hazards resulting from an earthquake may include a **tsunami**, which can have far reaching impacts spatially (for example Indian Ocean 2004), or **landslides** (for example Kashmir 2005) and **liquefaction** (for example Christchurch 2011), which are localised impacts. Landslides occur when earthquake (seismic) waves loosen rock or unconsolidated material on steep slopes; material loses cohesive strength and moves downwards under the influence of gravity. Landslides are more common in hilly or mountainous areas, and may hit small settlements within this topography, and block or destroy roads and railways, making it difficult to rescue people or supply emergency aid and rebuilding materials. Liquefaction occurs in lowland areas where shaking sorts the ground material to the point where it acts as a fluid.

CASE STUDY: Ground movement in the Christchurch earthquake

The Christchurch earthquake in New Zealand (6.3 M_w February 2011) had a shallow focus (5.95 km) and was interesting as the shockwaves were amplified by nearby solid rock (Port Hills). Contrasting resonances occurred in the horizontal soft layers beneath the city. An upper soft layer, about 30 m thick, slapped back down onto the layer underneath, sending renewed vibrations back to the surface, exaggerating the liquefaction. The shallow oblique, steeply dipping fault rupture (thrust) hit the centre of the city and the eastern hill suburbs very hard, with the strongest ever recorded ground motion in the world (mostly vertical). Areas built on finer deposits with a high water table were at greater risk, for example the Dallington suburb. In places waterlogged silt and sand lost its strength and cohesion and behaved like a fluid, including eruptions onto the surface as 'sand volcanoes'. Buildings lost their firm foundations and could no longer be supported, so subsidence or movement downslope (lateral spreading) were common, Fitzgerald Avenue, one of Christchurch's main roads, was damaged due to 1.5 m of lateral movement. The **Mercalli Intensity Scale** for this earthquake was VII (Table 1.6, page 40).

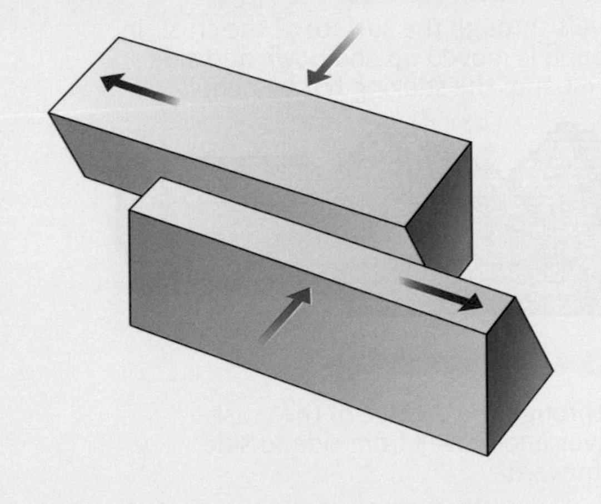

Oblique reverse faults indicate a shortening of the crust, with the land being pushed up as well as being moved horizontally.

Figure 1.12: Diagram of oblique reverse thrust mechanism experienced in the Christchurch earthquake of 2011

After the earthquake, cone penetrometer testing (CPT) was carried out to interpret soil characteristics across Christchurch and test for susceptibility to liquefaction. This is quick and inexpensive, but in some places boreholes were used to explore deeper. The aim was to identify the type of foundations necessary to protect buildings in the future, especially as 80 per cent of CBD buildings were damaged and had to be demolished. Many new buildings will have a grid of stone columns inserted into the soil to a depth of 10 m, with a geogrid of plastic matting and compacted sandy gravel above them. This method stabilises the ground and redistributes the weight of a building, reducing the stress on the ground and water pressure, and so reducing the risk of ejected sand. This method was tested by a 6.0 M_w earthquake in December 2011, and no

Synoptic link 

When urban areas are affected by a tectonic event it is important to understand the urban structure in order to fully assess the impacts. CBDs often have the most modern buildings and these may be earthquake proofed (for example Tokyo), but also this is where the most costly damage will be done. In low density suburbs damage is likely to be less but the possessions of individuals will be affected more. In cities of developing or emerging countries there may be vast shanty towns, with poorer people unable to help themselves (for example Port-au-Prince). (See page 50.)

Yellow = Players, Orange = Attitudes and actions, Purple = Futures and uncertainties

Figure 1.13: Earthquake damage in Christchurch

damage was caused where it had been used. Some new buildings will have base isolators. This is where buildings are not constructed directly onto the ground, but are put on flexible bearings or pads so that when the ground is moved by earthquake (seismic) waves the building does not move enough to cause significant damage or collapse (Figures 1.34 and 1.35 on page 60).

Historic buildings, such as the Clock Tower and College Hall, needed reinforcing. This was done by adding steel-reinforced concrete walls within existing wall cavities, bracing floors and roofs with plywood diaphragms, adding concrete roof braces to tie walls to roof trusses, putting vertical steel bars into stone columns and incorporating concrete into gable ends. In other buildings, pre-stressed laminated timber (Press-Lam) was used, with steel cables threaded through high grade timber walls and frames and then tightened. This system allows the walls and building frame to rock during an earthquake with the energy being absorbed by steel energy dissipaters, which can then be replaced after an event.

Synoptic link 

When considering the intensity of tectonic events, the stage of development of a country, and in particular its access to technology and finances, should be considered (for example the affordability of stronger housing and buildings). (See page 46.)

Tsunamis

Tsunamis are a major secondary hazard because of their potential destructive ability and the spatial area that they can cover. A tsunami usually consists of a sequence of waves with deep troughs in between, because water is drawn upwards into each wave. In deep water tsunamis may hardly be noticeable due to long wavelengths, but in shallow water the friction at the base of the wave slows them down, shortening the wavelength and allowing them to build in height (maybe 20 times higher than in open water). Tsunamis are created by **water column displacement**, mostly by undersea plate movements where part of the seabed is thrust upwards (or downwards) very quickly, or by explosive volcanic eruptions on a volcanic island (which may involve landslides or cone collapses into the sea), or underwater landslides such as at continental shelf edges. The Tohoku tsunami took 22 hours to cover the Pacific Ocean in a complex pattern due to reflection and refraction by islands and land masses, and the bathymetry of the ocean. There is a tsunami **intensity** scale (Soloviev 1978) (Table 1.6, page 40).

Synoptic link 

With tsunami waves it is important to understand local coastal processes, for example erosion may completely remove land areas that were previously occupied, as occurred in Banda Aceh, and integrated coastal zone management should consider tsunami impacts in regions likely to be affected. (See pages 160–161.)

ACTIVITY

1. Study the photograph of the 2011 Tohoku, Japan tsunami (Figure 1.14) and the cross-sectional diagram (Figure 1.15).

    a. Using the tsunami intensity scale (Table 1.6, page 40), suggest the level of intensity of this event.

    b. Describe and explain the physical characteristics of this tsunami event.

Figure 1.14: Photograph showing the Tohoku earthquake tsunami 2011

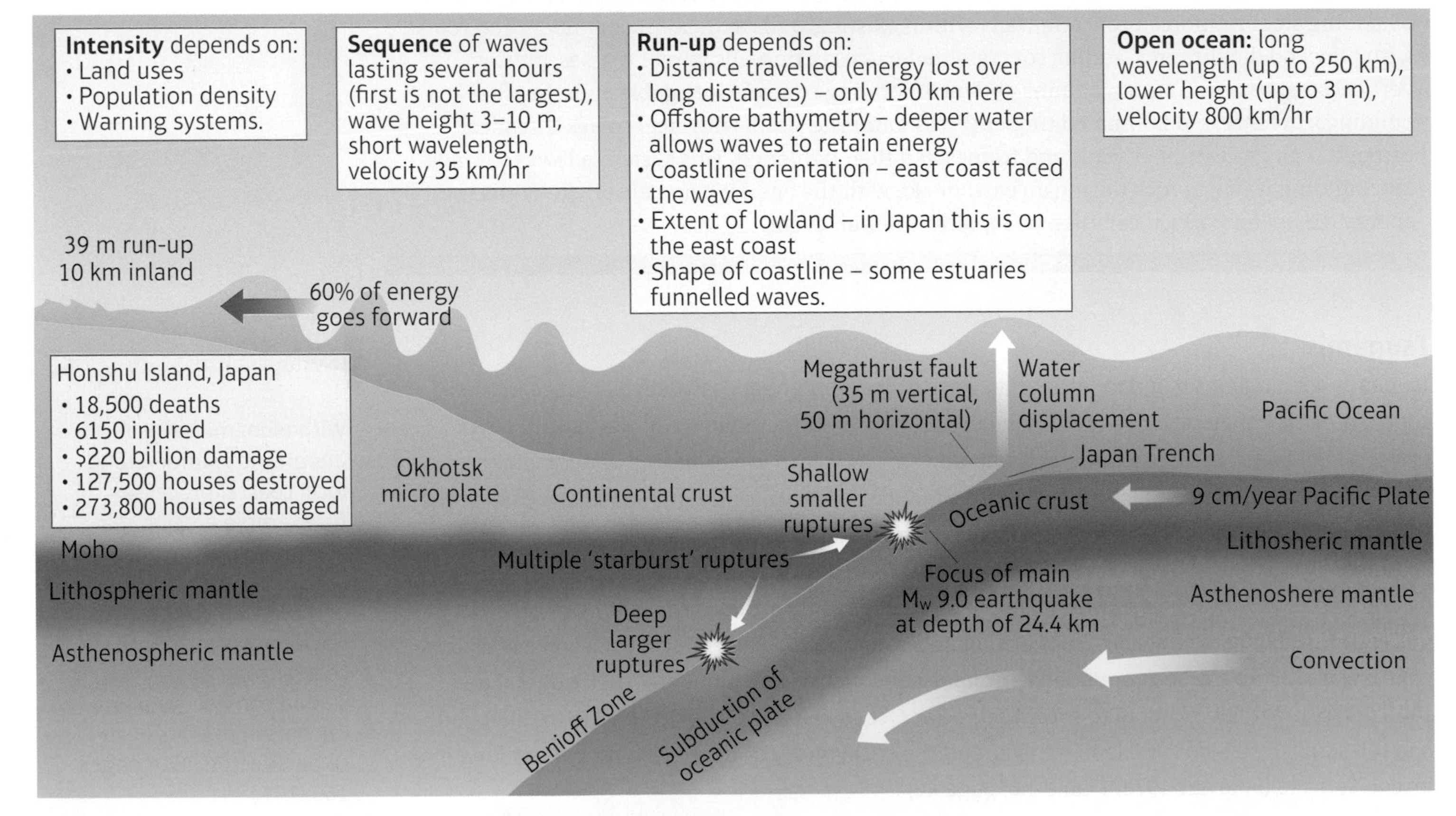

Figure 1.15: 2011 Tohoku earthquake and tsunami features and processes

Yellow = Players, Orange = Attitudes and actions, Purple = Futures and uncertainties

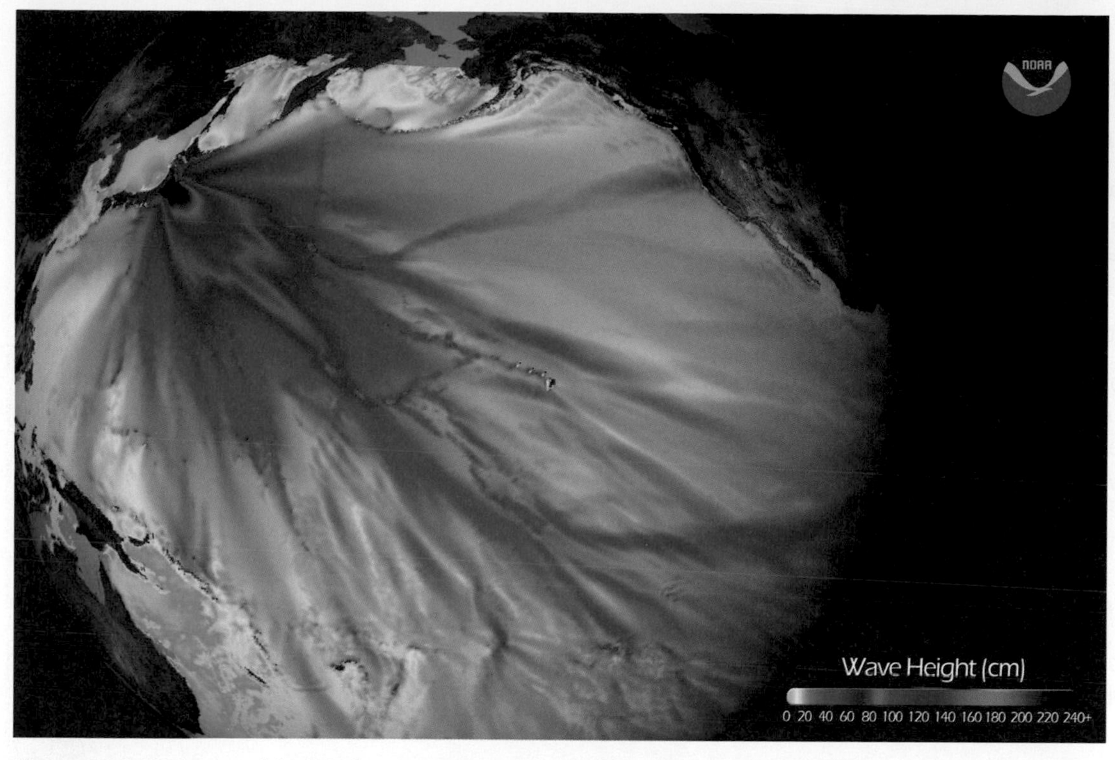

Figure 1.16: GIS image of the Tohoku tsunami direction and height

Extension

Study Figure 1.16.

- Describe the spread of the Tohoku tsunami. Consider direction, countries and wave heights.
- Investigate imagery of the Tohoku tsunami by viewing photographs and video clips.

Table 1.3: Selected data for the Tohoku tsunami wave in the Pacific Ocean

Location	Distance from epicentre (km)	Arrival time (minutes after earthquake)	Velocity of tsunami wave (km/hr)	Tsunami height at the location (m)
Midway Island	4100	302		1.27
Kahului, Hawaii	6345	486	783.33	1.74
Vanuatu	6345	596		0.69
Hilo, Hawaii	6535	503	779.52	1.41
Crescent City, California	7835	671	700.60	2.02
Port San Luis, California	8500	668	763.47	1.88

ACTIVITY

NUMERICAL AND STATISTICAL SKILLS

1. Study Table 1.3.
   a. Calculate the two missing velocities for the Tohoku tsunami wave.
   b. Describe and explain the patterns and trends shown by the data and in particular suggest reasons for any anomalies. You may wish to consult a map source to check the location of the named places as well as the GIS image (Figure 1.16).

Volcanic processes

Volcanoes are extrusive features found on the Earth's surface, ranging from gentle fissure eruptions to explosive composite cones. This range of volcano types reflects the amount of energy released during an eruption (Table 1.7, page 40). Supervolcanoes also exist, such as Yellowstone in the USA or Taupo in New Zealand, with large calderas and very long time periods between eruptions, but when they happen they may have a significant impact on the world. The Yellowstone eruption 2.1 million years ago was 6000 times larger than the Mount St Helens eruption of 1980, and 500 times greater than the Mount Pinatubo eruption of 1991. The shape of volcanoes is mainly related to the type of lava erupted, of which there are three main types: basalt, andesite and rhyolite (Table 1.4, page 34).

Table 1.4: Details of lava types

	Basaltic lava	Andesitic lava	Rhyolitic lava
Temperature	Hottest (1000 to 1200°C)	800 to 1000°C	Coolest (650 to 800°C)
Main minerals	Low silica (50%), water, gases and aluminium. High CO_2, iron and magnesium.	Intermediate silica (60%), gas content, magnesium and iron. High water and hydrochloric acid. Low SO_2.	High silica (70%), potassium, sodium, aluminium, and gas content. Low iron and magnesium.
Gas content	Low (0.5 to 2%)	3 to 4%	4 to 6%
Formed by	Melting of mantle minerals (e.g. olivine), mostly from upper zone but some from core-mantle boundary.	Subducted oceanic plate melts and mixes with seawater, lithospheric mantle and continental rocks.	Melting of lithospheric mantle and slabs of previously subducted plate.
Flow characteristics	Thin and runny (low viscosity, gases escape).	Slow (intermediate viscosity traps gases).	Thick and stiff (high viscosity traps gases).
Eruption energy	Gentle, effusive	Violent, moderately explosive	Very violent, cataclysmic
Location	Ocean hot spots, mid-ocean ridges, shield volcanoes (e.g. Mauna Loa).	Composite cone volcanoes (e.g. Chances Peak), subduction zones.	Supervolcanoes (e.g. Taupo), or composite cone volcanoes.

Volcanic hazards

Volcanic hazards increase with the magnitude of the event, and composite cones (Figure 1.17) are more dangerous than shield volcanoes (Figure 1.18). Close to volcanic activity, the hazards include lava flows and emissions of gases or steam (phreatic eruption). At greater distances, **pyroclastic flows** are perhaps the deadliest volcanic hazard and **ash** may be carried by atmospheric processes far from the eruption site (transboundary). Significant secondary hazards involve water (often from melted snow or ice) in the form of **lahars** and **jokulhlaups**.

Synoptic link

The lightest tephra (ash) emitted during a volcanic eruption may reach the upper troposphere or even enter the stratosphere. It is then affected by atmospheric processes, such as the general movement of air from west to east in mid-latitudes, the planetary wind belts, high altitude jet streams and the Quasi-Biennial Oscillation (QBO) in the stratosphere. These processes determine the spread of ash and the spatial scale of its impacts. (See page 72.)

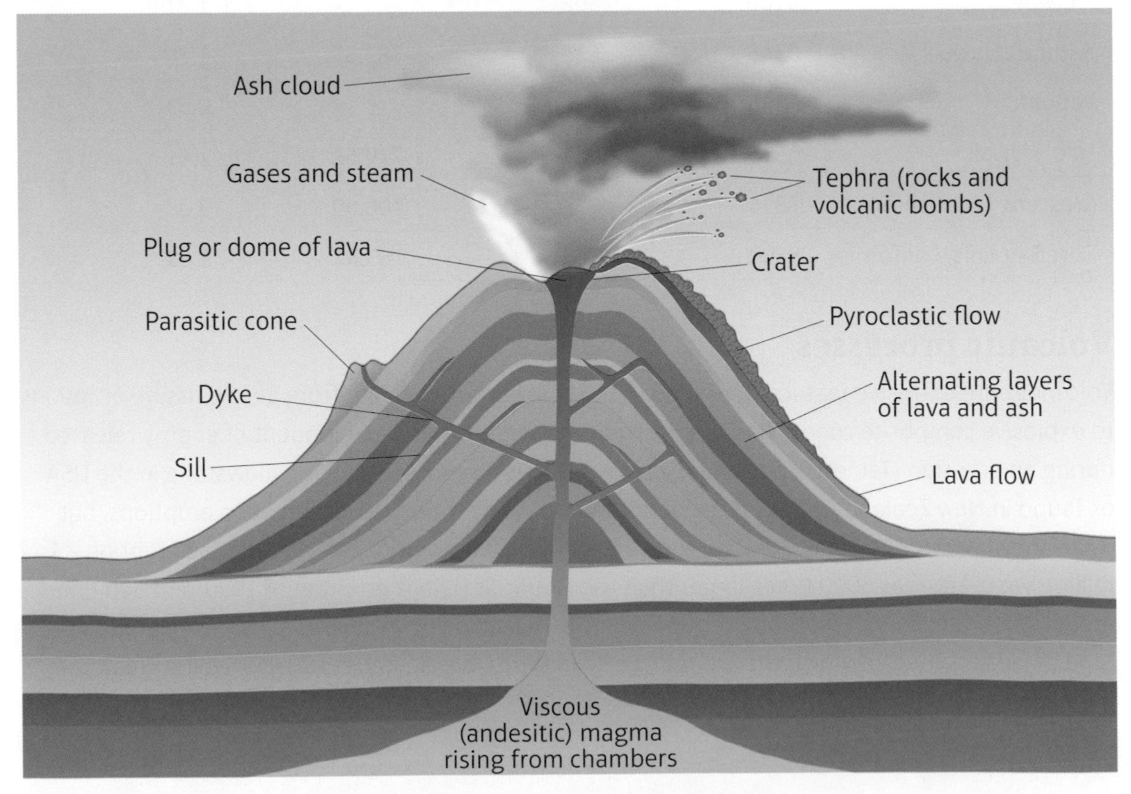

Figure 1.17: Cross-section through a composite cone volcano (with hazards)

ACTIVITY

Study Figures 1.17 and 1.18. Write a comparison of the characteristics of a composite cone volcano and a shield volcano.

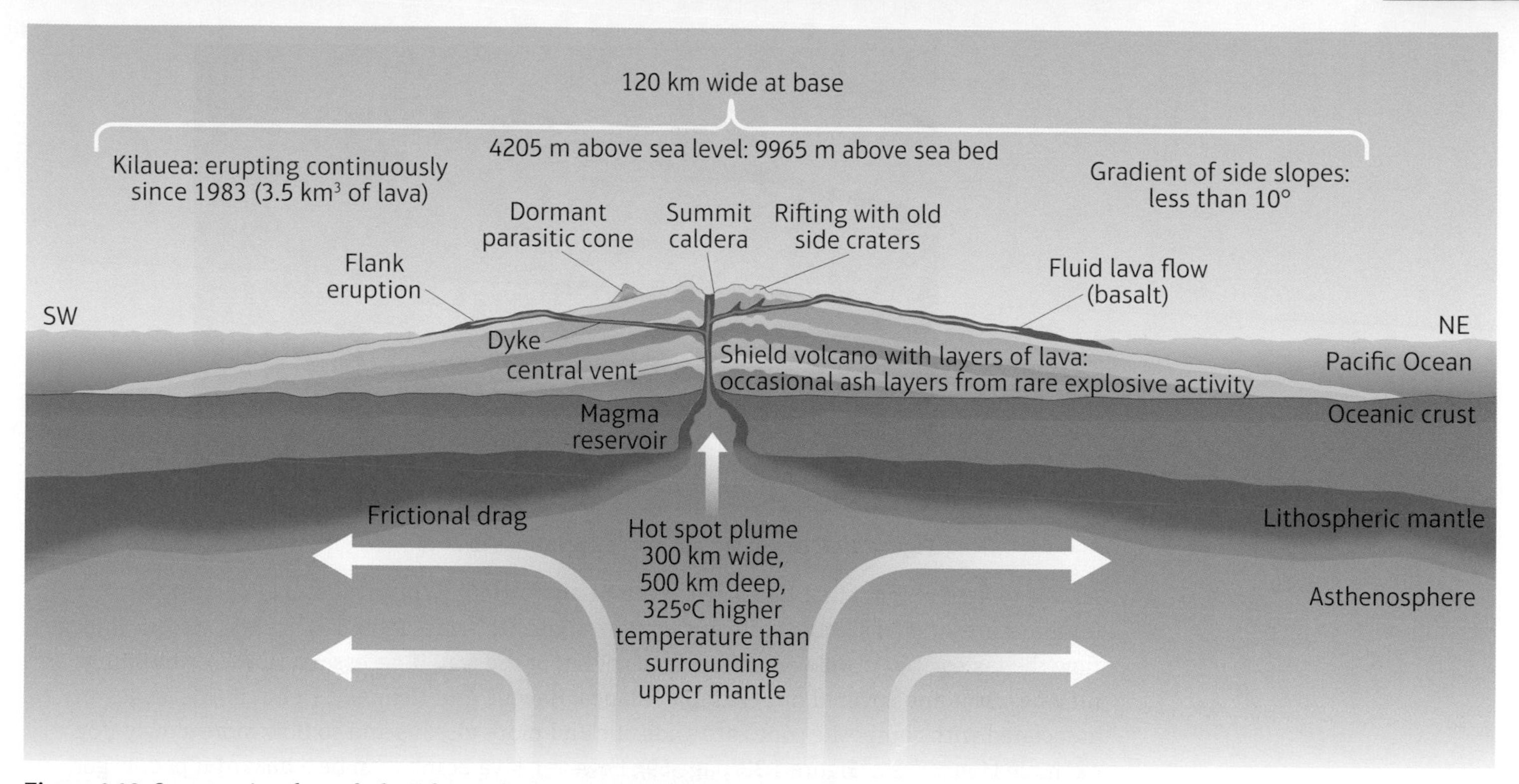

Figure 1.18: Cross-section through the Kilauea shield volcano in Hawaii

CASE STUDY: Volcanic hazards

A *jokulhlaup* is a flood of meltwater issuing from underneath an ice cap or glacier. This happens when a volcano erupts underneath the ice and melts the ice so that it forms a lake, either in a crater or at the crater side where it is dammed by the ice. Eventually the warm meltwater will have sufficient volume to lift the ice away from the bedrock. When this happens the water bursts out underneath the ice carrying with it glacial moraines and blocks of ice. This moraine then becomes a fluvial deposit (sandur) as the jokulhlaup deposits the material in lowland areas or valleys. The speed and volume of the event will wash away anything in its path and change the route of river channels. The best examples of jokulhlaups occur in Iceland, where the main road along the south coast has often been cut and bridges washed away. The Eyjafjallajökull (EFJ) eruption of 2010, which was infamous for its ash clouds disrupting aeroplanes, also created jokulhlaups (such as the Makarfljot river), but these were anticipated and the main road was closed and breaches made in the road embankment to allow water and deposits to pass through without causing major damage.

Pyroclastic flows are a dense mixture of superheated tephra (all types of ejected material) and poisonous gases (up to 1000°C) moving rapidly down the sides of a composite cone volcano (at speeds of up to 700 km/h). This severe hazard occurs with explosive eruptions of molten or solid tephra, or from the collapse of part of a volcanic cone or thick **lava flow**. Gravity causes larger particles, including boulders, to move more slowly at ground level while lighter particles appear as a turbulent fast moving cloud of ash. Anything in the path of a pyroclastic flow is likely to be destroyed, either by impact, burying, burning or poisoning. For example, Chances Peak on Montserrat in 1997 destroyed the capital town of Plymouth with a pyroclastic flow, and in 2010 350,000 people had to be rescued, some dramatically, from the slopes of Mount Merapi in Indonesia (Figure 1.19, page 36). As well as destruction, pyroclastic flows leave behind thick layers of volcanic deposits (ignimbrite), up to 200 m thick, and may cause secondary hazards such as river flooding if they temporarily dam them (for example Pinatubo, Philippines 1991).

Figure 1.19: Mount Merapi rescue operations in 2010

Lava is molten magma that has reached the Earth's surface. While it is still at very high temperatures it is in a liquid state and so flows. Basaltic lavas (Table 1.4) are less viscous and so flow the fastest, but people are still able to get out of the way if necessary. However, buildings may be burnt and covered (for example Kilauea, Hawaii, and Heimaey, Iceland). The lavas associated with composite cones are andesitic and more viscous and so flow more slowly (for example Mount Etna, Figure 1.32, page 59); however, lava bombs can be a hazard if people get too close. One lava **disaster** occurred on the flanks of the Nyiragongo volcano in 1977 when a flank fissure caused a lava lake to drain in less than an hour from the crater (at a rate of 6000 m^3/s). Fluid lava, with a velocity of between 20 and 60 km/h, flowed for 9 km, swamping several villages while people were asleep, unofficially killing 400 people.

An explosive volcanic eruption will eject solid and molten rock fragments into the air, these are known collectively as *tephra*. The smallest fragments (ash) are carried high into the atmosphere, where atmospheric systems such as jet streams and planetary winds carry the ash some distance from the eruption site. Most ash will fall locally and may cause roofs to collapse, choke machinery and electronics, cause breathing difficulties for people and animals and bury crops and vegetation. High in the atmosphere it can disrupt flight paths of aeroplanes because of the potential failure of jet engines and abrasion of fuselages. There is a network of Volcanic Ash Advisory Centres around the world to monitor ash clouds. An example was the 2010 eruption of Eyjafjallajökull (EFJ) in Iceland, where the ash cloud was present across the trans-Atlantic flight paths, disrupting aircraft movements to and from Europe for a week (Figure 1.20). There were also local evacuations (700 people) in Iceland, distribution of advice to people on avoiding inhalation of the ash, and the main road had to be closed due to poor visibility.

ACTIVITY

TECHNOLOGY AND ICT SKILLS

Describe the location and extent of the ash cloud shown in the GIS image Figure 1.20.

Investigate the spread of volcanic ash clouds and gases by viewing clips on the EUMETSAT and NASA websites.

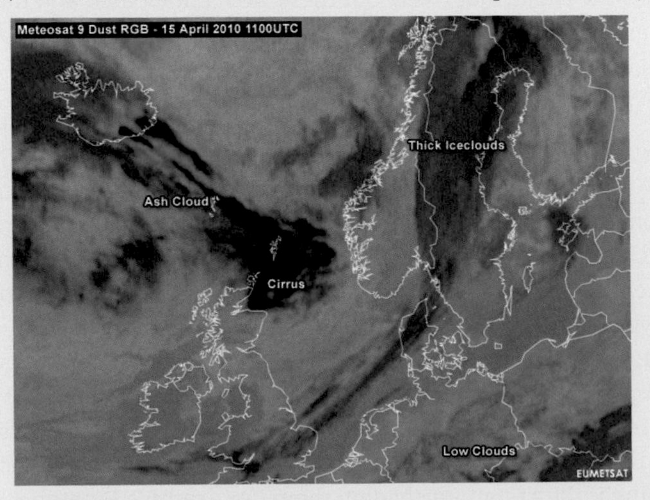

Figure 1.20: GIS image of ash cloud from the Eyjafjallajökull volcano in 2010

Yellow = Players, Orange = Attitudes and actions, Purple = Futures and uncertainties

The most common *volcanic gas* is water vapour (70–90 per cent), which can lead to heavy rainfall events (and possible lahars) at the location. Steam, sometimes containing other gases, escapes from geothermally active areas and can be seen in the form of geysers or fumaroles (Figure 1.21), attracting tourists who need to be careful where they walk. Another common gas is sulphur dioxide which forms fine particles that get carried as high as the stratosphere. These solid particles reflect radiation from the Sun back into space, so cooling the Earth. The eruption of Mount Pinatubo in 1991 produced a 20 million tonne sulphur dioxide cloud which rose to a height of 32 km and cooled the Earth for 3 years by about 0.6°C. Sulphur dioxide can combine with water to form 'acid rain' which may corrode human structures or affect vegetation. Fluorine gas can be poisonous in high concentrations and can contaminate farmland or water supplies. Tiny amounts of carbon dioxide are also emitted during eruptions, but it would take 700 large eruptions a year to match the anthropogenic emissions.

Figure 1.21: geysers in a geothermal area of New Zealand

Lahars consist of water mixed with volcanic deposits flowing rapidly along existing valleys. They are caused by either heavy rainfall, a result of humid air being seeded with volcanic ash which accelerates condensation, the formation of clouds and rain; or the emptying of a crater lake; or through the melting of snow and ice due to the heat from an eruption at high altitude; or a mudslide/landslide on the flank of the volcano. The dangers from lahars are their fast velocity, the amount of material carried (they can contain boulders 10 m in diameter), and the great distances they can travel. The town of Armero, Colombia, was buried by a lahar from the Nevado del Ruiz volcano in 1985 after pyroclastic flows melted 10 per cent of the snow and ice cover on the volcano. Several small lahars were funnelled into 6 major river valleys and grew in size with the addition of river water. The lahars travelled at 60 km/hr and in places were 50 m deep. Within 2.5 hours of the eruption, lahars had travelled 100 km and killed 23,000 people, mostly in the town of Armero which was at the exit of a canyon. The main pulse arrived at 11.35 p.m. so people were in their homes asleep and unable to escape. Buildings and people were swept away and buried by 2–5 m of deposits. A further 10 lahar surges prevented the escape of any who initially survived.

Literacy tip

Expand your vocabulary with wider reading. For example, do you know the meaning of anthropogenic, cloud seeding, particulates, transboundary, seismic, aseismic, magnitude and intensity?

CHAPTER 1

Why do some tectonic hazards develop into disasters?

Learning objectives

1.4 To understand that tectonic disaster occurs when the risk to people is increased by the magnitude of a hazard and vulnerability of people.

1.5 To understand that tectonic hazards can be compared through their profiles, to help understand contrasting impacts and the links to vulnerability and resilience.

1.6 To understand that levels of development and governance are important in determining disaster impact, vulnerability, and resilience.

Hazards, vulnerability, resilience and disasters

Defining risk, hazard and disaster

A natural tectonic event that does not affect people in any way is not a hazard. As soon as an event disrupts normal daily home or work routines then it becomes a hazard, and the level of severity may increase to cause the destruction of property and death. In a world that could be considered to be overcrowded, it is now rare for a tectonic event *not* to become a hazard in some way, and what is increasingly possible is that the hazard becomes a disaster. So when does a **natural hazard** event become a disaster or **mega-disaster**? This can be viewed quantitatively through numbers of deaths and costs of damage (**socio-economic impacts**) (Table 1.5), or qualitatively in human terms such as the amount of upset within a community. The United Nations suggests 500 or more deaths as a disaster, with a catastrophe (mega-disaster) being when there is:

- over 2000 deaths, or
- over 200,000 made homeless, or
- the GDP of a country is reduced by at least 5 per cent, or
- dependence on aid from abroad for a year or more after the event.

It is difficult to measure the grief of those affected by disasters, but it is certain that it can take families and communities years to overcome. Tohoku 2011 was a mega-disaster because there were over 20,000 deaths but only a 3.5 per cent fall in GDP. Haiti 2010 was a mega-disaster because it cost 100 per cent of GDP, had over 200,000 direct deaths and needed international aid, such as from MSF, for several years. Chile 2010 was a disaster with just over 500 deaths. Few volcanic hazards ever reach mega-disaster proportions due to their greater temporal and **spatial predictability** and smaller affected area (Table 1.5). However, a few have had a wider impact. For example, Tambora, Indonesia in 1815 (VEI 7), is estimated to have killed 92,000 people. Approximately 12,000 were killed by the blast, some by a tsunami, but most on Sumbawa and Lombok islands by starvation when ash covered crops and choked livestock. Farming in Europe and North America was also hit by global decreases in temperature.

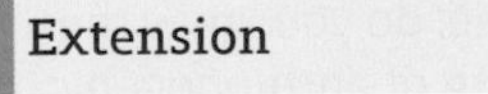

Extension

Assess the Bradford disaster scale (1990).

Yellow = Players, Orange = Attitudes and actions, Purple = Futures and uncertainties

Table 1.5: Table showing top five tectonic disasters by cost, fatalities and insured losses

Tectonic event	Cost (US$m)	Tectonic event	Estimated fatalities	Tectonic event	Insured losses (US$m) and % of total losses
(1) Earthquake and tsunami: Tohoku, Japan 2011	210,000	(1) Earthquake: Tangshan, China 1972	655,000	(1) Earthquake and tsunami: Tohoku, Japan 2011	40,000 (19%)
(2) Earthquake: Kobe, Japan 1995	100,000	(2) Earthquake: Haiti, 2010	222,570	(2) Earthquake: Christchurch, New Zealand 2011	16,500 (69%)
(3) Earthquake: Sichuan, China 2008	85,000	(3) Earthquake and tsunami: Indian Ocean 2004	220,000	(3) Earthquake: Northridge, California, USA 1994	15,300 (35%)
(4) Earthquake: Northridge, California, USA 1994	44,000	(4) Earthquake: Kashmir, Pakistan/ India 2005	88,000	(4) Earthquake and tsunami: Concepcion, Chile 2010	8,000 (27%)
(5) Earthquake and tsunami: Concepcion, Chile 2010	30,000	(5) Earthquake: Sichuan, China 2008	84,000	(5) Earthquake: Canterbury, New Zealand 2010	7,400 (74%)

ACTIVITY

Study Table 1.5. Analyse the differences between developed and developing (or emerging) countries shown by the data.

Magnitude and intensity scales

When judging the impact of a tectonic disaster there are several scales that can be used to put the events into context. There are two main scales for measuring the *magnitude* (energy released) of an earthquake. The Richter Scale (M_L) (1935) uses the arrival times of the P and S waves, the amplitude of the S wave, and distance from the epicentre; however, the Moment Magnitude Scale (M_w) (Kanamori 1977) is more accurate, as it uses the energy released by all shockwaves as well as the area of rupture and movement. Both scales are logarithmic, with each level being 10 times greater than the level below. In practice there is little difference between the measuring of earthquakes using either scale, except for the largest earthquakes, when the Moment Magnitude Scale is more accurate (Table 1.6, page 40). Scientifically, the M_w Scale should be used, although in the news the Richter Scale is still referenced.

There is also a scale for measuring the *intensity* of an earthquake; this uses the amount of damage caused by the earthquake (seismic) waves and is called the Mercalli Scale (Table 1.6, page 40). Usually the stronger the magnitude of an earthquake the higher the intensity, but local conditions may modify this correlation through secondary hazards or where the epicentre coincides exactly with human settlements.

ACTIVITY

Study Tables 1.6 and 1.7. Suggest the level required on these scales to create a disaster and the level for a mega-disaster. Explain your answer.

AS level exam-style question

Explain the geographical criteria that can be used to decide if a tectonic event is a hazard, disaster, or mega-disaster. (6 marks)

Guidance

As well as recalling definitions by organisations such as the UN, also consider the suggestions of impacts contained within the scales used for measuring volcanoes, tsunamis and earthquakes.

A Level exam-style question

Explain the correlation between the magnitude and intensity scales used for measuring earthquakes and their secondary hazards. (4 marks)

Guidance

Show clearly that you understand magnitude and intensity within your answer. Think about the similarities and differences between the levels by recalling the descriptors for selected levels, and remember the logarithmic nature of magnitude scales. Reference to examples of events would help to illustrate your answer with Mercalli and Moment Magnitude scales.

Table 1.6: Selected levels from magnitude and intensity scales

Moment Magnitude Scale		Mercalli Intensity Scale		Tsunami Intensity	
3.0	Minor; over 100,000 a year at this level; rarely causes damage; energy = 134.4 gigajoules (GJ)	IV	Felt indoors; people woken at night; windows rattle	III	2 m run-up height; flooding of lowest land; light vessels carried away; some damage to weak buildings; reversed river flow
5.0	Moderate; between 1000 and 1500 per year at this level; damage over a small region; energy = 134.4 terajoules (TJ)	VII	Slight to moderate damage of well-constructed buildings; aseismic designs negligible damage; walls and plaster cracks; people alarmed	IV	4 m run-up height; shore flooded to some depth; scouring of land; structures damaged; large vessels swept inland or out to sea; lots of debris
7.0	Major; 10 to 20 per year at this level; serious damage over large area; energy = 210 petajoules (PJ)	X	Wooden structures destroyed; masonry and frame buildings destroyed; rails bent; ground cracks; landslides on steep slopes	V	8 m run-up height; general flooding; significant structures damaged; soil scoured away; coast littered with floating debris; all vessels carried inland or out to sea; large bores in estuaries; harbours damaged and people drowned
9.0	Great; one every decade or more at this level; serious damage in a region of several hundred miles diameter; energy = 23.5 exajoules (EJ)	XII	Total destruction; lines of sight and level distorted; objects thrown into air; river courses altered; topography changed	VI	16 m plus run-up height; complete destruction of many buildings some distance from shore; coastal lowland flooded to great depth; large ships damaged; many deaths

There is a separate *magnitude* scale for volcanic eruptions called the Volcanic Explosivity Index (VEI), this also has a logarithmic scale (Table 1.7), and is a useful way of comparing the energy released and the type of eruption.

Table 1.7: Table of selected levels of the Volcanic Explosivity Index (VEI scale)

Scale	Description
1	Small gentle eruption, generally less than 10^6 m of ejected material; small fissure eruptions with intermittent bursts of activity. Hawaiian type.
3	Moderate to large eruptions, sometimes severe with up to 10^8 m of ejected material and an eruption column between 5.5 km and 10.5 km high; eruption phases lasting up to half a day. Strombolian type.
5	Very large eruption with 10^{10} m of ejected material and an eruption column perhaps reaching 28 km high; can be a cataclysmic event. Composite cone (Plinian or Peléean) type (e.g. Mount St Helens 1980).
7	Very large, colossal eruption with 10^{12} m of ejected material and a column height of over 47 km with eruption phases over half a day. Composite cone (ultra-Plinian) type (e.g. Tambora 1815).

Yellow = Players, Orange = Attitudes and actions, Purple = Futures and uncertainties

ACTIVITY

1. Read about the volcanic processes, volcanic hazards and the VEI scale on the previous pages before completing the following activities.

- **a.** Create a brief fact file for Mount Merapi and Eyjafjallajökull.
- **b.** Was the eruption of Mount Merapi in 2010 a hazard, a disaster, or a mega-disaster?
- **c.** Was the eruption of Eyjafjallajökull in 2010 a hazard, a disaster, or a mega-disaster?

Tectonic hazard profiling

Pressure and Release model and hazard risk equation

The risk to people can be represented by a simple equation: Risk (R) = Hazard (H) × Vulnerability (V). This is sometimes shown visually in a Venn diagram (Degg's Model). It is the combination of physical factors, such as magnitude, and human factors, such as lack of knowledge, that combine to create the level of risk. The **Pressure and Release model** (Figure 1.22) suggests what should be tackled in order to reduce the risk of a disaster, such as root causes, dynamic pressures and unsafe living conditions. The physical factors and processes are difficult to change, certainly earthquakes are impossible to predict. However, small things such as building barriers to divert lava flows may be possible as used on Mount Etna and at Heimaey. A similar model is the Disaster Crunch Model, which highlights that a disaster will only happen when a hazard event impacts on vulnerable people.

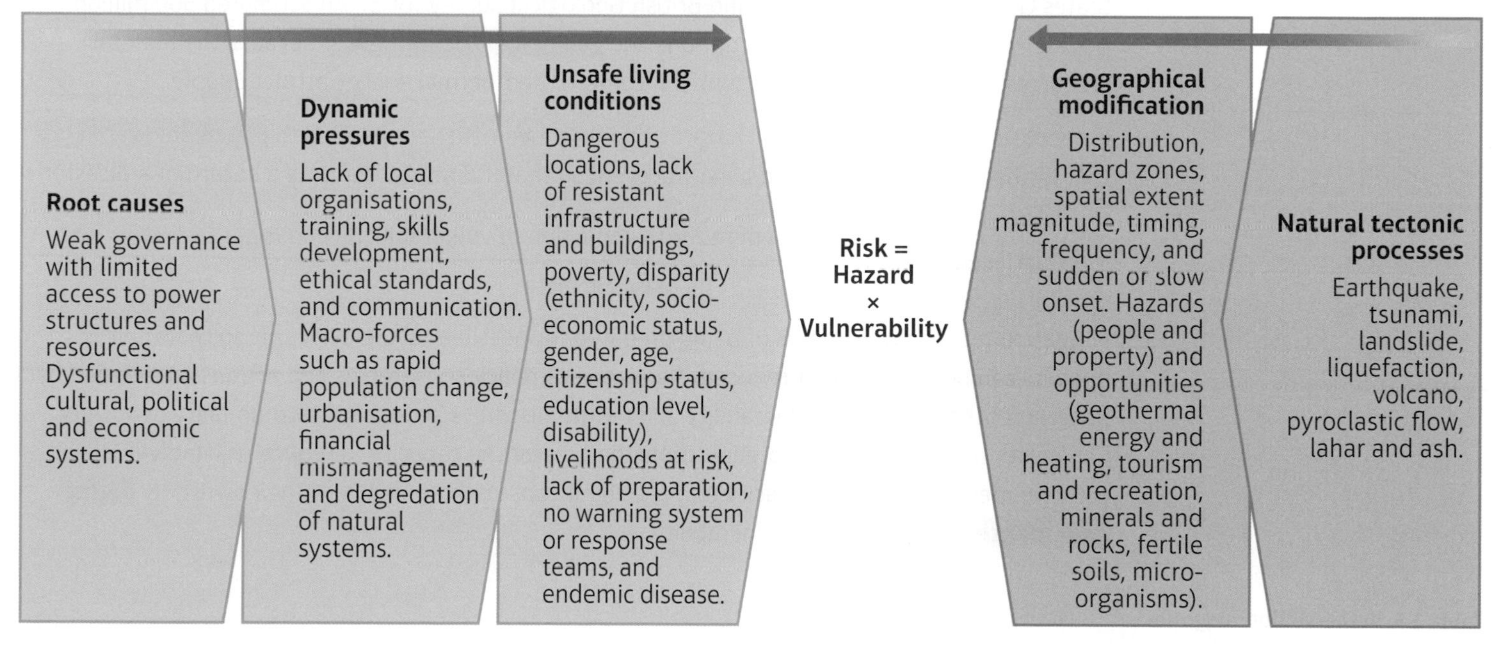

Figure 1.22: Pressure and Release (PAR) model

Risk (R) is the probability of harm or loss taking place. This includes deaths, injuries, trauma and upset, loss of livelihoods, damage or loss of property and disruption to economic activities. Some people also include the natural environment as a factor, and certainly damage to ecosystems may degrade areas so that they are less able to support people.

Risk can be viewed in terms of damage thresholds; at the upper extreme a magnitude 9 earthquake may cause a huge amount of damage, while a moderate volcanic eruption (VEI 3) is above the damage threshold and causes some losses, but a small volcanic eruption depositing new minerals in the soil will be within human tolerances and a benefit (Figure 1.22).

Hazard (H) is the earthquake or volcanic event itself, including relevant secondary hazards, with consideration of the character of the event (such as magnitude, **speed of onset**, spatial extent, frequency and duration).

Vulnerability (V) relates to the human geography characteristics, such as location of settlements, knowledge and understanding, the ability to react, **resilience**, **community adaptation** and **preparedness**, and technology for warning systems and protection (Table 1.8). There are contrasts around the world. For example, it has been estimated that only 0.06 per cent of the population was vulnerable in the Christchurch (2011) earthquake and 0.3 per cent in Tohoku (2011) earthquake, but 10 per cent in the Haiti (2010) earthquake and 85 per cent in the Bam (2003) earthquake.

In terms of reducing risk, it is clear that the factor that should be managed is vulnerability. This is because the energy of tectonic processes is beyond the ability of current technology to control. The prediction of earthquakes remains elusive and Charles Richter wrote that he believed that it would never be possible. One useful predictive tool is estimating the build-up of stress and strain by using the time gaps between earthquakes; this was scientifically successful for Haiti in 2010 and Nepal in 2015. But this method still does not give the precise location or precise time of an event. There are 700 active volcanoes on Earth, with about 50 erupting in any year. However, their locations are well known and there are precursors to eruptions, such as harmonic tremors (small earthquakes created as the magma forces its way to the surface), gas and steam emissions or bulges in the crater or sides of the volcano (for example Mount St Helens 1980). Many volcanoes are constantly monitored by Earth scientists and geologists, working for organisations such as United States Geological Survey (USGS) and British Geological Survey (BGS). An estimated 500 million people are at risk from volcanic eruptions and vulnerability is increasing because the mineral resources (including fertile soils), tourist income, or geothermal energy attract people.

If people are to live with tectonic hazards then improvements are needed to community resilience, that is, their ability to withstand a natural event so that it does not become a hazard or a disaster.

$$\textit{Risk reduction} = \frac{\text{mitigation of hazard} \times \text{reduction of vulnerability}}{\text{increase capacity}}$$

Mitigation means finding ways of being prepared for possible tectonic hazards so that their impacts can be prevented or reduced. Management policies, strategies and actions are needed by governments to minimise vulnerability and reduce disaster risk for all in a community or society. The better the strategies and organisation, the greater the capacity to reduce risk. Individuals or communities may lack the capacity to make significant improvements on their own, only better governance can help reduce vulnerability.

Table 1.8: Vulnerability categories

Physical vulnerability is when people live in hazard prone areas in buildings that offer little protection.
Economic vulnerability is when people risk losing their jobs, assets (for example their house) and money.
Social vulnerability is when a household or community is unable to support the disadvantaged people within it, for example political isolation may exist for the poor, females, elderly and rural residents.
Knowledge vulnerability exists when people lack education or training (and therefore understanding), and there are no warning or evacuation systems in place.
Environmental vulnerability exists where the area that people are living in has increased in hazard risk because of population pressure, forcing people into riskier areas.

Yellow = Players, Orange = Attitudes and actions, Purple = Futures and uncertainties

CASE STUDY: Tectonic vulnerability in the Middle East

A complicated fault area exists between the Eurasian and African continental plates (Figure 1.2, page 21). There is a long history of seismic activity and there have been some very damaging earthquakes, such as Bam (2003 (6.6 M_w, Mercalli IX) which killed over 31,000 people when unbaked clay houses collapsed, as well as destroying the UNESCO heritage citadel Arg-é Bam. There is high vulnerability as ordinary people are not involved with mitigating risks because they lack knowledge and understanding of the tectonic situation and how to be prepared. The absence of this has been a significant factor in many Middle East earthquake disasters.

Natural disasters have uneven effects on individuals within communities, both in terms of impacts and access to resources to help with recovery. In Turkey a culture of seismic mitigation had evolved in rural communities with aseismic building practices (earthquake resistant engineering methods). But this culture has changed in recent generations because rapid **urbanisation** (69 per cent urban in 1995, 73 per cent in 2014, 87 per cent by 2025) has replaced traditional building materials and methods with reinforced concrete for example. This allowed rapid cheap construction, but not safety from earthquakes.

- Rural community house building, with local knowledge and skills has been replaced by a dependence on contractors working to local and national government building regulations.
- As young adults have migrated away from rural areas, the opportunity to pass on knowledge and understanding of aseismic building practices is being lost.

After the 1999 Izmit (7.4 M_w, Mercalli X) earthquake which killed 17,000 people and badly damaged 245,000 buildings, there was a focus on training local people within communities in strategies to reduce the impacts of earthquakes; this raised community awareness of risks and led to simple mitigation steps. A leaflet with maps of Turkey's seismic risk areas and advice on making homes safer, such as fixing free standing furniture to the walls, was distributed. Turkey is particularly vulnerable (Figure 1.23) as the Anatolian Fault runs through the north of the country and a sequence of stress and pressure release earthquake pulses is making its way from east to west, getting closer to Istanbul, a megacity of 14 million people.

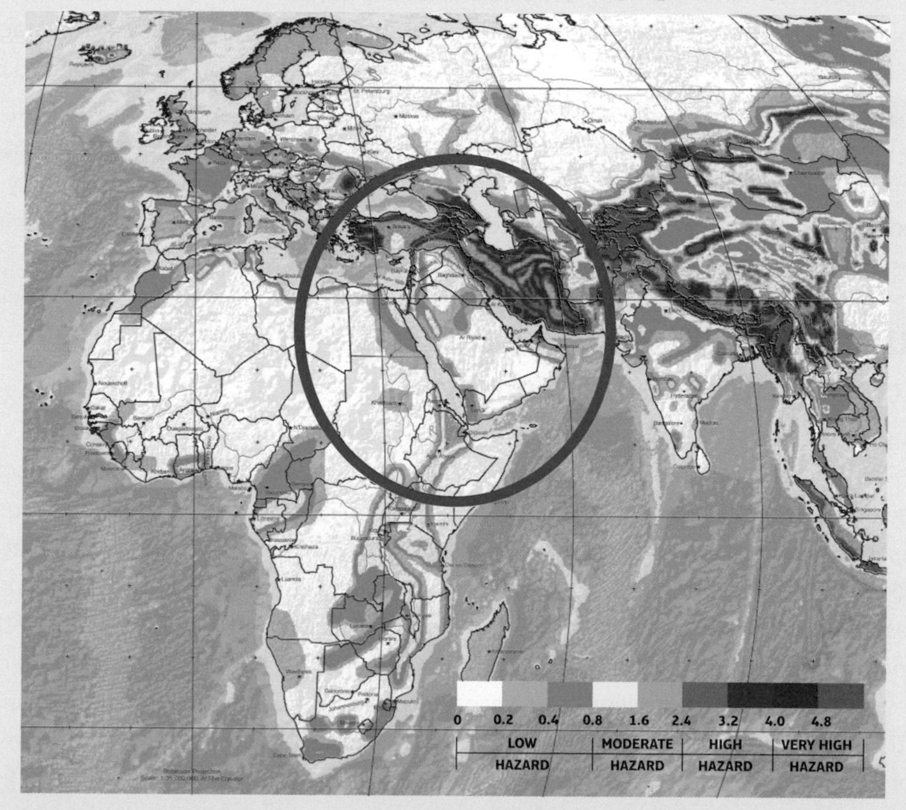

Figure 1.23: Peak ground acceleration probabilities in the Middle East

ACTIVITY

CARTOGRAPHIC SKILLS

Study Figure 1.23 and an atlas map of Iran. Analyse the seismic risk in Iran (reference to the Bam earthquake of 2003 should also be made).

ACTIVITY

Assess the role of governance in determining the vulnerability and resilience of a community to tectonic hazards.

In Egypt, ancient structures were partially earthquake proof, but recent attempts by Egyptian architects to adapt this knowledge to make strong cheap adobe housing (bricks made with clay rich soil and straw and baked by the Sun) for rural poor people met with resistance. This is because the frequency of earthquakes in Egypt is low and so people's perception of risk is poor, but also the social and educational structures in the country have not helped the distribution of knowledge and understanding. Providing accessible information to huge numbers of people at risk from earthquakes in the Middle East is a major task for the authorities. The Egyptian government produced a free booklet about earthquake disasters and how to prepare for them and distributed it through schools. It included useful scientific information supported by quotes from the Qur'an so that it would be acceptable to people. Egyptians gained knowledge on what to do if an earthquake occurs, but the emphasis was on emergency responses rather than action to mitigate the impacts. This could be seen in the Nile Valley villages where rebuilding after the 1992 Dahshur (5.5 M_w, Mercalli VIII) earthquake, which affected 8300 buildings, was completed by villagers without any technical guidance on how to 'build-back better'. So physical vulnerability still exists, and people living in informal housing on the outskirts of Cairo were excluded because officially they do not exist, and the people themselves do not want to attract attention.

For a culture of disaster prevention to be created it would appear to be important for governments to organise a flow of information from experts to ordinary people, and for the people themselves to be willing to take action based on this information.

Extension 

- Investigate the preparedness of Tokyo and Jakarta for a major earthquake (including a tsunami).
- Why is the preparedness of these two cities different?

Socio-economic impacts

Figure 1.24 shows the **hazard profiles** of selected earthquakes, tsunamis and volcanic eruptions for a developed country and an emerging/developing country. The USA, New Zealand and Japan have high Human Development Index (HDI) scores (0.914, 0.910 and 0.890 respectively in 2013) and high Gross Domestic Product (GDP) per capita ($55,904, $36,963 and $32,481 respectively in 2015), while Indonesia, Haiti and Montserrat have lower HDI scores (0.684, 0.471 and no data available respectively) and lower GDP per capita amounts ($3,416, $830 and no data available respectively). While Figure 1.24 shows that there are similarities in terms of physical characteristics (for example magnitude), there are notable differences in the impacts on the areas linked to levels of development. There are higher financial costs of damage in developed countries; but it must be remembered that even small costs may be significant to the people, communities and governments of developing/emerging countries (as shown by the GDP figures), which may create a dependency on international aid. Fatalities do appear to be higher in developing/emerging countries and this is linked to higher vulnerability.

Influence of development and governance

Impacts in different types of country

Increasing losses due to natural hazards around the world is largely a result of increasing vulnerability. It is not always those in financial poverty that are most at risk, but rather those with an overall disadvantage due to a combination of political, physical, social, cultural and economic factors (such as in the Middle East). The World Risk Report (WRR) of 2014 commented that the urban governments of rapidly growing cities face the major challenge of establishing planning measures to reduce vulnerability. Concern is increasing because by 2050, 66.4 per cent of the world's population will be urban (a total of 6.3 billion people). The World Risk Index (WRI) considers measures of exposure, susceptibility, coping capacity and adaptive capacity. Japan was ranked 17th and Haiti 21st (Table 1.9, page 46) so both countries have disaster risks, but for different reasons: Japan for its exposure to natural events and Haiti for its susceptibility and poor capacity. Insurance companies are concerned about the increasing risks and losses, especially in **multiple-hazard zones** (Table 1.9).

Yellow = Players, Orange = Attitudes and actions, Purple = Futures and uncertainties

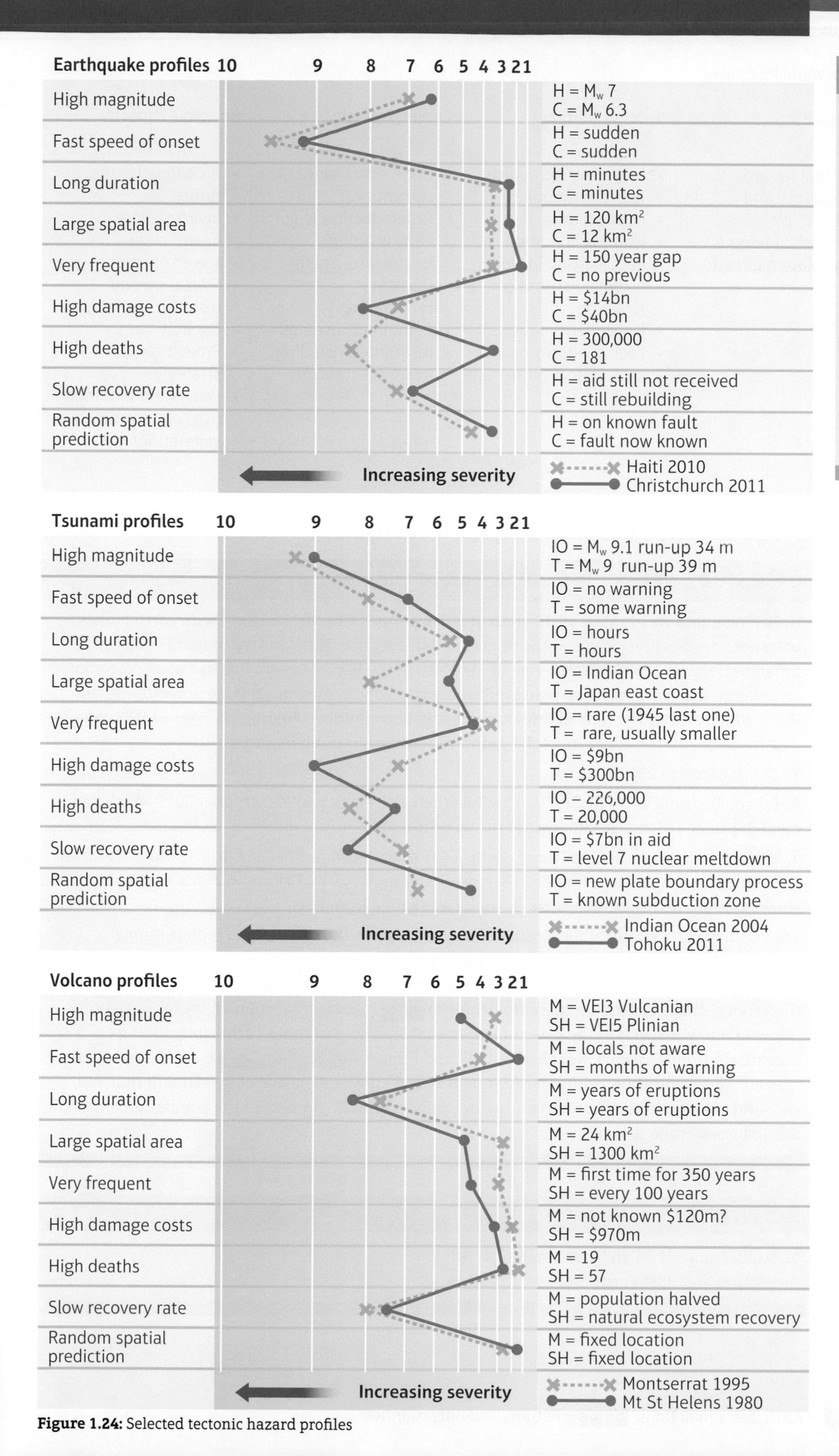

Figure 1.24: Selected tectonic hazard profiles

ACTIVITY

1. Study Figure 1.24.
   a. Compare the paired hazard profiles. Which tectonic hazard has the highest severity rating? Why? (Note that a log scale has been used).
   b. Compare the tectonic hazard impacts in developed countries with those in developing/emerging countries. Why are they similar or different?

Table 1.9: Factors considered by the World Risk Index

Top ten (2015) highest risk	Exposure	Susceptibility	Coping capacity	Adaptive capacity
• Vanuatu • Tonga • Philippines • Guatemala • Solomon Islands • Bangladesh • Costa Rica • Cambodia • Papua New Guinea • El Salvador	• Natural hazards • Frequency and probability • Number of people (population growth)	• Public infrastructure (e.g. sanitation) • Food security • Housing condition • People living in slums • Income distribution (poverty) • Dependency ratio • Lack of education • Lack of health care	• Good governance (no corruption) • Risk reduction (e.g. early warning) • Healthcare system • Disaster preparedness • Community strength • Insurance (including micro-insurance)	• Preparing for the future • Investment (e.g. health care) • Speed of response • Education and research levels • Equality • Projects and strategies • Ecosystem protection (e.g. water, protective vegetation)

ACTIVITY

1. Study the list of top ten countries on the WRI in Table 1.9, Figure 1.2, page 21, and an atlas.

a. Where are most of these countries located?

b. Does this location affect their exposure to tectonic hazards?

c. What must countries like Vanuatu do to increase their coping capacity?

Extension

Papua New Guinea is rated as the most susceptible out of the top ten riskiest countries; investigate why.

Inequalities, vulnerability and resilience

CASE STUDY: Vanuatu, a multiple-hazard and risk zone

In 2015 Vanuatu was ranked as the most at risk country (WRI) in the world with earthquakes, volcanoes, tropical cyclones and the impact of sea level rise. This country consists of 65 inhabited islands 1800 km north-east of Australia in the Pacific Ocean. It is in a zone of tectonic complexity with several micro-plates and subduction of the Australian Plate under the Pacific Plate, which creates the New Hebrides trench and volcanic arcs (Figures 1.2, page 21, 1.5, page 24 and 1.6, page 24). The exposure of the country to hazards is demonstrated by

1. Cyclone Pam in 2015,

2. Frequent strong earthquakes, for example the 7.1M_w earthquake in October 2015 which had a **focal depth** of 130 km (100 km east of the trench), and

3. Volcanic eruptions, such as Mount Yasur, a 405 m high composite cone, which was active from 2013 emitting ash and lava bombs in Strombolian eruptions, and Mount Gaua, a 979 m high shield volcano with a caldera 8x6 km, which was active from 2011.

The country is also vulnerable because of isolated islands (for example no mobile phone service) spread over a large area, women's groups lacking access to resources, poverty ($3,124 GDP per capita [129th position]), a stagnant Human Development Index of 0.616 (131st position), undernourishment and little space for people to move into safer locations. Coping capacity is low in many respects as the government is unable to help in times of disaster, relying on outside aid from Australia, New Zealand, UK and France. Adaptive capacity is low because there is insufficient education (adult literacy rate is only 53 per cent), a lack of investment in health care and poor sanitation infrastructure (for example only 57 per cent of the population have an adequate sewerage system).

ACTIVITY

1. Study Figures 1.25 and 1.26. Analyse which megacities are most at risk from earthquakes.

Yellow = Players, Orange = Attitudes and actions, Purple = Futures and uncertainties

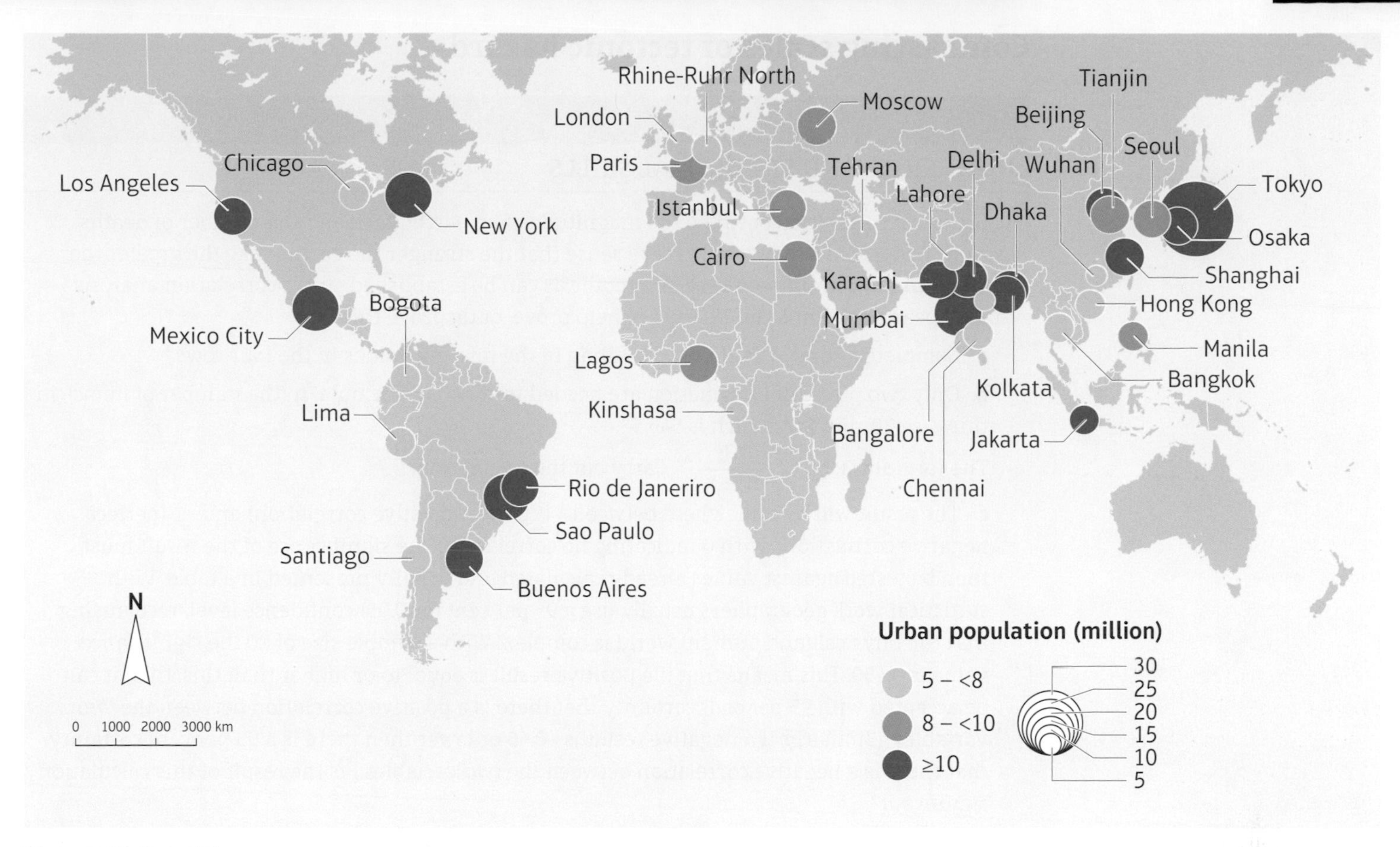

Figure 1.25: Megacities

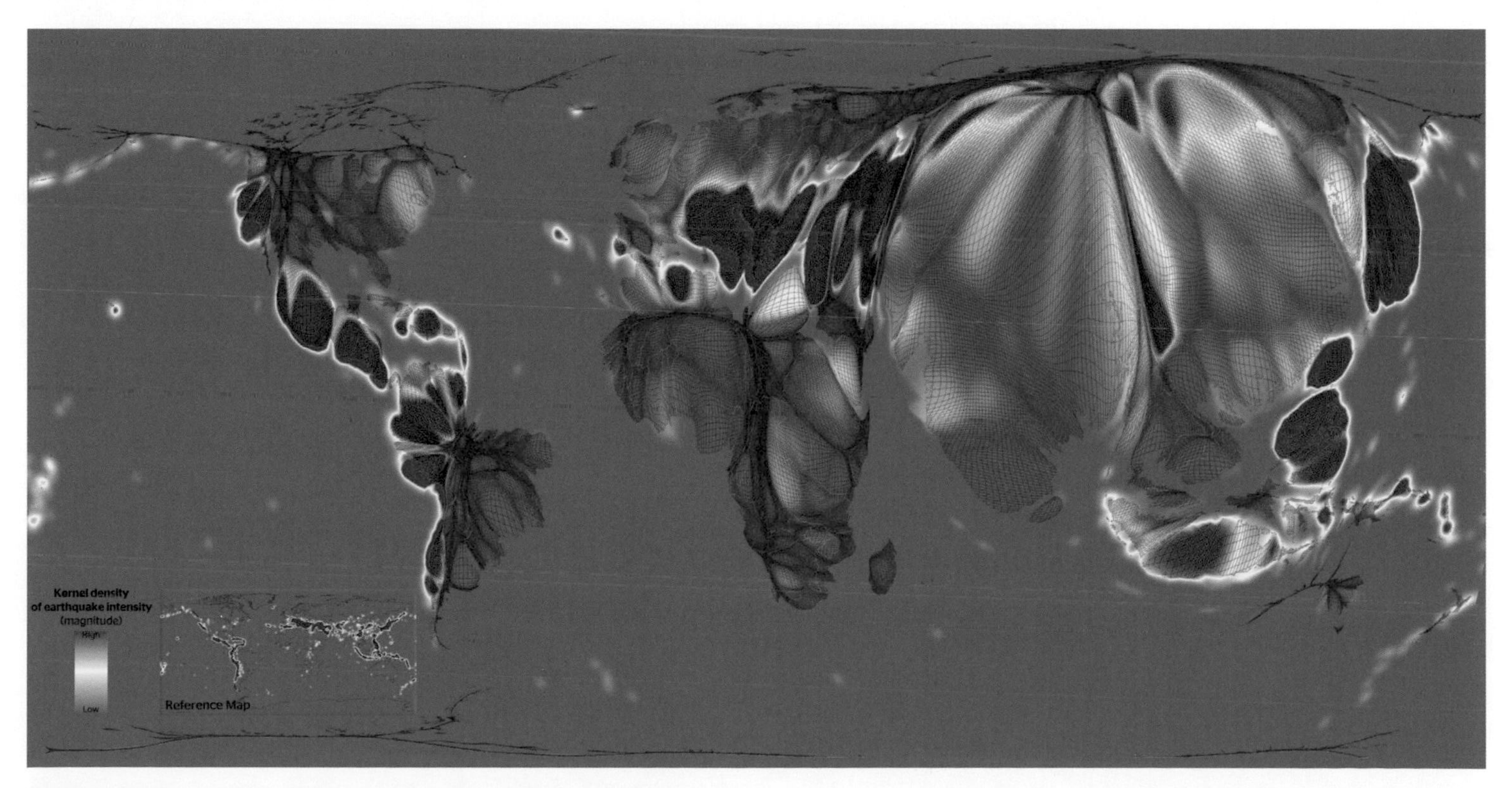

Figure 1.26: Global earthquake intensity and population distribution

Contrasting scales of tectonic hazard

ACTIVITY

NUMERICAL AND STATISTICAL SKILLS

1. Is there a correlation between the magnitude of an earthquake and the number of deaths caused? Geographically it would make sense that the stronger the earthquake the greater the damage and loss of life would be. A hypothesis can be established, and a correlation analysis called Spearman's Rank can be used to help prove, or disprove this.

a. Complete a copy of Table 1.10 by filling in the missing values in the four rows.

b. Only two pieces of information are needed to solve the formula, **n** (the number of items), in this case 20, and $\mathbf{\Sigma d^2}$, which is 546.5.

The formula is $r_s = 1 - \frac{6\Sigma d^2}{n^3 - n}$ Carry out the calculation.

c. The result will be somewhere between +1 (perfect positive correlation) and –1 (perfect negative correlation), with 0 indicating no correlation. The significance of the result must then be tested against values already calculated and usually presented in a table. With statistical work geographers usually use a 95 per cent (or 0.05) confidence level, recognising that the physical (and human) world is complex. With a sample size of 20 the significance value is 0.450. This means that if a positive result is equal to or higher than this, then it can be accepted with 95 per cent certainty that there is a positive correlation between the two variables. (Similarly, if a negative result is –0.45 or lower then there is a 95 per cent certainty that there is a negative correlation between the two variables.) Is the result of this calculation significant?

d. Write a geographical explanation of the result. Consider factors such as the acceptance of the hypothesis, the location of the countries involved, the types of plate boundary involved, population densities, the level of development of the countries, magnitude and intensity, why the correlation was not even greater.

Maths tip

When carrying out the Spearman's Rank correlation analysis there are two common errors to avoid:

- When ranking values that are equal, it is necessary to allocate an average (mean) rank to each. For example, in this data set there are six earthquakes with a Richter magnitude of 8.3; these occupy rank positions 12 to 17. However, as the rank values are crucial to the calculation they cannot all be called 12th=. The total ranks they occupy must be added up and divided by 6 (as there are six of them): So 12+13+14+15+16+17 = 87 ÷ 6 = 14.5. Each of these must be given the rank 14.5. The next rank would then be 18.
- When solving the formula do not forget the 1 – at the front of it.

This statistical technique can be very useful with fieldwork data to show whether two things are correlated or not.

Table 1.10: The 20 most powerful earthquakes since 1960 (Deaths and injuries include any tsunami impacts.)

Earthquake	Magnitude	Rank	Deaths and injuries	Rank	d	d^2
Chile 1960	9.5	1	7223	4	3	9
Chile 1960	8.2		0			
Chile 2010	8.8	5	12,558	3	2	4
Chile 2015	8.3	14.5	28	12	2.5	6.25
Indonesia, Banda Sea 1963	8.3	14.5	0	17	3.0	9
Indonesia, Sumatra 2004	9.1		228,900			
Indonesia, Sumatra 2005	8.6	7.5	1653	6	1.5	2.25
Indonesia, Sumatra 2007	8.4	10.5	186	10	0.5	0.25
Indonesia, Sumatra 2012	8.6		22			
Japan, Hokkaido 1994	8.3	14.5	393	9	5.5	30.25
Japan, Hokkaido 2003	8.3	14.5	755	7	7.5	56.25
Japan, Honshu 1968	8.2	19	434	8	11	121
Japan, Honshu 2011	9.0	4	26,113	2	2	4
Peru 2001	8.4	10.5	2816	5	5.5	30.25
Russia, Kuril Islands 1963	8.5	9	0	17.5	8.5	72.25
Russia, Kuril Islands 2006	8.3	14.5	0	17.5	3	9
Russia, Kurilskiye 1969	8.2	19	0	17.5	1.5	2.25
Russia, Severo Kurilskiye 2013	8.3	14.5	0	17.5	3	9
USA, Alaska 1964	9.2		139			
USA, Alaska 1965	8.7	6	17	14	8	64
n=20						$\sum d^2 = 546.5$

Synoptic link 

The Lloyd's City Risk Index estimates insurance losses from 18 threats; these are mostly man-made risks, but also include natural risks. The Index identifies that most of the man-made risks are the result of globalisation, involving financial systems, cyber-attacks, human pandemics, and plant epidemics, as well as solar storms and power outages. (See page 53.)

Table 1.11: Table showing selected cities and risk

City	Population size 2014 (million)	Population size 2030 (million)	Main natural hazards (according to Lloyd's City Risk Index)	Lloyd's City Risk Index (2015–25)	Socio-economic resilience to earthquakes	Earthquake physical vulnerability
Tokyo-Yokohama, Japan	37.8	37.2	Tropical cyclone, earthquake, flood, volcano, tsunami, drought, freeze, heatwave	$153.3bn	1 = very strong	1 = very strong
Beijing, China	19.5	27.7	Earthquake, flood, drought, freeze, wind storm	$46.5bn	5 = very weak	3 = moderate
Karachi, Pakistan	16.1	24.8	Earthquake, wind storm, volcano, flood, heatwave	$18.3bn	5 = very weak	4 = weak
Cairo, Egypt	18.4	24.5	Earthquake, flood, heatwave	$25.1bn	4 = weak	4 = weak
Mexico City, Mexico	20.8	23.9	Tropical cyclone, volcano, earthquake, flood, freeze, heatwave	$60.7bn	4 = weak	3 = moderate
Osaka, Japan	20.1	20.0	Tropical cyclone, flood, earthquake, volcano, tsunami, drought, heatwave	$79.3bn	1 = very strong	1 = very strong
Manila, Philippines	12.8	16.8	Tropical cyclone, earthquake, volcano, flood, drought, tsunami	$101.1bn	4 = weak	3 = moderate
Istanbul, Turkey	14.0	16.7	Earthquake, flood, drought, heatwave	$82.5bn	3 = moderate	3 = moderate
Jakarta, Indonesia	10.2	13.8	Earthquake, volcano, flood, drought, tsunami, tropical cyclone	$48.2bn	4 = weak	3 = moderate
Los Angeles, USA	12.3	13.3	Earthquake, flood, drought, wind storm, tsunami, heatwave (wildfire)	$90.3bn	1 = very strong	1 = very strong
Tehran, Iran	8.4	10.0	Earthquake, flood, drought, wind storm	$64.1bn	4 = weak	4 = weak
Seoul, South Korea	9.8	10.0	Tropical cyclone, flood, drought, freeze, volcano	$103.5bn	2 = strong	3 = moderate
Taipei, Taiwan	6.9	7.0	Tropical cyclone, earthquake, flood, volcano, drought	$181.2bn	2 = strong	3 = moderate
Port-au-Prince, Haiti	3.2	4.3	Tropical cyclone, earthquake, flood, drought	$0.5bn	5 = very weak	5 = very weak
Islamabad, Pakistan	1.4	2.3	Earthquake, drought, volcano	$0.67bn	5 = very weak	4 = weak
Wellington, New Zealand	0.2	0.3	Earthquake, flood, volcano, wind storm, tsunami, freeze	$1.2bn	1 = very strong	1 = very strong

ACTIVITY

GRAPHICAL SKILLS

1. Using the data in Table 1.11.
   - **a.** Plot a two-axis scattergraph to show the 2030 population size and the amount of financial risk (City Risk Index) for all cities.
   - **b.** Identify and explain any trends and patterns by annotating the graph.

CHAPTER 1

How successful is the management of tectonic hazards and disasters?

Learning objectives

1.7 To understand that the differential impact of tectonic disasters can be understood through their trends and patterns.

1.8 To understand that the prediction, impact and management of tectonic hazards can be understood through models and cycles.

1.9 To understand that the effectiveness of mitigation and adaptation to the impact of tectonic hazards varies according to the strategies used.

Tectonic disaster trends and patterns

Disaster trends and context

In terms of **disaster trends,** Figure 1.27 shows that the number of tectonic disaster events has fluctuated; with peak years in 1997 and 2000, lower years in the early 1980s, and 2012 was another notable lower year. The number of volcanic events has not changed significantly and the data for landslides (dry mass movements) has only just been introduced to the data base, showing that care must be taken when examining data.

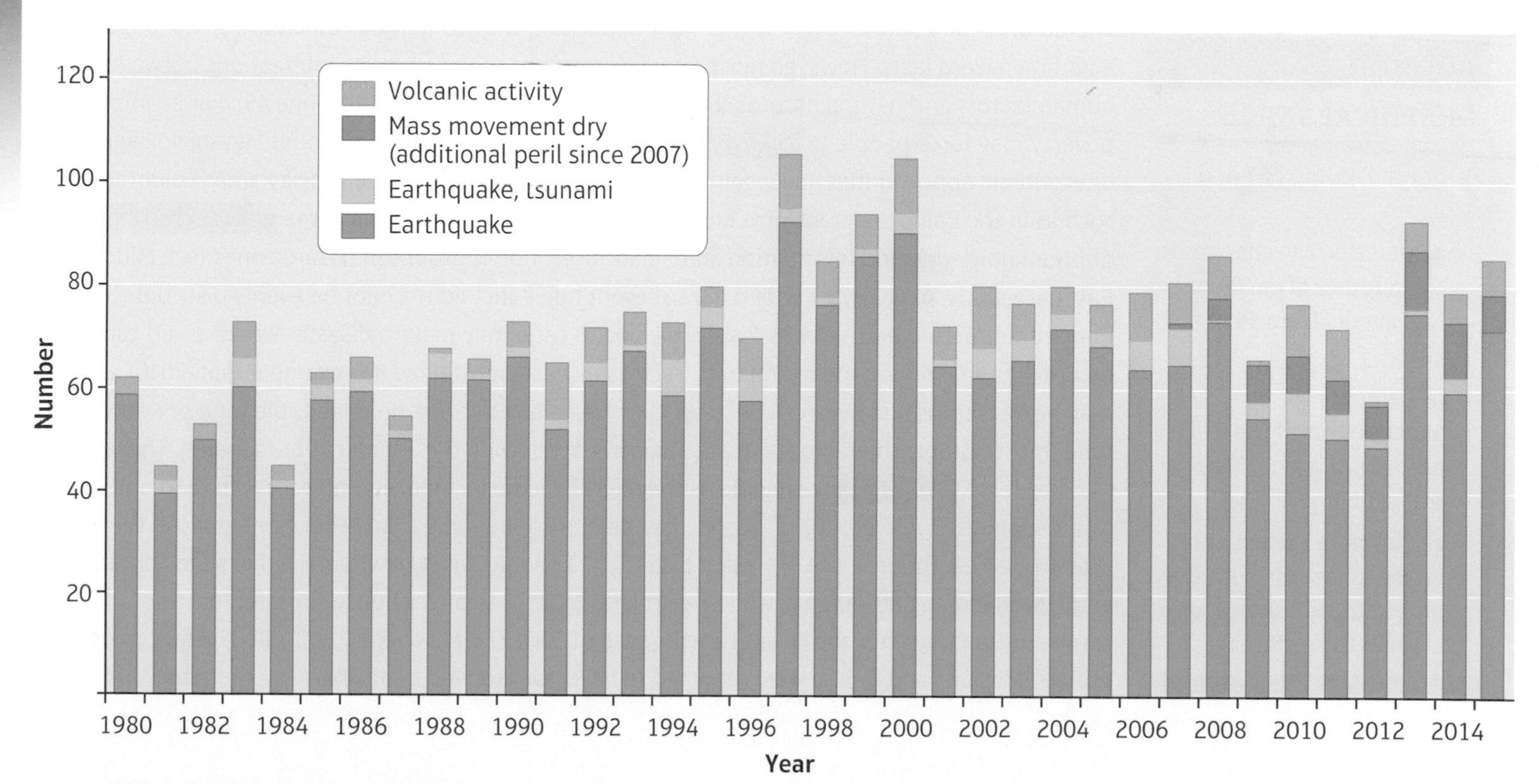

Figure 1.27: Graph showing tectonic disaster trend 1980 to 2015.

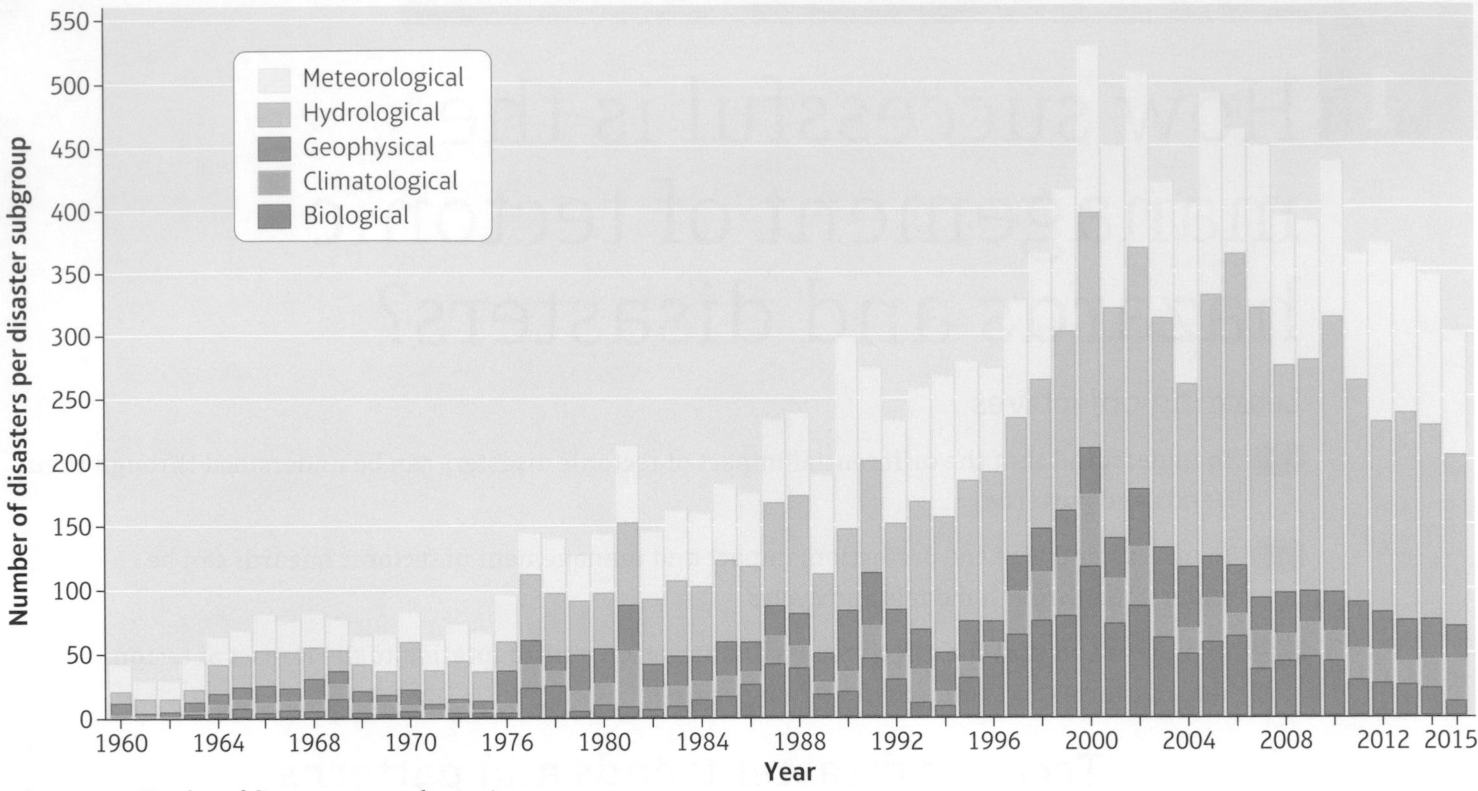

Figure 1.28: Number of disasters reported over time.

Significance of disasters

It has been estimated that in the 20th century only 2.2 per cent of human fatalities from natural hazards were the result of earthquakes and their secondary hazards, and volcanic eruptions were only responsible for 0.1 per cent. The majority of fatalities were due to **slow onset** natural events such as drought and famine, rather than the **rapid onset** events such as earthquakes. The duration of a natural event is therefore important; with vulnerability increasing the longer a natural hazard lasts. However, most disasters result from the complex interactions between human factors and natural hazards (Figure 1.22, page 41, and Table 1.10, page 49). For example, poverty may force people to live in risky areas, such as on the slopes of Mount Merapi volcano, or economic opportunities and wealth may encourage people to move to risky areas voluntarily, such as in the California transform fault zone. A key vulnerability factor is the uneven distribution of prior knowledge and information among socio-economic groups in hazard zones (see Middle East case study, page 43), or after a hazard event relief and aid may not be evenly distributed. Poor infrastructure and services (especially health care) may make a disaster worse as aid cannot be distributed, and secondary impacts, such as disease may follow (for example Haiti 2010). Jobs may be lost, crops and livestock destroyed, or areas may become uninhabitable for a period of time (most notably the radioactivity at Fukushima following the Tohoku 2011 tsunami). These factors create further risk for people as they may have no income, and no access to food and safe water supplies. Some groups of people may be more vulnerable than other. For example, the 1976 Guatemalan earthquake (7.5 M_w) killed over 23,000 people, most of whom lived in slum conditions and Mayan Indian settlements, while wealthier people in stronger houses in safer locations survived. Similarly after the Gujarat earthquake of 2001 (7.6 M_w, over 20,000 killed) ethnic minority groups (Muslims and Dalits) were reported to have not received a fair share of emergency aid.

ACTIVITY

GRAPHICAL SKILLS

1. Study Figures 1.27 and 1.28.
   a. Describe the changes to the number of earthquake loss events from 1980 to 2015.
   b. Compare the effectiveness of the graphs shown as Figure 1.27 and Figure 1.28.
   c. Compare the trend of tectonic hazards with the other hazards shown, and explain the differences.

Yellow = Players, Orange = Attitudes and actions, Purple = Futures and uncertainties

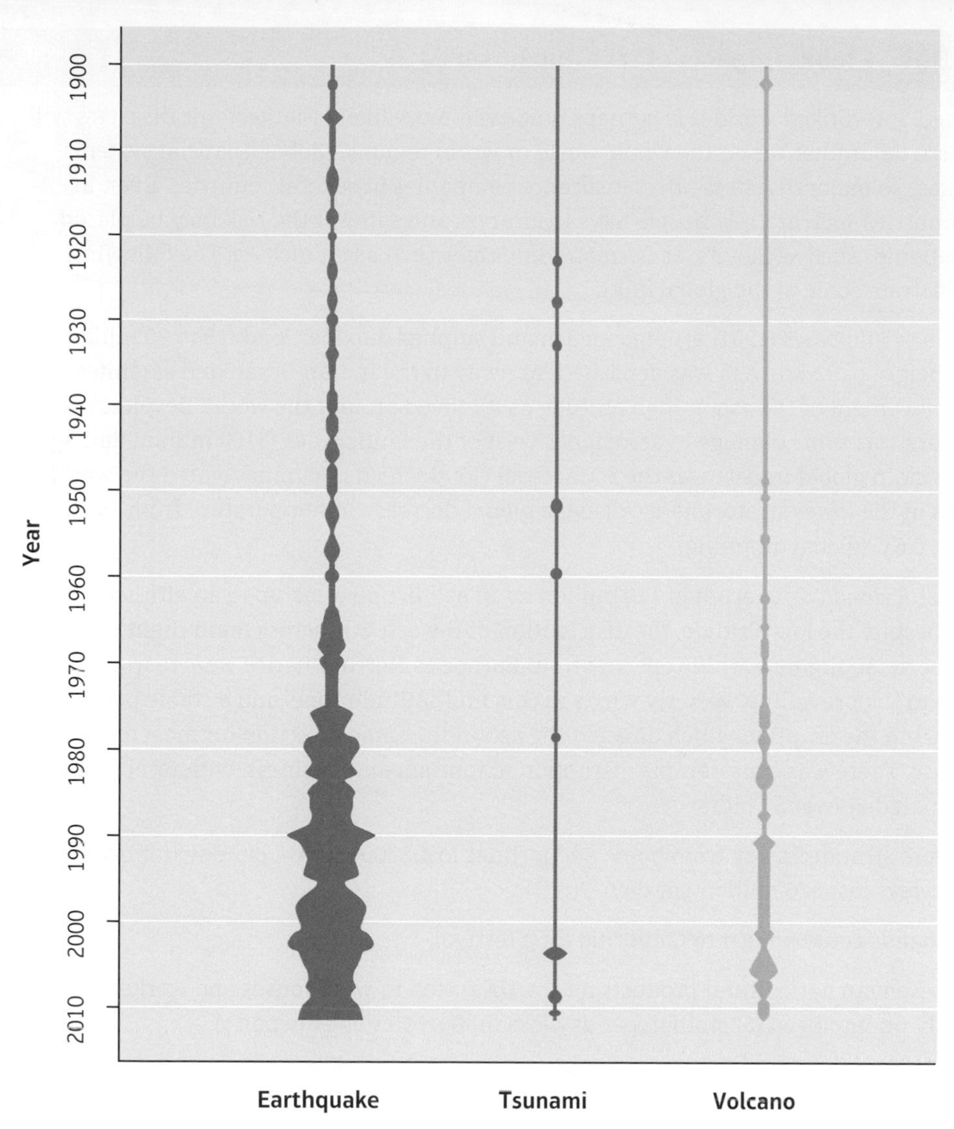

Figure 1.29: Number of natural disasters reported 1900 to 2012, by sub-category.

Reports of hazards and disasters may be biased over time because there is now more reporting of events by the media; much of this is instantaneous due to the global communications network, including social media. Media focus tends to be on 'sudden onset' events which are more dramatically newsworthy, perhaps giving the impression that there are actually more tectonic hazard events now than in the past, which is not true according to long-term trends. Many of the most hazardous countries are in Asia, but then it is the largest continent with the largest area and population. The number of geophysical events, which includes tectonic hazards, has remained about the same over recorded history, but the number of **hydro-meteorological** events (such as river flooding and severe weather events) have been increasing, most probably due to the effects of climate change. However, the number of people being affected has increased in line with population increase and urbanisation rates, and the costs of damage are also increasing as economic development takes place in more countries. Insurance companies, such as Munich Re, Swiss Re and Lloyd's, are very aware of this and conduct considerable research of their own into tectonic hazards trends and patterns (as well as other natural hazards). In a globalised world, not only is the information about tectonic hazards spread very quickly, but also the impacts of tectonic hazards are affecting world regions or the whole world (see case studies).

Synoptic link

Worldwide population growth and urbanisation will continue this century. This places more people at risk from natural hazards, not just tectonic ones, but, for example, hydro-meteorological (such as river flooding and wind storm events) as well. The UN Millennium Development Goals (up to 2015) placed tackling poverty as a major target, but with limited success in some world regions, a significant proportion of the world's population remain unable to reduce their vulnerability to hazards. The Sustainable Development Goals (2015 to 2030) are the latest initiative to improve the lives of people, especially in emerging or developing countries. (See page 189.)

Extension

Investigate the United Nations' Sustainable Development Goals and in particular 11.5 and 11.b.

ACTIVITY

After reading the case studies demonstrating the possible global impacts of tectonic hazards, produce a summary diagram to show the links between the physical and human processes involved.

Extension 

Choose one source of disaster data (for example, EM-DAT, CRED, Munich Re, NOAA) and investigate how accurate this data may be for the tectonic hazards.

CASE STUDIES: Global impacts of tectonic hazards

In a globalised, interlinked world it is perhaps now even more likely that tectonic disasters will have major spatial influences on the whole world or world regions. Global financial systems are interlinked, so major disasters affect insurance companies in several countries. Even in emerging countries such as Chile people have insurance, and some of the risk may be placed with UK companies such as Lloyd's, or German companies such as Munich Re. The following examples illustrate some of the global links:

Mount Pinatubo, Philippines (1991) erupted an ash and sulphur dioxide cloud (15 to 20 million tonnes) to a height of 32 km. Ash was deposited far away in the Indian Ocean and satellites tracked the movement of the ash at high altitude as it moved around the world. Despite warnings, there was some damage to aeroplanes west of the Philippines ($100 million damage). However, the main global impact was the SO_2 aerosol cloud which circumnavigated the world several times in the lower stratosphere causing a global decrease in temperature from 1991 to 1993 of about 0.6°C (global dimming).

Eyjafjallajökull, Iceland (2010) erupted 110 million m^3 of ash in one week up to an altitude of about 9 km. Despite the low altitude, the distribution of the ash was across main flight paths and over airports, so flights over Europe and from Europe to North America were disrupted. This was due to the prevailing westerly winds in this mid-latitude zone, and a stable polar front jet stream during the eruption which directed the ash in the same direction for most of the eruption phase. There was considerable disruption to tourism and business, with total economic impacts estimated at over $3 billion.

- Tourists were stranded away from home and airlines lost $200 million per day (for example, British Airways lost $26 million per day).
- UK music bands could not get to California for a festival.
- Perishable Kenyan agricultural products for the UK rotted in warehouses and workers were temporarily unemployed ($2 million per day loss in Kenyan flower exports).
- Fresh fish from Iceland had to be cold stored.
- Car parts did not reach European factories (for example, BMW production was down by 7000 vehicles in one week).

Sports were affected too, with European football teams (for example, FC Barcelona) having to travel by coach rather than air for Champions League matches, and a MotoGP race in Japan was postponed. However, greater use was made of video and tele-conferencing facilities via the internet, boosting that global link, and an improvement in air quality near European airports was recorded.

The *Tohoku, Japan* (2011) tsunami affected places around the Pacific Ocean, destroying docks, boats, and killing one person in Crescent City, California. The wave was 2 m high in Chile, 17,000 km from the source and it caused calving of icebergs from the Sulzberger Ice Shelf in Antarctica. So many buildings were destroyed in Japan that chemicals from the debris were released into the atmosphere, affecting stratospheric ozone and global warming. There was a decline in Japan's contribution to world industry, particularly the supply of semi-conductors and other high-tech products, and also vehicle manufacturing. Debris was carried throughout the northern Pacific, reaching the coast of North America, along with radioactive seawater. Perhaps the most significant global impact was to encourage countries to think about the safety of nuclear energy, due to the damage to the Fukushima reactors. Despite several reviews of nuclear reactor procedures in earthquakes over recent decades, with automatic shutdown part of the design, a combination of circumstances led to a major incident (now classified as a level 7 nuclear accident). Three reactors at Fukushima did shut down, but then a 15 m tsunami wave flooded the reactor buildings to a depth of 5 m, shutting off the emergency electrical supplies

Yellow = Players, Orange = Attitudes and actions, Purple = Futures and uncertainties

to generators that powered the cooling systems. Three reactors overheated, melted and released radioactivity into the air during explosions, and also into groundwater which eventually reached the Pacific Ocean. Germany has had a long historical empathy with Japan, and this accident made nuclear energy a sensitive political issue in Germany, which has strong environmental and anti-nuclear weapons pressure groups. Following the Fukushima accident Germany decided to phase out its nuclear energy by 2022, closing 17 reactors, and a major German company, Siemens, decided not to continue producing equipment for nuclear power stations. Germany is already the world leader in solar energy, and is capable of finding alternative energy technologies to replace the lost capacity. Other European countries, such as Italy, Switzerland and Spain, confirmed their anti-nuclear energy stances after this accident. Another global impact was to change the Earth's wobble by 17 cm and change the rate of spin by 1.8 microseconds, but these are extremely tiny compared with cosmic influences.

The *Indian Ocean* tsunami (2004) affected countries around the Indian Ocean, with deaths in Indonesia, Thailand, Sri Lanka, India and even Tanzania in Africa. But there were also fatalities from 46 other countries due to tourism, including European countries. Scandinavian countries, for example Sweden, in particular were affected, because it was more likely that communities knew someone who had been killed (Table 1.12). Similarly the Nepal earthquake of 2015 affected foreign climbers and trekkers visiting the Himalayas, with the USA experiencing a few deaths.

Table 1.12: Indian Ocean tsunami (2004): Deaths by selected European countries

Country	Population in 2005 (in millions)	2004 tsunami deaths	Deaths per million inhabitants
Sweden	9.0 m	543	60.3
Finland	5.2 m	179	34.4
Norway	4.6 m	84	18.3
Switzerland	7.5 m	106	14.1
Austria	8.2 m	86	10.5
Germany	82.4 m	539	6.5
United Kingdom	60.5 m	143	2.4

Multiple-hazard zones

CASE STUDY: Philippines, a multi-hazard country

The Philippines is a country of over 7000 islands; eleven large, such as Luzon, and approximately 1000 are settled by people. With all of these islands there is a long coastline on which many people have settled to get access to fishing. The population is about 102 million, with a density of 332 people per km^2. It is located on the western side of the Pacific Ocean on part of the Ring of Fire where there is subduction of the Philippine Plate and the Sunda Plate. It has 37 volcanoes (18 currently active) and experiences numerous earthquakes. The most significant earthquake was in August 1976 in the Gulf of Moro (8.0M_w and Mercalli VII) which happened in the middle of the night and caused 6000 deaths, 85 per cent of which were due to a tsunami. In 2015 the WRI placed the country as the 3rd most at risk country overall, mainly because of an exposure to risk rating of 52.5 per cent (3rd highest in the world). It is a multi-hazard country and hazards may combine to create riskier conditions in the densely populated country. It has been estimated that 60 per cent of the land area is exposed to multiple hazards and 74 per cent of the population is exposed to two or more hazards (World Bank). For example, the Autonomous Region in Muslim Mindanao has higher risk than other islands due to poverty. Between 1960 and 2015 the Philippines experienced 555 major hazard events, an average of 10 a year, and 6 times more than Taiwan (Table 1.13, page 56), with annual damage costs estimated at 0.5 per cent of GDP.

ACTIVITY

GRAPHICAL SKILLS

1. Using the data in Table 1.13.

a. Produce a suitable graph to show the information for both countries.

b. Using your graph(s), compare the countries in terms of their exposure to tectonic hazards and multiple hazards.

Table 1.13: Number of major natural disasters 1960 to 2015: Philippines and Taiwan

Hazard	Philippines	Taiwan
Tropical cyclone (typhoon)	278	67
River flood	94	10
Other storm	67	0
Landslide	33	1
Earthquake	23	9
Volcano	22	0
Epidemic	17	2
Coastal flood	11	0
Drought	8	0
Tsunami	1	0
Wildfire	1	0
Total	555	89

In 1990 a 7.8 M_w earthquake 100 km from Mount Pinatubo caused a landslide on the volcano, and 8 months later it showed signs of erupting after 500 years of dormancy, with harmonic tremors and phreatic eruptions that blasted away parts of its north flank. Due to monitoring, many thousands of people were evacuated and the number of deaths was relatively small (about 800), but ash mixed with rainfall caused the collapse of roofs, and monsoon rains and typhoon rain (meteorological factors) mixed with the ash and pyroclastic deposits months later to cause lahars, which were more damaging than the eruption itself (costs were estimated at $211 million).

In 2009 the Philippine government passed the Disaster Risk Management Act to help increase resilience and reduce risk. This included removing illegal settlers from areas prone to hydro-meteorological hazards. There is an advisory system for meteorological events and monitoring of volcanoes, mapping of fault systems in Manila, the capital, has taken place, and there is a comprehensive 2011 to 2028 National Disaster Risk Reduction and Management Plan for the whole country. A National Disaster Response Pillar provides information (such as updates and contacts) and links all government and relief agencies, and volunteers for co-ordination.

Synoptic link

The World Risk Index shows that many countries are multiple-hazard zones, with not just tectonic events but also hydro-meteorological events. The frequency of tectonic events, across the range of magnitudes, has remained constant over historical time (Figure 1.27, page 51), while the number and magnitude of weather/climate related events has increased. The main reason for this increase is linked to climate change. Higher temperatures mean that there is more heat energy in the atmosphere, which alters the heat budget and strengthens atmospheric movements (such as faster jet streams), increases evaporation, and in combination gives more energy to weather systems such as extra-tropical cyclones (depressions) and tropical cyclones (such as hurricanes). For volcanic eruptions, more water in the atmosphere may mean more lahars, and perhaps more widespread ash. (See page 143.)

Yellow = Players, Orange = Attitudes and actions, Purple = Futures and uncertainties

Tectonic hazard management frameworks

Responses to hazards

Common to all major disasters is the help and aid donated by governments and non-governmental organisations, often supported by citizens, once an event is reported as a disaster. While there may be some criticism of the amounts donated, and the possible corruption involved with distribution of aid, this is undoubtedly a major global link. Major sources of disaster aid (emergency to reconstruction) include the European Union, World Bank, United Nations, Asian Development Bank, Asian Development Fund, individual countries (such as the USA, UK and Germany), international NGOs such as Red Cross and Red Crescent, Médecins sans Frontières (MSF), religious groups, celebrities, trans-national corporations and online communities, as well as individuals donating to emergency appeals. Expertise is required to provide disaster relief essentials such as food, water, shelter, medical care, search and rescue, and military assets such as helicopters and planes. Kashmir, Pakistan (2005) attracted a lot of aid, partly due to emigration creating links to wealthier countries through family connections (for example UK to Pakistan). Haiti (2010) had $4.5 billion of aid given or pledged, and the countries affected by the 2004 Indian Ocean tsunami had a collective total of $10 billion given or pledged. It was estimated that Nepal (2015) needed $415 million in aid, and 22 countries sent Urban Search and Rescue teams (USAR), creating 54 teams out of 1719 people and 147 search dogs (these teams ranged from Israel, Turkey, South Korea, and most from India).

Stages of hazard management

The United Nations Office for Disaster Risk Reduction (UNISDR) emphasises the importance of developing resilience to natural hazards within communities. Resilience can be considered to be the ability to resist, cope with, adapt to, and then recover from the impacts of a natural hazard (for example a volcanic eruption), in an efficient manner and in a reasonable amount of time. This recovery sequence is illustrated well by Park's **response curve** model (Figure 1.30). However, resilience does require action before a natural event as shown by the disaster management cycle (Figure 1.31), and as the PAR model (Figure 1.22, page 41) suggests, this may include making changes to social, political, and economic structures.

ACTIVITY

1. Consider Figure 1.30 (Park's model), information on examples of tectonic events included within this chapter and your own research.

a. Suggest *one* example of a tectonic hazard where recovery has returned a place to normal and explain the evidence for your judgement.

b. Suggest *one* example where recovery has been below the previous quality of life level and explain the evidence for your judgement.

c. Has there been an example of a tectonic hazard where recovery has led to a better quality of life? Explain your answer.

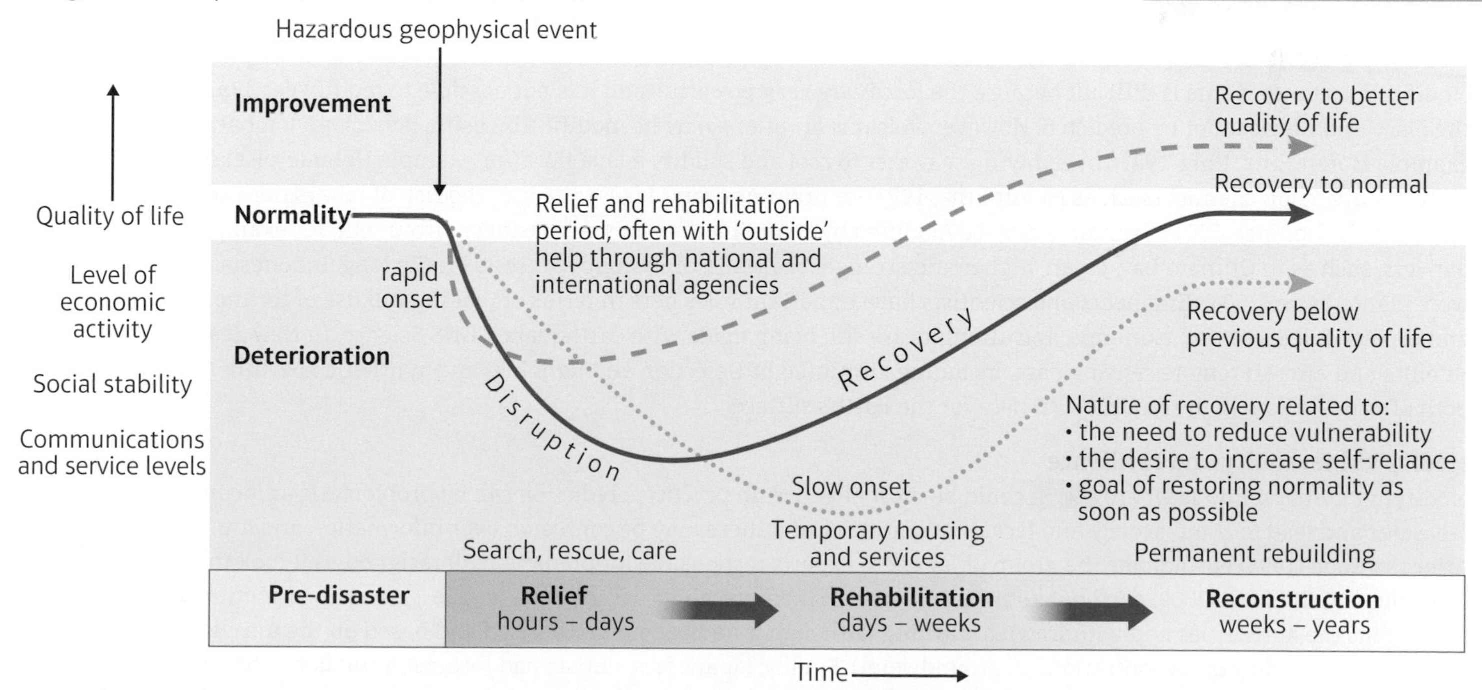

Figure 1.30: Modified Park's model showing quality of life changes over time after a hazard event

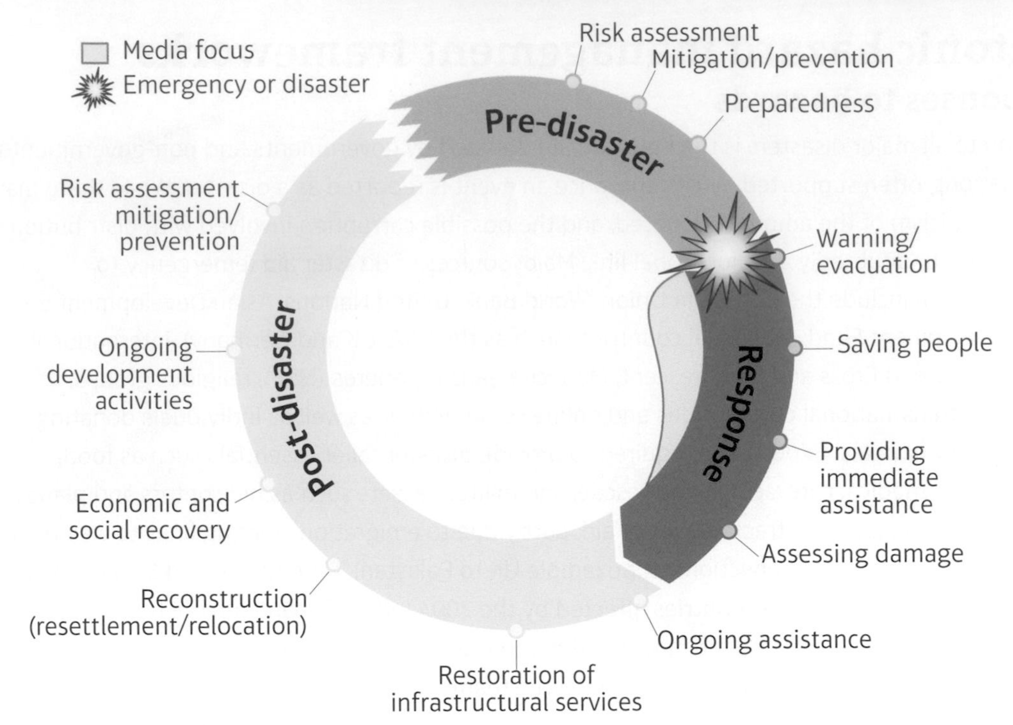

Figure 1.31: Hazard management cycle

Mitigation and adaption strategies

Mitigation is a term used to describe the actions and interventions that a community may take to reduce vulnerability in advance of a tectonic hazard event. Adaptations are ways in which communities may be able to live with a tectonic hazard by making adjustments, to help the community survive by reducing risk.

CASE STUDIES: Strategies for reducing impacts

Modify the event

Modifying tectonic events is difficult because the forces are very powerful, and it is not possible to modify earthquakes because the place and time cannot be predicted. However, volcanic eruptions may be modified by using lava diversion barriers (for example Mount Etna, Italy, 1983), by spraying seawater to cool and solidify a lava flow (for example Heimaey, Iceland, March 1973), cutting a diversion channel (such as Mount Etna, 1993), or draining crater lakes to reduce the risk of lahars (for example Mount Pinatubo, Philippines, 2001). Tsunamis may be modified by changing the offshore coastal environment, for example with offshore barriers, such as in Ofunato Bay, Japan, higher and stronger sea walls, or mangrove forests (at Gle Jong, Indonesia, 70,000 trees were planted after 2004). Engineers and scientists have important roles here in terms of making full use of technology to monitor and predict volcanoes and tsunamis, and attempts are still being made with earthquakes. GNS Science, in New Zealand, use satellite and aircraft remote sensing data, including LiDAR (Light Detection and Ranging) and synthetic aperture radar (or SAR), both of which create high resolution 3D data for the Earth's surface.

Modify vulnerability and resilience

Modifying vulnerability is where action could be most effective. In practice, prediction can be problematic as it can make people feel safer and lead to complacency and lack of preparation, and there may be confusion over information and announcements. After the Kobe (1995) earthquake the group of Japanese experts responsible for prediction all resigned as it took them by surprise. After the L'Aquila (Italy 2009) earthquake six scientists had their warnings described as 'vague, generic, and ineffective' in a court of law, plus the authorities had withdrawn a warning (although it had been unauthorised and based on the unreliable technique of measuring radon gas concentrations in groundwater). Finally, Japanese scientists had forecast a smaller earthquake and tsunami for northern Honshu at some point, but not the large one of 2011.

Yellow = Players, Orange = Attitudes and actions, Purple = Futures and uncertainties

Before an event risk mapping can be completed, such as predicting liquefaction areas or lahar routes. This could be used to produce **land use zoning** and strict planning laws to remove people and property from areas at risk from volcanic eruption, such as Chances Peak, Mount Merapi, or Mount Etna (Figure 1.32), or from lowland coasts facing possible tsunamis. Land reform may help by relocating ownership to less risky areas, or encouraging an increase in food production and stockpiling, or a diversification of the economic base so that not all jobs are lost in a disaster. Technology can be used to establish monitoring and warning systems, for example the tsunami warning system in the Pacific gives people time to evacuate. Technology can also help construct buildings that are more **hazard resistant** (aseismic) **design**, such as the counterweight near the top of Taipei 101 (Taiwan), or the base isolators under New Zealand's old parliament building in Wellington, backed up by enforced building codes (Figure 1.34).

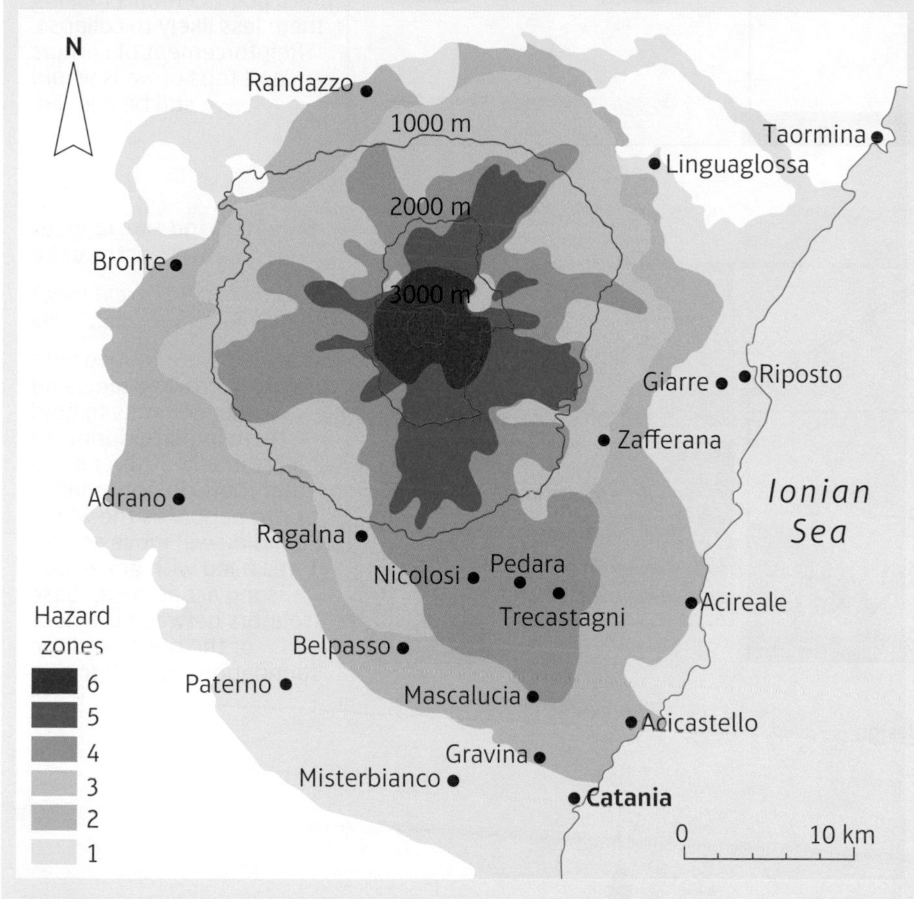

Figure 1.32: Hazard zoning map for Mount Etna

Extension

Assess the attitudes and roles of the different groups of people that would be involved with a major eruption of Mount Merapi in Indonesia. Consider groups such as villagers farming on the sides of the volcano, the religious beliefs of the village elders, the local/regional government, the national government, the USGS, international rescue services and humanitarian groups. Consider the types of volcanic hazard that may be faced, jobs and economy of the local area, the costs of preparedness and evacuation, and the level of technology available.

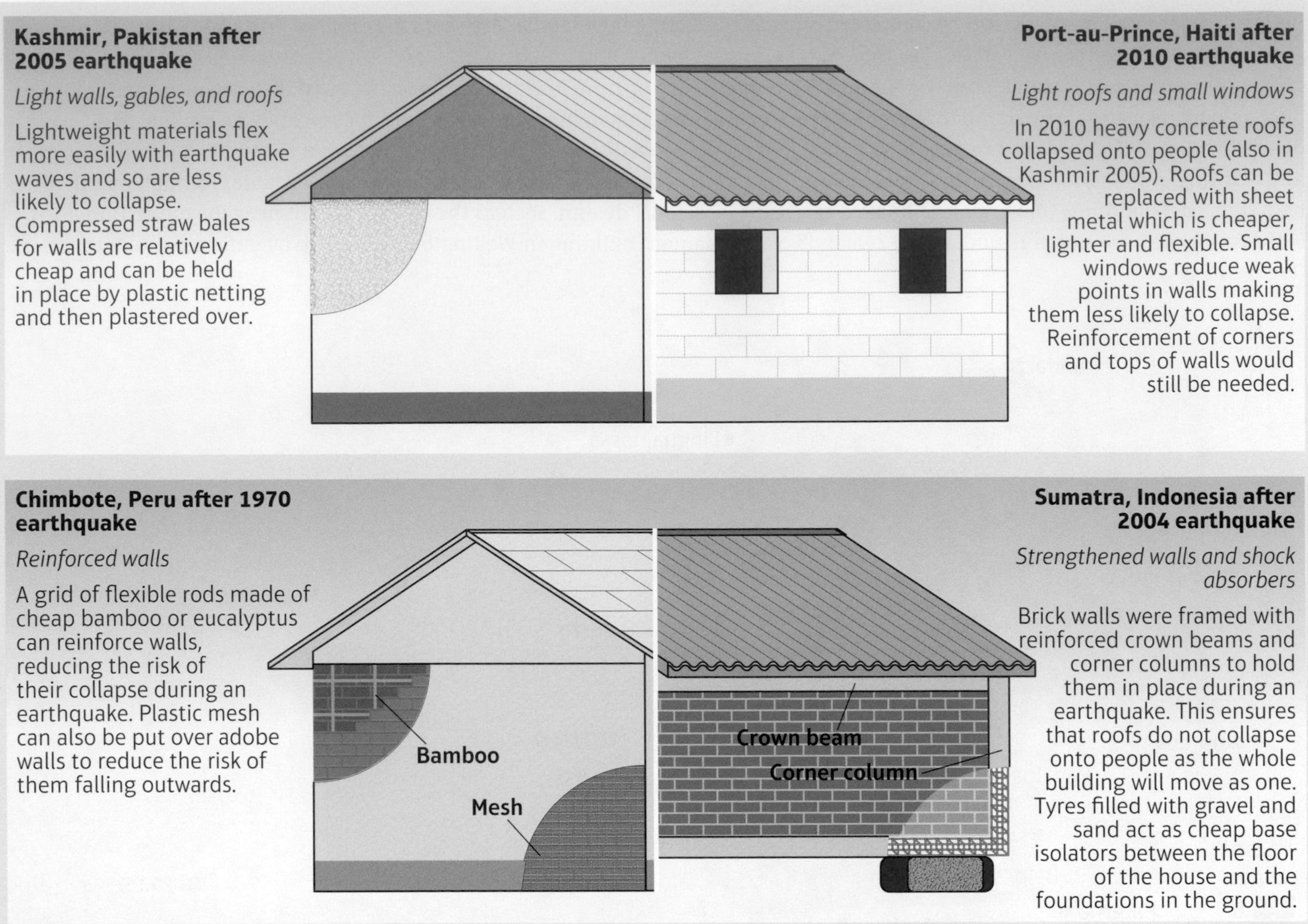

Kashmir, Pakistan after 2005 earthquake
Light walls, gables, and roofs
Lightweight materials flex more easily with earthquake waves and so are less likely to collapse. Compressed straw bales for walls are relatively cheap and can be held in place by plastic netting and then plastered over.
Port-au-Prince, Haiti after 2010 earthquake
Light roofs and small windows
In 2010 heavy concrete roofs collapsed onto people (also in Kashmir 2005). Roofs can be replaced with sheet metal which is cheaper, lighter and flexible. Small windows reduce weak points in walls making them less likely to collapse. Reinforcement of corners and tops of walls would still be needed.
Chimbote, Peru after 1970 earthquake
Reinforced walls
A grid of flexible rods made of cheap bamboo or eucalyptus can reinforce walls, reducing the risk of their collapse during an earthquake. Plastic mesh can also be put over adobe walls to reduce the risk of them falling outwards.
Bamboo
Mesh
Sumatra, Indonesia after 2004 earthquake
Strengthened walls and shock absorbers
Brick walls were framed with reinforced crown beams and corner columns to hold them in place during an earthquake. This ensures that roofs do not collapse onto people as the whole building will move as one. Tyres filled with gravel and sand act as cheap base isolators between the floor of the house and the foundations in the ground.
Crown beam
Corner column

Figure 1.33: Inexpensive tectonic resistant (aseismic) design

Location of tuned mass damper in building
Close up of tuned mass damper within building
Movement of building during earthquake and action of tuned mass damper
Base isolator

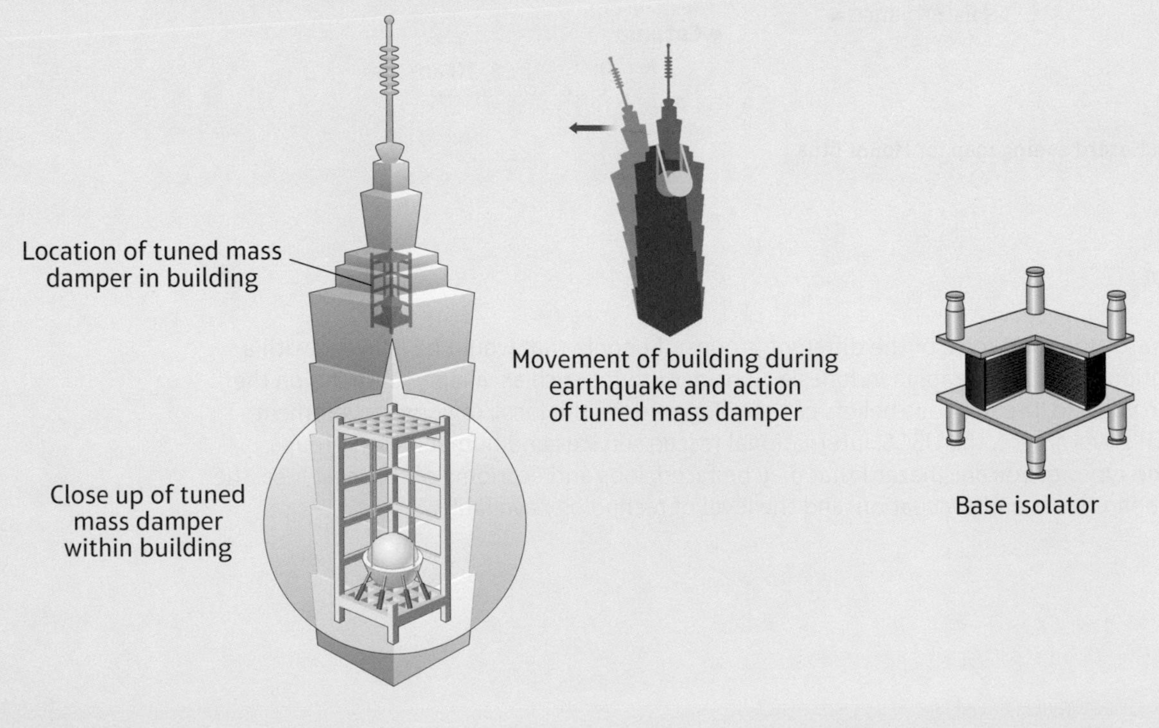

Figure 1.34: High-tech tectonic resistant (aseismic) design

Yellow = Players, Orange = Attitudes and actions, Purple = Futures and uncertainties

Investment in health care and education helps recovery and provides knowledge of what to do in an emergency, including practice drills and storing emergency supplies in public open spaces (such as in Japan). Improvements to communications routes, especially roads, make it easier to establish evacuation routes and help rescue efforts and the distribution of aid, for example in 2011 San Francisco established a Regional Catastrophic Earthquake Mass Transportation/Evacuation Plan. Research and study of tectonic hazards can lead to better understanding, which enables knowledge and understanding to be passed on to communities and distributed through community networks. After a hazard event lessons are frequently learned, and 'build-back-better' is a common target, for example in Christchurch and Haiti, and this principle is part of the 2015 Sendai framework.

A new high-speed tsunami warning system was established for Indonesia from 2008, enabling accurate warnings to be given via 11 regional centres across the country. There is a network of sirens, and a few evacuation centres with open ground floors to let water pass under. However, during a major earthquake in 2012 and the issuing of a tsunami warning, it was found that there were several problems with the system. For example, in Banda Aceh, one of the places hit hard in 2004, warnings did not reach vulnerable people, and officials had not educated local people enough about what to do. So people rushed about to get family and belongings rather than evacuating immediately, evacuation centres were locked, and people operating the sirens fled. Fortunately, the plate movement in 2012 was horizontal and not vertical, so there was no tsunami (Figure 1.15, page 32). Afterwards it was also discovered that people did not know about the evacuation centres or where the evacuation routes were, again showing lack of information from the authorities. Effective warning systems are desperately needed because huge numbers of people still live by the Indian Ocean coast, despite the mega-disaster of 2004.

The USGS set up the Prompt Assessment of Global Earthquakes for Response (PAGER) in 2007. This system analyses earthquake events within 30 minutes of their occurrence, providing information to governments, scientists and relief agencies, so that they can decide what their responses need to be. CalTech and other universities and institutions use the latest technology to monitor, map and produce research models of tectonic events. This technology includes teleseismic data, GPS data, radar measurements of ground deformation, optical ground and satellite imagery, and accelerogram data. The Christchurch earthquake, which had the fastest ground acceleration ever recorded, was analysed using InSAR, a satellite radar system which can measure ground displacement.

Modify loss

During and after a tectonic event the immediate response by the authorities is important. This includes fully equipped rescue efforts followed by relief aid (water, blankets, food, shelter and sanitation) to prevent further deaths from exposure (Kashmir 2005 and Nepal 2015), hunger, thirst or disease (Haiti 2010). Developing countries may depend on external assistance from international non-governmental organisations (NGOs) such as CAFOD, MSF and Red Crescent, who will organise relief after events. Governments and NGOs may help with long-term reconstruction, for example after the Kashmir 2005 earthquake the Pakistan government set up the Earthquake Rehabilitation and Reconstruction Authority (ERRA) to 'build-back-better', although progress was slow. Volunteers within affected communities are common, and where communications break down they are essential, for example Kobe (1995) had 1.2 million volunteers. Some cultures may find comfort in prayer or their beliefs, such as villagers around Mount Merapi. Some people may rely on insurance to recoup losses, including the growing trend of micro-insurance. In some cases whole settlements may be moved away from risky areas and rebuilt in safer positions, for example Balakot was moved 23 km westwards after the 2005 Kashmir earthquake.

ACTIVITY

Which is the best strategy to reduce the impacts of tectonic hazards – modifying the event, modifying vulnerability, or modifying the loss? Explain your answer.

Extension

The Centre for Research on the Epidemiology of Disasters (CRED) has assembled data and advice on health for over 30 years, and it links with other organisations such as USAID and the World Health Organisation (WHO). Investigate the CRED website and read one of the publications listed below, or a more recent publication linked to tectonic hazards: Credcrunch 39 (CRED 2015), Credcrunch 35 (CRED 2014), or The Andaman Nicobar earthquake and tsunami 2004 (CRED 2005).

Synoptic link

Mortality is a more specific indicator than death rate when studying population change, as it can be related to the cause of death or specific groups within the structure of a population, such as maternal mortality or natural hazard mortality. (See page 20.)

AS level exam-style question

Explain how emergency planners and engineers may help to modify the impacts of a tectonic hazard. (6 marks)

Guidance

Think about what the two groups of people can do, describe what these actions are and how each action may reduce the impact of an earthquake, tsunami or volcanic eruption.

A level exam-style question

Explain why insurance companies may be interested in encouraging the accurate prediction of, and effective preparation for, a tectonic hazard. (4 marks)

Guidance

Think about the proportion of financial costs that may be insured, and recall information from Munich Re and Lloyd's data for example. Explain how prediction may help reduce losses. Explain how preparation may reduce losses. Think also about population, urbanisation, and losses trends.

There was an international decade for national disaster reduction (IDNDR) in the 1990s which emphasised the need to combine 'top down' technological and government approaches with 'bottom up' community and NGO efforts. This strategy may form part of a **hazard management cycle** to prepare a place for a natural hazard or to learn lessons after each hazard event or disaster. The overall aim is to reduce vulnerability for the future. All those involved with planning for an emergency have a key role in reducing the impacts of a tectonic hazard.

The Sendai framework (2015 to **2030**) recognises that **emergency planners** have a key role in reducing the impacts of tectonic hazards within its priorities. This framework for reducing risk includes the targets of reducing (a) disaster mortality, (b) numbers affected, (c) economic losses, and (d) damage, and set targets of increasing (i) the number of disaster risk reduction strategies, (ii) cooperation between developed and developing countries, (iii) the number of warning systems, and (iv) the information flow to people. There are four priorities for action: (1) understanding disaster risk, (2) strengthening disaster risk governance, (3) investment in making places more resilient and (4) strengthening disaster preparedness and actions during the recovery phase.

Yellow = Players, Orange = Attitudes and actions, Purple = Futures and uncertainties

Summary: Knowledge check

Through reading this chapter and by completing the tasks and activities, as well as your wider reading, you should have learnt the following, and be able to demonstrate your knowledge and understanding of Tectonic processes and hazards (Topic 1).

a. Where do earthquakes, tsunamis and volcanic eruptions occur?

b. What types of plate boundary are there and what are their main characteristics?

c. Explain the plate tectonic theory and the evidence that supports it.

d. Explain the processes operating at plate margins and intra-plate locations?

e. What are the features of earthquake, tsunami and volcanic processes?

f. What are a hazard, disaster, and mega-disaster, and how do these link to vulnerability and resilience of people and communities?

g. Explain the social and economic impacts of tectonic events in countries at different stages of development?

h. How is the magnitude and intensity of earthquakes, tsunamis and volcanoes measured?

i. Explain why the characteristics of hazard profiles differ?

j. Explain the different types of vulnerability and how human factors influence the scale of a disaster?

k. Explain tectonic disaster trends over time and whether these trends are accurately reported?

l. What is the global significance of mega-disasters?

m. Explain multiple-hazard zones, and how hydrological and meteorological processes interlink with tectonic hazards?

n. How can tectonic hazards be managed effectively at the different stages of an event?

o. Which specific strategies can be used to modify a tectonic event, vulnerability and resilience, or loss?

As well as checking your knowledge and understanding through these questions you also need to be able to make links and explain ideas, such as:

- There is a variation in tectonic activity according to the type of plate boundary.
- The mantle is a complex layer of the Earth and the role of large mantle plumes is still not fully understood.
- A disaster occurs when the risk to people is increased by the magnitude of an event and the vulnerability of people; these combine to increase the severity of impacts.
- Hazard profiles demonstrate that physical factors are important in countries at any level of development, but that vulnerability is greater in developing or emerging countries.
- Vulnerability is increased by a combination of poverty, poor access to knowledge and understanding, poor access to technology, lack of community involvement in decision-making, and high population density.
- Tectonic mega-disasters have a range of significant global impacts due to interlinked natural and human systems.
- There are variations in prediction, forecasting, response, recovery, mitigation, preparedness, and the modelling of responses according to local human factors and levels of development.
- The effectiveness of strategies to modify tectonic events, vulnerability and loss vary considerably, and are particularly difficult with earthquakes.

Preparing for your AS level exams

Sample answer with comments

Assess the reasons why earthquakes create more disasters than volcanic eruptions. (12 marks)

A disaster occurs when there is a significant loss of life or high financial costs or disruption of socio-economic conditions. A mega-disaster or catastrophe is a more extreme event affecting people. While earthquakes and volcanoes are both mostly found on or near plate boundaries, their characteristics create differences in their impacts, although this can be modified by the human factors at any location.

Sound understanding demonstrated and good use of terminology. However, a clearer definition of what makes a natural hazard a disaster was possible with reference to UN criteria.

Earthquakes cannot be predicted in terms of their exact location and exact time, but volcanoes are usually very visible and there are often warning signs that an eruption is going to happen before it does, such as gas emissions, harmonic tremors and bulges. This difference means that people cannot be as prepared for an earthquake as they can for a volcanic eruption, and therefore there may be a greater loss of life and more damage from an earthquake. A volcanic eruption phase can last a lot longer (years) than an earthquake, which can be measured in minutes at most, and can affect areas over a long period of time. For example, some historical eruptions have had global impacts, such as starvation due to crop failures resulting from global dimming.

Again, great use of terminology and correct understanding. But what about secondary hazards as well to extend the range of ideas? Also some place detail through case studies would greatly assist to increase the level of the answer.

Earthquakes happen more frequently than volcanic eruptions, for example the Philippines, and their intensity depends on the location of the epicentre. There are major cities on transform fault plate boundaries, for example San Francisco, but it is also true that there are significant settlements on or near volcanoes, for example Naples on Mount Vesuvius. The energy released by an earthquake through its seismic waves is centred on a relatively small area and the shaking affects buildings, which, when they collapse, kills people and costs money to rebuild. People are therefore more vulnerable during an earthquake, especially if it happens at a time of day when they are inside. In some countries buildings are not earthquake proofed and they collapse easily. The most dangerous volcanic hazards are unexpected pyroclastic flows and lahars, but hazard mapping and land use zoning can reduce problems. So, earthquakes create more disasters than volcanoes.

The candidate continues to demonstrate good understanding and terminology within a coherently written answer and at last names a couple of places. There is also no substantial conclusion, it appears that the candidate may have been under time pressure and needed to move on.

Verdict

This is an average answer despite the sound and relevant understanding shown, because there is a lack of detailed knowledge included within the answer. There are some logical connections and partial interpretation of the question with a good range of factors. This answer needed specific reference to actual recent volcanic and earthquake events with data on losses, perhaps using the idea of hazard profiles as a structure to provide an assessment of the significance of factors, and provide a reasoned suggestion for any difference.

Preparing for your A level exams

Sample answer with comments

Assess the relative importance of the physical characteristics of volcanic eruptions in creating risk for people. (12 marks)

Volcanoes have a range of characteristics that affect the risk to people, these include the magnitude, duration and the type of activity and hazards created. However, the risk equation R=HxV suggests that it is not just the characteristics of the hazard that should be considered, but also the human factors linked to vulnerability.

Concise introduction showing understanding of the complexities of the question. Possible expansion points could have been the Pressure and Release model. But not much space in a 12-mark question to generalise too much in an introduction.

Volcanic hazards vary according to the type of volcano and the explosivity of the eruption (VEI scale); effusive eruptions are gentle with lavas flowing slowly onto the landscape and therefore there is low risk to people. Further up the VEI scale there are explosive eruptions, among the most violent type are Plinian which blast ash high into the atmosphere. The Iceland volcano of 2012 spread ash all over Europe and disrupted flights and closed airports, so creating risks for people.

Candidate understands magnitude and describes relative risk in a fair general discussion. However, there is a lack of terminology and factual material in this section (for example name the Icelandic volcano and get the year of the eruption correct – Eyjafjallajökull or at least EFJ, 2010). The types of risk could also have been explored more factually.

Composite cone volcanoes, for example Mount St Helen's, may create various hazards during their eruptions, perhaps the most dangerous of which are pyroclastic flows – these are superheated clouds of ash and gas travelling at high speeds, usually created by the collapse of an eruption column. These create high risk as they burn or poison or suffocate anybody or anything in their way and they cannot be stopped. Lahars are also dangerous as water mixed with volcanic material rush down the sides of volcanoes and can reach some distance away, such as at Armero.

Understanding is still present in the answer, but there is a lack of precision and detail. Where is the evidence from recent case studies such as Mount Merapi or Mount Pinatubo?

In conclusion, it is clear that the physical characteristics of an eruption do affect the level of risk, but volcanoes are predictable if monitored, and so people can be evacuated so reducing their vulnerability and risk. This happened with Mount St Helen's in 1980 where just 57 people were killed because the whole area had been evacuated, and those killed were because lahars travelled some distance. Vulnerability is also an important factor and may be more important than the physical characteristics.

A partially supported conclusion with the lack of development of the vulnerability concept compared to the physical geography side of the risk equation.

Verdict

This is an average answer with mostly relevant knowledge and understanding of risk and volcanic hazards, and a partial interpretation of the question through logical connections. There is good coverage of the types of hazards created by volcanic eruptions and some assessment of their relative dangers. The candidate appears to understand the complexity of the question, but then there is a lack of expansion of ideas; in particular how important are the physical characteristics when compared to vulnerability factors? The other main weakness is the limited range of factual evidence.

THINKING SYNOPTICALLY

Read the following extract carefully and study Figure A. Think about how all the geographical ideas link together or overlap. Answer the questions posed at the end of the article. These extracts are from the New Scientist online sources dated 5 and 11 February 2015.

MELTING ICE SPELLS VOLCANIC TROUBLE

LESS ice, more lava? Earth's crust falls and rises as ice caps grow and melt – each raising the risk of volcanism.

Kathleen Compton at the University of Arizona and her team analysed data from GPS receivers attached to rocks in Iceland since 1995. They show that areas where five of the country's largest glaciers are melting have been rising by around 3.5 cm a year. Rates elsewhere were much lower (*Geophysical Research Letters*, doi.org/zzk).They think the loss of ice is relieving pressure on rocks beneath and allowing them to spring up. Some fear that this could trigger more volcanoes. During the last great melt 12,000 years ago, volcanic activity on Iceland was up to 50 times greater than over the past century, says Bill McGuire at University College London.

Another study reinforces the link between climate and volcanism. John Crowley at the University of Oxford and his team found that huge eruptions in the Southern Ocean coincided with ice ages. By locking up much of the world's water in ice sheets on land, they reduced the ocean's pressure on the seabed enough to allow magma to escape from Earth's mantle (*Science*, doi.org/zzm).

So it seems glaciation can trigger submarine eruptions, while deglaciation may lead to magma outflows on land. Both studies reinforce the idea that redistribution of water caused by climate change can elicit volcanic eruptions, says McGuire. Melting ice is causing the land to rise up in Iceland – and perhaps elsewhere. The result, judging by new findings on the floor of the Southern Ocean, could be a dramatic surge in volcanic eruptions.

Last week, researchers at the University of Arizona in Tucson showed that a recent dramatic uplift of the Earth's crust in parts of Iceland coincided with the rapid melting of nearby glaciers.

Kathleen Compton's team used data from GPS receivers that have been attached to rocks since 1995 to show that some parts of south-central Iceland, where five of the country's largest glaciers are melting fast, have been rising by around 3.5 cm a year. Away from the glaciers, the rates of land rise were much lower. Their explanation is that the disappearance of the ice is relieving pressure on rocks beneath and allowing them to spring up.

Rapid rebound

It has long been known that the Earth's crust falls and rises as ice caps grow and melt. But the speed of the rebound is surprising, says Compton. Richard Katz of the University of Oxford finds the discovery 'very exciting'. 'The measurements show that there is a response even at a very short time-scale of 30 years,' he says.

The land uplift could be handy to protect some coastal areas from rising sea levels as the melting ice flows into the oceans. But there is a growing fear among geologists that climate-induced changes to water and ice levels could trigger more dangerous events, such as volcanic eruptions.

The evidence is mostly from the past. For instance, during the last great melt 12,000 years ago, volcanic activity on Iceland was up to 50 times greater than the activity observed over the past century, says Bill McGuire, a volcanologist at University College London. Iceland has suffered three major volcanic eruptions in the past five years – although no one has shown a certain link with climate change.

Seabed clue

And today, fresh evidence of climatic cues for volcanic eruptions emerges from an analysis of thousands of kilometres of ridges on the floor of the Southern Ocean. The ridges were created by huge eruptions.

Through detailed analysis of the topography of the seabed, John Crowley of the University of Oxford shows that the eruptions coincided with phases of orbital wobbles known as Milankovitch cycles, which trigger ice ages. He concludes that glaciations caused the increase in eruptions. By locking up much of the world's water in ice sheets on land, they lowered sea levels and so reduced the ocean's pressure on the seabed enough to allow magma to escape from the Earth's mantle. 'Our work reinforces the link between climate change and volcanism,' says Katz, a co-author of the Crowley paper.

So it seems that glaciation can trigger submarine eruptions, while deglaciation may lead to magma outflows on land.

'Both these studies reinforce the idea that the wholesale redistribution of water that accompanies major climate change elicits a significant response from the solid earth in the form of potentially hazardous phenomena such as earthquakes and volcanic eruptions,' says McGuire. 'We saw this very dramatically at the end of the last ice age, and we are seeing it again today in Iceland and elsewhere.'

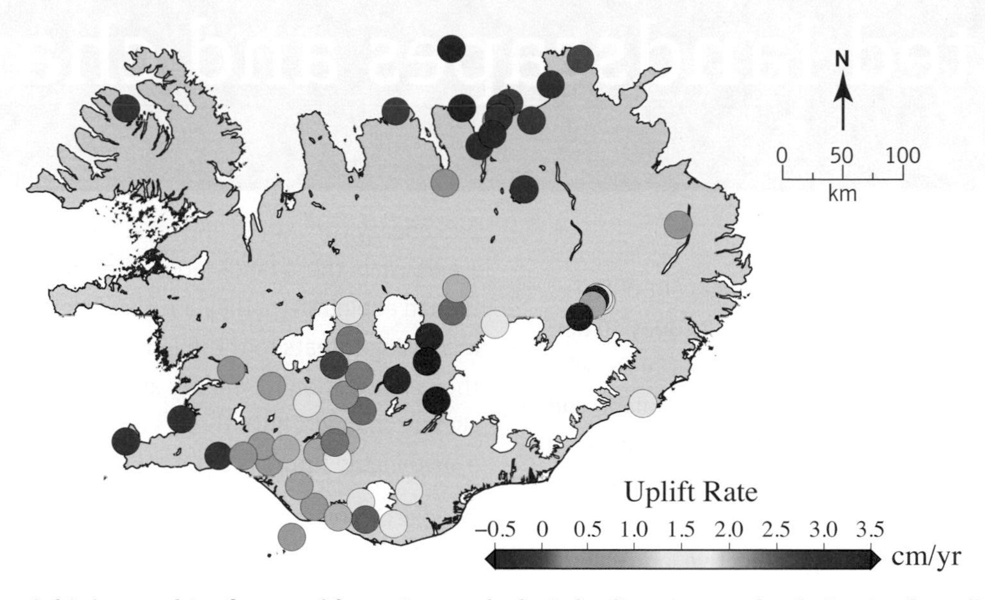

Figure A: Iceland's glaciers (white) are melting faster and faster. As a result, the Icelandic crust near the glaciers is rebounding at an accelerated rate. Credit: Kathleen Compton/University of Arizona Department of Geosciences

ACADEMIC SKILLS

1. How would you rate the validity of information provided in this article? Explain why.
2. What is meant by circumstantial evidence? To what extent may some of the evidence presented in this article be of this type?
3. To what extent is the research from the different universities supportive of each other's findings?
4. How is evidence from the past being used to indicate possible futures?
5. Study Figure A. Evaluate the strengths and weaknesses of this map and suggest how you would improve it.

ACADEMIC QUESTIONS

6. **a.** Explain how the researchers believe that melting ice can cause volcanic eruptions.
   b. Explain how the researchers believe that lower sea levels can cause volcanic eruptions.
7. Using what you have learnt about coasts or glaciation, explain fully why Richard Katz is 'very excited' about the new evidence of isostatic change.
8. Suggest how an increase in earthquake activity could also be a result of changes to the distribution of water on Earth?
9. Using evidence from this article, assess the extent to which places such as Iceland should be concerned about links between climate cycles and tectonic activity.

Dynamic landscapes

Glaciated landscapes and change

Introduction

We are living in an ice age. For more than 90 per cent of the Earth's history, the poles have been ice-free, but today 10 per cent of the Earth's land surface is covered with glacial ice, which includes glaciers, ice caps and the ice sheets of Greenland and Antarctica. Glaciers are one of nature's bulldozers – massive bodies of ice, capable of sculpting and carving spectacular landscapes. The glacial processes of erosion, transport and deposition are responsible for creating distinctive landscapes in both upland and lowland areas. By studying these processes in today's polar and alpine glacial environments, we can understand the relict glacial landscapes formed by past glacier advance.

Glaciated landscapes are constantly changing, not just because of ongoing physical processes, but also because of human activities. These activities involve a diverse range of stakeholders – such as indigenous peoples, tourists, conservationists, mining and energy companies and local and national governments – and, while these landscapes offer many opportunities for human exploitation, they are also fragile. Human activities pose increasing threats to these landscapes, for example through their exploitation for energy and mineral resources, and through climate change. These threats create management challenges, if a sustainable future is to be achieved.

Periglacial environments cover approximately 25 per cent of the Earth's land surface and yet, due to the harsh physical conditions in these environment, they are home to less than 0.3 per cent of the world's population. However, the value and importance of these environments is increasingly recognised: not only do they offer a wealth of economic resources, including minerals and fossil fuels, but the permafrost is also an enormous store of carbon and methane. Large-scale thawing of the permafrost could make a significant contribution to climate change in the future.

In this topic

After studying this chapter, you will be able to discuss and explain the ideas and concepts contained in the following enquiry questions, and provide information on relevant located examples.

- How has climate change influenced the formation of glaciated landscapes over time?
- What processes operate within glacier systems?
- How do glacial processes contribute to the formation of glacial landforms and landscapes?
- How are glaciated landscapes used, threatened and managed today?

Figure 2.1: Powerful glaciers erode spectacular landscapes such as the Cordillera Huayhuash in Peru. Why do you think this landscape is culturally, economically and environmentally valuable?

Synoptic links

The topic 'Glaciated landscapes and change' is relevant to the three synoptic themes: players; attitudes and actions; and futures and uncertainties. There are many players in glacial and periglacial landscapes at a range of scales; for example, international governmental organisations, transnational corporations, national governments, pressure groups and indigenous peoples. These players have different attitudes towards the exploitation or conservation of glacial and periglacial landscapes, and their actions can cause environmental degradation. The future of many cold environments is uncertain because of the threats posed by both global climate change and increasing economic exploitation. As the threats escalate, so does pressure for global action to achieve a sustainable future for glacial and periglacial landscapes.

Useful knowledge and understanding

During your previous studies of Geography (KS3 and GCSE), you may have learned about some of the ideas and concepts covered in this chapter, such as:

- physical and glacial processes in glaciated upland landscapes
- the effect of the past and current UK climate on processes in glaciated upland landscapes
- the role of glacial erosion and deposition in the development of landforms
- the human activities that change glaciated upland landscapes
- the advantages and disadvantages of development in glaciated upland landscapes
- the physical and human processes responsible for the formation of a distinctive, glaciated upland landscape in the UK.

This chapter will reinforce this learning and also modify and extend your knowledge and understanding of glaciated landscapes. Remember that the material in this chapter features in both the AS and A level examinations.

Skills covered within this topic

- Comparison of past and present distributions of glacial landscapes, using global and regional maps.
- Graphical analysis of global changes in glacier mass balance.
- Use of GIS and OS maps to identify glacial landforms in active and relict landscapes, and to reconstruct past ice extent and ice-flow direction.
- Cirque orientation analysis using OS maps and rose diagrams.
- Till fabric analysis using rose diagrams.
- Use of measures of central tendency, dispersion (standard deviation) and Student's *t*-test to analyse changes in sediment size and shape in outwash plains.
- Drumlin morphometry and orientation surveys requiring statistical and cartographical analysis.
- Calculations of mass balance and rates of recession.
- Fieldwork skills.

CHAPTER 2

How has climate change influenced the formation of glaciated landscapes?

Learning objectives

2.1 To understand the causes of long-term and short-term climate change leading to icehouse–greenhouse changes

2.2 To understand the changing distribution of ice cover in the Pleistocene epoch

2.3 To understand the periglacial processes that produce distinctive landforms and landscapes

From icehouse to greenhouse: climate changes

Current geological evidence suggests that the Earth is 4.6 billion years old, and that throughout its history the planet's climate has been fluctuating between two dominant states: the **greenhouse Earth** and the **icehouse Earth**. A greenhouse Earth occurs when there are no continental glaciers on the planet as a result of warming processes such as higher levels of greenhouse gases in the atmosphere. The cause of this warming may be increased volcanic activity. An icehouse Earth is a global ice age, when large ice sheets are present on the Earth. During this time, the climate fluctuates between cooler **glacials**, when ice advances, and warmer **interglacials**, when ice retreats.

There are five known ice ages in the Earth's history, the most recent being the Quaternary Ice Age. The Quaternary started approximately 2.6 million years ago and extends up to and including the present day. It is divided into two epochs:

- The **Pleistocene**, which lasted until 10,000 years ago
- The **Holocene**, which began 10,000 years ago and continues today

The Pleistocene epoch is often known as the Ice Age, as it was characterised by over 50 glacial–interglacial cycles (Figure 2.2), and glaciers reached their maximum extent during this time. The last glacial maximum is known as the Devensian, which occurred approximately 18,000 years ago (Figure 2.3). The last glacial advance in the UK is known as the **Loch Lomond Stadial**, which occurred between 12,000 and 10,000 years ago, marking the end of the Pleistocene epoch.

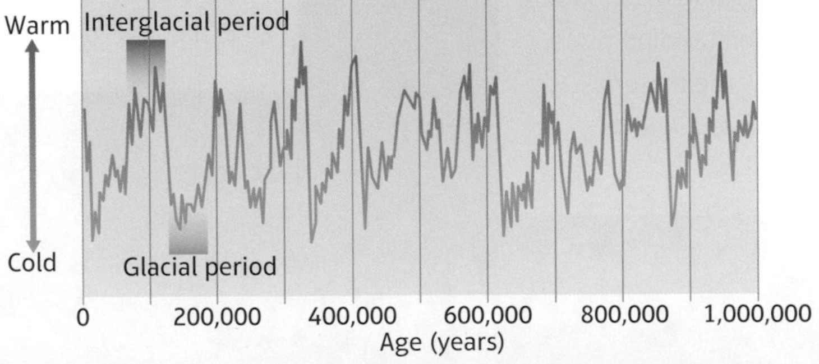

Figure 2.2: A chronology of multiple glacial and interglacial cycles during the Pleistocene epoch

Yellow = Players, Orange = Attitudes and actions, Purple = Futures and uncertainties

The long-term causes of climate change

The start of the Quaternary Ice Age has been linked to the changing position of the continents, a result of plate tectonics. Three million years ago the North and South American tectonic plates collided, creating the Panama Isthmus (the narrow land bridge that joins the two continents). This re-routed ocean currents, so that the warm Caribbean waters that had once flowed through the gap between the Americas were forced northwest towards Europe, creating the Gulf Stream. Scientists think that the formation of the Gulf Stream transported extra moisture into the Arctic atmosphere and this fell as snow, triggering the build-up of the Greenland Ice Sheet, which in turn may have kick-started the last ice age.

Literacy tip

This topic involves a large number of specialist geographical terms, which you must use fluently in your writing. Make a glossary list for each enquiry question and practise using the terms in all written work.

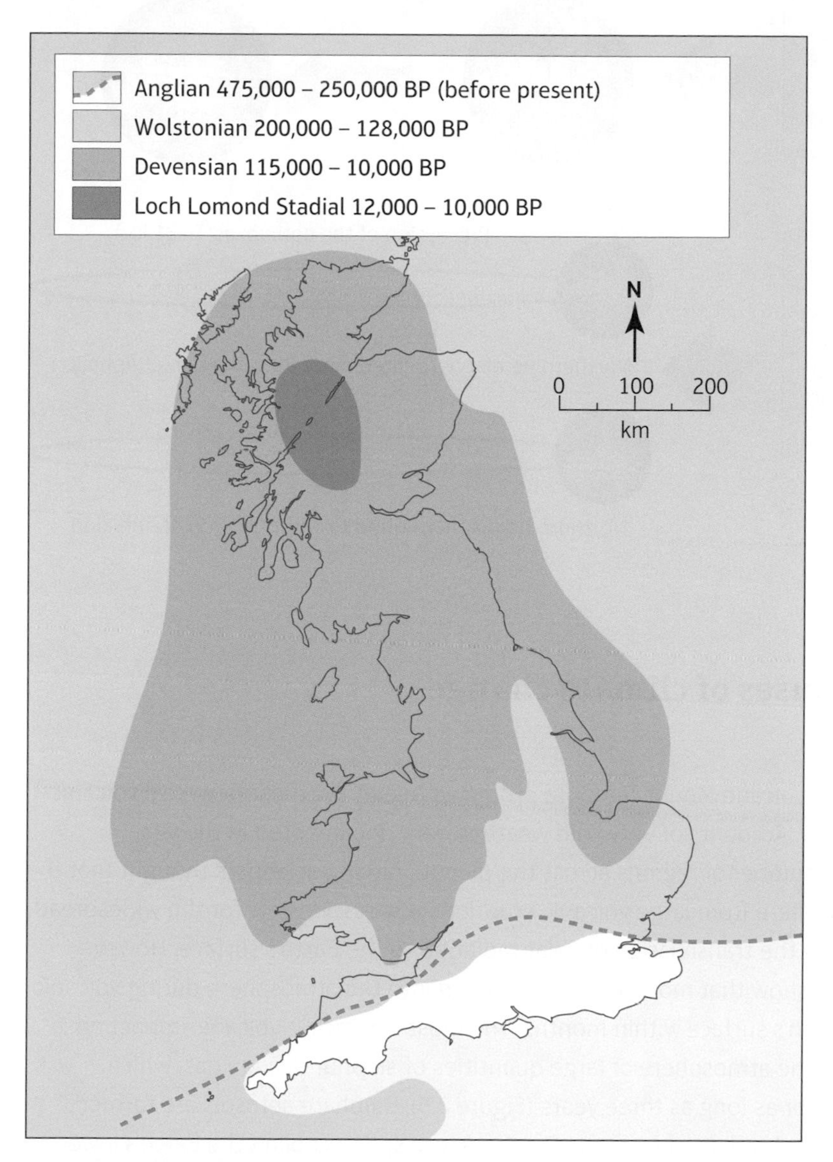

Figure 2.3: Pleistocene glacial limits in the UK

Astronomical theory: Milankovitch cycles

In 1920 the geophysicist Milankovitch proposed that the glacial–interglacial cycles were caused by variations in the amount of solar radiation received by the Earth. This is a result of three cyclical changes in the orbit and axis of the Earth (Figure 2.4). Evidence from coral reefs in Barbados reveals that there is a strong correlation between the timing of the interglacial periods of the past 160,000 years and eccentricity cycles, indicating that **Milankovitch cycles** are a primary driver of climate change.

ACTIVITY

TECHNOLOGY/ICT SKILLS

1. a. Using an internet search engine, search for animations of Milankovitch cycles.
   b. Watch the animations to help you understand how the three cycles cause climate change.
2. Draw three diagrams to show each Milankovitch cycle and annotate your diagrams to explain how climate change is caused.

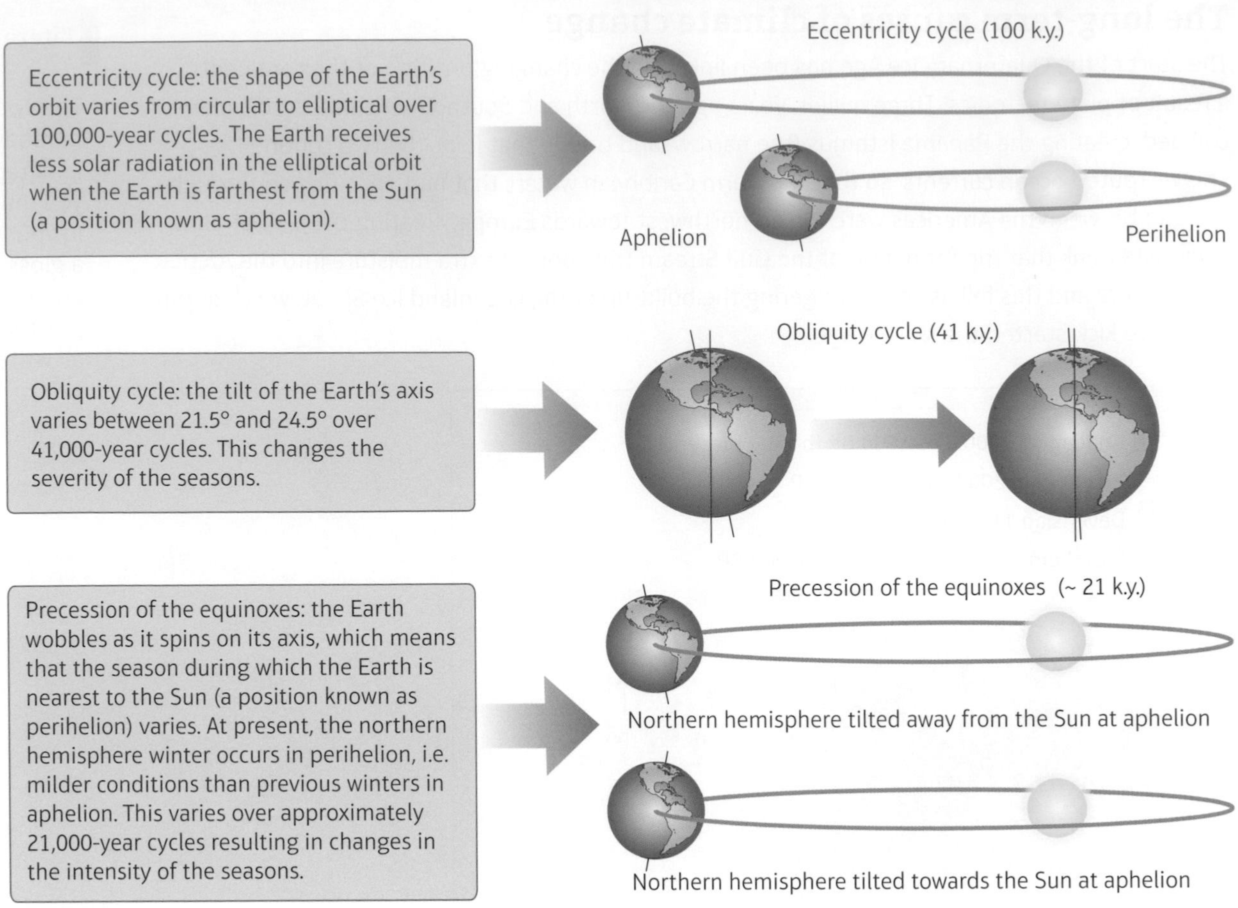

Eccentricity cycle: the shape of the Earth's orbit varies from circular to elliptical over 100,000-year cycles. The Earth receives less solar radiation in the elliptical orbit when the Earth is farthest from the Sun (a position known as aphelion).

Eccentricity cycle (100 k.y.)

Aphelion

Perihelion

Obliquity cycle: the tilt of the Earth's axis varies between 21.5° and 24.5° over 41,000-year cycles. This changes the severity of the seasons.

Obliquity cycle (41 k.y.)

Precession of the equinoxes: the Earth wobbles as it spins on its axis, which means that the season during which the Earth is nearest to the Sun (a position known as perihelion) varies. At present, the northern hemisphere winter occurs in perihelion, i.e. milder conditions than previous winters in aphelion. This varies over approximately 21,000-year cycles resulting in changes in the intensity of the seasons.

Precession of the equinoxes (~ 21 k.y.)

Northern hemisphere tilted away from the Sun at aphelion

Northern hemisphere tilted towards the Sun at aphelion

Figure 2.4: Milankovitch cycles

The short-term causes of climate change

Volcanic emissions

In April 1815 the Indonesian volcano Mount Tambora produced one of the most powerful volcanic eruptions in recorded history. Accounts of very cold weather were documented in the year following this eruption in a number of regions across the planet. Initially, scientists thought that the ash emitted into the atmosphere from large volcanic eruptions was responsible for the widespread cooling, by partially blocking the transmission of solar radiation to the Earth's surface. However, more recent measurements show that most of the ash thrown into the atmosphere during volcanic eruptions returns to the Earth's surface within months. The most significant volcanic impact on climate is the injection into the atmosphere of large quantities of sulphur dioxide gas, which remains in the atmosphere for as long as three years (Figure 2.5). Sulphate aerosols are formed, which increase the reflection of radiation from the Sun back into space, cooling the Earth's lower atmosphere.

Figure 2.5: The eruption of Mount Pinatubo, 15 June 1991

Variations in solar output

For hundreds of years, scientists have regularly counted the number of dark patches on the face of the Sun. These sunspots are caused by intense magnetic activity in the Sun's interior. An increase in the number of sunspots means that the Sun is more active and giving off more energy, so sunspot numbers indicate levels of solar output, and they appear to vary over an 11-year cycle (Figure 2.6). The climate has fluctuated during the Holocene epoch, with cooler temperatures occurring between 1300 and 1870, a time known as the **Little Ice Age**. The causes of the Little Ice

Yellow = Players, Orange = Attitudes and actions, Purple = Futures and uncertainties

Age cooling are still debated, but there is evidence to suggest that it may have been triggered by volcanic emissions and variations in solar output. Observations of the Sun during the latter part of the Little Ice Age (1650 to 1750) indicate that very little sunspot activity was occurring on the Sun's surface (a period known as the Maunder Minimum) and during this time Europe and North America experienced colder than average temperatures.

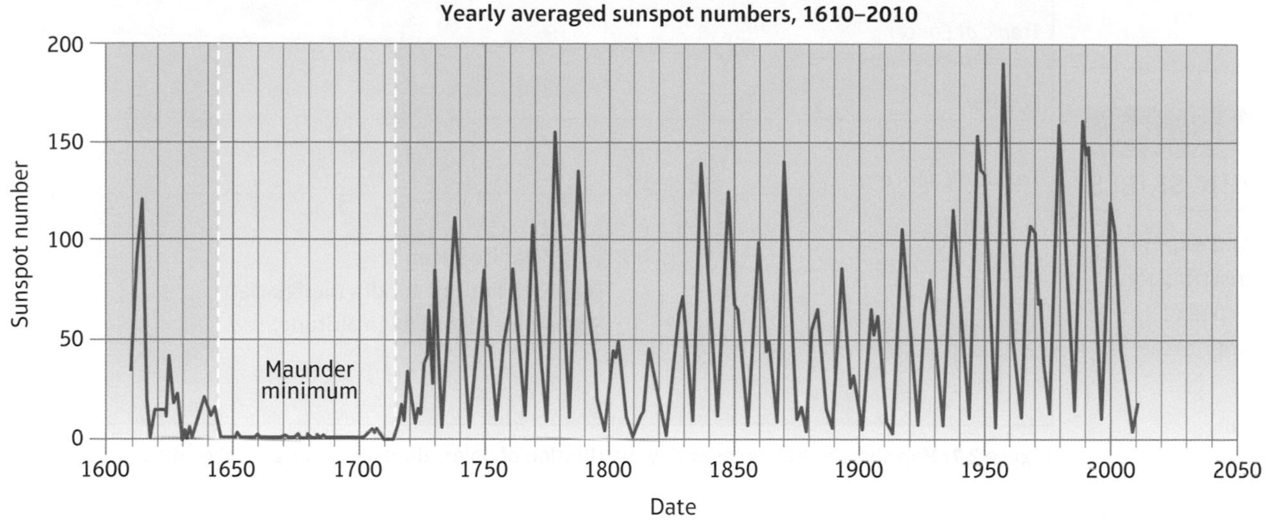

Figure 2.6: Observations of sunspot numbers showing considerable variations, including the Maunder Minimum, during the Little Ice Age

Ice cover in the Pleistocene epoch

Cryosphere comes from the Greek word for cold, *kryos*. It refers to the cold environments of our planet, where water is in its solid form of ice. This includes sea ice, lake ice, river ice, snow cover, glaciers, ice caps, ice sheets, and frozen ground (**permafrost**). The present-day distribution of cold environments is shown in Figure 2.7.

- **Polar** glacial environments are found in the high latitudes of the Antarctic and Arctic. They are characterised by extremely cold temperatures (mean annual temperatures of –30 to –40°C) and low levels of precipitation.
- **Alpine** glacial environments are found at high altitudes in mountain ranges in the mid to low latitudes, for example the European Alps, the Himalayas and the Andes. They are characterised by high levels of precipitation and a wide temperature range, with frequent freeze–thaw cycles.
- **Glaciers** are slow-moving bodies of ice in valleys, which shape the landscape in both polar and alpine environments.
- **Periglacial** environments do not feature glaciers, but they are usually found next to glacial areas. They are characterised by **permafrost** (permanently frozen ground) and occur in high-latitude or high-altitude areas where seasonal temperatures vary above and below freezing point. There are extensive periglacial environments across Siberia, Alaska and northern Canada.

Extension

On 15 June 1991 the Philippine volcano Mount Pinatubo produced one of the largest eruptions of the 20th century. Research the impact of this eruption on the Earth's climate, and use statistics to show the scale of change.

Synoptic link 

Volcanic eruptions can have a regional to global impact on the Earth's climate, and can affect glacial and periglacial environments. An understanding of volcanic processes may help your understanding of glacial advances and retreats. (See page 33.)

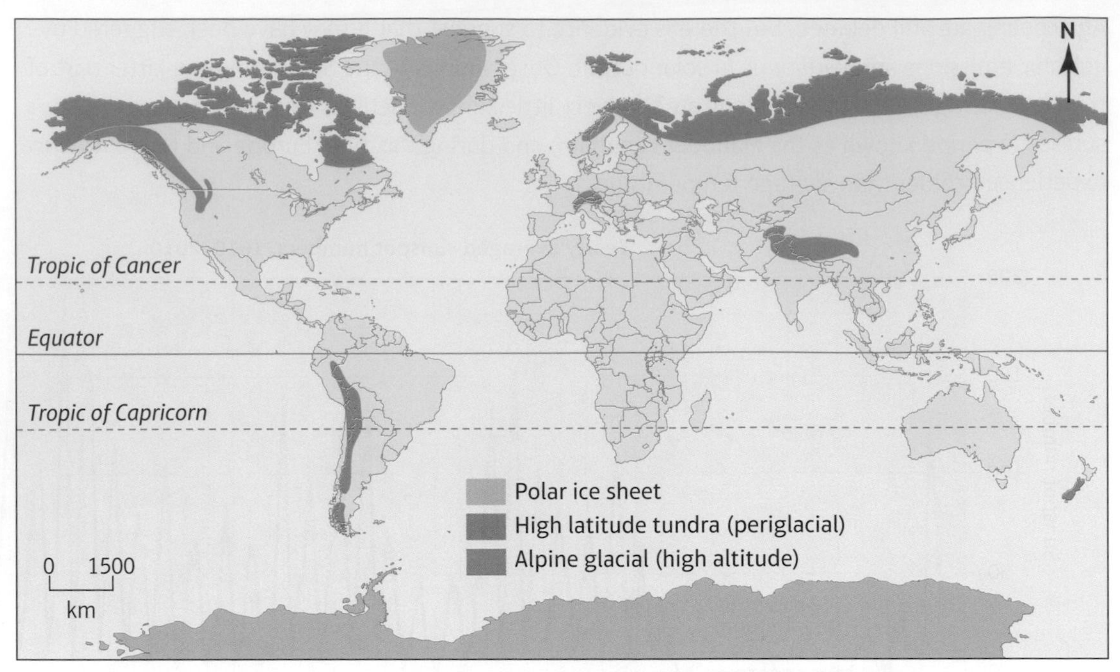

Figure 2.7: Map showing the present-day distribution of polar, alpine and periglacial cold environments

ACTIVITY

CARTOGRAPHIC SKILLS

1. **a.** Describe and explain the present distribution of the three types of cold environment (see Figure 2.7).

   b. Describe the global extent of the last glaciation in the Pleistocene epoch (see Figure 2.8).
2. **a.** Study Figure 2.3 and describe the distribution of relict glacial environments in the UK.

   b. The Pleistocene ice did not extend over the whole of the UK, but the area beyond the ice limit experienced periglacial conditions. Describe the location of the maximum ice extent in the UK, and identify the areas of the UK that experienced only periglaciation (Figure 2.3).
3. Why is it important to study the cryosphere?

The global distribution of glacier ice changed during the glacial–interglacial cycles of the Pleistocene, as the ice advanced and retreated in response to climate change (Figure 2.8). Periglacial environments existed at the margins of the ice and shifted with the ice movement, replacing glacial environments as the ice retreated. Today, much of Britain is a **relict** glacial environment, which means that it no longer experiences active glacial processes, but it does display geomorphological evidence of the Pleistocene glaciation.

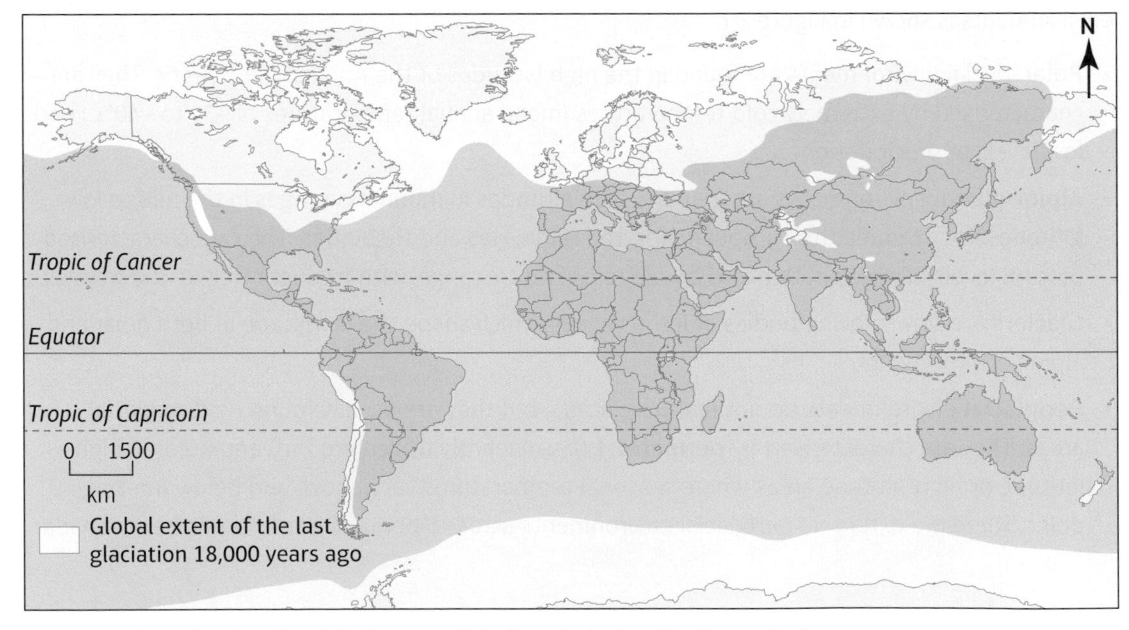

Figure 2.8: Map showing the global extent of the last glaciation (the Devensian)

Yellow = Players, Orange = Attitudes and actions, Purple = Futures and uncertainties

Classifying ice masses by scale

Glaciers are large bodies of ice that flow downhill under the influence of gravity. They form in areas where temperatures are sufficiently low to allow snow to persist from year to year, enabling thick layers of snow to accumulate and be slowly compressed to form glacial ice. Snowflakes have an open feathery structure that traps air as they accumulate on the ground, a process known as **alimentation**. With continued alimentation the lower snowflakes are compressed and gradually lose their edges, to become tiny grains separated by air spaces. Meltwater seeps into the gaps, expelling the air, and then freezes. This process increases the compaction of the snow, and further accumulation increases pressure. The result is a mass of tightly packed, randomly orientated ice crystals separated by tiny air passages, which is known as **firn** or **névé** if it lasts from one year to the next. The transition from firn to ice occurs when the interconnecting air passages are sealed, isolating bubbles of air. The duration of this transition depends on the location of the glacier, and may range from 30 to 40 years in some Alpine locations, to over 1000 years in parts of Antarctica.

The smallest ice masses are niche glaciers, which are patches of ice in hollows and gullies on north-facing mountain slopes in the northern hemisphere. With continued snowfall and ice formation, a niche glacier will grow to form a **cirque** (or **corrie**) **glacier**, which forms in an armchair-shaped hollow in the mountains. Many cirque glaciers spill out of their hollows and flow downhill to join or form **valley glaciers**, which resemble rivers of ice contained within steep-sided (U-shaped) valleys (Figure 2.9). Valley glaciers move downslope until they reach a flat plain, where they spread out into large lobes known as piedmont glaciers.

The largest masses of ice are **ice sheets** (over 50,000 km^2); the two ice sheets on Earth today cover much of Greenland and Antarctica, with the Antarctic ice sheet covering an enormous area of almost 14 million km^2. Ice sheets bury the whole landscape except for the highest mountain summits, called **nunataks**. Where ice sheets extend out over the sea, they form expanses of floating ice, known as ice shelves – one example is the Larsen Ice Shelf in Antarctica – and these may calve to form detached floating ice masses, called icebergs. **Ice caps** are dome-shaped ice masses (less than 50,000 km^2) that cover mountain peaks and plateau areas. An example is the Vatnajökull ice cap in Iceland. **Icefields** are similar to ice caps, but are typically smaller, and icefield topography is determined by the shape of the surrounding land.

Figure 2.9: The Franz Josef Glacier in New Zealand – an example of a valley glacier

ACTIVITY

TECHNOLOGY/ICT SKILLS

1. Use an internet search engine to find photographs of different types of glacier or ice mass: niche, cirque, valley, piedmont, ice field, ice cap, ice sheet, ice shelf and iceberg.

2. Save and annotate the photographs to show the features of each type of glacial ice, and include a named, located case-study example.

AS level exam-style question

Explain **two** short-term causes of climate change. (4 marks)

Guidance

For each cause, a logical sequence of processes is needed to explain why the climate changes.
Include an example for each cause as evidence of the climate change.

A level exam-style question

Explain one approach to classifying ice masses. (6 marks)

Guidance

Use data and named examples in your answer to show the differences between the ice masses, either by scale or location.

Classifying ice masses by location

Polar (cold-based) glaciers

The glaciers in polar glacial environments such as Greenland and Antarctica are also called cold-based glaciers. These glaciers are frozen onto the bedrock below, and melting occurs only at the surface during the short summer season.

Temperate (warm-based) glaciers

Temperate or warm-based glaciers occur in alpine glacial environments such as the European Alps, Norway and New Zealand. The temperature of the ice is often close to zero, and mild summer temperatures will cause melting.

Periglacial processes, landforms and landscapes

The prefix 'peri' means near or around, and the term periglacial has traditionally been used to refer to cold environments on the margins of glaciers. Today the term is used more widely to refer to non-glacial cold environments, which are characterised by periods of extreme cold, dry conditions with frequent freeze–thaw cycles and the development of permafrost. Periglacial regions are also referred to as tundra, a term that describes the treeless vegetation of dwarf shrubs, sedges and grasses, mosses, and lichens (Figure 2.10). Tree growth is hindered by mean annual temperatures of less than 3°C and minimum temperatures as low as –50°C in the coldest periglacial locations. There is also a short growing season due to the long, dark winters in the high latitudes. The coldest inhabited place on Earth is the village of Oymyakon in the Russian tundra, where the lowest recorded temperature was –71.2°C.

Figure 2.10: The periglacial tundra landscape of the Arctic National Wildlife Refuge, Alaska

There are a number of processes shaping periglacial landforms and landscapes (Table 2.1). Some are unique to periglacial areas, such as **pingos** and **patterned ground** (the collective term for stone and ice-wedge polygons and stone stripes). A periglacial landscape describes the collection of periglacial landforms in an area (see Figure 2.13).

Yellow = Players, Orange = Attitudes and actions, Purple = Futures and uncertainties

Table 2.1: Periglacial processes and landforms

Periglacial process	Explanation	Landforms produced by the process
Freeze–thaw weathering	When water freezes in the cracks and joints of rock, it expands by up to 10 per cent of its volume, weakening the rock and causing disintegration through repeated freeze–thaw cycles. This process is also known as frost action or frost shattering.	An accumulation of angular, frost-shattered rock fragments – known as a blockfield when on a flat surface, and as scree when on a slope.
Solifluction	This is the downslope movement of the saturated active layer under the influence of gravity (known as gelifluction when it occurs over impermeable permafrost).	A tongue-shaped feature at the foot of a slope – known as a solifluction lobe.
Nivation	A combination of processes weakens and erodes the ground beneath a snow patch. These processes include freeze–thaw weathering, solifluction and meltwater erosion.	Rounded nivation hollows formed in upland areas.
Frost heave	The freezing and expansion of soil water causes the upward dislocation of soil and rocks. As the ground freezes, large stones become chilled more rapidly than the soil. Water below such stones freezes and expands, pushing the stones upwards and forming small domes on the ground surface.	On flat ground, stone polygons formed as the large stones settle around the edges of the domes; on slopes, stone stripes formed as the stones move downhill.
Groundwater freezing	Where water is able to filter down into the upper layers of the ground and then freeze, the expansion of the ice causes the overlying sediments to heave upwards into a dome, which may rise as high as 50 m.	An ice-cored dome known as a pingo.
Ground contraction	When dry areas of the active layer refreeze, the ground contracts and cracks. Ice wedges will form when meltwater enters the crack during the summer and freezes at the start of winter. Repeated thawing and refreezing of the ice widens and deepens the crack, enlarging the ice wedges.	Large-scale polygonal patterns on the ground surface, known as **ice-wedge** polygons.
Aeolian action	Due to limited vegetation cover, the wind is able to pick up and transport the fine, dry sediment from the ground surface.	Extensive accumulations of wind-blown deposits, known as **loess**.
Meltwater erosion	During the short summer, thawing creates meltwater, which erodes stream or river channels. Refreezing at the onset of winter causes a reduction in discharge and sediment deposition in the channel.	Braided streams with multiple channels separated by islands of deposited material.

The distribution and types of permafrost

Permafrost is permanently frozen ground where subsoil temperatures remain below 0°C for at least two consecutive years. The distribution of permafrost gives a good indication of the main periglacial regions (Figure 2.11). Today, permafrost is concentrated in the high latitudes, but it is also found in lower latitudes at high altitudes. During Pleistocene glacials, large areas of the present temperate mid-latitudes experienced periglacial conditions because of their proximity to ice sheets. Evidence suggests that an additional 20–25 per cent of the Earth's land surface would have experienced permafrost and/or intense frost action during Pleistocene glacials. There are several terms to describe the different types of ground in the permafrost zone (Figure 2.12).

- **Continuous permafrost** is found at the highest latitudes, where virtually all the ground is permanently frozen and there is very little, if any, surface melting. It may extend to depths of several hundred metres.

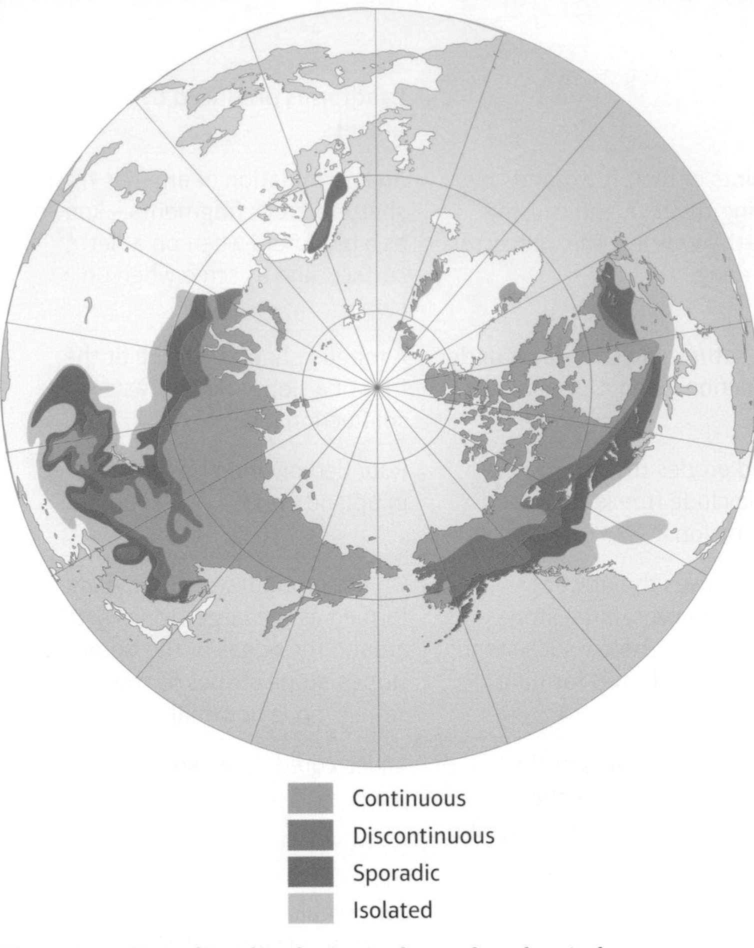

Figure 2.11: Permafrost distribution in the northern hemisphere

- **Discontinuous permafrost** is shallower and the permanently frozen ground is fragmented by patches of unfrozen ground, or **talik**. The surface layer of the ground melts during the summer months.
- **Sporadic permafrost** occurs where the mean annual temperature is only just below freezing and permafrost cover amounts to less than 50 per cent of the landscape.
- **Isolated permafrost** occurs when less than 10 per cent of an area is affected.
- The **active layer** is the upper part of the ground that regularly thaws during the summer months. Unlike the permafrost below, it is highly mobile as a result of frequent freeze–thaw cycles and meltwater saturation, caused by the impermeable nature of the permafrost. On slopes, the saturated active layer will move downslope under the influence of gravity, slowly shaping landforms and the landscape.

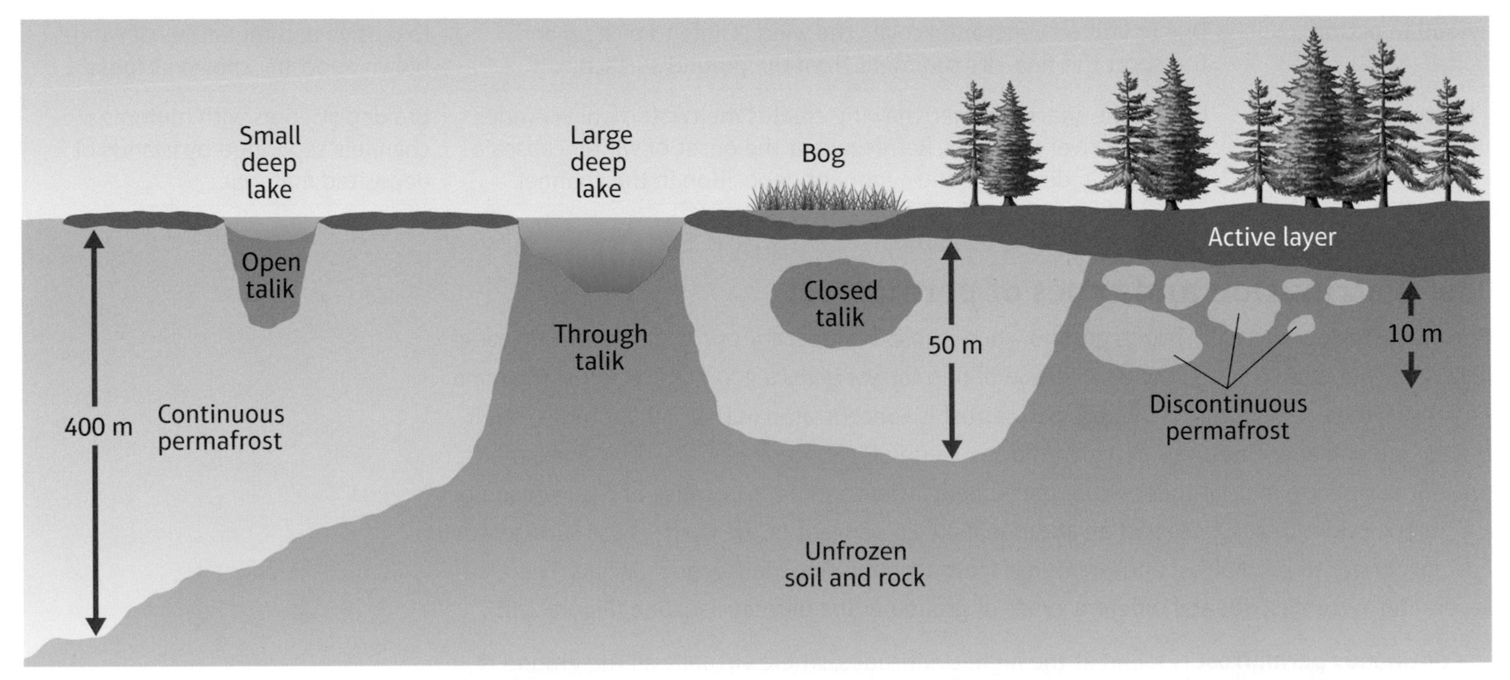

Figure 2.12: Types of permafrost, the active layer and talik

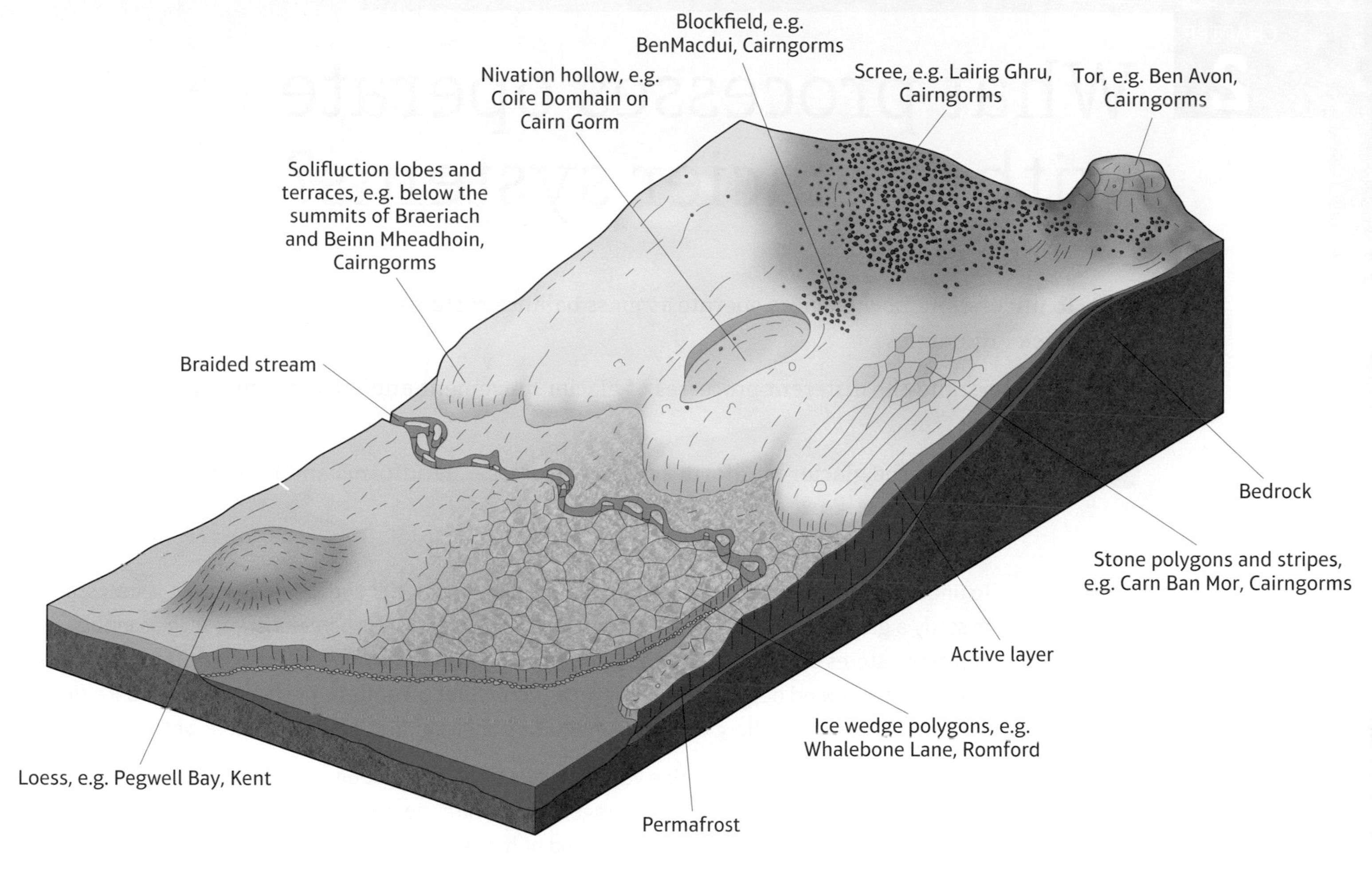

Figure 2.13: A periglacial landscape with relict UK case-study examples

ACTIVITY

GRAPHICAL SKILLS

Landform sketches or diagrams are useful revision tools that can be used in exams to help improve explanations.

1. Use Figure 2.13 to draw a simple diagram of each landform and annotate your diagram to explain how it is formed. You could also use an internet search engine to find photographs of each of the periglacial landforms to help you with your drawings.

2. Explain how the scale of changes in the distribution of ice cover during the Pleistocene

3. Research a periglacial landscape to find a located case-study example for each landform. Examples include the Cairngorms National Park in the UK and the Arctic National Wildlife Refuge in Alaska.

CHAPTER 2

What processes operate within glacier systems?

Learning objectives

2.4 To understand how glaciers operate as mass balance systems and why this is important in glacier dynamics

2.5 To understand the different processes of glacier movement and factors controlling the rate of movement

2.6 To understand the glacier landform system – processes, landforms and landscapes

Glaciers as systems

To understand the whole glacial landscape, including the glacier and its landforms, it is necessary to study a glacier as a system with interrelated components and characteristics, such as inputs, processes, stores and outputs (Figure 2.14). The input of snow and ice to a glacier – by precipitation, avalanches and wind deposition – is known as **accumulation** and the output of snow and ice from a glacier – by melting, **iceberg calving**, **sublimation** and evaporation – is known as **ablation**.

Rates of accumulation and ablation vary with climate. Accumulation rates increase where there are high levels of precipitation, low average temperatures, low levels of insolation and low wind speeds. The highest rates are therefore found at higher altitudes on slopes with a poleward aspect. It is possible to identify different zones on a glacier depending on the balance of accumulation and ablation. The upper part of the glacier, where accumulation is greater than ablation, is known as the accumulation zone, and the lower part of the glacier, where ablation exceeds accumulation, is known as the ablation zone (Figure 2.15). The boundary between the two zones, where accumulation equals ablation, is known as the equilibrium line.

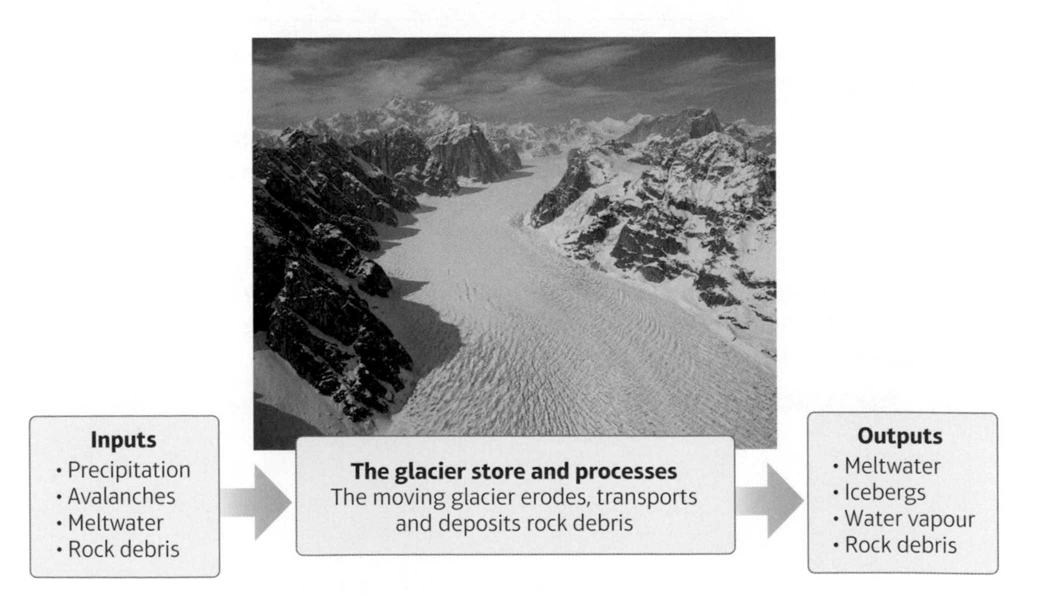

Figure 2.14: The glacier system: inputs, outputs, processes and stores

Yellow = Players, Orange = Attitudes and actions, Purple = Futures and uncertainties

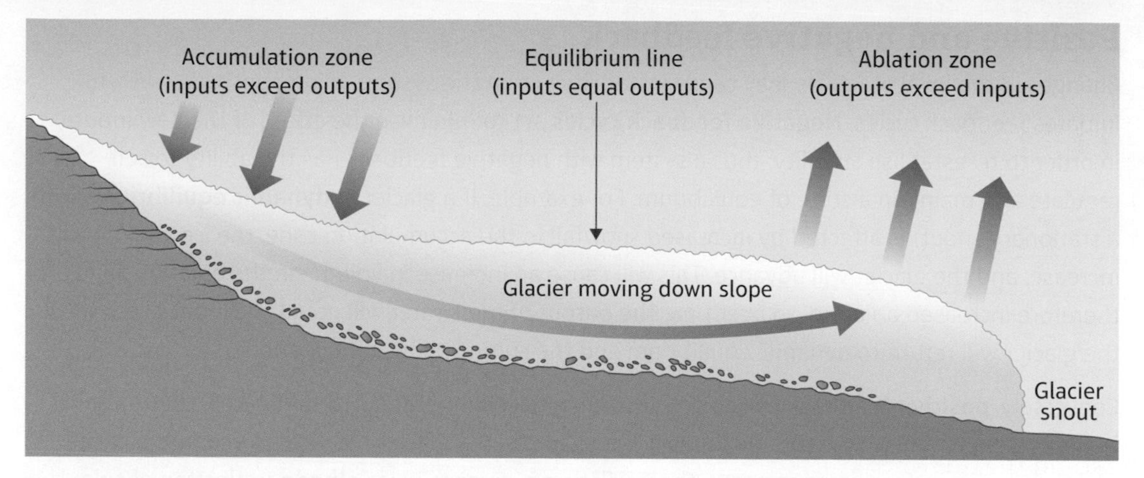

Figure 2.15: The glacier system: the accumulation zone, ablation zone and equilibrium line

Glacier mass balance

Glacier mass balance (or annual budget) is calculated by subtracting the total ablation for the year from the total accumulation (Figure 2.16). When total annual accumulation is greater than ablation, there is a positive mass balance and the glacier will advance. When total annual ablation exceeds accumulation, there is a negative mass balance and the glacier will retreat. When accumulation equals ablation, the glacier snout will be stationary (in dynamic equilibrium). Glacier mass balance has been measured since the 1950s, and is widely used as an indicator of climate change, and to assess the glacier contribution to run-off and sea-level rise. The World Glacier Monitoring Service annually compiles the mass balance measurements from around the world (Figure 2.16).

ACTIVITY

GRAPHICAL SKILLS

1. Describe and explain the global variations in glacier mass balance shown in Figure 2.16.
2. Explain the importance of studying glacier mass balance in order to understand glacial systems.

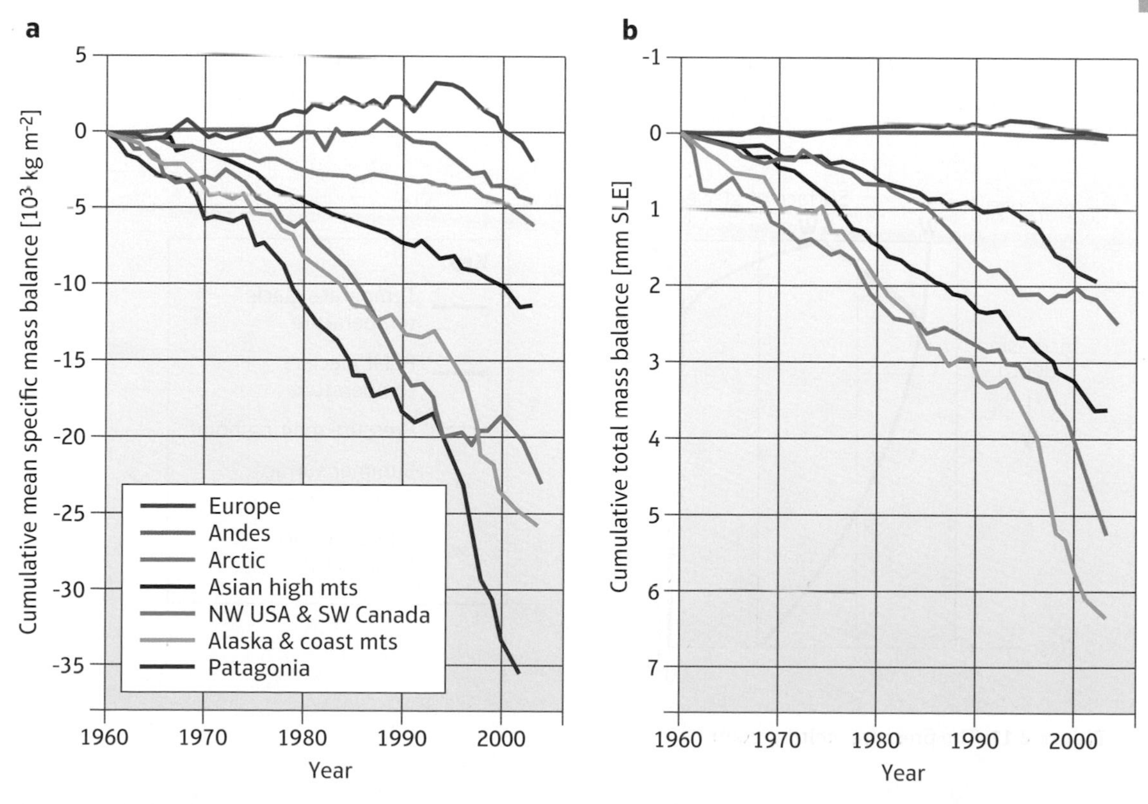

Figure 2.16: Global trends in glacier mass balance, 1960–2005. a. Mean specific mass balance shows the strength of climate change in the respective region. b. Total mass balance is the contribution from each region to sea-level rise

Positive and negative feedback

Changes in the level of inputs may cause instability within the system. In response, the system initiates feedback cycles. **Negative feedback cycles** act to minimise the effect of the new inputs, in order to re-establish stability; thus a system with negative feedback has the ability to self-regulate and maintain a state of equilibrium. For example, if a glacier in **dynamic equilibrium** (with a stationary snout) is affected by increased snowfall in the accumulation zone, the ice volume will increase, and the glacier will advance. This will cause an increase in volume in the ablation zone and therefore increased ablation, so over time the output of meltwater will equal the input of snowfall, the glacier will return to dynamic equilibrium and the snout will become stationary.

Conversely, **positive feedback cycles** amplify the initial change, and may ultimately cause a shift in the system to a new state of equilibrium. For example, if a glacier has a positive mass balance and the glacier surface area increases, there will be an increase in ice **albedo** (reflection of solar radiation). This will cause a further reduction in air temperature, thus increasing accumulation and initiating a positive feedback cycle in which the glacier will continue to advance.

Extension

Use the World Glacier Monitoring Service website to research changes in glacier mass balance over different timescales (for example, over the last five years, 25 years and 50 years), and to compare and contrast trends in different parts of the world.

Glacier movement

The process and rate of glacier movement will depend on the temperature of the ice and whether the pressure melting point is reached. At the surface of the glacier the melting point is 0°C, but with increased ice depth the melting point is fractionally lowered by the pressure of the overlying ice (Figure 2.17). Most temperate (warm-based) glaciers reach the pressure melting point, and therefore produce a great deal of meltwater, which lubricates and increases the rate of movement by a number of processes collectively known as **basal slip** (Figure 2.18). Polar (cold-based) glaciers are too cold to reach the pressure melting point, meaning that they are frozen onto the bedrock and movement only occurs by a process called **internal deformation** (Figure 2.18).

ACTIVITY

GRAPHICAL SKILLS

1. Draw a simple diagram or sketch to show each of the different processes of glacier movement in Figure 2.18.
2. Annotate your diagram to explain how the glacier moves by each process.

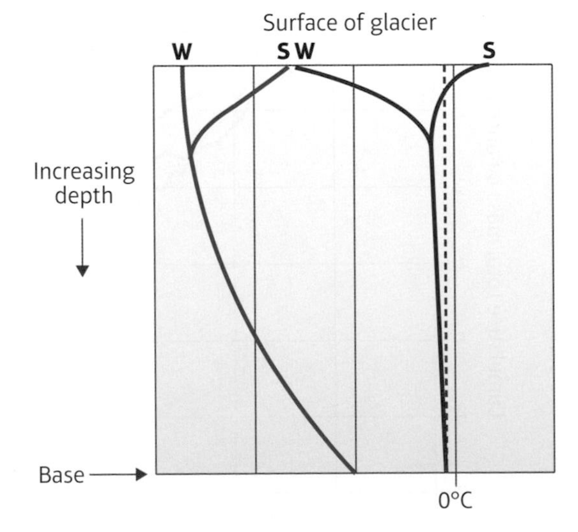

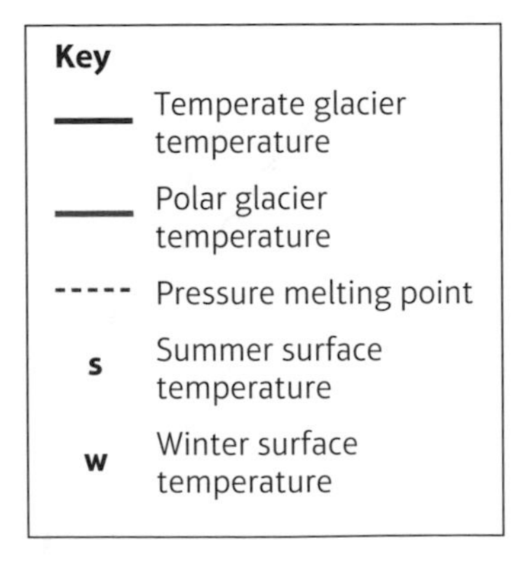

Figure 2.17: The pressure melting point

Yellow = Players, Orange = Attitudes and actions, Purple = Futures and uncertainties

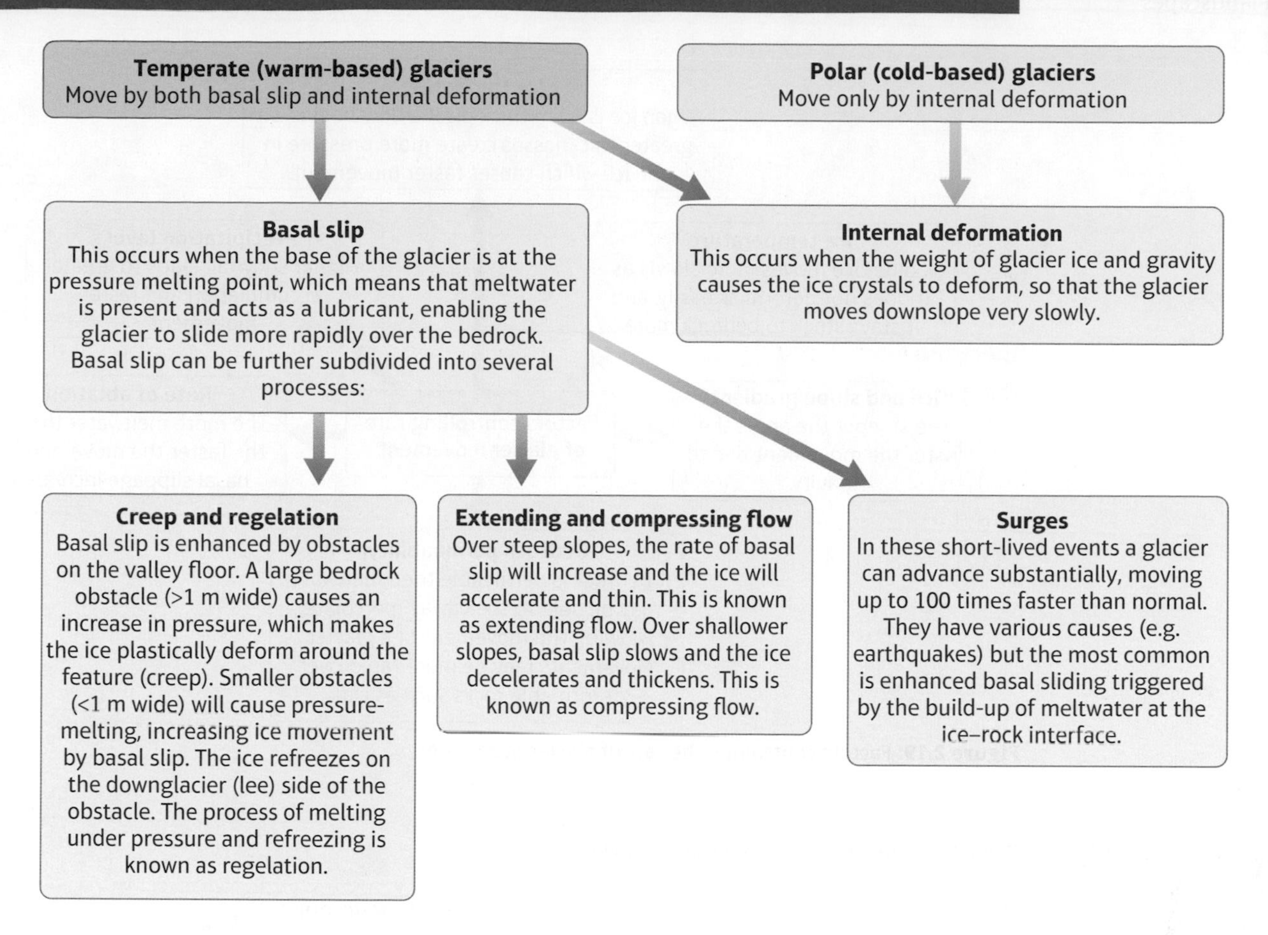

Figure 2.18: Different processes of temperate and polar glacier movement

Factors controlling the rate of glacier movement

A number of factors control the rate of glacier movement (Figure 2.19). In 2014, it was reported that the Jakobshavn Glacier in Greenland was moving at a rate of 17 km per year. The rate of movement varies not only between glaciers (see Table 2.2) but also over time for individual glaciers, as a result of changes in inputs and feedback cycles. For example, an increase in basal meltwater will increase basal slip, which could promote a positive feedback cycle by generating increased frictional heat and therefore a further increase in the rate of basal slip. Alternatively, negative feedback can occur; for example an increase in ice thickness may increase pressure melting and basal slip so the ice moves faster, but in turn this will reduce ice thickness and therefore lead to reduced pressure melting and basal slip.

AS level exam-style question

Explain the factors that cause variations in the rate of ablation. (6 marks)

Guidance

Consider the factors affecting different forms of ablation – melting, iceberg calving, sublimation and evaporation – and the role of feedback mechanisms.

A level exam-style question

Explain the role of pressure melting in glacier movement. (6 marks)

Guidance

This can be a difficult concept to explain, so check that you understand Figure 2.17 before you attempt this question.

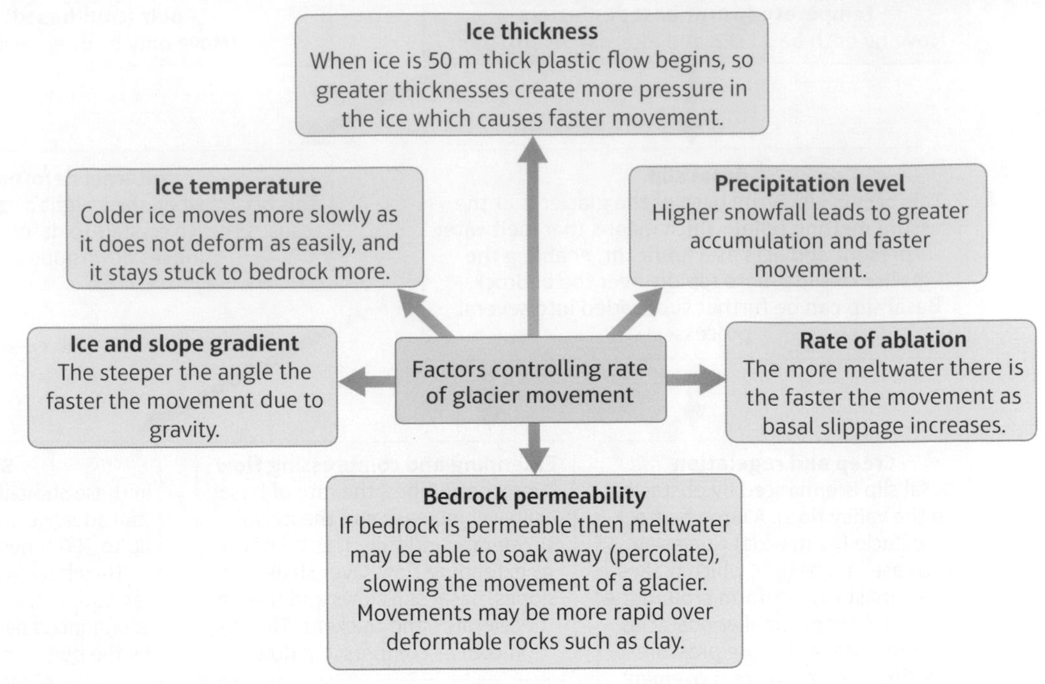

Figure 2.19: Factors controlling the rate of glacier movement

ACTIVITY

1. Suggest the possible reasons for the differences in the rates of movement shown in Table 2.2.

2. Explain the importance of the glacial mass balance concept to the understanding of glacier dynamics.

Table 2.2: Rate of movement of selected glaciers

Glacier	Average rate of movement (m per year)	Year
Jakobshavn, Greenland	12,600	2003
Pine Island Glacier, West Antarctica	2075	2007–2008
Nigardsbreen, Norway	204	2001
Fox Glacier, New Zealand	182	1991
Trapridge, Glacier, Canada	8.5	2005

Glacier landforms and landscapes

Glaciers alter landscapes through a number of processes, and these processes create landforms at a range of scales and in a range of different environments (Figure 2.20). Many landforms are uniquely formed by glaciers and produce distinctive landscapes, which will remain following ice melting and retreat. Relict glacial landscapes can help develop an understanding of past glaciations, including the rate and direction of ice movement and the extent of ice cover. Figure 2.20 summarises and links glacial processes, glacial landforms, glacial environments and glacial landscapes (see pages 86–99). For example, the plucking and abrasion by a glacier erodes a glacial trough, which is a major landform during and after glaciation, such as the Lauterbrunnen valley in Switzerland. This landform has smaller features within it, such as terminal moraines where the ice margin used to be, and ribbon lakes, for example, Buttermere, in the Lake District, where depressions were eroded by subglacial processes. These macro- and meso-landforms may be found in areas where glaciers are still active, such as Greenland, or in relict glacial landscapes, such as the Alps.

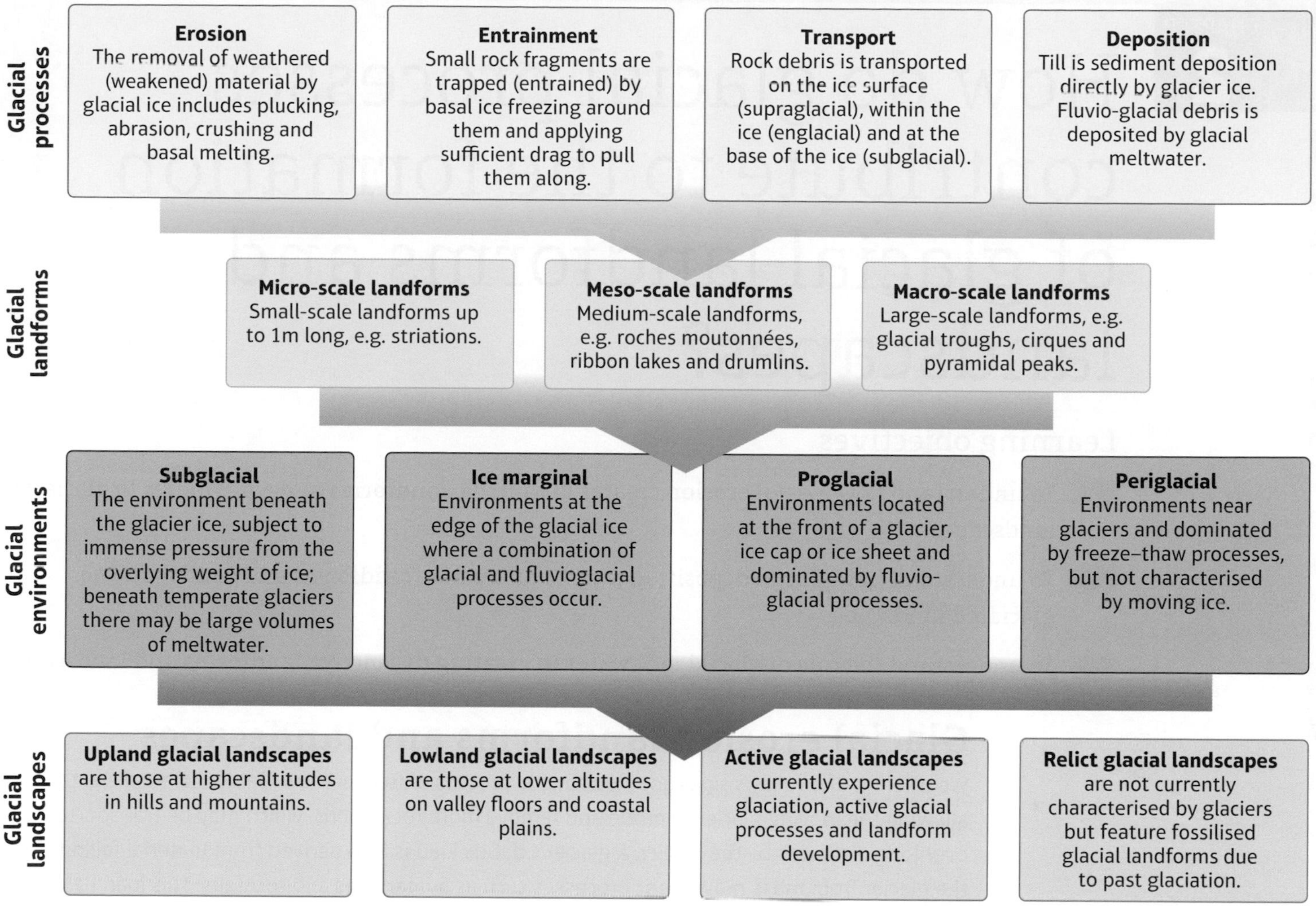

Glacial processes	**Erosion** The removal of weathered (weakened) material by glacial ice includes plucking, abrasion, crushing and basal melting.	**Entrainment** Small rock fragments are trapped (entrained) by basal ice freezing around them and applying sufficient drag to pull them along.	**Transport** Rock debris is transported on the ice surface (supraglacial), within the ice (englacial) and at the base of the ice (subglacial).	**Deposition** Till is sediment deposition directly by glacier ice. Fluvio-glacial debris is deposited by glacial meltwater.
Glacial landforms	**Micro-scale landforms** Small-scale landforms up to 1m long, e.g. striations.	**Meso-scale landforms** Medium-scale landforms, e.g. roches moutonnées, ribbon lakes and drumlins.	**Macro-scale landforms** Large-scale landforms, e.g. glacial troughs, cirques and pyramidal peaks.	
Glacial environments	**Subglacial** The environment beneath the glacier ice, subject to immense pressure from the overlying weight of ice; beneath temperate glaciers there may be large volumes of meltwater.	**Ice marginal** Environments at the edge of the glacial ice where a combination of glacial and fluvioglacial processes occur.	**Proglacial** Environments located at the front of a glacier, ice cap or ice sheet and dominated by fluvio-glacial processes.	**Periglacial** Environments near glaciers and dominated by freeze–thaw processes, but not characterised by moving ice.
Glacial landscapes	**Upland glacial landscapes** are those at higher altitudes in hills and mountains.	**Lowland glacial landscapes** are those at lower altitudes on valley floors and coastal plains.	**Active glacial landscapes** currently experience glaciation, active glacial processes and landform development.	**Relict glacial landscapes** are not currently characterised by glaciers but feature fossilised glacial landforms due to past glaciation.

Figure 2.20: Glacial processes, landforms, environments and landscapes

CHAPTER 2

How do glacial processes contribute to the formation of glacial landforms and landscapes?

Learning objectives

2.7 To understand how glacial erosion creates distinctive landforms and contributes to glaciated landscapes

2.8 To understand how glacial deposition creates distinctive landforms and contributes to glaciated landscapes

2.9 To understand the role of glacial meltwater in creating distinctive landforms and landscapes

Glacial erosion landforms and landscapes

Weathering of a valley's sides and floor by the process of freeze–thaw (Table 2.1) weakens the rock, allowing the moving glacier to erode and remove more rock debris, which may be transported over long distances by the glacier. A glacier's debris load is also derived from material falling onto the glacier from **mass movement** processes such as **avalanches** and rockfalls. This load has an important role in increasing the rate of glacial erosion.

Processes of glacial erosion

There are several processes of glacial **erosion**: plucking, abrasion, crushing and fluvio-glacial erosion processes.

Glacial plucking

Plucking will occur where rocks are well jointed and weakened and where meltwater is present at the base of the glacier due to pressure melting. The meltwater penetrates into joints and around blocks and then freezes onto the rock. As the ice moves, it exerts an immense pulling force onto the attached rock, which may fracture and be plucked from its position (Figure 2.21). Plucking leaves a very jagged landscape.

Glacial abrasion

Material plucked from the bedrock is frozen into the glacial ice, and as the glacier moves downslope this material rubs against the valley sides and floor, wearing them away by a process similar to sandpapering (Figure 2.21). Coarse rock debris will scrape the rock surface, forming scratches called striations, and fine material smoothes and polishes the rock, creating a fine rock 'flour'. Striations can be mapped in relict glacial landscapes to reveal the extent of former glacial ice and the direction of ice movement. The rate of **abrasion** will be highest under thick, fast-moving ice with large amounts of coarse, angular basal debris, particularly where the debris is hard rock and the bedrock is less resistant. A high basal water pressure and/or a large amount of fine rock debris may reduce rates of abrasion by protecting the bedrock.

Glacial crushing

This is the direct fracturing of weak bedrock by the weight of ice above it. The greatest recorded thickness of ice on Earth is 4780 m, measured in Antarctica by radio echo soundings from a US research aircraft. The bedrock must first be weakened, either by intense freeze–thaw weathering or repeated glacier advance and retreat causing dilation (rock fracture due to the removal of overlying glacier weight). Bedrock crushing typically produces large, angular blocks of rock.

Fluvio-glacial erosion

Basal ice melting beneath temperate glaciers can produce large volumes of meltwater, which causes fluvial erosion processes such as abrasion, hydraulic action, attrition and corrosion. Attrition refers to the collision of rock fragments in the meltwater, which breaks them into smaller, more rounded fragments. **Subglacial** meltwater often travels very fast and under high pressure, because it is confined beneath the glacier. The force of the water may dislodge and remove rock debris through the process of hydraulic action.

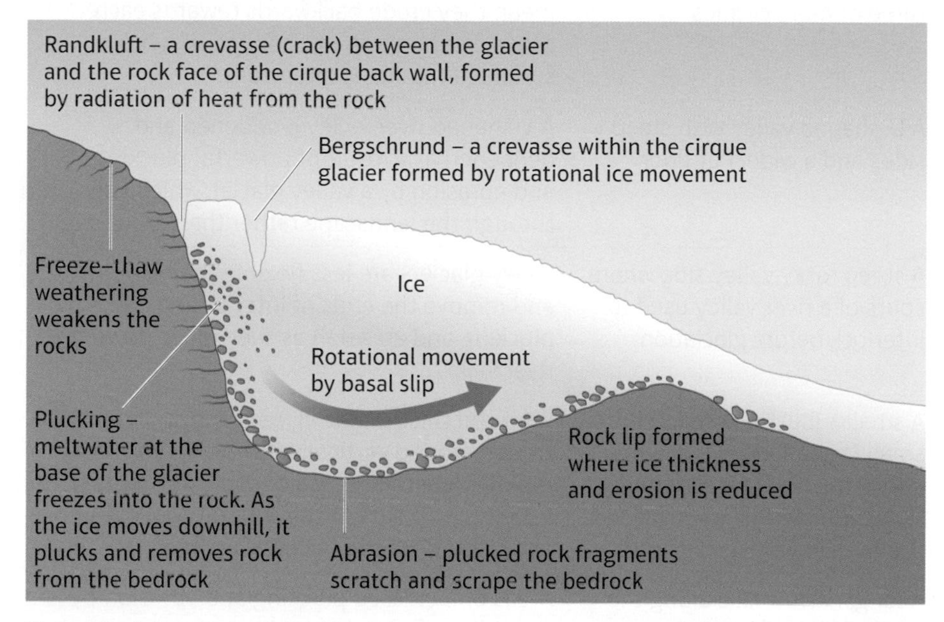

Figure 2.21: Processes of glacial erosion

Extension

Investigate the location of upland glaciated areas, past and present, in each continent. Compare these locations (e.g. latitude, coastal orientation, mountains). Why did glaciation processes, especially erosion, take place in these locations?

Landforms of glacial erosion

A niche glacier will slowly erode a small rounded hollow in a mountainside and, as more ice accumulates, a cirque glacier will form and erode a large armchair-shaped hollow, known as a cirque or corrie. Multiple cirques around a mountain summit create landforms known as **arêtes** and a pyramidal peak. If the cirque glacier overflows, it will form a valley glacier, which will erode a glacial trough (a U-shaped valley). Glacial troughs have a number of **meso-scale** and **micro-scale** erosional landforms, including those formed by **ice sheet scouring** or erosion of the valley floor (Table 2.3 and Figure 2.22). Ice is a powerful erosion agent and cuts away at the landscape leaving large scars – the cirques and glacial troughs. When these are shaped, other prominent features are left in between, such as aretes and truncated spurs. Then, as ice powers its way through and over the landscape, it will ride over obstacles creating features such as roches mountonnee (Figure 2.22). The larger features can be identified from detailed Ordnance Survey maps.

Table 2.3: Landforms of glacial erosion

Landform	Description	Process of formation
Cirque (corrie or cwm)	An amphitheatre-shaped depression in a mountainside with a steep back wall and a rock lip. The name given to this landform depends on its location, e.g. a cwm is found in Wales.	A large rounded hollow high on a mountainside is eroded and deepened by plucking and abrasion due to the rotational ice movement of a cirque glacier.
Arête	A narrow, knife-edged ridge between two cirques	Plucking and abrasion on the back wall of two cirques on a mountainside mean they erode backwards towards one another, creating a narrow ridge. Freeze–thaw action is also important.
Pyramidal peak	A pointed mountain peak with three or more cirques	The erosional processes within nearby cirques mean they erode backwards towards each other, creating a sharp, pointed mountain summit. Plucking is important.
Glacial trough	A U-shaped valley with steep sides and a wide, flat floor	A V-shaped river valley is widened and deepened as a result of powerful plucking and abrasion by a valley glacier, which goes through the landscape rather than around it.
Truncated spur	A steep rocky valley side where spurs of a river valley used to interlock before glaciation	Valley glaciers are less flexible than rivers and remove the ends of interlocking spurs by plucking and abrasion as they move down the river valley.
Hanging valley	A small tributary V-shaped or small U-shaped valley high above the main glacial trough floor, often with a waterfall as the river flows over the edge	Powerful thicker glacial ice in the main glacial trough eroded vertically downwards more rapidly than thinner ice or rivers in tributary valleys. The floors of the tributary valleys are left high above the main valley floor.
Ribbon lake	A long, narrow lake along the floor of a glacial trough	Areas of increased plucking and abrasion by the valley glacier deepen part of the valley floor, as a result of either the confluence of glaciers or weaker rocks. Sometimes the lake forms behind a terminal moraine after glaciation.
Roche moutonnée	A mass of bare rock on the valley floor with a smooth stoss (up-valley side) and a steep jagged lee (down-valley side)	A more resistant rock outcrop causes ice movement by creep and regelation around it. As the ice slides over the rock, it scours and smoothes the stoss, while refreezing on the lee causes plucking.
Knock and lochan	A lowland area with alternating small rock hills (knock) and hollows, often containing small lakes (lochan)	Scouring at the base of a glacier excavates areas of weaker rock, forming hollows that fill with meltwater and precipitation following ice retreat.
Crag and tail	A very large mass of hard rock forms a steep stoss with a gently sloping tail of deposited material	A large mass of hard rock is resistant to ice scouring and creates a steep stoss. Reduced glacier velocity on the lee protects softer rock and allows deposition, but the sheltering effect diminishes with distance, creating a sloping tail.

ACTIVITY

1. For each of the glacial landforms shown in Table 2.3, search the internet for a photograph or satellite image of an example.
2. Save these to a document and label the images to show the dimensions of each feature. Add annotations using the information given in Table 2.3.
3. Consider the size of these features in relation to landscape features near to where you live (i.e. are they bigger or smaller? How much bigger or smaller?).

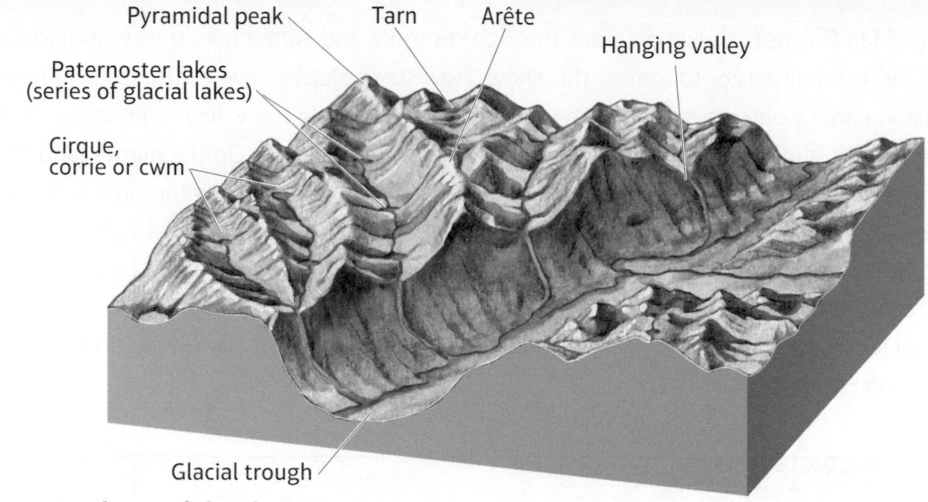

Figure 2.22: Landscape of glacial erosion

ACTIVITY

CARTOGRAPHIC SKILLS

Use Figure 2.23 to construct a cross-section through the glacial trough. Place a plain piece of paper over the map from a 750 m contour to the bottom of the map. On the edge of the paper, mark with a small pencil dash where each contour disappears under the paper. Write the correct contour heights next to these marks. Take a piece of graph paper and create a vertical axis to represent height above sea level (perhaps 2 mm = 25 mm). Place the marked edge of the piece of paper along a blank horizontal axis. On the graph paper, plot the correct heights directly above the lines on the paper. Once plotted, draw a line between the plots to show the land surface. Label the landforms in the correct place on your cross section. Describe the shape of the glacial trough.

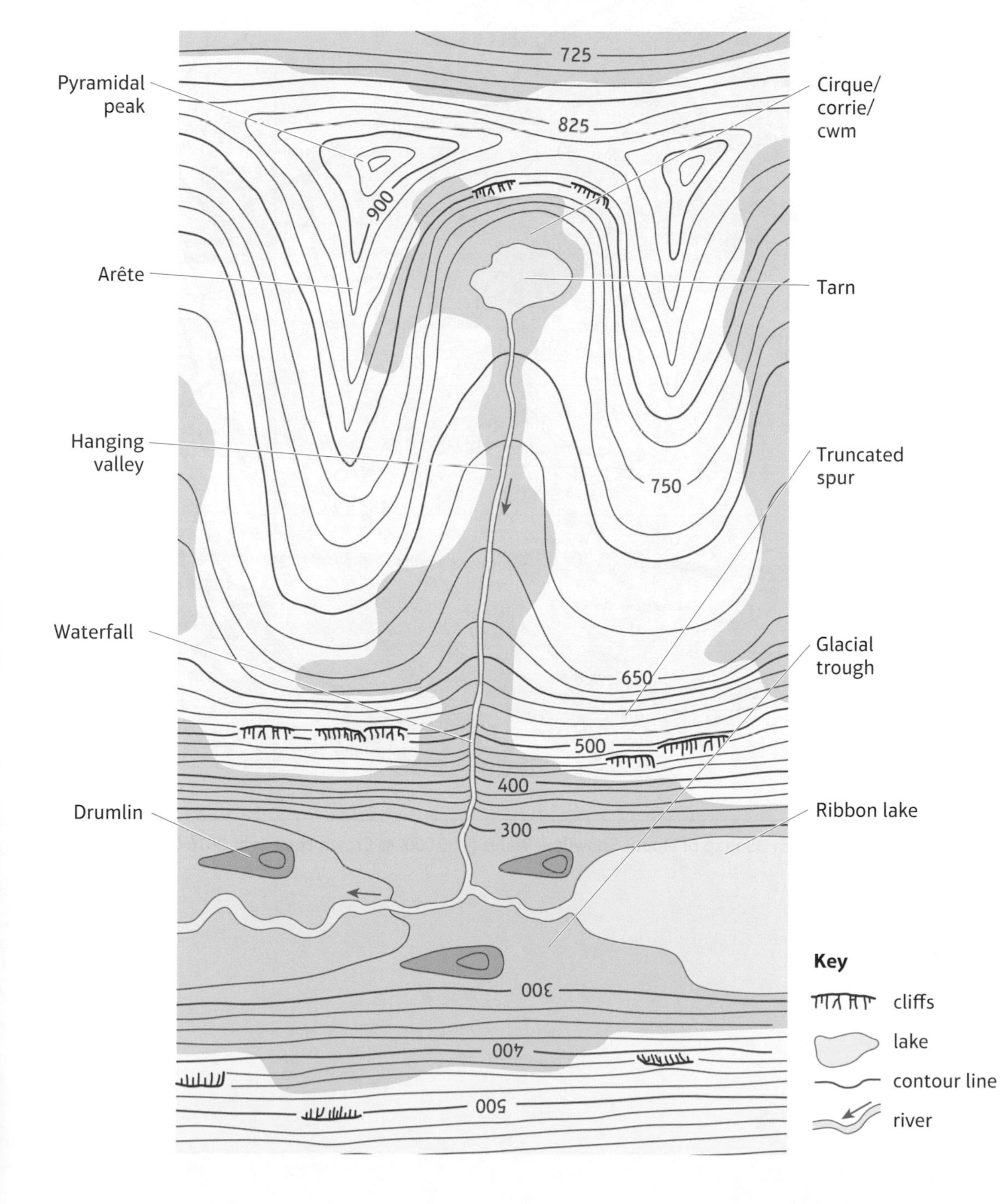

Figure 2.23: The use of contour lines to identify glacial landforms

ACTIVITY

CARTOGRAPHIC SKILLS

Figure 2.24 is an OS map of Mount Snowdon in Wales. It is a relict glacial environment.

1. Mount Snowdon is a pyramidal peak with multiple cirques (cwms) and arêtes radiating around the summit. Can you identify them? Record their names as case-study examples.
2. At the peak of glaciation, 18,000 years ago, the cirque glaciers on Mount Snowdon overflowed to create valley glaciers, which eroded glacial troughs and left ribbon lakes. Identify these features with names and grid references and record them as case-study examples.
3. Today, hanging valleys sit high above the main glacial troughs, with rivers that descend as waterfalls to the valley floor. Identify these features with names and grid references and record their location.
4. **a.** Identify and locate evidence of human activities in this glaciated environment.

   b. Suggest what problems for the physical landscape may arise because of these activities.

The shape of landforms is shown on maps through contours and sometimes the symbol for rock. Figure 2.23 is a simplified contour map; this shows how steep slopes have contours close together with occasional rocky outcrop symbols. The absence of contours in the valley floor shows that it is almost flat. This is a relict glacial environment as there is no ice shown. On the higher land, there is a circular area with a lake, with higher land on three sides (a cirque with a tarn in it showing that there is depth to the hollow). There are triangular patterns to the contours in places which represent the typical three sides of a glaciated mountain peak. Study Figures 2.22 and 2.24 together to see what the features look like on 3D and on a map. Figure 2.24 is an OS map of a real relict glacial landscape and is therefore more complicated, but look for the similar arrangement of contours as on Figure 2.23.

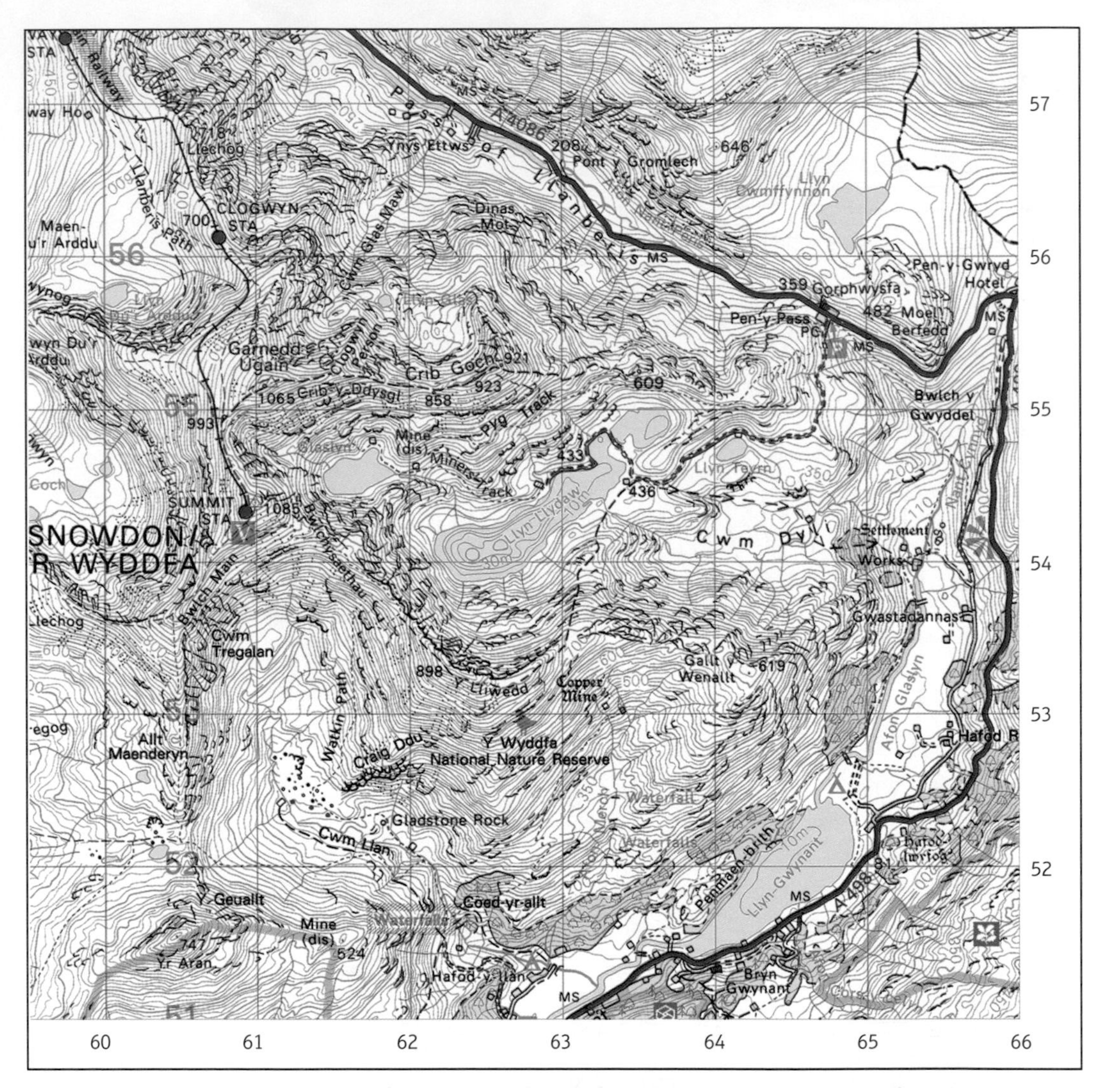

Figure 2.24: Extract from OS map of Mount Snowdon, Wales (1:50,000) © Crown copyright 2016 OS 100030901

Glacial deposition landforms and landscapes

Glaciers generally deposit their load when velocity is reduced or they become overloaded with debris, or when ablation increases, causing retreat and deglaciation. Material deposited directly underneath ice is known as **till** or boulder clay, and it could have been transported over hundreds of kilometres and deposited in areas with different geology. These contrasting rock fragments deposited by ice are known as **erratics** and they are valuable indicators of the direction of past ice movement, as the rock types can be traced back to their source. For example, microgranite erratics from the small Scottish island of Ailsa Craig are found in Pembrokeshire, South Wales, so they were transported by ice over a distance of approximately 400 km.

Moraine is the collective term for rock material carried by a valley glacier and deposited to form a range of landforms (Table 2.4 and Figures 2.25 and 2.26).

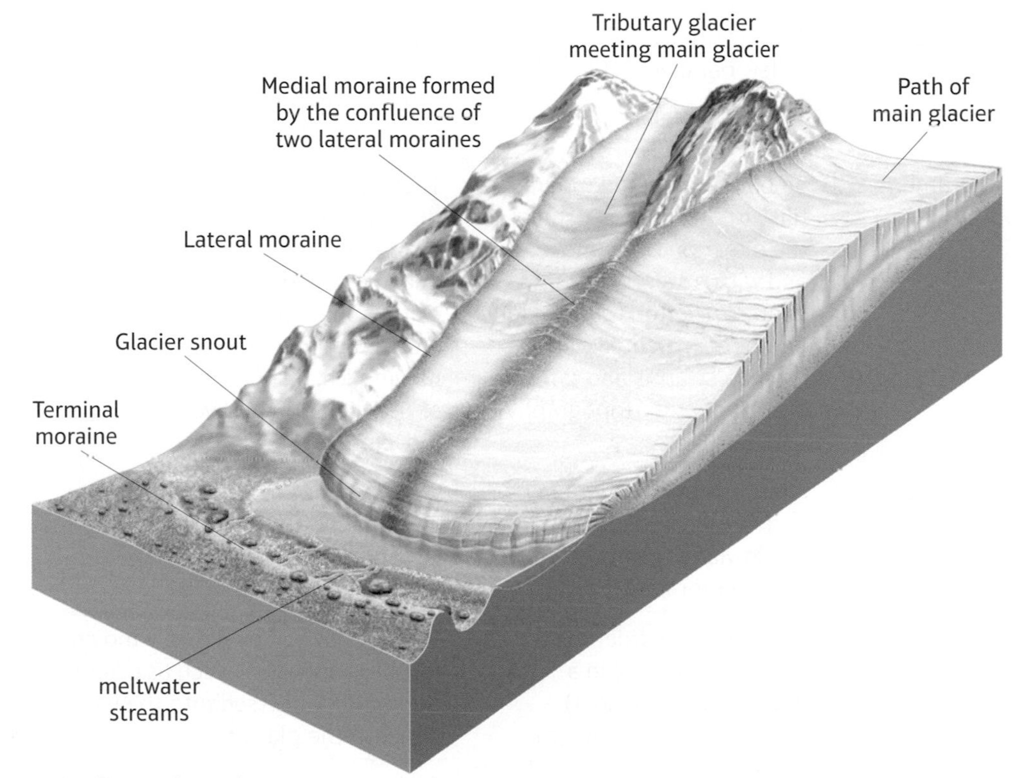

Figure 2.25: Types of moraine

Extension

Find a key for Ordnance Survey 1:50000 scale maps. Refresh your skills of map reading, for example, scale and use of grid references.

AS level exam-style question

Choose one glacial landform from Figure 2.24. Explain two processes that have shaped the landform. (6 marks)

Guidance

Link each process to the part of the landform that it shapes; for example plucking forms the steep back wall.

A level exam-style question

Explain the formation of landforms by ice sheet scouring. (6 marks)

Guidance

Study Table 2.3 carefully to determine the landforms created by this process.

ACTIVITY

TECHNOLOGY/ICT SKILLS

1. **a.** Search online to find a detailed satellite image of an active ice margin, next to an ice sheet such as in Greenland, ice cap such as in Iceland, or glacier such as in New Zealand.

   b. Save the image to a document and identify the ice contact landforms shown in Table 2.4 that are present.

   c. Annotate the image with relevant details from Table 2.4.

2. **a.** Visit the British Geological Survey website and view the time lapse presentation on Virkisjokull glacier.

   b. Which depositional features are visible?

   c. What influenced the changes that took place over the time period covered?

Table 2.4: Ice-contact and lowland landforms of glacial deposition

	Landform	Description	Process of formation
Ice-contact features	Lateral moraine	A ridge of till deposited along the valley sides	Debris from freeze–thaw weathering of the valley sides falls onto a glacier; it is then transported and deposited at the edge of the glacier when it melts.
	Medial moraine	A ridge of till deposited in the middle of the valley (parallel to the valley sides)	At the confluence of two tributary glaciers, the lateral moraines join to form a medial moraine, which is deposited during ice melt and retreat.
	Terminal moraine	A high ridge of till extending across a valley at right angles to the valley sides	Debris is deposited at the maximum extent (limit) of a glacier, especially if the glacier is in equilibrium and the snout is stationary, and will be continuously supplied to the same place.
	Recessional moraine	A lower ridge of till across the valley, parallel to the terminal moraine	Debris is deposited during interruptions in the retreat of glacier ice, when the glacier remained stationary long enough for a ridge of material to build up.
	Drumlins	Smooth, elongated mounds of till, with a long axis parallel to the direction of ice movement and with a steep stoss and gentle lee; often found in large numbers in an area, called drumlin swarms and forming 'basket of eggs' topography.	There is controversy over the origin of drumlins. A popular view is that they are formed by deposition when glacier ice becomes overloaded with debris when exiting an upland area. The deposits are streamlined and shaped by the moving ice. They are valuable indicators of the direction and velocity of past ice movement.
Lowland depositional features	Till plains	A large, relatively flat plain or undulating landscape of till in a lowland area (also called ground moraine)	When a sheet of ice retreats, large amounts of material are deposited over a sizeable area due to melting.
	Lodgement till	Deposits of angular rock fragments (clasts) in a fine matrix (rock flour) – unstratified (not layered), unsorted (mixture of clast sizes) and containing erratics (mixed geology)	Till is deposited by actively moving ice, forming landforms such as drumlins. It is lodged or pressed into the valley floor beneath the glacier.
	Ablation till		Till is deposited by melting ice from stationary or retreating glaciers, forming landforms such as terminal and recessional moraines.

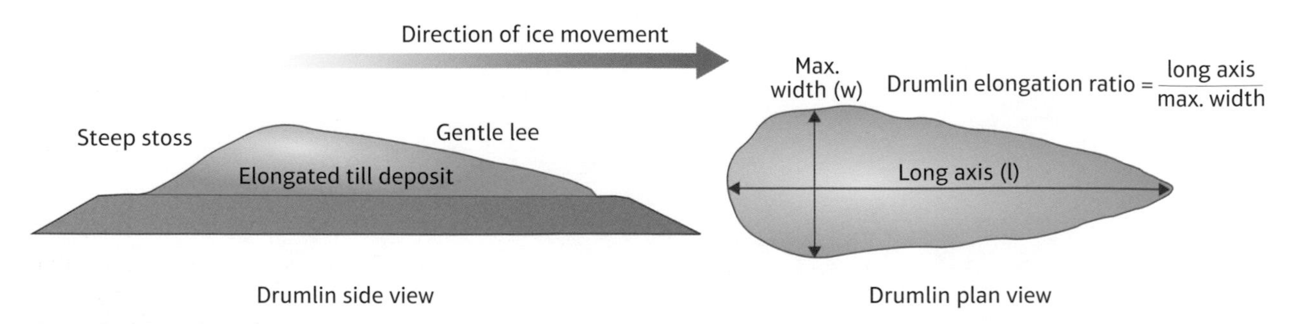

Figure 2.26: Drumlin formation

Yellow = Players, Orange = Attitudes and actions, Purple = Futures and uncertainties

Reconstructing past glaciation using relict glacial landscapes

The glacial landforms of a relict landscape can be studied to reconstruct former ice extent, direction and rate of ice movement and ice origin. For example, we can determine ice extent by studying the distribution of ground moraine and mapping the position of the terminal moraines. A number of erosional and depositional landforms can be mapped to indicate the direction of ice movement, including striations, **roche moutonnées**, **glacial troughs**, **crag and tail** and **drumlins**. Furthermore, the elongation ratio of drumlins can be calculated to indicate the rate of ice movement; it is thought that greater elongation is related to faster ice movement (Figure 2.26).

It can be difficult to identify these landforms in the field, because they may be modified and confused by multiple glaciations and post-glacial weathering and erosion, as well as human activities. Therefore several landforms must be studied for accurate reconstruction. Till fabric analysis is a valuable fieldwork technique for studying both the origin and the route of former glaciers using glacial deposits (Figure 2.27).

The provenance of glacial deposits is important for reconstructing the movement of ice in the past. This usually involves working out the origin or source of rock fragments (or clasts) by studying the geology. For example, the lower layers of the till matrix in the cliffs at Happisburgh (North Norfolk) have local clasts mixed with rock fragments from north and north east England (dolerite and limestone), southern and central Scotland (metamorphic, basalt and granite), and from Oslo fjord in Norway.

Till fabric analysis and cirque (corrie) orientation

As a glacier moves, it turns the larger till rock fragments (clasts) to point in the direction of the ice movement. Therefore it is possible to study the orientation of the clasts to understand the former ice movement. A compass can be used to measure the orientation of a large sample of clasts using an appropriate sampling technique (see the fieldwork example at the end of the chapter). The geology of the clasts and the topography will also indicate the direction from which the ice was coming. A rose diagram can be used to plot the orientation of clasts (Figure 2.27) or cirque orientation in a glaciated mountain area.

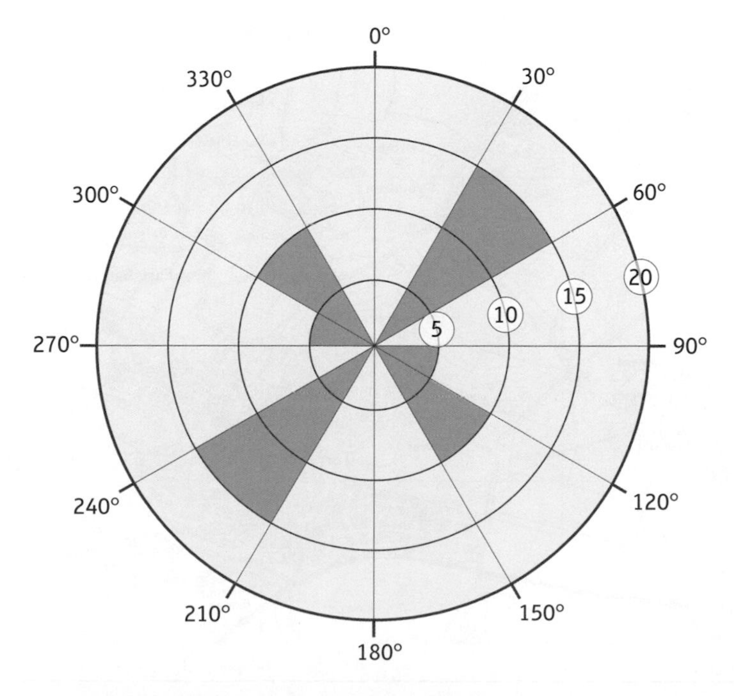

Figure 2.27: Till fabric analysis – clast orientation

ACTIVITY

GRAPHICAL AND CARTOGRAPHIC SKILLS

1. a. How many clasts would need to be sampled to ensure accurate analysis of orientation?

   b. What would be an appropriate sampling technique for till fabric analysis?

2. Describe and explain the till fabric analysis data shown in Figure 2.27.
3. Table 2.5 shows data for the Glyderau cwms in Snowdonia. Plot the data on a rose diagram.
4. Suggest reasons for the Glyderau cwm orientation shown by your rose diagram.

ACTIVITY

CARTOGRAPHIC AND STATISTICAL SKILLS

Figure 2.23 shows how drumlins can be identified on OS maps using contour lines. There is an extensive drumlin field (or swarm) in and around the Scottish city of Glasgow, and studying their orientation and shape can help us understand past ice dynamics.

1. Figure 2.28 shows part of the Glasgow drumlin field. Place tracing paper over the map and trace the outline of an appropriate sample of drumlins. Draw a line along the long axis and number each drumlin with its six figure grid reference.

2. Table 2.6 shows data for ten drumlins in this sample area. Draw a similar table for your data: use the map scale and compass to work out the length, width and orientation of each of your drumlins.

3. Calculate the elongation ratio of each drumlin in your sample and complete Table 2.6. Calculate the mean value of the sample.

4. Drumlin elongation varies with the 'normal' range, commonly between 2.5 to 4.0. What does your drumlin orientation and elongation analysis tell you about past ice dynamics in this area?

Table 2.5: Glyderau cwm orientation data

Glyderau cwms, Snowdonia	Orientation (°)
Marchlyn Bach	48
Marchlyn Mawr	32
Ceunant	40
Graianog	42
Perfedd	51
Bual	59
Coch	60
Cywion	48
Clyd	73
Idwal	38
Cneifio	20
Bochlwyd	25
Tryfan	40
Gwern Gof	41
Y Gors	43

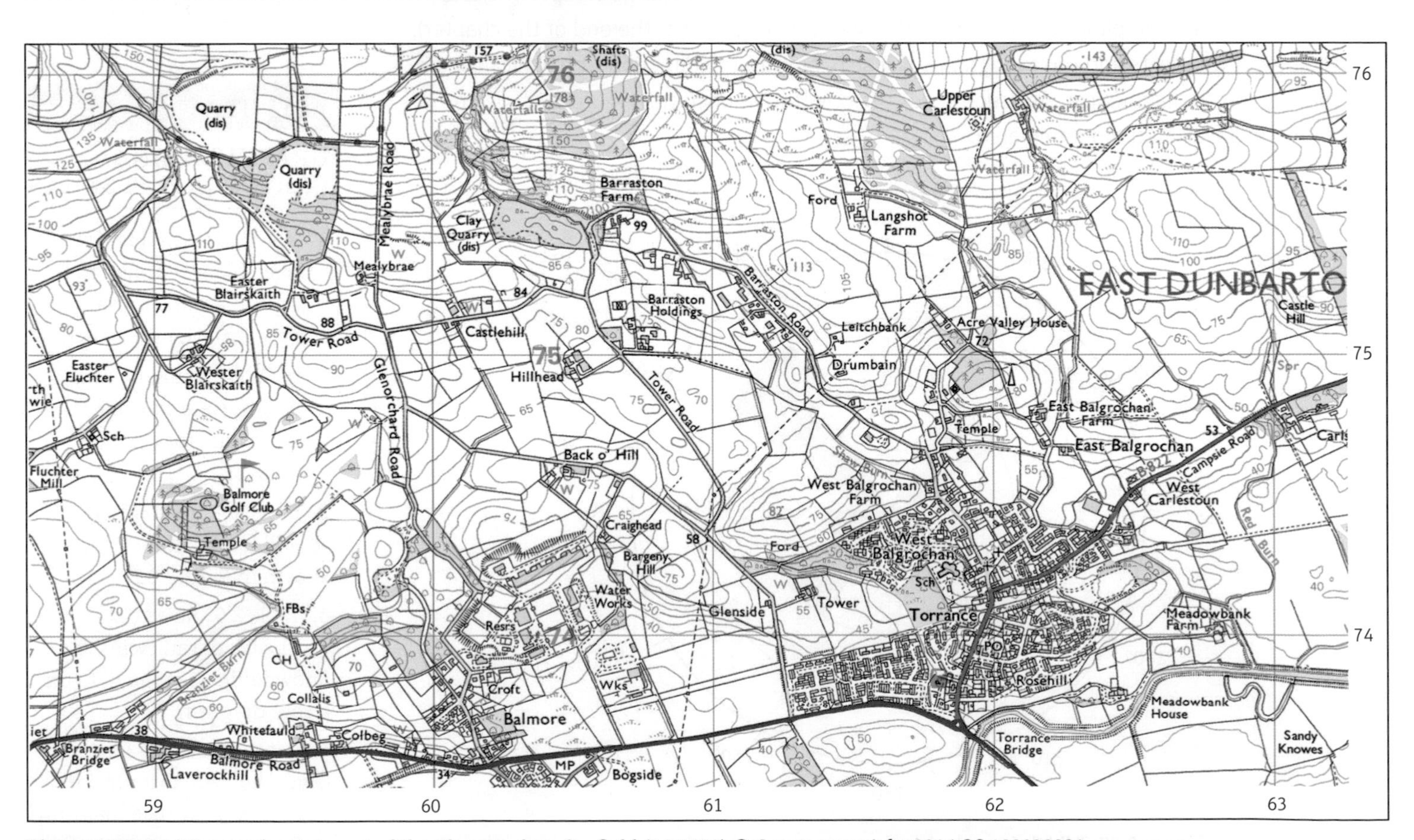

Figure 2.28: An OS map showing part of the Glasgow drumlin field (1:25,000) © Crown copyright 2016 OS 100030901

Table 2.6: Data for ten drumlins in the Glasgow drumlin field

Drumlin number	Length (m)	Width (m)	Elongation ratio	Orientation (°)
1	700	431	1.62	278
2	730	477	1.53	258
3	734	460		262
4	895	600		258
5	637	482		257
6	900	568		271
7	775	657		266
8	1085	661		269
9	625	450		271
10	625	450		259

Extension

- Research the elongation ratio of drumlins in a different location, for example Cumbria, Ireland or Canada.
- How does this data compare to the Glasgow drumlin data? What does the data tell you?

Glacial meltwater landforms and landscapes

Meltwater moves through the glacial system in a number of ways (Figure 2.30) and is most abundant in retreating temperate glaciers. Meltwater stream discharge is highly variable, both diurnally and seasonally, depending on temperature fluctuations. When temperatures rise above zero, large volumes of turbulent meltwater erode the landscape. Significant fluvio-glacial deposition then occurs when temperatures and the volume of meltwater discharge and velocity decrease. Fluvio-glacial deposits have very different characteristics from glacial till, due to their different processes of erosion, transport and deposition (Table 2.7 and Figure 2.29), and they are shaped into a number of different landforms (Figure 2.30 and Table 2.8).

Table 2.7: The characteristics of till and fluvio-glacial deposits

Classification	Definition	Till	Fluvio-glacial debris
Clast shape	Clasts (rock fragments) may be angular or show evidence of rounding.	Clasts are frozen into ice, limiting their movement and maintaining their angular shape.	The process of attrition in meltwater makes clasts more rounded.
Imbrication	The clasts have a preferred orientation and dip caused by a strong current.	Clasts are aligned in the direction of ice movement but often horizontal rather than dipping, unless part of a push moraine.	Clasts are aligned in the direction of flow and often dip upstream.
Stratification and grading	The deposit is layered, with coarse sediments at the base, grading upward into progressively finer ones.	Unstratified – clasts are dumped chaotically by the glacier.	Rock fragments are stratified and graded by seasonal variation in meltwater discharge. A layer of fine grains is deposited in spring and summer when discharge is high; a layer of coarse grains is deposited when discharge falls in autumn and winter.
Sorting	Sorted sediment has a common grain size.	Unsorted – ice has enough energy to transport a wide range of grain sizes, from fine rock flour to large boulders.	The seasonal variation in stream discharge sorts the grains into layers of consistent size.

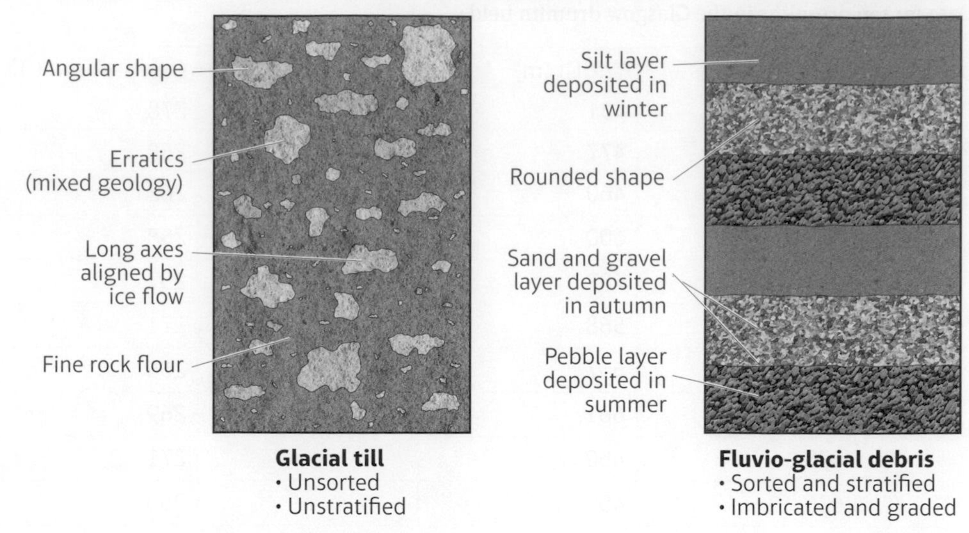

Figure 2.29: The different characteristics of glacial and fluvio-glacial sediments

Table 2.8: Landforms of fluvio-glacial deposition

	Landform	Description	Process of formation
Ice-contact features	Kame	An undulating mound of fluvio-glacial sand and gravel deposited on the valley floor near the glacier snout	As meltwater streams emerge onto the outwash plain or proglacial lake at the glacier snout, their velocity suddenly falls and sediment is deposited.
	Kame terrace	A flat, linear deposit of fluvio-glacial sand and gravel deposited along the valley sides	During the summer the valley sides radiate heat, melting the edge of the glacier and forming meltwater streams, which deposit sediment. When the glacier retreats, the sediment will fall to the valley floor, forming a kame terrace.
	Esker	A long, narrow, sinuous (winding or meandering) ridge of fluvio-glacial sand and gravel	Subglacial streams can carry large amounts of rock debris due to their high hydrostatic pressure inside tunnels. The streams often meander beneath the glacier. When the glacier retreats, the debris load is deposited at a consistent rate and forms a ridge.
Proglacial features	Sandur (outwash plain)	A flat expanse of fluvio-glacial debris in front of the glacier snout	As meltwater streams emerge from the glacier and enter lowland areas, they gradually lose their energy and deposit their debris load. The coarse gravels are deposited first, nearest the glacier, then the sands, and finally clay, farthest from the glacier.
	Kettle hole	A circular depression, often forming a lake in an outwash plain	As the glacier retreats, detached blocks of ice remain on the outwash plain. Meltwater streams flow over the ice, covering them in deposits of fluvio-glacial debris. Eventually the ice melts and the debris subsides to form a depression, which often fills with meltwater to form a kettle-hole lake.
	Proglacial lake	A lake formed in front of the glacier snout	A proglacial lake is often formed by the damming action of a terminal or recessional moraine during the retreat of a melting glacier, or because hills block the escape of meltwater. It can also be formed by meltwater trapped against an ice sheet as a result of isostatic depression of the crust around the ice.
	Meltwater channel	A narrow channel cut into bedrock or deposits, either underneath or along the front of an ice margin	Meltwater can erode deep channels, even gorges, as a result of the high hydrostatic pressure within the glacier and their high sediment load. They have some unique characteristics: under hydrostatic pressure beneath the glacier, they are able to flow uphill and they are often larger than post-glacial streams; and braiding of proglacial meltwater channels is common, due to seasonal variations in discharge.

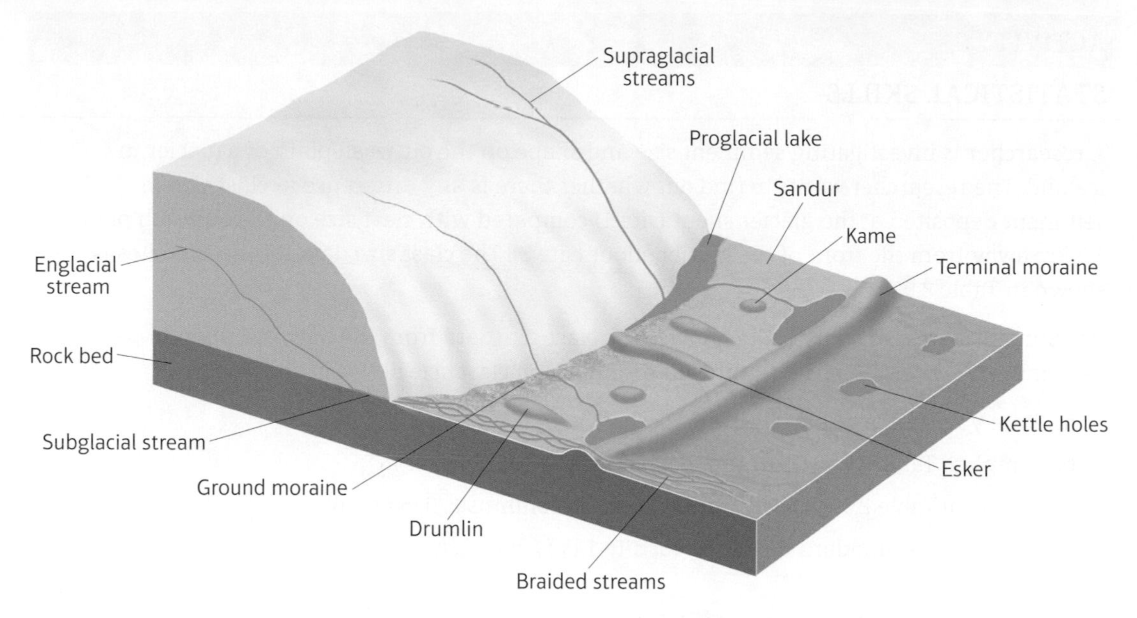

Figure 2.30: Meltwater movement within the glacial system and landforms of glacial and fluvio-glacial deposition

ACTIVITY

TECHNOLOGY/ICT SKILLS

1. **a.** Search online to find photographs of each fluvio-glacial deposition landforms shown in Table 2.8.

   b. Create a photo-album document and annotate each photo with name and location of the example and detail of the processes that formed each feature.

2. Visit the British Geological Survey website and investigate the information on Blakeney Esker Explored.

Glacial sediment size and shape

Examining the sediment size and shape of glacial deposits can help us understand past glacial processes and identify whether these deposits are till or fluvio-glacial. Sediment size can be determined by measuring three axes, and sediment shape can be analysed using the Cailleux Index (Figure 2.31). A Cailleux value of 1 is a perfectly rounded sphere; the higher the number, the flatter or more angular is the clast. Therefore glacial till deposits with angular clasts are likely to have higher Cailleux Index values than more rounded fluvio-glacial deposits.

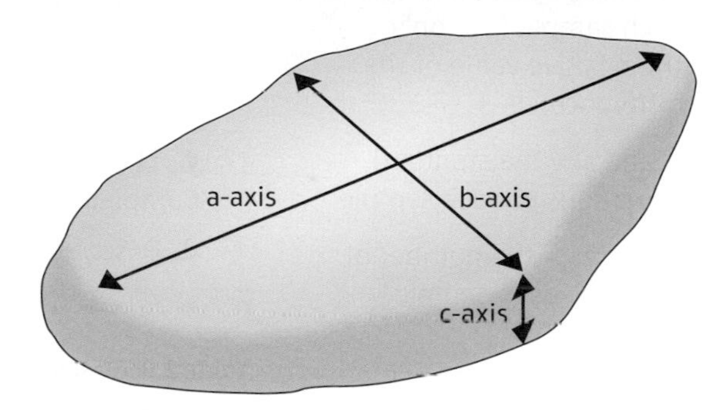

$$\text{Cailleux Index} = \left(\frac{a+b}{2c}\right)$$

Figure 2.31: Sediment size measurements and the Cailleux Index

Maths tip: Analysis of central tendency and dispersion

It is possible to compare sediment samples from different locations by calculating the central tendency using mean, median or mode, and measures of dispersion such as standard deviation.

- The mean is the total of the values of all observations (Σx) divided by the total number of observations (n).
- The median is the middle value when all values are placed in ascending or descending order.
- The mode is the value that occurs most frequently.
- Standard deviation is a measure of dispersion, or how much each value differs from the mean. A large number for the standard deviation usually means that there is a wide spread of values around the mean, whereas a small number for the standard deviation implies that the values are grouped close together around the mean. However, this depends on the actual scale of the values in the sample. Standard deviation can be best used to compare different samples.

The formula for standard deviation is:

$$\sigma = \sqrt{\frac{\Sigma(x-\bar{x})^2}{N}}$$

Student's *t*-test

It is also possible to further test the difference between two samples using Student's *t*-test. The null hypothesis is always that two sample means are the same, i.e. there is no difference between samples. The alternative hypothesis is that there is a significant difference between the sample means.

The formula for Student's *t*-test is:

$$t = \frac{(\bar{x}_1 - \bar{x}_2)}{\sqrt{\frac{s_1^2}{n_1} + \frac{s_2^2}{n_2}}}$$

where:

$\bar{x}_1$ and $\bar{x}_2$ = the means of each sample (use only the positive value of the difference)

s_1 and s_2 = the standard deviations of each sample

n_1 and n_2 = the number of values in each sample

ACTIVITY

STATISTICAL SKILLS

A researcher is investigating sediment size and shape on the outwash plain of a glacier in Iceland. The researcher wants to find out whether there is any difference in clast size in sediment deposited at the glacier snout (site 1) compared with clast size on the outwash plain 1.5 km away from the front of the glacier snout (site 2). The clast size data for the two sites is shown in Table 2.9.

Use Student's *t*-test to statistically analyse the clast size data from the outwash plain to determine whether there is a significant difference in clast size between sites 1 and 2.

1. **a.** State your null and alternative hypotheses.
   b. Complete Table 2.9 to show the mean clast size for site 2 ($\bar{x}_2$).
   c. Complete Table 2.9 with the missing data in columns 2, 3, 5 and 6.
   d. Calculate the standard deviation for site 1 (s_1) (the calculation for site 2 has already been completed).
   e. Using the formula, complete Student's *t*-test for this data.

The value of t may be either positive or negative. If it is negative, the minus sign is ignored when *t* is compared with the critical values in the relevant significance table. The critical value for a sample size of 9 at the 0.05 significance level is 2.12. If the value of *t* is greater, the null hypothesis can be rejected.

1. **a.** Is the result of Student's *t*-test statistically significant?
   b. Describe and explain the variations in clast size on the outwash plain.
   c. The researcher also collected data to analyse differences in clast shape (Table 2.10). Complete the table to show the average Cailleux Index value for site 2.
   d. Describe and explain the variations in clast shape on the outwash plain.

Table 2.9: Clast size data for two sites on an outwash plain in Iceland

Column 1 Site 1 – clast long-axis in mm (x_1)	Column 2 ($x_1 - \bar{x}_1$)	Column 3 $(x_1 - \bar{x}_1)^2$	Column 4 Site 2 – clast long-axis in mm (x_2)	Column 5 ($x_2 - \bar{x}_2$)	Column 6 $(x_2 - \bar{x}_2)^2$
30	−10	100	12	-2	4
51	11	121	10		16
35		25	13		1
48		64	14	0	0
35		25	16	2	
29	−11		18	4	
33	−7		15	1	1
50	10	100	16	2	4
49	9	81	12	-2	4
$\Sigma x_1 = 360$		$\Sigma (x_1 - \bar{x}_1)^2 = 686$	$\Sigma x_2 = 126$		$\Sigma (x_2 - \bar{x}_2)^2 = 50$
$N_1 = 9$		$s_1 =$	$N_2 = 9$		$s_2 = 2.36$
$\bar{x}_1 = 40$			$\bar{x}_2 =$		

Table 2.10: Clast shape data for two sites on an outwash plain in Iceland

Site 1				Site 2			
a, b, c axes in mm	a + b	2c	Cailleux Index values	a, b, c axes in mm	a + b	2c	Cailleux Index values
30, 14, 2	44	4	11.00	12, 10, 9	22	18	
51, 21, 4	72	8	9.00	10, 10, 8	20	16	
35, 21, 5	56	10	5.60	13, 12, 11	25	22	
48, 27, 7	75	14	5.36	14, 14, 13	28	26	
35, 27, 4	62	8	7.75	16, 15, 13	31	26	
29, 14, 3	43	6	7.17	18, 17, 14	35	28	
33, 9, 4	42	8	5.25	15, 13, 11	28	22	
50, 25, 6	75	12	6.25	16, 15, 12	31	24	
49, 23, 14	72	14	5.14	12, 12, 9	24	18	
			$\Sigma x_1 = 62.52$				$\Sigma x_2 =$
			$n_1 = 9$				$n_2 =$
			$\bar{x}_1 = 6.95$				$\bar{x}_2 =$

ACTIVITY

TECHNOLOGY/ICT SKILLS

1. Use the GIS software 'Google Earth' to investigate the landforms in an active glacial landscape (such as the Unteraargletscher in Switzerland). Use the different layers and images to identify and record named examples of different glacial and fluvio-glacial landforms.

2. Explain the fluvio-glacial processes responsible for creating ice-contact features.

3. Explain the use of different glacial landforms to reconstruct past glacial environments.

CHAPTER 2

How are glaciated landscapes used and managed today?

Learning objectives

2.10 To understand the intrinsic cultural, economic and environmental value of glacial and periglacial landscapes

2.11 To understand the threats facing fragile active and relict glaciated upland landscapes

2.12 To understand how threats to glaciated landscapes can be managed using a range of approaches

The value of glacial and periglacial landscapes

Table 2.11 illustrates the vital importance of both glacial and periglacial landscapes, culturally, economically and environmentally, and at a range of scales.

Table 2.11: Examples to show the cultural, economic and environmental value of glacial and periglacial landscapes

Cultural value examples	Economic value examples	Environmental value examples
• Scientific research, e.g. ice core analysis • Native peoples with distinctive cultures • Leisure and recreation opportunities, e.g. skiing • Spiritual and religious inspiration	• Hunting and fishing for food and clothing • Pasture, e.g. for reindeer herding • Forestry • Tourism • Minerals and metals, e.g. gold • Fossil fuels – coal, oil, gas • Renewable energy resources, e.g. HEP • Freshwater resources	• Fragile ecosystems • Endemic species • Carbon cycling • Water cycling • Carbon sequestration • Genetic diversity • Climate control • Weather system control

Threats to glaciated upland landscapes

As the global population and economy have grown, the economic value of glacial and periglacial landscapes has been increasingly recognised. Developments in technology, such as deep-sea drilling, mean that these environments are becoming more and more accessible for exploitation. However, human activities pose a number of threats to the natural environment and to the native peoples of these regions (Table 2.12).

Yellow = Players, Orange = Attitudes and actions, Purple = Futures and uncertainties

Table 2.12: Natural and human threats facing active and relict glacial landscapes

Examples of natural threats	Examples of human threats
• Avalanches and mass movements • Glacial outburst floods (*jökulhlaup*) • Thick, unstable till deposits • Natural climate change • Extreme cold temperatures • Thin soils • Fragile ecosystems • Steep rugged terrain • Seasonal extremes (e.g. large temperature range)	• Footpath trampling by tourists • Introduction of invasive (alien) species • Visual pollution, e.g. litter, mining waste • Water and air pollution, e.g. oil spills, sewage, vehicle emissions • Noise pollution, e.g. from mining or tourist crowds • Overfishing • Construction, e.g. infrastructure, hotels, power stations, ski lifts • Resource exploitation, e.g. mineral, metal, fossil fuel mining • Human-induced climate change disrupting natural cycles

CASE STUDY: Greenland – a glacial landscape

Greenland, the world's largest island, has about 80 per cent ice cover (Figure 2.32). Most of its small population lives along the ice-free coast, particularly in the southwest, where the population density is only 0.14 people per km^2. Greenland was granted self-government in 1979 by the Danish parliament, although it remains a Danish territory. The population is 88 per cent Inuit and 12 per cent of European descent, mainly Greenland Danes. The glaciated landscape offers many valuable opportunities but is increasingly under threat from a range of processes.

Figure 2.32: Greenland's glacial landscape

Greenland's environmental value

Although mostly covered by ice sheet, the land and waters of Greenland support a fragile biodiversity of endemic plants, large mammals such as polar bear, reindeer, Arctic fox and whale, and a diverse range of fish and birds. These organisms and ecosystems offer important opportunities for scientific research, wilderness recreation, cultural identity and economic exploitation.

The Greenland ice sheet contains approximately 10 per cent of the total global ice mass and it therefore plays an important role in the global **water cycle**. In 2014 the Intergovernmental Panel on Climate Change (IPCC) reported that the average rate of ice loss from the Greenland ice sheet had increased from 34 gigatonnes per year between 1992 and 2001 to 215 gigatonnes per year between 2002 and 2011 (Figure 2.33). (One gigatonne is one thousand million tonnes.) This means that the Greenland ice sheet contributed 0.33 mm per year to global mean sea level rise between 1993 and 2010. The IPCC calculates that if the whole Greenland ice sheet melted it would cause a global mean sea level rise of up to 7 m. This would mean that over future centuries many cities such as Boston, Los Angeles, London, New York and Shanghai would flood. The US state of Florida would be mostly under water, and some countries would disappear completely, such as Bangladesh and the Maldives..

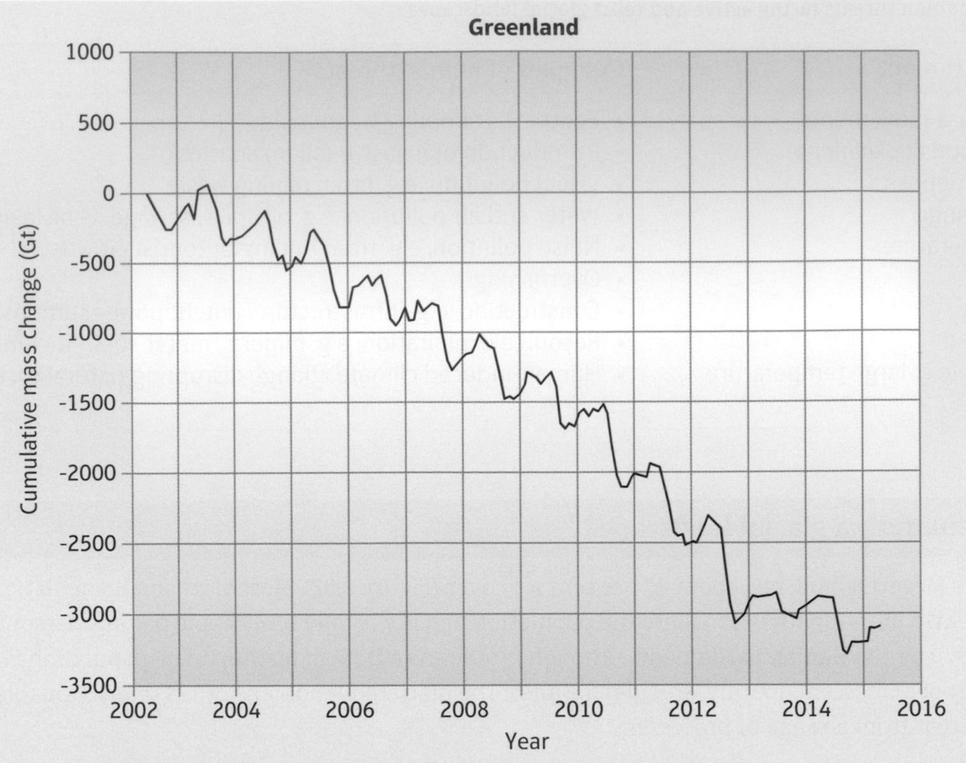

Figure 2.33 Greenland ice sheet mass balance changes, 2002–13

As well as regulating global sea levels, Greenland ice also plays a vital role in regulating global climate via feedback cycles. Ice has a high albedo, which means it reflects more solar radiation than land. However, current ice loss in Greenland is reducing surface albedo, increasing the amount of solar energy absorbed and further increasing air temperatures. This is a positive feedback mechanism.

Greenland's cultural value

Although many Greenlandic Inuit now live in towns, with modern homes and appliances, and work in the growing mining and tourism industries, they still utilise the glacial and periglacial landscape to maintain their traditional lifestyle of hunting, fishing and herding. However, Inuit hunting culture is increasingly threatened by modern culture, by conservationists and by climate change. Pressure from environmental groups has led to hunting limits for most species, and the loss of sea ice is reducing the size of hunting grounds. Nonetheless, there is an active movement among indigenous people to pass on their traditional knowledge, skills and native languages to the younger generation.

Greenland ice possesses immense value for scientific research. For example, the Greenland ice sheet contains a unique record of the Earth's climate history, as it is made up of layers of snow and ice that formed over millions of years. The layers contain trapped gases, dust, pollen and water molecules that scientists can use to study past climates by drilling deep ice cores.

Greenland's economic value

The glacial landscape of Greenland offers a wealth of economic opportunities. Meltwater provides huge potential for hydroelectric power – the country is thought to have the world's biggest unexploited hydropower capacity. Furthermore, as the ice retreats it is revealing highly valuable deposits of oil, gas, metals and minerals, including rubies and gold. Tourism also offers another area of economic growth for Greenland, with increasing numbers of cruise liners now operating in the island's western and southern waters during the peak summer tourism season (Figure 2.34).

Yellow = Players, Orange = Attitudes and actions, Purple = Futures and uncertainties

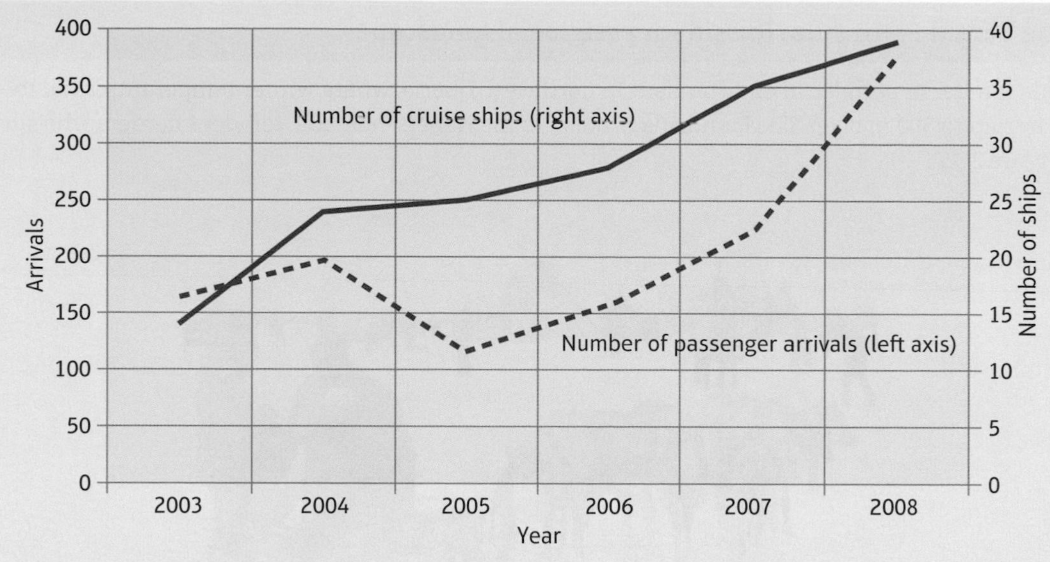

Figure 2.34 Cruise ship arrivals in Greenland, 2003–8

Threats to the fragile glacial landscape

Climate change poses a significant threat to the natural glacial environment, as a result of both ice loss and the opening up of areas for economic exploitation. Increased industrial activity, such as oil and gas exploration and marine shipping, will bring stresses to the environment. For example, as the sea ice retreats, new commercial shipping routes could open in the Arctic, such as the Northwest Passage connecting the Pacific and Atlantic Oceans. Increased shipping increases the risk of marine pollution. Overfishing and the discharge of ballast water into Arctic seas may introduce invasive species that may outcompete and displace resident species.

In 2014 protests by conservationists in Greenland's capital Nuuk followed a government decision to reverse a ban on radioactive uranium mining. The decision means that areas in the south of Greenland could be opened up to large-scale mining projects for uranium and rare earth metals. The conservationists feared that the impact of mining extensive deposits in pristine natural environments would be widespread and irreversible. They claimed that radioactive waste from the mining process would endanger fisheries and farmland in the region, both of which are vital for local communities. Southern Greenland is the only part of the island capable of supporting farming.

Polar bears now face an uncertain future in Greenland, threatened by climate change and environmental pollution. As the sea ice melts, the polar bear's hunting grounds are reduced. POPs (persistent organic pollutants which can impair the ability to reproduce) have been discovered in high concentrations in polar bears from East Greenland and Svalbard.

The fragile glacial landscape is also threatened by political disputes and conflict. In 2007, Russia claimed part of the Arctic seabed at the North Pole, when polar explorers descended through the icy waters to plant a titanium Russian flag on the seabed. This caused conflict with other Arctic nations such as Denmark and Canada, which have now staked their own claims to sections of the Arctic seabed.

CASE STUDY: The Yamal Peninsula, Russia – a periglacial landscape

The Yamal Peninsula is a harsh periglacial environment in northwest Siberia, where winter temperatures fall to –50°C and permafrost penetrates up to 300 m deep. The Peninsula is home to the Nenets, nomadic reindeer herders who survive on the tundra pastures (Figure 2.35).

Figure 2.35: Nenets on the Yamal Peninsula

The Yamal Peninsula's environmental value

The biodiversity of the tundra biome is low but it has global value, particularly for birds, as it provides a summer home for many migratory species and thus plays a role in worldwide food webs. The permafrost also has global value as a large-scale carbon sink, storing an immense amount of carbon and methane (twice as much carbon as in the atmosphere). In the warming climate, permafrost is expected to melt, and these stored gases will be released to add further warming (positive feedback cycle).

The Yamal Peninsula's cultural value

The Nenets' understanding of the harsh climate and fragile ecosystem has enabled them to live sustainably in this inhospitable land. The reindeer provide the Nenets with transport, clothing, hides for tents, meat and income. The Nenets and their reindeer migrate seasonally to avoid the extreme cold winter temperatures in the north and to prevent the overgrazing of pastures.

The Yamal Peninsula's economic value

The Nenets' herder economy is driven by the reindeer meat they sell. Reindeer herding supports more than 10,000 nomads, who herd over 300,000 domesticated reindeer on the pastures of the Peninsula, 80 per cent of which is privately owned by the herders. However, the economic value of reindeer herding cannot compete with the natural resources that lie beneath the pastures. The Yamal Peninsula contains the biggest gas reserves on the planet, holding almost a quarter of the world's known gas reserves. There is now increasing investment and infrastructure development for large-scale exploitation of this valuable energy resource.

Threats to the fragile periglacial landscape

Climate change poses a significant threat to the tundra and to the Nenets. Earlier spring melts and delays to the autumn freeze are affecting the reindeers' and herders' ability to cross the frozen tundra, threatening their survival. In July 2014, large sinkholes were discovered on the Yamal Peninsula, which Russian researchers believe to have been caused by methane released as the permafrost thawed. This change to the **carbon cycle** is a threat to the global climate system, as well as to local culture and biodiversity.

Early attempts to exploit gas on the Yamal Peninsula have posed considerable threats to both the natural environment and the Nenets. In the 1980s the infrastructure constructed to exploit the Bovanenkovo gas field destroyed pasture, forcing overgrazing of the tundra and disruption to migration routes, causing conflict with the Nenets. There had been no environmental impact assessment or consultation with the native people. Eventually, work on this gas field was suspended due to its extremely high costs.

The area is now part of the 'Yamal megaproject' being developed by the majority state-owned energy giant Gazprom. The aim of the megaproject is to exploit and bring to the market the vast natural gas reserves in the Yamal Peninsula, so there will be ongoing cultural and environmental concerns. Several ambitious infrastructure projects are included, such as a 572-km railway line, a gas pipeline (Figure 2.36) and several bridges. These have already led to the eviction of more than 160 reindeer herders,

Yellow = Players, Orange = Attitudes and actions, Purple = Futures and uncertainties

and loss of tundra to concrete. According to Gazprom, the company is planning to build housing, kindergartens, hospitals and both fish- and venison-processing factories, and they are providing financial compensation to the Nenets. Many question whether the Nenets will be able to maintain their traditional culture, as these changes are forcing them into modern settlements and jobs. Without the Nenets, the Yamal ecosystems will lose their main custodians.

Figure 2.36: Gas pipeline construction in Russia

Synoptic link 

A number of different players are involved in glacial landscapes, at different scales and with different attitudes and different levels of influence. Economic players, including both transnational corporations (such as oil companies) and local businesses (such as the ski industry), generally favour exploitation, whereas others, such as environmental pressure groups (such as Greenpeace), favour conservation. The world is globalised and the influences on glacial environments can come from far away. (See pages 106–107.)

Managing threats to glaciated landscapes

There is a wide range of stakeholders with diverse interests in glacial landscapes; consequently there are many potential conflicts at different scales (see Table 2.13, page 106).

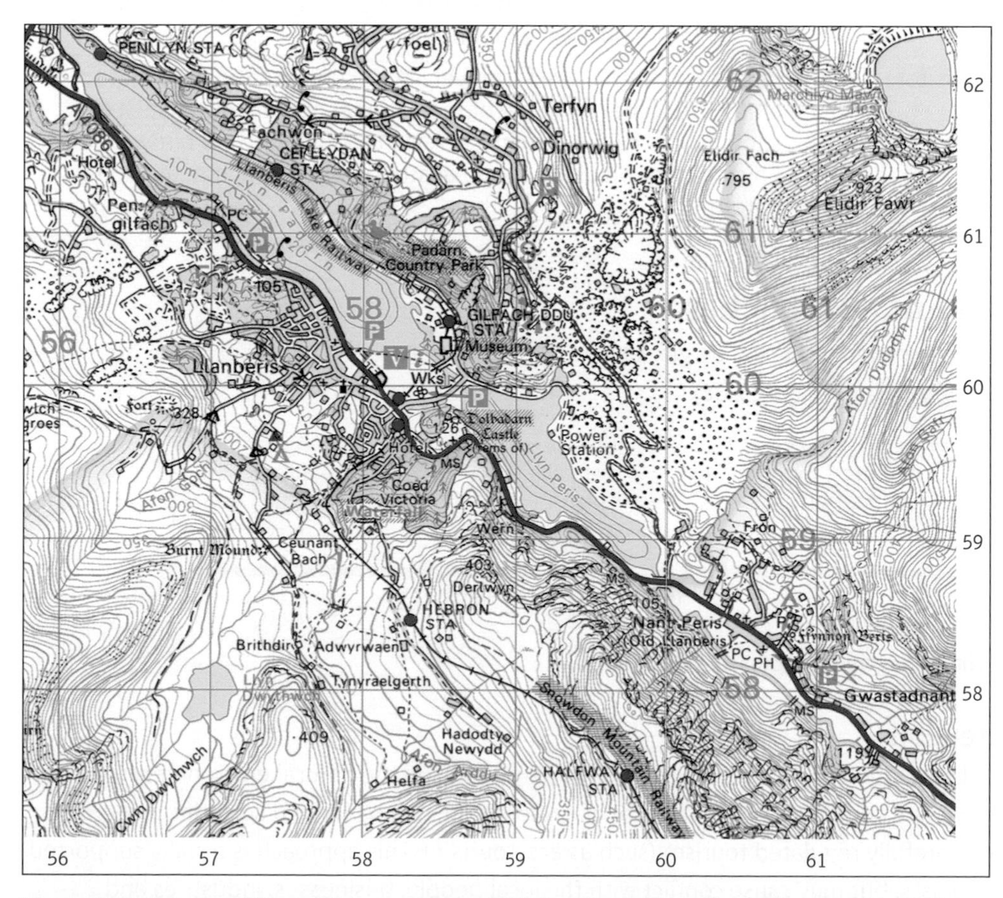

Figure 2.37: OS map of Llanberis, a town situated below Mount Snowdon in Snowdonia, a relict glacial landscape (1:50,000) © Crown copyright 2016 OS 100030901

AS level exam-style question

Using examples, explain why periglacial landscapes play an important role in the carbon cycle. (6 marks)

Guidance

Explain the positive feedback cycle using accurate geographical terminology.

A level exam-style question

Explain the economic importance of glacial landscapes. (6 marks)

Guidance

Use named located case studies to provide evidence to support a range of economic (money, business) reasons.

ACTIVITY

CARTOGRAPHIC SKILLS

Figure 2.37 is an OS map of Snowdonia, which is a relict glacial landscape. It shows the glacial trough and ribbon lakes of Llanberis, near Mount Snowdon.

1. Study the map and suggest reasons why this relict glacial landscape is economically, culturally and environmentally valuable.

2. Using evidence from the map, suggest the natural and human threats facing this relict glaciated landscape.

Table 2.13: Stakeholders in glacial landscapes

Stakeholder	Interest in glacial landscapes
International governmental organisations, e.g. UNEP	These organisations aim to promote international cooperation, including protection of the natural environment to achieve global sustainability. For example, Antarctica is protected by an international treaty, written in 1959.
Transnational corporations, e.g. Royal Dutch Shell, ExxonMobil	TNCs aim to utilise economic resources to maximise profits and meet the needs of their shareholders, industries and consumers.
Global, national and local pressure groups	Pressure groups try to influence public policy in the interest of a particular cause. For example, conservationists such as Greenpeace, who campaign to protect fragile glacial landscapes from resource exploitation. There are also groups who seek to change legislation to allow resource exploitation in protected areas e.g. ANWR.org, who are campaigning for oil exploitation in the Arctic National Wildlife Refuge in Alaska.
National and local governments	These aim to establish appropriate management strategies to balance economic, social and environmental concerns to meet the needs of society. This can be very difficult to achieve where there are valuable economic resources in fragile natural environments, and therefore they must manage conflict.
Local businesses, e.g. farming, fishing, ski hire	These need to use local economic resources to maintain livelihood and quality of life, often in areas with few alternatives.
Native peoples, e.g. Inuit	Often depend on the natural environment for survival, including the provision of food, shelter, clothing and transport. Often nomadic, and traditionally migrate within a region, which may conflict with new stakeholders that move in. Have deep spiritual and cultural links with the physical environment.
Tourists and visitors	Require infrastructure for travel and accommodation in order to enjoy the scenery, wildlife or recreational opportunities such as snow sports, whale-watching, trekking, climbing, and birdwatching. Environment provides strong aesthetic value and appreciation of nature.

Management approaches

The conflicts and challenges posed by glacial landscapes are managed using a spectrum of management approaches (Figure 2.38).

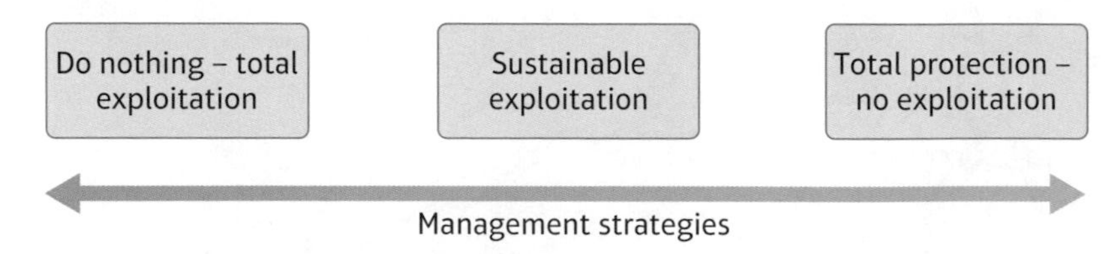

Figure 2.38: The management approaches spectrum

Extension

Is there legislation to protect the landscape in Greenland and/or the Yamal Peninsula? Research and explain your findings.

Total protection

The aim of this management approach is to completely conserve the natural environment, maintaining pristine conditions; the only form of exploitation may be limited scientific research and, possibly, carefully regulated tourism (such as eco-tourism). This approach is usually supported by conservationists, but may cause conflict with the local people, businesses, industries and governments who wish to increase economic productivity through resource exploitation. Therefore

Yellow = Players, Orange = Attitudes and actions, Purple = Futures and uncertainties

this approach is most feasible in remote locations, which are either uninhabited or sparsely populated, for example Antarctica, the North East Greenland National Park and the Arctic National Wildlife Refuge in Alaska. It is often enforced by national or international legal frameworks, such as the Antarctic Treaty.

Total exploitation

At the opposite end of the management approach spectrum there is maximum economic exploitation without protection of the natural environment. This is often the preferred approach of business and industry where there are large reserves of minerals, metals or fossil fuels that require opencast mining or large-scale infrastructure. It causes significant conflict with conservationists and with native people, who often depend on the natural landscape and live in harmony with it. There are concerns that the Russian government will favour total exploitation of the Yamal Peninsula for gas (see the case study above), which could result in considerable environmental and cultural degradation.

Sustainable exploitation

This management approach aims to find a balance between the need for resource exploitation and economic growth, and the need to conserve the natural environmental and indigenous cultures. It aims to take into account the interests of all stakeholders to reduce conflict, but this means it relies on compromise, which is often difficult to achieve. The Alpine Convention, for example, is a **legislative framework** that aims to achieve **sustainable management** of the European Alps.

ACTIVITY

1. Compare and contrast the scale of the threats facing the landscape in Greenland and the Yamal Peninsula.

2. Compare the interests of the different stakeholders in Greenland and on the Yamal Peninsula. You could draw a table similar to Table 2.13 for each case study.

3. For each case study, research and explain how the different stakeholders are involved in managing the challenges posed by the landscape.

CASE STUDY: The Antarctic Treaty

The Antarctic Treaty came into force in 1961, and by 2015 there were 53 signatory nations. This international agreement has been extended several times in order to regulate international relations on the Earth's only continent without a native human population. It sets aside Antarctica as a scientific preserve, establishes freedom of scientific investigation and bans military activity on the continent. The signatory nations meet regularly to review issues and add additional agreements or protocols, such as the Madrid Protocol. This protocol came into force in 1998, prohibiting any exploitation of Antarctic mineral resources for 50 years, except for scientific research. The treaty is heralded as one of the most successful international agreements. It has allowed nations to peacefully cooperate to further scientific research and understanding of the Earth. For example, environmental monitoring in Antarctic led to the discovery of the stratospheric ozone hole in the early 1980s.

The Antarctic Treaty recognises tourism as a legitimate activity in Antarctica and it regulates the industry (Figure 2.39). Tourism companies are required to have a permit to visit the continent, and guidance is given to visitors so that they are aware of their responsibilities when in the Antarctic Treaty area. Commercial tourism using both ships and aircraft has increased steadily since the first commercial expeditions in the 1950s, and the total number of tourists visiting Antarctica peaked in the 2007/08 season at around 46,000.

Figure 2.39: Tourism in Antarctica, regulated by the Antarctic Treaty

CASE STUDY: The Alpine Convention

The Alpine Convention is an international treaty between the Alpine countries (Austria, France, Germany, Italy, Liechtenstein, Monaco, Slovenia and Switzerland) and the European Union. The aim is to achieve sustainable development in the Alps by protecting the natural environment while promoting economic development. The aim is to balance the needs of 14 million residents and 120 million tourists per year (Figure 2.40). It entered into force in 1995 and consists of a number of protocols that provide specific measures required to achieve sustainability. Not all parties have ratified all the protocols. For example, Switzerland has not ratified any protocols yet, which means that they have not passed national legislation to support the protocol. The Swiss central government found that the Convention was opposed by most of the mountain cantons, which considered that responsibility for their local area would be taken out of their hands. They fear that the Convention focuses too much attention on protection of the environment and therefore threatens their livelihoods and economy.

Figure 2.40: The area covered by the Alpine Convention

CASE STUDY: Zermatt, Switzerland – a sustainable ski resort?

Zermatt is best known for its spectacular pyramidal peak, the Matterhorn, which towers above the Swiss town (Figure 2.41). This **active upland glacial landscape** faces a number of human threats arising from its popularity as a ski resort, with over 2 million visitors per year. Zermatt has a resident population of almost 6000 people, which reaches more than 35,000 at the height of the ski season. This creates a huge demand for energy and water resources and threatens environmental degradation due to urbanisation, increased noise and vehicle emissions and the expansion of ski areas. For example, the preparation of pistes damages the fragile ecosystem and soil. The WWF (World Wide Fund for Nature) reports that winter ski tourism is one of the most ecologically devastating leisure activities in the Alps. Irreparable damage to the landscape results from the construction of ski runs and the increasing use of energy-intensive snow cannons, which may apply large amounts of water, chemical and biological additives to the slopes.

Yellow = Players, Orange = Attitudes and actions, Purple = Futures and uncertainties

Figure 2.41: Zermatt, with the Matterhorn (a pyramidal peak) in the background

However, Zermatt markets itself as a sustainable ski resort, and both buildings and companies have won sustainability prizes. The town is car-free, with tourists encouraged to go on foot or use bikes, electric buses or electric taxis. A new high-altitude restaurant, the 'Matterhorn Glacier Paradise', has won the European Solar Prize for its innovative design and use of renewable energy. The building is designed to reduce energy consumption, and waste water is collected and purified before being reused for sanitation. Zermatt's new youth hostel uses solar energy and several hotels use geothermal energy.

Zermatt Bergbahnen AG, the company responsible for managing the ski pistes and lifts, states that they have invested up to 1 million Swiss francs per year in environmental projects, including ongoing repair of past environmental damage. They are conducting tests to identify the ideal types of plant to revegetate high-altitude slopes, and they claim that more than 85 per cent of priority areas are now repaired. The company requires a qualified environmental expert to supervise all construction work and they enforce the laws protecting conservation areas.

Zermatt has six protected forests and ten areas designated as wildlife sanctuaries, including seven endemic plants. Legally enforced sanctuary areas are either closed entirely to the public during winter, or only designated routes may be used, with infringements punishable by law.

However, critics claim that there is still much more to be done to make Zermatt sustainable, including limiting both the number of tourists and future construction. In 2007 there was controversy over a proposal to build a tower on the Klein Matterhorn with a hotel and restaurant, to raise the mountain to over 4000 m in height. While this may have brought significant economic benefit to the tourist industry, environmentalists feared further degradation of the wild natural landscape. In 2010 a proposal to build a car park for 2000 cars in Zermatt also caused conflict with local environmental groups such as the WWF and Pro Natura, which claimed that 9000 m^2 of forest would be lost to the construction.

Management challenges posed by climate change

The case studies illustrate the need for management of glacial landscapes at a range of scales, from continental-scale management in Antarctica, to regional-scale management in the European Alps and local-scale management in Zermatt in Switzerland. Global climate change is a **context risk**, meaning that it poses a widespread, global hazard, including many threats to glacial landscapes. Successful management of these unique and fragile landscapes is set to become even more challenging in the future, and will require coordinated approaches at the global, national and local scales.

ACTIVITY

1. Explain why climate warming is considered to be a 'context risk'.
   a. Explain why successful management of glaciated landscapes needs coordinated approaches at global, national and local scales.
   b. Using case-study examples, explain how glaciated landscapes can be managed at different scales – global, national and local.

CASE STUDY: Climate change management strategies

The member states of the Alpine Convention have agreed that the time has come to act collectively, and on a large scale, to make the Alpine region a model for tackling climate change issues. This has prompted a number of mitigation initiatives across the Alpine states to reduce carbon emissions.

- In Slovenia, financial support and subsidies are granted to families and public bodies for using renewable energy sources (biomass boilers, solar collectors, heat pumps for heating).
- In Embrun (France), an association called Le Gabion is offering seminars to professional and individuals in eco-construction to reduce energy consumption and carbon emissions, as well as in protecting heritage. They make the most of local materials, such as walls and building frames made out of wood, straw, soil, hemp, stones, plaster and lime.
- In the town of Gap (France), buses are free for its 39,000 inhabitants, with a city-centre shuttle every ten minutes.
- Arosa (Switzerland) has set up a scheme to offset the carbon emissions generated by tourists to the mountain resort. The tourism office asks participants how they travelled to the resort, where they stayed, and the activities they participated in. Hiking, for example, generally leaves a smaller carbon footprint than alpine skiing. Arosa then spends part of the money it collects through a local tourism tax to buy carbon credits from a German biogas plant. The resort also provides a range of eco-friendly activities such as free use of buses, electric bikes, ski lifts, cable cars and pedalos.

The Alpine Convention also promotes adaptation strategies to cope with increased air temperatures and increased winter precipitation and storm frequency. For example, in Bavaria (Germany) protective measures against floods have been implemented, including enlarged reservoirs to store floodwaters, restoration of natural peatlands and wetlands and afforestation schemes.

Synoptic link

Direct actions by many players often reduce the resilience of glacial landscapes, leading to ecological degradation. For example, the creation of ski pistes often involves regrading and deforestation, causing considerable disturbance to habitats and food webs. The effects of indirect actions – for example, the negative impact of human-induced climate change on glacier mass balance and the melting of ice sheets – take time to become apparent, but may be greater. These in turn may cause sea levels to rise, leading to coastal flooding and erosion. (See page 147.)

Extension

Research a range of climate change adaptation and mitigation strategies suitable for glaciated landscapes.

Yellow = Players, Orange = Attitudes and actions, Purple = Futures and uncertainties

Fieldwork exemplar: Glaciated landscapes and change

Enquiry question

What is the pattern of landform morphology and orientation in the Devensian glacial deposit at Abermawr?

Fieldwork hypotheses

Null hypothesis (H_0): The glacial deposits are randomly orientated.

Alternative hypothesis (H_1): The glacial deposits are orientated in the direction of ice movement. The glacial deposits are orientated in the direction of ice movement.

Locating the study

This is a remote beach, exposed at low tide, with tree stumps buried by sea level rise 8000 years ago. It is therefore a relatively undisturbed site for fieldwork with both coastal and glacial features. It is within the Pembrokeshire Coast National Park and accessible from the Pembrokeshire Coast Path, and is managed by the National Trust who ensure that nature and natural processes have priority. The nearest larger settlement is Fishguard 20 km to the north east.

Methodology

Secondary research was undertaken to locate the glacial deposits in the area. The British Geological Survey 1:50,000 Solid and Drift map for St David's was used, along with the British Geological Survey online 'Geology of Britain Viewer'. Information on the date and extent of past glacial advances in West Wales was collected from books and journals published by local experts and academics, such as the chapter on Pembrokeshire by the geographer Dr Brian John in the book *The Glaciations of Wales and Adjoining Regions*. The Field Studies Council website was used to research possible fieldwork techniques for till fabric analysis. These are academic sources and therefore considered reliable.

Primary fieldwork was conducted to measure the clast orientation and clast shape within the glacial deposit. On arrival at Abermawr, a field sketch was drawn including a grid-referenced location, orientation, scale and detailed description and interpretation of the deposit and landscape features. The deposit was also photographed, with a metre ruler included in the photograph for scale. The field sketch and photographs will be used to support subsequent data analysis and interpretation.

A random sampling technique was used to sample 100 clasts; a 100 m tape measure was randomly placed along the deposit and a random number generator was used to generate 20 random numbers. The clasts positioned on the tape measure at those numbers were sampled and, once complete, the tape measure was placed in another random location and the process was repeated. This should ensure that the results are not affected by bias and that a large sample was obtained to make the results reliable.

A compass was used to measure the orientation of the clasts by placing the long side of the compass along the long axis of each clast and recording the orientation of each end (180° apart) of the clast in degrees; there are therefore two recordings for each clast. The data was plotted on a rose diagram (Figure 2.27) and the mean, mode and standard deviation were also calculated to understand the trend. The clasts must be undisturbed for orientation analysis, so it is important that it is done before any clasts are removed for shape analysis. All of the data were recorded in standardised tables to ensure clarity for later statistical analysis.

There were a number of risks involved in the primary fieldwork. Abermawr is on the coast and therefore subject to strong winds, high tides, large waves and frequent intense precipitation and cold temperatures during the winter. Therefore appropriate thermal and waterproof clothing was required to prevent hypothermia in the winter. Sun cream was worn to prevent sunburn in the summer. It is a remote site, so group work was essential for safety and mobile phones were carried for emergency communication. Gloves were worn for protection when removing clasts from the deposit. Tide charts and weather forecasts were consulted to select the safest time to visit the site.

Data presentation

Clast orientation data was plotted on a rose diagram to show whether there is a strong common trend to indicate the direction of ice movement (Figure 2.42). The mean and standard deviation were calculated to determine both the average orientation and the amount of dispersion.

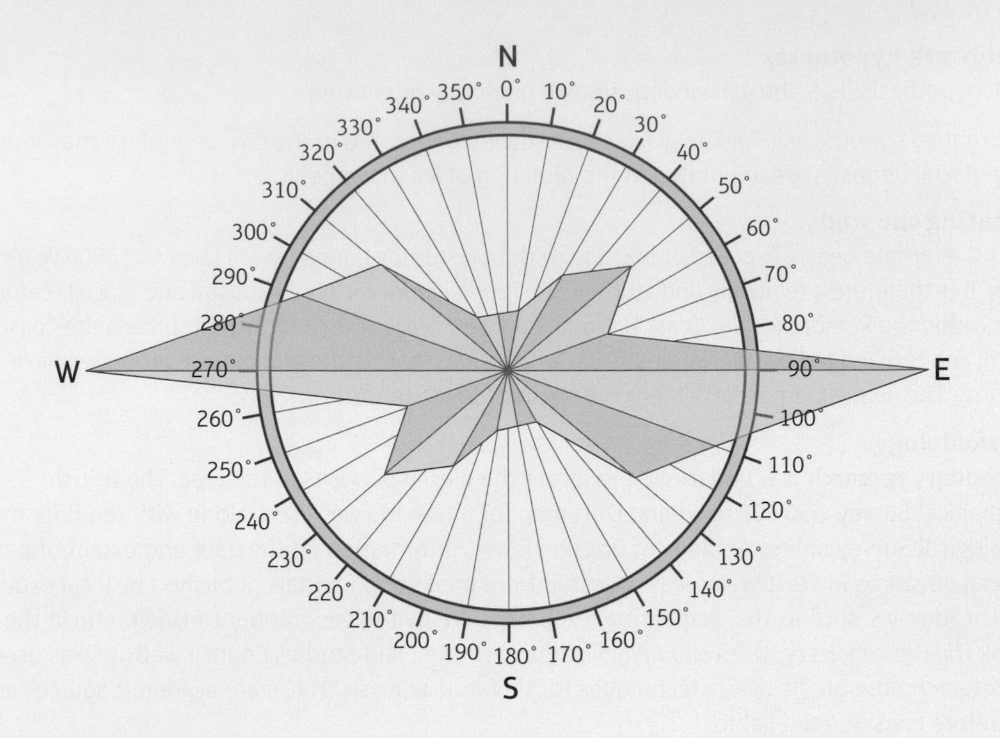

Figure 2.42: Till fabric analysis – clast orientation in a till sample taken from Abermawr, Pembrokeshire, Wales

Analysis and conclusion

The orientation analysis revealed that there is a trend with a mean orientation of 275° to 95°, which is an approximately west-to-east orientation. The standard deviation was 51°, which indicates that there is some dispersion about the mean; this is likely to be the result of post-glacial modification of the deposit by mass movement processes under the influence of gravity. The secondary research revealed that the rock fragments are lodgement till deposited by the Irish Sea glaciation during the Devensian, approximately 18,000 years ago. The till contains angular erratics and fragments of sea shells, indicating that the till was picked up from the Irish Sea floor by the ice and subsequently deposited at the site. The glacial ice therefore moved from west to east at this location, scraping deposits from the Irish sea floor and depositing them on land as lodgement till.

From this data it is possible to accept the alternative (H_1) hypothesis.

Evaluation

The random sampling technique reduced the risk of bias, and the large sample size facilitated statistical analysis. It was often difficult to ensure that the orientation of the longest axis of the clast was measured, as part of each clast was buried in the cliff. An even larger sample of 200 clasts for orientation and shape analysis, or measurements at a different site, would further improve the accuracy of the results. Further primary data could have been collected on clast sizes and shapes using the Cailleux Index and rock type analysis of the fragments to show their origin. The primary and secondary data produced complementary and conclusive results to show that Devensian ice had extended over Abermawr, as well as the direction of ice movement, but the data is only for one locality in West Wales. The British Geological Survey maps indicate extensive till deposits across Wales, which would need to be studied to understand Devensian ice extent and patterns of ice movement from other directions and sources.

Summary: Knowledge check

By reading this chapter and completing the tasks and activities, as well as reading more widely, you should have learned the following and be able to demonstrate your knowledge and understanding of the topic of glaciated landscapes and change (Topic 2a).

- **a.** What is the difference between icehouse climate and greenhouse climate?
- **b.** What are the Pleistocene, Devensian, Loch Lomond Stadial, Holocene and Little Ice Age?
- **c.** Explain the long-term causes of climate change?
- **d.** Explain the short-term causes of climate change?
- **e.** What are the differences between the three environments: polar glacial; alpine glacial; periglacial?
- **f.** How can ice masses be classified?
- **g.** What is permafrost and where is it distributed?
- **h.** Which periglacial processes and landforms are distinctive (unique) to periglacial landscapes and why?
- **i.** Explain the different components of a glacier system?
- **j.** Why do glaciologists study glacier mass balance?
- **k.** Explain the similarities and differences in the movement of polar and temperate glaciers?
- **l.** Which of the glacial landforms are more common in upland areas and which are more common in lowland areas? Why?
- **m.** How can the features of a relict glacial landscape be studied to reconstruct former ice extent, movement and provenance?
- **n.** Why are glacial and periglacial landscapes valuable?
- **o.** Why is a spectrum of management approaches needed to manage the threats to glacial landscapes?

As well as checking your knowledge and understanding through the above questions, you also need to be able to make links and explain ideas such as these:

- The distribution of polar glacial, alpine glacial and periglacial environments has changed considerably during the Pleistocene and Holocene as a result of different causes of climate change.
- There are several different periglacial landforms, which are formed by a range of different periglacial processes creating distinctive landscapes.
- Glacier mass balance varies due to long-term and short-term causes of natural climate change and human-induced climate change.
- Positive and negative feedback mechanisms affect glacier mass balance, global climate change and periglacial processes, increasing the complexity of the systems and creating uncertainty.
- A range of factors controls the rate of glacier accumulation, ablation and ice movement and the processes of erosion, transport and deposition.
- Glacial and fluvio-glacial processes of erosion and deposition create a number of landforms in different locations and at different scales, which contribute to glaciated landscapes.
- The assemblage of landforms can be used to reconstruct former ice extent, movement and origin.
- Glacial and periglacial landscapes are culturally, economically and environmentally, valuable and they are increasingly threatened by a range of activities and processes.
- There is a spectrum of strategies to manage glaciated landscapes, and a number of different stakeholders are involved at different scales.

Preparing for your AS level exams

Sample answer with comments

Assess the extent to which threats to an active glaciated landscape can be managed sustainably. (12 marks)

Active glaciated landscapes are characterised by moving bodies of ice called glaciers or ice sheets and they face both natural and human threats. These are events or processes that are likely to cause damage to the natural balance of processes. It is possible to assess the relative importance of such threats by considering the geographical scale of impact. The most important threats will have a global impact, whereas less important threats will be localised.

Good understanding of the key terms in the question; some criteria given to assess importance and recognition of the need to consider both natural and human processes. The student could also consider the importance of threats over different timescales and how sustainability may be achieved. It is essential not to simplify judgements, even in an introduction – which should clarify your thoughts about the question ideas.

Natural threats to sustainability include a range of natural hazards. For example, Alpine areas such as Zermatt face the winter threat of avalanches, extreme cold temperatures (below –10ºC) and freeze–thaw processes. In Iceland, at the snout of Breiðamerkurjökull, the glacial landscape is changed by *jökulhlaups* or glacial outburst floods, as a result of volcanic activity beneath the ice, and there are large volumes of thick deposits. However, all of these natural threats are localised as they only affect and change small glaciated landscapes, and therefore may be considered less important than the more globalised human processes that cause change.

Suitable use of named locations linked to natural processes causing short-term change, supported by geographical terminology. However, there is a danger that the judgement is oversimplified; for example, are these natural processes being accelerated or exaggerated by human processes such as climate change?

Many human activities pose threats to glaciated landscapes. For example, the increasing number of tourists visiting Antarctica (46,000 in 2007/08) threaten the fragile ecology by trampling footpaths, introducing pollutants such as litter, noise and waste (e.g. sewage, oil and exhaust emissions) and introducing alien species. However, these are all localised threats, which are being effectively managed by a legislative framework called the Antarctic Treaty. The Antarctic Treaty started in 1961 and the number of member countries has increased over time. It ensures that the Antarctic is maintained for science, which is vital because important discoveries have been made by scientific research on the continent, such as the discovery of the ozone hole. The Madrid Protocol makes it illegal to mine for any resources on the continent. A more important human threat to glacial landscapes is human-induced climate change because it is a global threat impacting all glacial landscapes.

A good example with plenty of relevant factual information. The climate change process is only mentioned at the end, but many would argue that this is the key factor for an answer to this question. Perhaps the candidate ran out of time, in which case better planning at the start of this mini-essay was required.

In conclusion, the most important threat to the sustainability of active glaciated areas is human-induced climate change because it is a global threat, whereas the natural threats and many of the human threats are localised and can be effectively managed.

A well-structured answer, but the conclusion is almost non-existent and climate change appears almost as an afterthought. The candidate would have benefited from making a plan in the form of a list.

Verdict

This is an average answer. The candidate could have improved it by explaining a wider range of threats to sustainability or fully explaining the human-induced climate change threat, rather than giving extensive but less relevant case study detail such as management strategies. The candidate could have considered criteria to assess the importance of the threats. The global threats such as natural climate change have not been recognised in this answer.

Preparing for your A level exams

Sample answer with comments

Evaluate the usefulness of evidence from relict glacial landscapes to the reconstruction of past ice extent and movement. (20 marks)

Glacial landforms are features of the Earth's surface that have been formed by moving bodies of ice called glaciers or ice sheets. A number of landforms can be used to show past ice movement.

Till deposits can be analysed to show former ice direction because the long-axes of the clasts are aligned in the direction of ice flow. The orientation of a large sample of clasts is plotted on a rose diagram and the trend will reveal ice direction. For example, a till deposit at Abermawr in Pembrokeshire shows that the Devensian Irish Sea ice sheet moved across the area from west to east.

Drumlins can be analysed to show the direction of past ice movement as, similar to till clasts, their long-axis is also aligned by the ice flow. It is also possible to calculate the elongation ratio of the drumlins (length/maximum width) to understand the rate of movement, as the elongation ratio increases with ice velocity. For example, a swarm of drumlins near Glasgow reveals that former ice was moving west–east. However, many drumlins are altered or destroyed by human activity through farming, construction of infrastructure and urbanisation, reducing the value of this technique. For example, Glasgow city is built on a drumlin field.

Striations are scratches in the bedrock formed by glacial abrasion. As the glacier or ice sheet moves over the bedrock, rock fragments frozen into the base of the ice scratch the bedrock and fine material (rock flour) polishes the rock. Therefore striations reveal the direction of past ice movement. However, they are often very hard to see, especially if there is a lot of vegetation or joints in the rock.

Finally, the direction of glacial troughs shows past ice movement. This is a U-shaped valley carved by a valley glacier and they can be mapped to show the movement of glaciers. For example, there are a number of glacial troughs below Mount Snowdon in Snowdonia, which reveal the movement of ice away from this pyramidal peak.

In conclusion, there are a number of landforms including till clasts, striations, glacial troughs and drumlins that can be mapped to reconstruct past ice movement. However, the value of some of these landforms, e.g. drumlins, is reduced if they are likely to be eroded or modified by subsequent human activity.

This conclusion is too short for A level: it should return to the question and ensure that it has been fully answered. While movement of ice has been considered quite well, there is no expansion into the ice extent or provenance and only some evaluation.

Avoid definitions at the start of long essays; this is far too brief and simplistic. The candidate needed to demonstrate understanding of the complexities of the question, such as which areas of the world are available for investigation, how glaciers and ice sheets move, what sorts of evidence they leave behind, and whether this evidence is useful.

Shows relevant knowledge and understanding supported by case study evidence. However, there is a lack of evaluation: for example, till deposits are only valuable when undisturbed by post-glacial change. They also only show local movement and suggest little else about the characteristics of the glacier or ice sheet.

A stronger paragraph: the candidate considers past ice velocity as well as direction and evaluates the value of the landform. However, a simple sketch would confirm the candidate's knowledge and understanding and improve the answer.

A further relevant piece of evidence, but not located. Another limitation is that striations may only show local movement unless there is an extensive exposed area of bedrock.

The candidate adds a further piece of evidence but appears to be running out of ideas; this point could have been extended with an analysis of cirque glaciers and orientation of fjords.

Verdict

This answer is slightly below average because it does not cover the whole question. There are 20 marks here! The candidate does demonstrate relevant knowledge and understanding of some landforms to reconstruct past ice movement, but should have covered landforms that help reconstruct past ice extent (considering deposits such as moraines, outwash, erratics). The question requires an evaluation of the usefulness of these landforms, which should have been a major theme of the conclusion.

Coastal landscapes and change

Introduction

More than 200 countries and island states have a coastline. It has been estimated that 12 per cent of the world's population live in lowland coastal zones (less than 10 m above sea level), so it is not surprising that the coast plays an important role in the lives of many people.

Coastal zones are dynamic places, subject to constant change. What determines these changes is the effect of the power of wave motion in the sea on the rocks that make up the land. The UK is perhaps fortunate as its geology is such that the hardest, most resistant rocks (igneous and metamorphic) are on the western side of the country and so, even though the prevailing westerly wind creates powerful waves, erosion rates are low. The softer rocks on the eastern side of England would not last long if pounded by large Atlantic waves, but most of the time the wind blows waves away from the eastern shore. When waves do approach from the east across the North Sea, they are smaller than in the Atlantic, so erosion rates are less than they might have been. If the geology had been the other way around, the UK would be only half its present size!

Wave action and rock types combine to produce some spectacular features and landscapes, from sandy beaches to high cliffs, and unique natural environments such as salt marshes. Physical processes threaten these landscapes and the people who live in them: in some places erosion rates are very high, and lowland areas are always at risk from coastal flooding. Human activities may also cause damage, not just through global warming but also because people build in vulnerable locations and carry out activities in the sea such as dredging, or building offshore wind farms. Managing physical processes and human activities at the coast is not easy.

In this topic

After studying this chapter, you will be able to discuss and explain the ideas and concepts contained within the following enquiry questions, and provide information on relevant located examples:

- Why are coastal landscapes different, and what processes cause these differences?
- How do characteristic coastal landforms contribute to coastal landscapes?
- How do coastal erosion and sea-level change alter the physical characteristics of coastlines and increase risks?
- How can coastlines be managed to meet the needs of all players?

Figure 3.1: Waves are very powerful during storm conditions, such as here at Dawlish in Devon in February 2014. How does the destruction shown in the photograph demonstrate the power of the sea and the rate of possible change?

CHAPTER 3

Synoptic links

There are many players (stakeholders) at the coast, such as government departments responsible for managing the natural environment and human activities, local councils, engineers responsible for building coastal defences, retired people, poor people in developing countries, tourist businesses, transport businesses, fishing businesses, shipping businesses, offshore wind energy businesses and port managers, as well as geographers, geologists and environmentalists. Often there is little co-ordination between these players, which emphasises the need for integrated coastal zone management. According to United Nations Environment Programme (UNEP), 33 per cent of coastal regions are at risk of degradation, mainly from infrastructure development and pollution. Then there is climate change, which is increasing storm activity and the energy in waves, as well as the risk of coastal flooding.

Useful knowledge and understanding

During your previous studies of Geography (KS3 and GCSE), you may have learned about some of the ideas and concepts covered in this chapter, such as:

- Rock types, including igneous, sedimentary and metamorphic.
- Erosion processes such as abrasion and hydraulic action.
- Constructive and destructive waves.
- Places that experience rapid coastal erosion and retreat.
- Weathering processes such as oxidation and hydrolysis.
- Coastal sediment transport by longshore drift.
- Coastal defences such as groynes, sea walls and beach nourishment.
- Case studies of storm surges and coastal flooding.
- Coastal erosion features, including caves, arches, stacks and stumps.
- Coastal deposition features such as beaches and spits.
- Coastal settlements such as ports and seaside resorts.
- Coastal management strategies and policies.

This chapter will reinforce, but also modify and extend, your knowledge and understanding of coastal landscapes. Remember that the material in this chapter features in both the AS and A level examinations.

Skills covered within this topic

- Use and interpretation of maps at different scales, including GIS images of coastlines.
- Interpretation of satellite images to identify types of coastline.
- Use of measures of central tendency to classify wave types.
- Use of Student's *t*-test to investigate changes in particle sizes and shapes on a beach.
- Interpretation of photographs (including aerial) and diagrams to identify coastal features and processes, especially showing evidence of past sea-level change and recession rates.
- Use of diagrams and field sketches to interpret contrasting coastal landscapes.
- Interrogation of GIS and other land-use information mapping to inform choice of management strategies.
- Use of photographic evidence to assess environmental impact of schemes.
- Use of X^2 (chi-square) to assess succession in a sand dune or salt marsh environment.

CHAPTER 3

Why are coastal landscapes different, and what processes cause these differences?

Learning objectives

2B.1 To understand that the coast, and wider littoral zone, has distinctive features and landscapes.

2B.2 To understand how geological structure influences the development of coastal landscapes at a variety of scales.

2B.3 To know that rates of coastal recession and stability depend on lithology and other factors.

Coastal features and landscapes

The littoral zone

The littoral zone contains many coastal sediments (pebbles and sand particles). Waves, currents and tides move these sediments around in a zone along the coast called the **littoral zone**, which is from the highest sea-level line (linked to high tides and storm waves) to shallow offshore waters (where the base of a wave first encounters friction with the sea bed). This zone is subdivided into **backshore**, **foreshore**, **nearshore** and **offshore** zones (Figure 3.2).

ACTIVITY

1. Study Figure 3.2 and consider the processes operating in the littoral zone.
   a. Suggest which subdivision of the littoral zone experiences the most natural processes.
   b. Suggest which subdivision of the littoral zone experiences the most human activity.
   c. Outline the natural and human processes that may cause rapid change within the littoral zone.

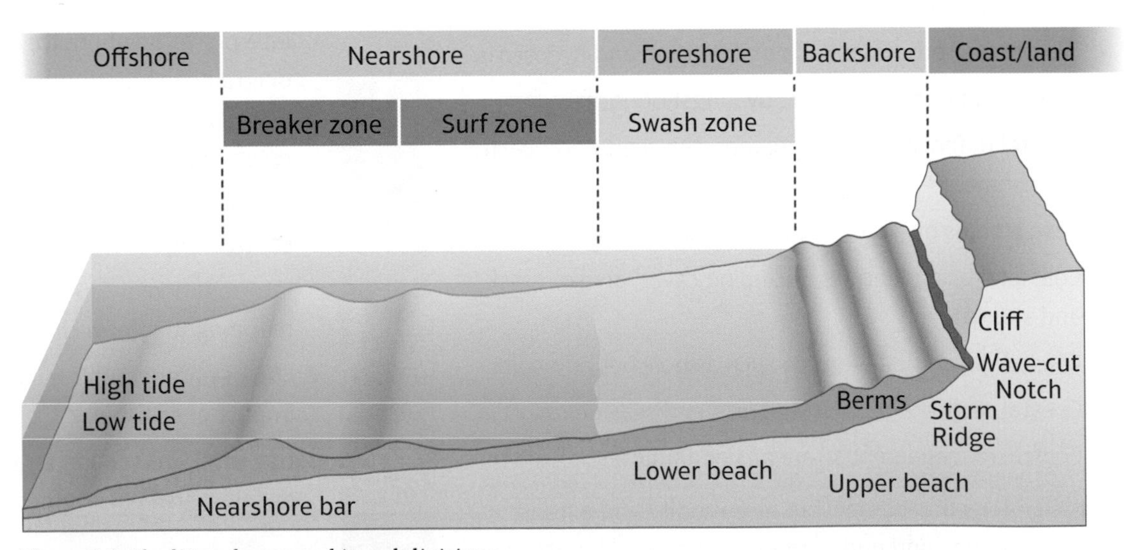

Figure 3.2: The littoral zone and its subdivisions

Yellow = Players, Orange = Attitudes and actions, Purple = Futures and uncertainties

The littoral zone is one of **dynamic equilibrium**, due to the wide range of natural processes that interact within it:

- There are inputs of sediments from the sea and currents from rivers flowing off the land.
- Weathering and mass movement occur on the backshore.
- Constructive and destructive waves occur on the foreshore and nearshore, causing deposition and erosion.
- The tidal range affects all parts by determining where wave action takes place.
- Offshore currents and longshore drift may move sediments some distance along a coast.

Human activities can interfere with these natural processes, for example:

- dredging of rivers to make them deeper for shipping
- dredging of offshore areas to get sand and gravel for construction
- the building of coastal defences against erosion and flooding.

Rapid change can take place when there is increased energy in the natural processes or when the impacts of human activities are not carefully considered. In England and Wales, the coastline has been divided into 11 sediment (littoral) cells, recognising lengths of coast where processes are closely linked, with the aim of improving coastal management.

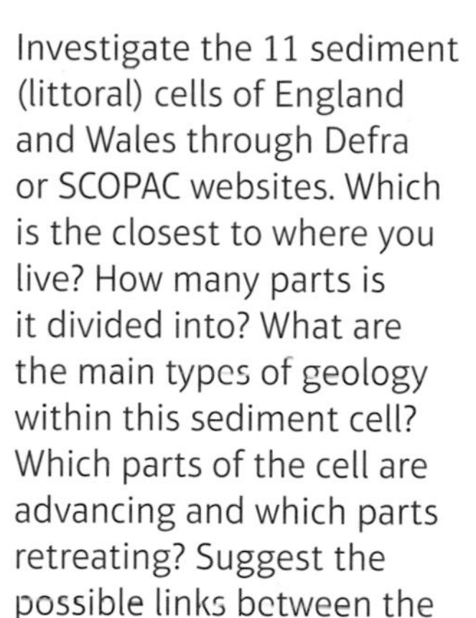

Extension

Investigate the 11 sediment (littoral) cells of England and Wales through Defra or SCOPAC websites. Which is the closest to where you live? How many parts is it divided into? What are the main types of geology within this sediment cell? Which parts of the cell are advancing and which parts retreating? Suggest the possible links between the sediment cell parts, geology and advance or retreat.

Classification of coasts

Sections of coast can be classified into different types using a variety of criteria, such as geology, sea-level rise and fall, and land-level rise and fall, which cause long-term changes, and erosion and deposition, which cause short-term changes. A classification of types of coast that are either advancing or retreating is summarised in Figure 3.3.

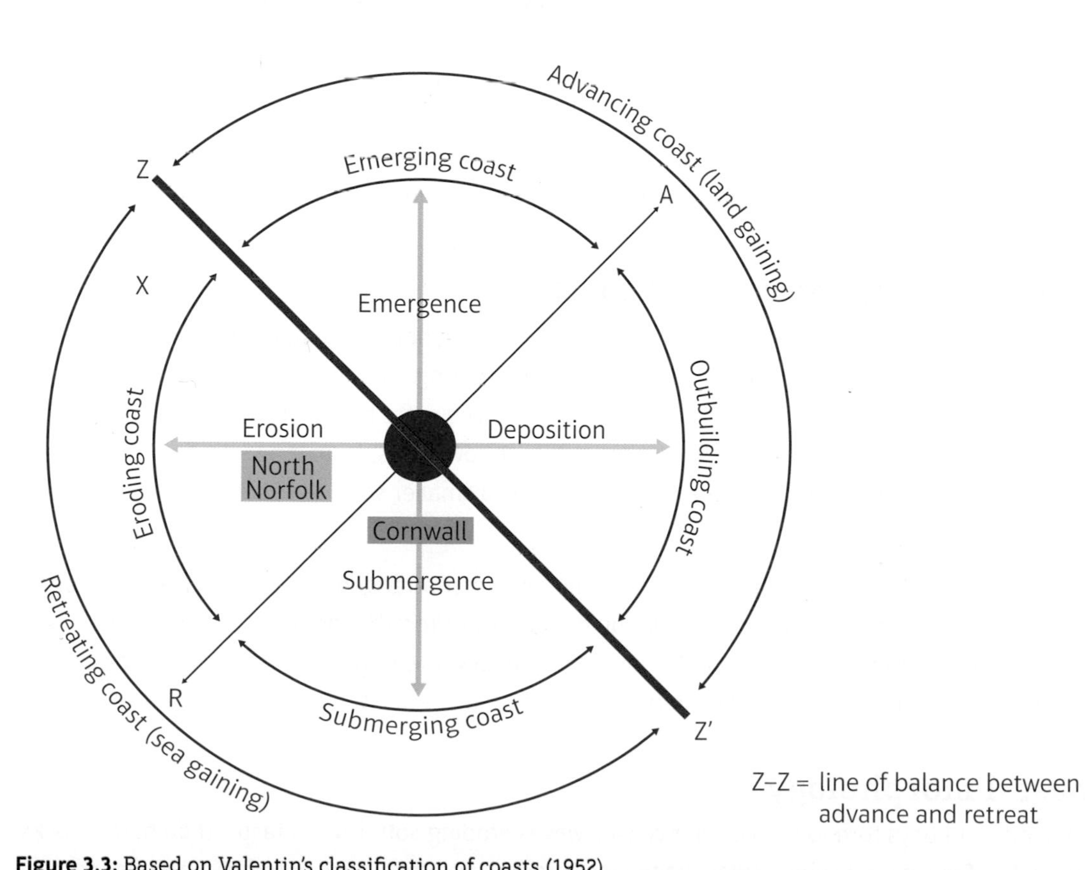

Figure 3.3: Based on Valentin's classification of coasts (1952)

ACTIVITY

GRAPHICAL SKILLS

1. Investigate the large-scale factors affecting the coastlines of the Netherlands, Norway and New Zealand's North Island. Use the government websites of these countries first.

2. Study Figure 3.3 carefully and understand the axes. Some named coastlines have been plotted on to the graph. On a copy of Figure 3.3, use the axes to plot the position of the UK coastline nearest to where you live, and the coastlines of the Netherlands, Norway and the North Island of New Zealand.

Literacy tip

When researching websites based in other countries, such as the USA, Australia, and New Zealand, be aware of different spellings and terminology. Always translate these back into UK English!

High-energy and low-energy coasts

The amount of energy in the coastal environment will determine the landscape of the coast (Table 3.1). This energy is mostly provided by waves, but also to some extent by the weather (rain, wind, and temperature), rivers, and large and small-scale sea currents.

Table 3.1: Characteristics of low-energy and high-energy coasts

	Low-energy coast	High-energy coast
Waves	• Less powerful (constructive) waves • Calmer conditions • Short fetches	• More powerful (destructive) waves • Storm conditions • Long fetches
Processes	• Deposition and transport • Sediments from rivers, longshore drift and nearshore currents	• Erosion and transport • Sediments from eroded land, mass movement and weathering, offshore currents
Landforms	• Beaches, spits, salt marshes, sand dunes, bars, mudflats	• Cliffs, wave-cut platforms, arches, sea caves, stacks
General location	• Sheltered from large waves • Lowland coasts • Coastal plain landscape	• Exposed to largest waves • Highland and lowland coasts • Rocky landscape
Example locations	• Mediterranean Sea coasts • East Anglian coast	• Atlantic coasts of Norway and Scotland • Pacific coasts of Alaska and Canada

Geological structure at different scales

The shape of the coast is largely determined by geology and wave action. Various aspects of geology should be considered, such as resistance to erosion and weathering, which relates to the degree of consolidation (how well the rock is stuck together), the number of joints or bedding planes, and the amount of folding. Wave characteristics include energy, which is related to size and shape, length of fetch, the prevalent and dominant wind direction, and the depth of nearshore and offshore water.

ACTIVITY

CARTOGRAPHIC SKILLS

1. Using an atlas and online mapping sites, investigate south-west Ireland and the east coast of the Adriatic Sea to understand the shapes of concordant and discordant coastlines.

2. Draw labelled sketch maps of these two coastline types, based on your investigations.

Concordant and discordant coasts

Concordant (or Dalmatian) coastlines occur where the folding or arrangement of rock types on a large scale is parallel to the coast. This means that alternating hard and soft rocks, or folds in the rock layers, lie in the same direction as the coastline. The more resistant rocks, or upfolds, form elongated islands, while the less resistant rocks, or downfolds, form long inlets or coves. The eastern coast of the Adriatic Sea is one example, and on a smaller scale the south-facing coast of Dorset is another.

Discordant (or Atlantic) coastlines occur where the folding or arrangement of rock types on a large scale is at right angles to the coast. Alternating hard and soft rock bands create headlands where there is more resistant rock, or an upfold, and bays or inlets where there is less resistant rock, or a downfold. South-west Ireland is an example, and on a smaller scale the east-facing coast of Dorset.

Coastal morphology

Headlands and bays form on discordant coasts, waves eroding softer rocks faster than harder rocks (Figure 3.4). Bays have a semicircular shape, once eroded, and the waves entering a bay dissipate, lose energy and deposit a bay-head beach. The resistant rock forms rugged rocky headlands that stick out into the sea. Once this stage is reached, the shape of the coastline causes waves to refract as they approach the headlands, and this concentrates wave energy on them, increasing the rate of erosion. In theory, if there are no large-scale forces such as tectonic uplift, the erosion and deposition patterns will eventually create a reasonably straight coastline again.

Haff coastlines form in **low-energy environments** where there is deposition of muds and sands. Large lagoons are found behind the deposits parallel to the shoreline. The Baltic Sea coastline of Poland has these features, such as Vistula Haff, which is a large lagoon (91 km long, 9 km average width and average depth of 2.7 m) that has formed behind Vistula spit and has deltaic (river) sediments building up within it.

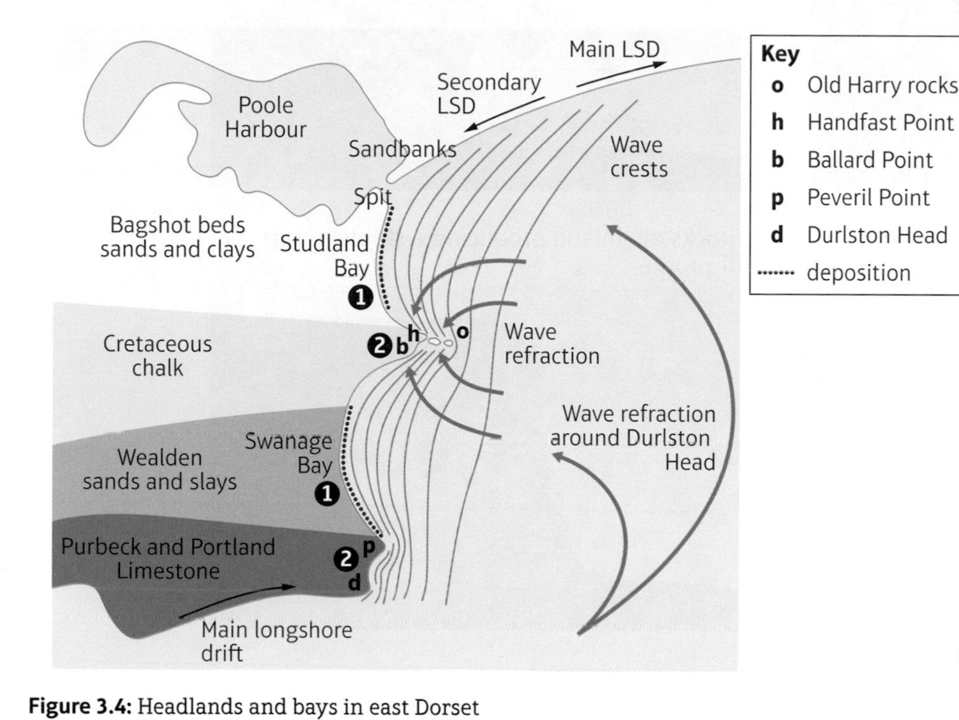

Figure 3.4: Headlands and bays in east Dorset

Synoptic link 

Tectonic activity can have an important influence on the coast. For example, it may push up sedimentary rocks from the seabed, such as the limestones in Dorset, and fold them through plate collisions, for example Stair Hole at Lulworth in Dorset, creating weak points for waves to exploit. Magmas also rise towards the surface and cool to form igneous rocks, or change other rocks into metamorphic types, which, when exposed after removal of the layers above them, resist wave erosion. In contrast, undersea volcanic activity may form temporary islands of tephra, which quickly erode in powerful waves, such as happened near Tonga in 2009 and 2015. (See pages 24–25.)

Geological structure has a major influence on coastal morphology (shape) and erosion rates. Tectonic forces may cause rock layers to fold into complex formations. For example, the Stair Hole 'crumple' at Lulworth in Dorset is thought to have been created by the collision of the African and Eurasian plates. During the process of rock formation, pressure and cooling may crack a rock to form crevices, joints or larger faults. Joints are made as sedimentary rock is formed by pressure, or as igneous and metamorphic rocks cool. In sedimentary rocks there are also usually bedding planes between the different layers of deposition. Faults, joints and bedding planes all create points of weakness where weathering and erosion processes can start to wear away the rock.

The dip of a rock layer is the angle of tilt from the horizontal. If a rock layer dips towards the sea, it will be possible for blocks of rock to slide under the influence of gravity – a mass movement. If the angle of dip is landward, weathering and erosion processes may attack exposed bedding planes and joints, creating an irregular profile.

Cliff profiles may include many protrusions and indentations, as well as small features such as caves, blowholes and geos (Figure 3.12). A weak point in a rock will be weathered or eroded (see page 129) so that a joint is enlarged; eventually this may become large enough to form a sea cave.

Further erosion along the jointing system can then create an opening to the surface at the top of the cliff. When large waves break into the cave, water and spray are then squirted out of this blowhole. Continued erosion may lead to the collapse of the roof of the cave and tunnel system, and the formation of a small narrow inlet known as a geo.

a. Uniform horizontal strata produce steep cliffs

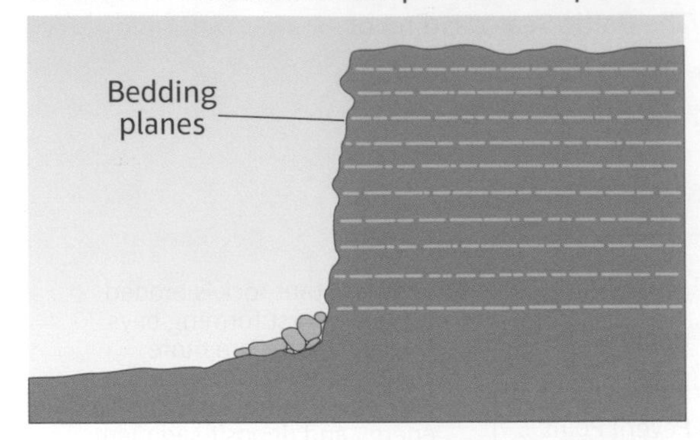

b. Rocks dip gently seawards with near-vertical joints

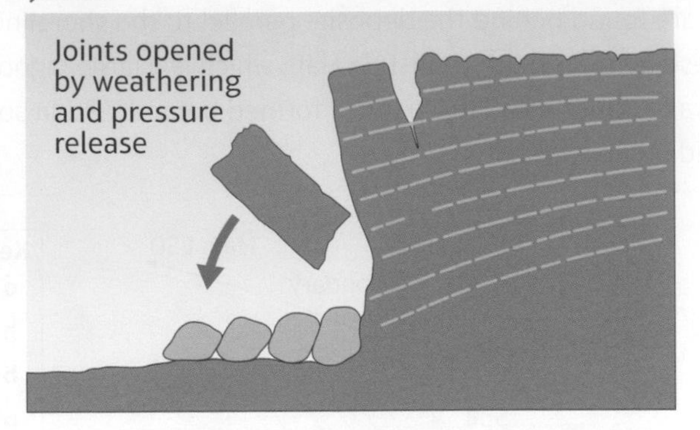

c. Steep seaward dip

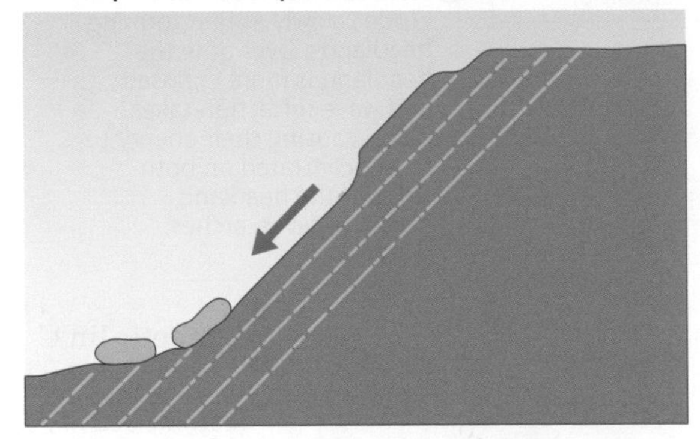

d. Rocks dip inland producing a stable, steep cliff profile

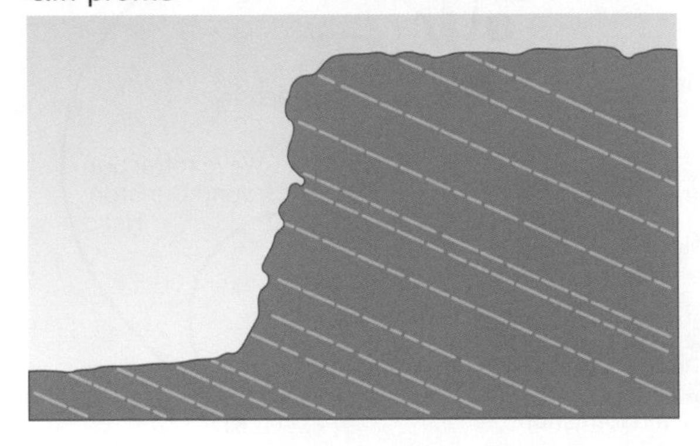

e. Rocks dip inland but with well-developed joints at right angles to bedding planes

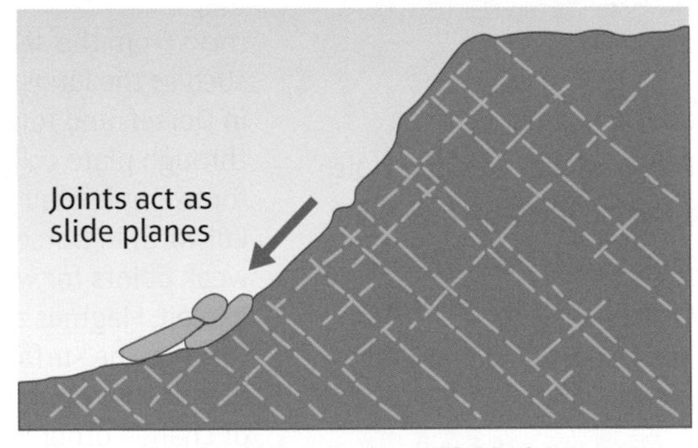

f. Slope-over-wall cliffs

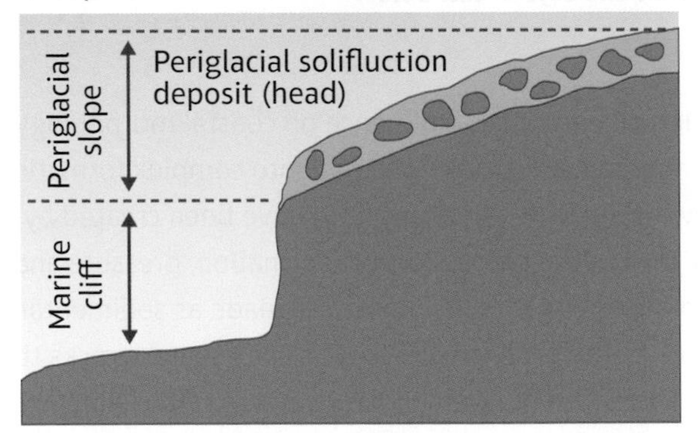

Figure 3.5: Cliff profiles linked to geological structure

Factors affecting coastal recession

Lithology

Bedrock **lithology** is important in determining rates of **coastal recession** (retreat); these are the rocks that make up the base of the land. Igneous rocks consist of cooled magmas (granites, basalts) and are more resistant to marine erosion and weathering than sedimentary rocks, which were formed underneath oceans (limestones, sandstones). For example, a typical annual recession rate for granite is 1 mm, while for thinly bedded limestone it is about 2.5 cm. Metamorphic rocks,

ACTIVITY

Explain how geology is an important influence on the shape (morphology) and features of a coastline.

Yellow = Players, Orange = Attitudes and actions, Purple = Futures and uncertainties

such as gneiss and marble, are formed from igneous and sedimentary rocks that were subjected to intense heat and pressure, which makes them more resistant. On top of bedrock there may be thick surface deposits left by rivers, wind or ice. For example, glacial till, a mixture of clay and stones, was left behind over much of eastern England by retreating ice sheets, creating vulnerable cliffs along the Holderness coast of Yorkshire and the North Norfolk coast, which in places can recede at a rate of 1 m a year. Recession is fastest of all on new islands formed as a result of volcanic activity, with rates of around 40 m a year. Where there are multiple types of rock at the same location, the cliff profile can become very complex (Figure 3.6).

Differential erosion

Rocks with many cracks, joints and bedding planes, such as limestone, are **permeable**, which means that water can pass through the rock. Rocks with spaces, or pores, within them, such as chalk, are porous and also allow water to pass through. Rocks that do not have spaces or many joints, such as granite, are **impermeable** and do not let water into the rock. The presence or absence of water may affect weathering and mass movement (collectively known as **subaerial processes**). For example, if there is a porous rock above an impermeable rock, such as a weakly consolidated sandstone above a layer of clay, water percolating through will collect in the upper layer, creating an instability that often leads to slipping and slumping (Figure 3.21). (See also page 139.)

Some types of rock contain soluble minerals, and so are vulnerable to chemical weathering such as carbonation or hydrolysis. Limestone is soluble in rainwater, which becomes a dilute carbonic acid by absorbing carbon dioxide from the air. Limestone rocks contain calcium carbonate, which is slowly dissolved into calcium bicarbonate, which is carried away in solution. Seawater also absorbs carbon dioxide from the atmosphere, and when waves splash on to limestone rocks the water dissolves the limestone, to create small pits and pinnacles. In hydrolysis, water chemically combines with some minerals, such as feldspar (in granite), to produce clays and salts. The feldspar mineral in the granite cliffs of Cornwall and Devon may turn into kaolin (china clay), contributing some of the salts that are in seawater.

Coastal recession rates can be determined by comparing historical maps with contemporary ones, comparing aerial photographs taken at different times, or using modern technology such as LIDAR (Light Detection and Ranging). Not all recession is due to marine erosion: sea-level rise and sinking land must also be considered (see page 142).

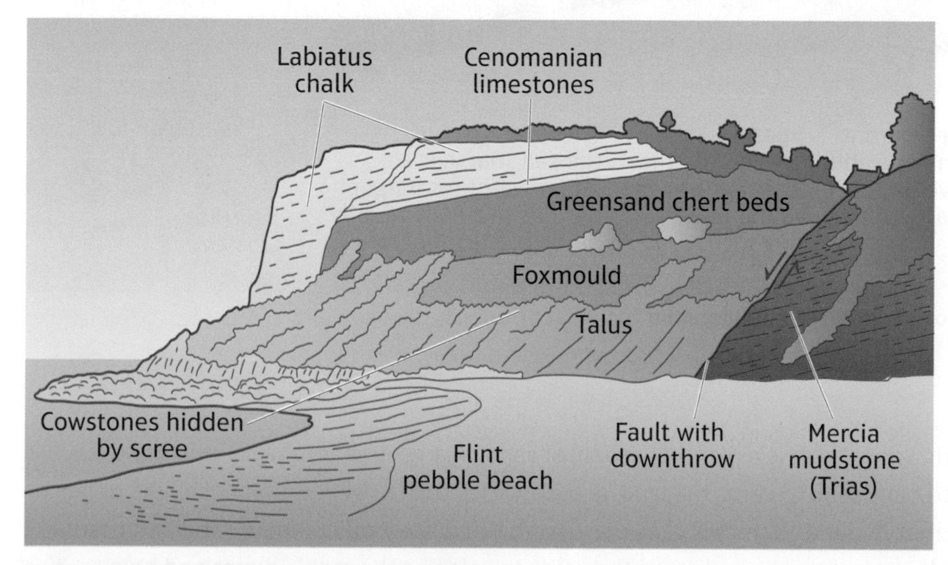

Figure 3.6: Cliff profile with multiple rock types

Extension

1. Investigate the chemical reactions (and formulas) between rain and limestone rock.
2. Some researchers are doubtful about the ability of seawater to dissolve rock minerals, because it is already saturated with salts. Investigate this to reach your own conclusion.

ACTIVITY

1. Investigate the changes to the coastline at Happisburgh, North Norfolk, over time. Look at websites such as the Coastal Concern Action Group, Eastern Daily Press or other mapping and satellite imagery sites.

2. Find a sequence of aerial photographs over time. Save these and write a detailed description of what has happened over the period chosen, perhaps by annotating the photographs.

3. Investigate the geology and rock characteristics at Happisburgh. Suggest how this coastline may change in the future, using information in this chapter.

The role of vegetation at the coast

Ecosystems at the coast include coral reefs, mangrove forests, salt marshes and sand dunes. The latter two are common in lowland UK areas. Vegetation is important for stabilising soft sediment low-energy coastlines based on sand and mud. As in all ecosystems, there is **plant succession** over time, with pioneer plants colonising bare areas first. As they change the conditions of the soil by adding humus (decayed vegetation), retaining moisture and stabilising loose sand or mud, new plants are able to establish themselves and take over. This keeps on happening until there is a balance between all the natural factors and the vegetation, when a climax (i.e. final) vegetation type establishes itself.

Sand dunes are known as a psammosere, where xerophytic plants such as marram grass can survive with little fresh water. Salt marshes are known as haloseres, with halophytic plants such as samphire that can survive in salty conditions. Both these natural environments are important buffer zones between the sea and the land, with salt marshes absorbing coastal flooding and sand dunes taking the impact of storms.

Salt marshes

Salt marshes are common in the low-energy environments of estuaries and sheltered bays. Here, tidal conditions bring seawater and sediments in and out, and rivers bring fine muds and silts and deposit them at the sides of the estuary. The tiny clay particles stick to one another, a process known as flocculation, and once deposited they are colonised by algae. Salt marshes can then be found in the creeks and at the edges of estuaries, as well as behind a spit which sometimes forms across the mouth of an estuary. The pioneer plants have to survive being covered with slightly salty (brackish) water twice a day. Over time, the plants change the conditions by trapping more sediment, which builds the salt marsh up to a higher level, so that other plants can then colonise. The muds contain many invertebrates such as lugworm, and many plants that can be grazed by migrating wildfowl such as Brent geese and waders such as oystercatchers.

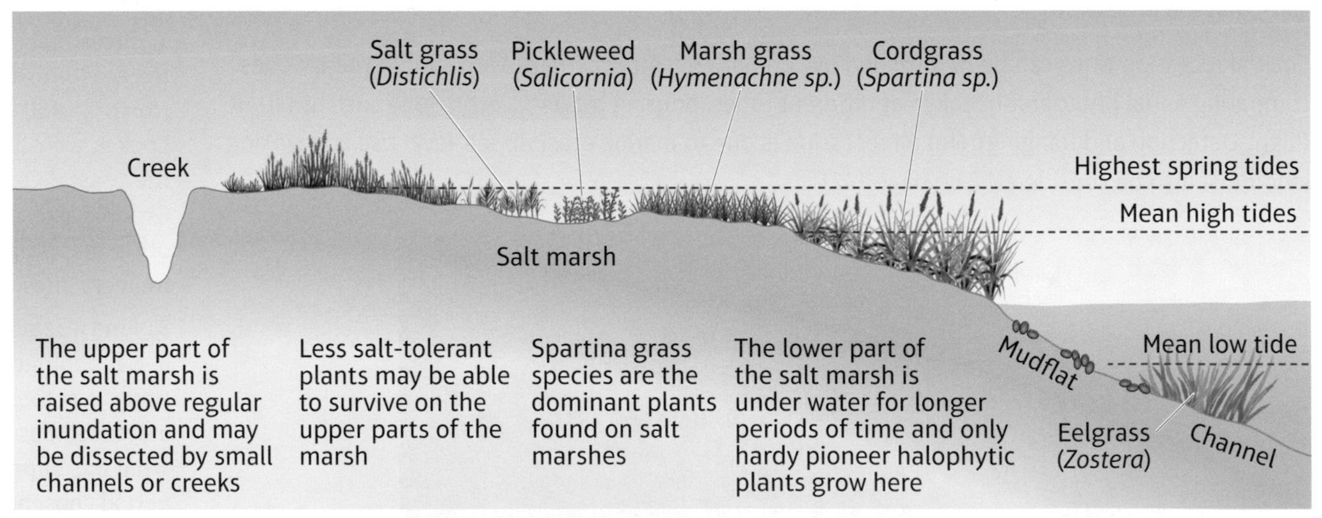

Figure 3.7: Salt marsh cross-section showing features and succession

Sand dunes

Sand dunes form where there is a plentiful supply of sand, a large area for the sand to dry out, onshore winds to blow sand towards the land, and obstacles such as vegetation or shingle ridges to trap the sand. Embryo dunes form first and, once established, they collect more sand and become larger. Pioneer plants such as sea rocket are then able to colonise the stable dunes and help to hold the sand together and trap more sand. Between 50 and 100 years may be enough time for a significant dune sequence to develop, with the oldest dunes closest to the land and the youngest closest to the sea.

Yellow dunes tend to be the highest, and may form a ridge near the dune front with marram grass. These are not fully vegetated and are subject to alteration by the wind and waves (with 'blow-outs' forming gaps in the line of dunes). At high tide or under storm conditions, seawater may reach the dips in the sand dunes, called slacks, allowing other plants such as marsh orchids to grow here. Mature dunes are known as grey dunes because their humus content is greater, and the climax vegetation is either pine forest (where acidity is high) or oak forest (where shell deposits neutralise the soil a little).

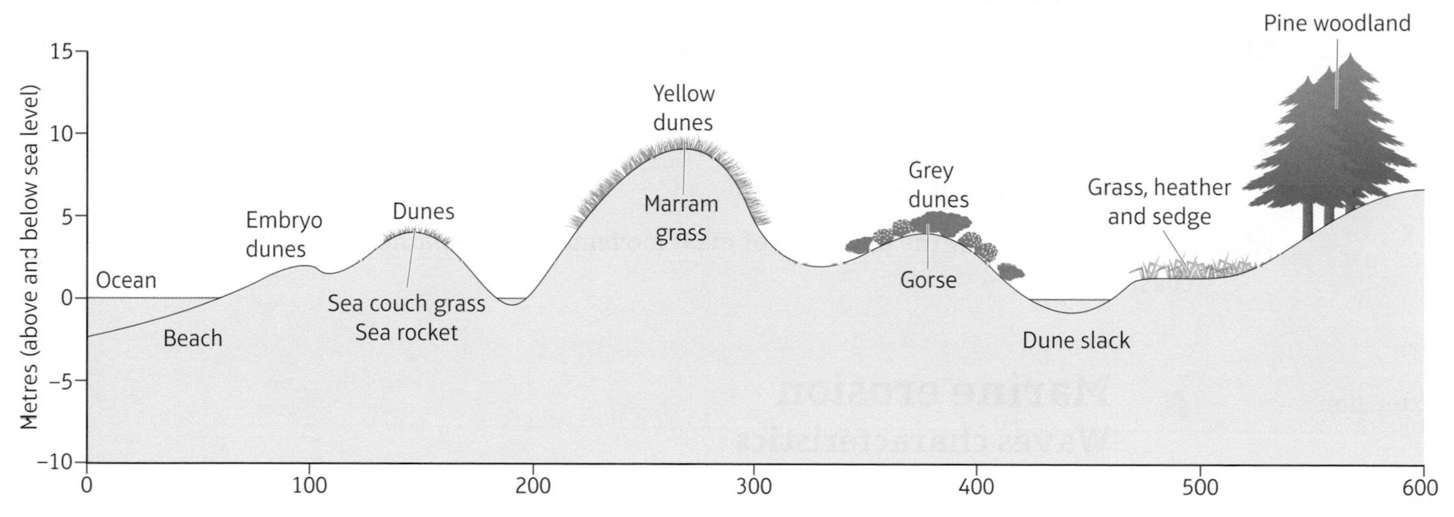

Figure 3.8: Sand dune cross-section showing features and succession

AS level exam-style question

Explain how vegetation can bring stability to low-energy coastlines. (6 marks)

Guidance

The answer needs to expand on how sand dune and salt marsh ecosystems can protect a coastline directly in the short term and indirectly in the longer term. Remember to explain – not just describe.

A level exam-style question

Explain the differences between low-energy and high-energy coastal environments. (6 marks)

Guidance

Be careful *not* to write two separate parts to the answer, one about low energy and one about high energy. The question requires you to compare the two, so combine statements about both and then add the reason(s) for each difference.

CHAPTER 3

How do characteristic coastal landforms contribute to coastal landscapes?

Learning objectives

- **2B.4** **To understand how marine erosion creates distinctive landforms and contributes to coastal landscapes.**
- **2B.5** **To understand how sediment transport and deposition create distinctive landforms and contribute to coastal landscapes.**
- **2B.5** **To understand how the processes of mass movement and weathering alter the shape of a coastline.**

Extension 

The highest ever recorded wave is 34 m (from trough to crest) in the western Pacific.

Where are the longest fetches in the world? Where are the strongest storms in the world? Which coasts will therefore be affected by the largest waves?

Marine erosion

Waves characteristics

Waves result from friction between the wind and the sea surface. The stronger the wind, the more powerful the waves will be when they break on the shore. The length of fetch is also important: the farther a wave can travel uninterrupted, the more powerful it will be when it reaches the shore. Dominant winds are those that produce the largest and most damaging waves at any given point on the coast, while prevailing winds are the ones that blow most frequently at a place. For the west coasts of the UK, the prevalent and dominant wind directions are the same – from the west, across the Atlantic Ocean – but for the east coast, the dominant winds, and therefore wave direction, are north-easterly.

During a storm, the wind creates sea waves, which continue under their own momentum to become swell waves. Different winds create waves of different wavelengths (distance between wave crests), and sometimes these coincide to make a larger wave. The movement of water particles within a wave in deep water is circular, the movement getting smaller with depth (Figure 3.9). The wave motion, not the water itself, moves forward, and the time it takes for successive waves to pass a point is known as wave frequency (or wave period).

The potential energy of a wave is proportional to its height. The forward movement of the water in a wave at the shore releases this energy. This forward movement does not start until the sea is shallow, defined as when the depth is less than the wavelength. This is when the water particle movement becomes elliptical, because of friction with the seabed at the base of the wave (Figure 3.9).

Destructive waves (plunging breakers)

With destructive waves, the crests move forwards and downwards at an angle of about 120 degrees, often briefly enclosing an air pocket before breaking. They are common when slower waves approach a steep shingle beach. The wave frequency is about 13 to 15 per minute. The motion of the water in the wave remains circular because of a deeper nearshore zone, so that the

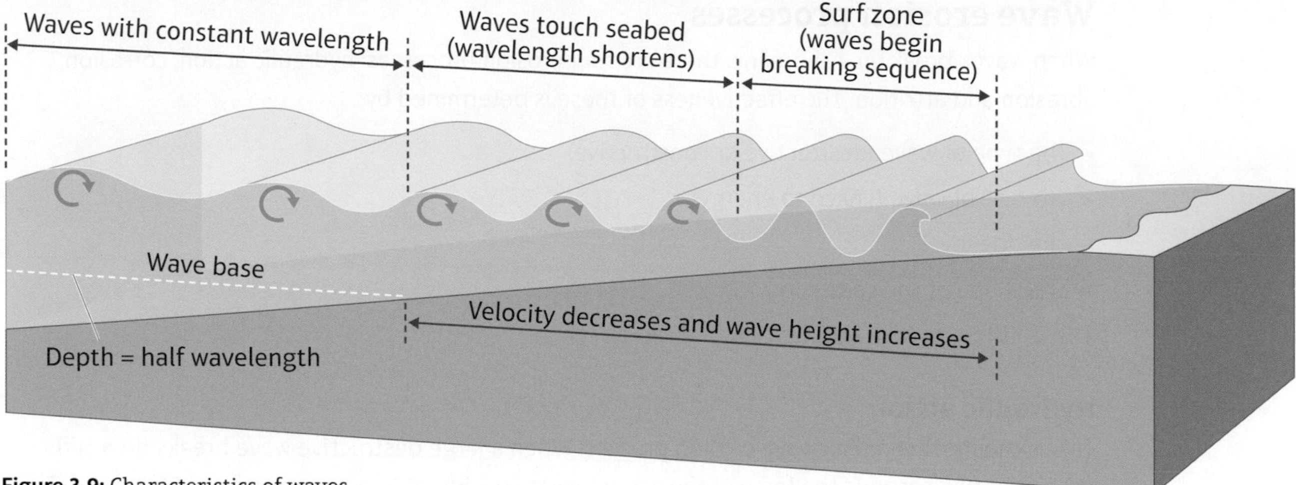

Figure 3.9: Characteristics of waves

mass of the wave is directed downwards at the beach – or a cliff at high tide. These waves have a stronger backwash than swash, the backwash of the previous wave making the swash of the wave that follows it weaker. These waves remove sediment from a beach or cliff, carrying away the smallest particles first, although the impact of a destructive wave can throw material up a beach to form a berm or storm ridge (see Figure 3.15, page 131), changing the **beach morphology**.

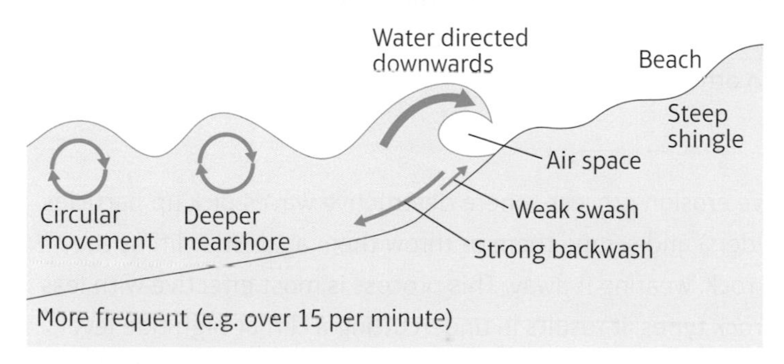

Figure 3.10: Cross-section of a destructive wave

Constructive waves (spilling breakers)

Constructive waves are usually found where nearshore depths are shallow and there are gently sloping sandy beaches. The swash is more powerful than the backwash, and sediment is moved up a beach, making it wider. Waves break with a frequency of about six to eight per minute, which allows each wave to complete its cycle and so the swash is not interrupted. The motion of water in the wave is elliptical, giving a strong forward movement. The energy of the backwash is reduced as the water percolates through the beach material, rather than running off the surface. If approaching at an angle to the beach, these waves may push larger sediments into semicircles, creating a feature known as beach cusps (see Figure 3.15, page 131).

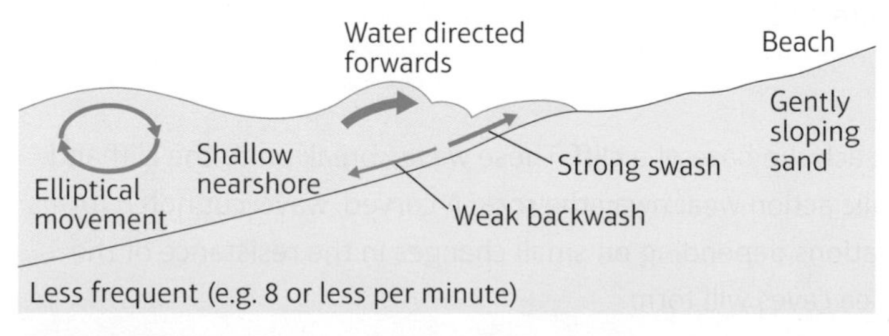

Figure 3.11: Cross-section of a constructive wave

ACTIVITY

CARTOGRAPHIC SKILLS

1. **a.** Using an atlas, measure the distance over which waves can travel to reach the coast of the British Isles, from ten different directions.

   b. Combine your measurements with those of other students and record the results on a circular graph to create a rose diagram based on compass directions.

2. Compare the geology of the British Isles with the results of the investigation of fetches. What significant patterns are there?

Wave erosion processes

When waves break on a shoreline, there are four erosion processes: hydraulic action, corrosion, abrasion and attrition. The effectiveness of these is determined by:

- the type of wave (destructive or constructive)
- the size of wave (linked to energy)
- the tide level
- the shape of the coastline
- the lithology (characteristics of the rocks).

Hydraulic action

This is the most complex wave erosion process. When a large destructive wave breaks on a cliff, for one or two seconds the force of the water exerts considerable pressure, around 50 kg/cm^2. This pressure may loosen pieces of rock. Air may also be trapped in the circular shape of the breaking wave and compressed into cracks and joints in the rock. Then, when the water from the wave suddenly falls away, this pressure is released explosively, which can shatter the rock around the joint or crack.

Corrosion

Seawater and salt spray from waves may react with rock minerals and actively dissolve them. Rock minerals are then carried away in solution within the seawater. Some rock types such as limestone are more susceptible to corrosion than others.

Abrasion

This is perhaps the most effective wave erosion process, where destructive waves pick up particles of sand and pebbles (sometimes boulders) and scrape them or throw them against a cliff as they break. This abrades (or scratches) the rock, wearing it away. This process is most effective with less resistant rocks. In most consolidated rock types, it results in undercutting a cliff at high tide level, especially under stormy conditions, leaving a curved wave-cut notch and sea caves where joints and cracks are widened.

ACTIVITY

Explain how waves influence the shape (morphology) and features of a coastline.

Attrition

Attrition occurs when boulders, rock particles, pebbles and sand are continually moved around by waves, especially in the breaker or surf zone (see Figure 3.2). As these sediments move, they collide with one another, and any projecting angular corners are knocked off. The process is slower with harder rocks, such as flint, which eventually form pebble beaches. The rocks are broken down into smaller and smaller pieces until only quartz grains (sand) are left, and even these are rounded.

Landforms of coastal erosion

Erosion creates distinctive coastal landforms: wave-cut notches, wave-cut platforms, cliffs and the cave-arch-stack stump sequence (Figure 3.12).

Wave-cut notch

At high tide, destructive waves may reach the base of a cliff. These waves break on to the cliff and the processes of abrasion and hydraulic action wear away the rock. A curved, wave-cut notch forms along the length of the cliff, with variations depending on small changes in the resistance of the rock. For example, at weaker points, sea caves will form.

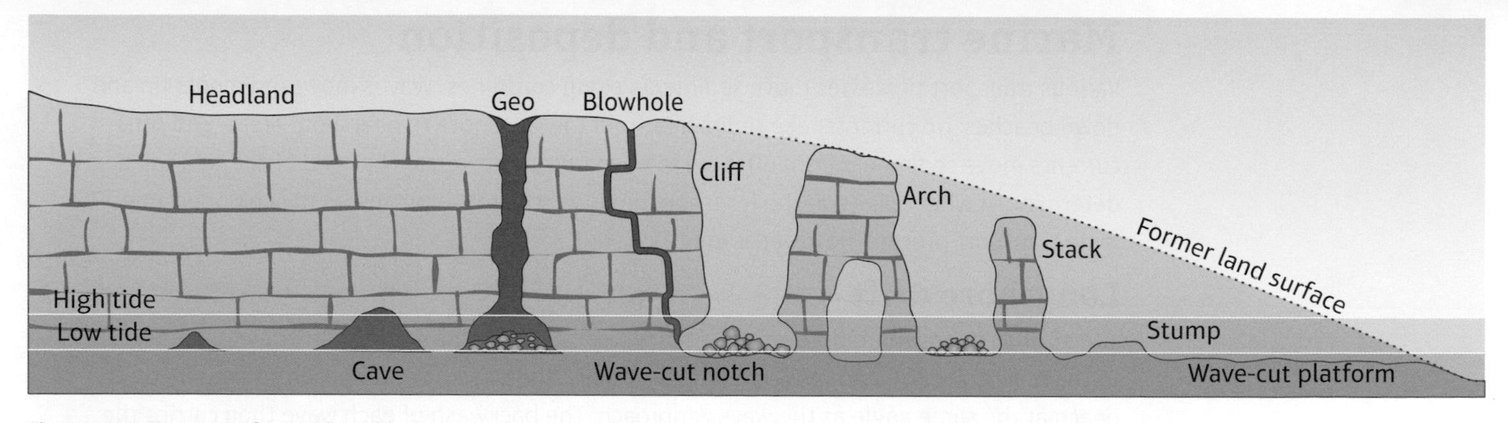

Figure 3.12: Features of coastal erosion

Wave-cut platform

As a cliff is eroded at its base, the rock above is left unsupported and eventually collapses under the influence of gravity. In this way the cliff retreats (recession), and leaves behind a flat or slightly sloping area of rock between the high- and low-tide levels where the land used to be. Sometimes this feature, also known as a shore platform, is not just the result of wave action but also of weathering processes. The surface is irregular, with rock pools or small ridges where the rock was either less or more resistant.

Cliffs

Cliffs are vertical or near-vertical slopes caused by waves undercutting the land at high tide, or constantly if there is no beach. As the cliffs are undercut, gravity is able to cause mass movement in the unsupported rock. Cliffs exist in both more resistant and less resistant rocks, so the type of mass movement depends on their geology (see Figure 3.5, page 122).

Caves, arches, stacks and stumps

At a headland or smaller promontory sticking out into the sea, waves will be refracted, so that the full energy of erosion processes is concentrated on weak points on the sides. Hydraulic action and abrasion will at first form caves; often these caves meet from opposite sides, to form a tunnel above which is a rock arch. The arch becomes more prominent with further erosion and weathering, and eventually the top collapses, leaving a pillar of rock, called a stack, at the seaward end of the headland. Waves continue to erode the base of the stack, especially during stormy conditions, cutting notches in several sides. In this way the stack becomes unstable and collapses, leaving just the base, known as a stump, which will be visible at low tide but may be submerged at high tide.

Figure 3.13: Arch and stack at Cathedral Cove, New Zealand

Marine transport and deposition

Various transport processes move sediments along coastlines. Waves move sediments up and down beaches; rip currents take sediments from the foreshore to nearshore areas; and other currents move sediments from offshore to onshore, or vice versa. The level of the tide will also determine at what height on the foreshore or backshore these processes may be operating. The main transport process, however, is longshore drift.

Longshore drift

The strongest longshore drift occurs when waves approach the coast at an angle of 30 degrees to the beach. The forward movement of each wave, the swash, moves beach sediments up the beach at the same angle as the waves approach. The backwash of each wave then carries the same sediments back down the beach at right angles to the shore, under the influence of gravity (Figure 3.14). In this way, sediments slowly move in small zigzag steps along the beach until they meet a natural obstacle such as a bay or an estuary, or an artificial barrier such as a groyne.

The dominant wind and wave direction will determine the longshore drift direction at any section of coast. Along the south coast of England it is from west to east because the dominant waves come from the Atlantic Ocean. However, wave refraction around headlands may reverse the drift, and in some locations there may be two longshore drift directions, such as at Dungeness in Kent (see page 134).

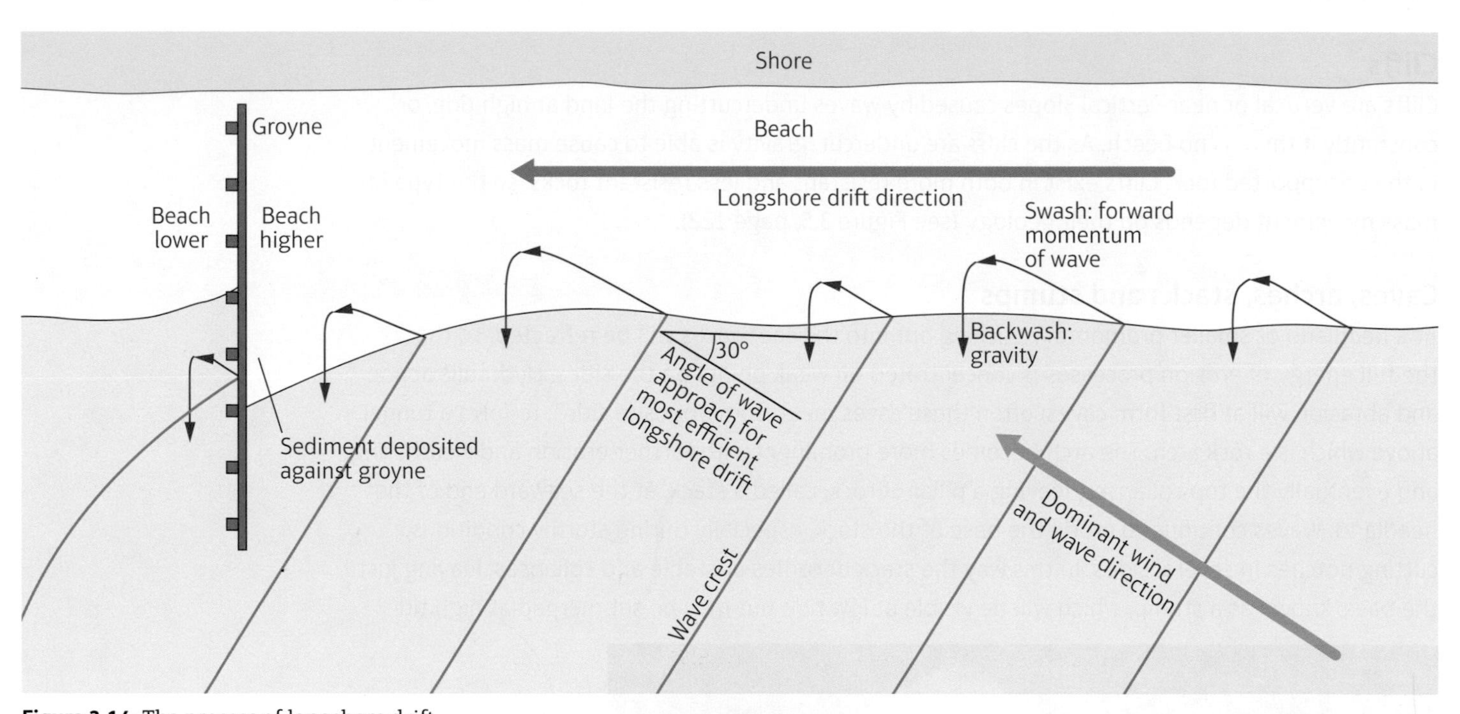

Figure 3.14: The process of longshore drift

Landforms of coastal transport and deposition

Beaches are common coastal depositional features, made of rounded sand and pebbles (Figure 3.15). Sandy beaches have a gentle angle, while pebble beaches are steeper because of the angle of rest of the particles and the wave energy shaping them. Beach sediment is deposited mostly by constructive waves, which have a strong swash, and the breaking action of these waves may form beach cusps and berms. Longshore drift also brings beach material from erosional locations, which is deposited by waves as they move material along the beach. Groynes are constructed to trap sediments to make a beach wider and higher, so that destructive waves crash on to the beach and not the shoreline. A wider beach also helps to attract tourists to seaside resorts.

Yellow = Players, Orange = Attitudes and actions, Purple = Futures and uncertainties

Destructive waves shape a beach by creating storm ridges at the backshore during high tide, throwing pebbles forwards up the beach by the force of their impact, or scouring sediments away with strong backwash. Tides determine where waves strike a beach, and rip currents take away beach material at certain points. Finally, subaerial processes of mass movement and weathering working on a cliff or shore supply sediment to a beach, for example debris from mass movement falling on to the beach.

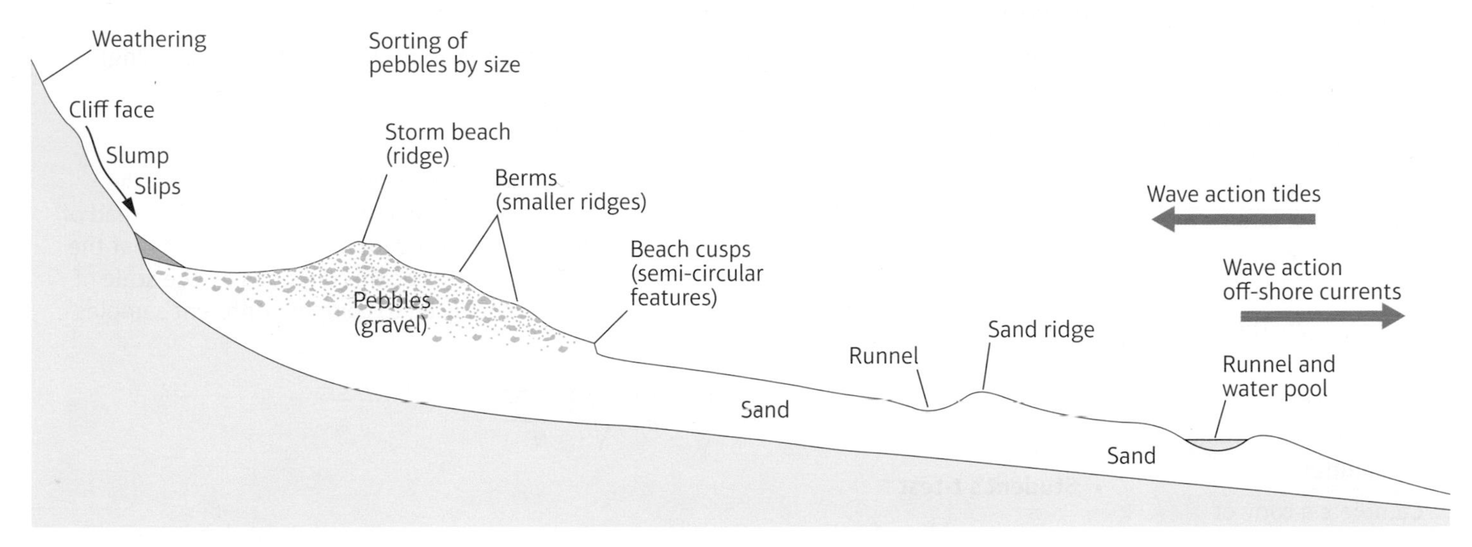

Figure 3.15: The features of a beach

An offshore bar is a long ridge of sand and pebbles found a short distance out to sea. It forms in shallow water, where destructive waves break before reaching the beach. These waves scour the seabed and throw material forward into a heap. Currents in the sea will also supply sediments for this deposition process.

Offshore bars are sometimes exposed at low tide, but when submerged at high tide can be a hazard to shipping. These shallow areas have been recently used as sites for wind farms, such as Scroby Sands in Norfolk off the East Anglian coast. They are also sometimes used to provide sand for beach nourishment, or dredged for building material. In the natural world, seals often use offshore bars for basking.

Barrier beaches are common on the east coast of the USA, such as Carolina, where there is a plentiful supply of sand, shallow nearshore and offshore areas, waves and winds with enough energy to move sand, and a rising sea level so that waves move the sand towards the shore. Sea levels have risen since the end of the last ice age and have accelerated recently with thermal expansion due to **climate change**, so conditions have become suitable for barrier beach formation.

In the UK, Chesil Beach in Dorset started as an offshore bar but was moved towards land by rising sea levels to become a barrier beach. Large barrier beaches form islands, which protect the mainland from storms. The sheltered water area behind them is a low-energy environment where salt marshes and other natural ecosystems may develop. Barrier beaches are dynamic (always changing) but this has not prevented some people from building houses on them, which creates a high hazard risk.

Nearshore bars are similar to barrier beaches, but smaller. They commonly form in the surf zone where storm waves break, scooping up sediments and adding them to onshore transport to pile them up in a long ridge, parallel to the coast. Bars may be sinuous or even attached to the shore in some places, and are gently shaped under swell wave conditions.

ACTIVITY

TECHNOLOGY/ICT SKILLS

Search online for satellite images of each coastal deposition feature.
Save these images and annotate their features and processes, using the information provided in these pages to help you.

Extension 

Investigate Chesil Beach and the theories behind its formation, and the contemporary threats to this depositional feature.

ACTIVITY

A student investigating sediment size wants to find out if there is a difference between the pebble sizes on the upper (site 1) lower beaches (site 2) (data is shown in Table 3.2).

Use the Student's *t*-test to statistically analyse the pebble size data to determine if there is a significant difference between the pebble sizes (calculations to two decimal places).

1. **a.** State the null and alternative hypotheses.

   b. Complete a copy of Table 3.2 to show the mean pebble size for site 2 ($\bar{x}_2$) (column 4).

   c. Complete a copy of Table 3.2 with the missing data in columns 2–6.

   d. Calculate the standard deviation for site 1 (s_1) (the calculation for site 2 has been done).

   e. Using the formula, complete the Student's *t*-test for this data.

The value of **t** may be either positive or negative. If it is negative, the minus sign is ignored when **t** is compared with the critical values in the relevant significance table. The degree of freedom is 18 – 2 = 16. Using a pre-prepared table of critical values, this means that the critical value in this case at the 0.05 significance level is **2.12**. If the value of **t** is greater than this, the null hypothesis can be rejected.

2. **a.** Is the result of the *t*-test statistically significant?

   b. Using this result, describe and explain the pebble size variations on this beach.

Maths tip

- **Analysis of central tendency and dispersion**

It is possible to compare coastal sediment samples from different locations by calculating the central tendency using mean, median or mode, and measures of dispersion such as standard deviation.

a. The *mean* is the total of the values of all observations (Σx) divided by the total number of observations (n).

b. The *median* is the middle value when all values are placed in ascending (or descending) order.

c. The *mode* is the value that occurs most frequently.

d. *Standard deviation* is a measure of dispersion, or how much each value differs from the mean. A large number for the standard deviation usually means that there is a wide spread of values around the mean, whereas a small number for the standard deviation implies that the values are grouped closely together around the mean. However, this depends on the scale of the values in the sample. Standard deviation can be best used to compare different samples.

The formula for standard deviation is

$$\sigma = \sqrt{\frac{\Sigma(x-\bar{x})^2}{N}}$$

- **Student's *t*-test**

It is also possible to further test the difference between two samples using Student's *t*-test. The null hypothesis is always that there is no difference between samples. The alternative hypothesis is that there is a significant difference between the sample means.

The formula for Student's *t*-test is

$$t = \frac{\bar{x}_1 - \bar{x}_2}{\sqrt{\frac{s_1^2}{n_1} + \frac{s_2^2}{n_2}}}$$

where:

$\bar{x}_1$ and $\bar{x}_2$ = the means of each sample (use only the positive value of the difference)

s_1 and s_2 = the standard deviations of each sample

n_1 and n_2 = the number of values in each sample.

Student's *t*-test is suitable for small sample sizes. In statistical tests, geographers usually work to the 95% (or 0.05) confidence level. For the *t*-test the 'degrees of freedom' need to be calculated: this is the total sample size minus 2.

Spits form when:

- there is a dominant main longshore drift direction
- there are plenty of sediments from mass movement and erosion
- there is a gap in the coastline, such as an estuary or bay.

Longshore drift moves sediment (pebbles and sand) along a coast over a long period of time. When the sediments reach a gap in the coastline, they are carried for a short while in the same direction until they are deposited on the seabed. Over time, so much of this material is deposited that it breaks through the surface of the sea to form a narrow strip of land across part of the bay or estuary. This strip of land is called a spit (Figure 3.16). It shelters the area of seawater behind it so that sediments from the sea and from rivers are deposited in the calm area. Eventually, salt-tolerant plants grow in this area to form mudflats and a salt marsh (see page 124). As the spit grows longer, the tides, river currents and other wave directions turn the end of the spit into a hook (recurved spit). Sometimes the remains of old hooks are visible, representing the growth and former positions of the spit.

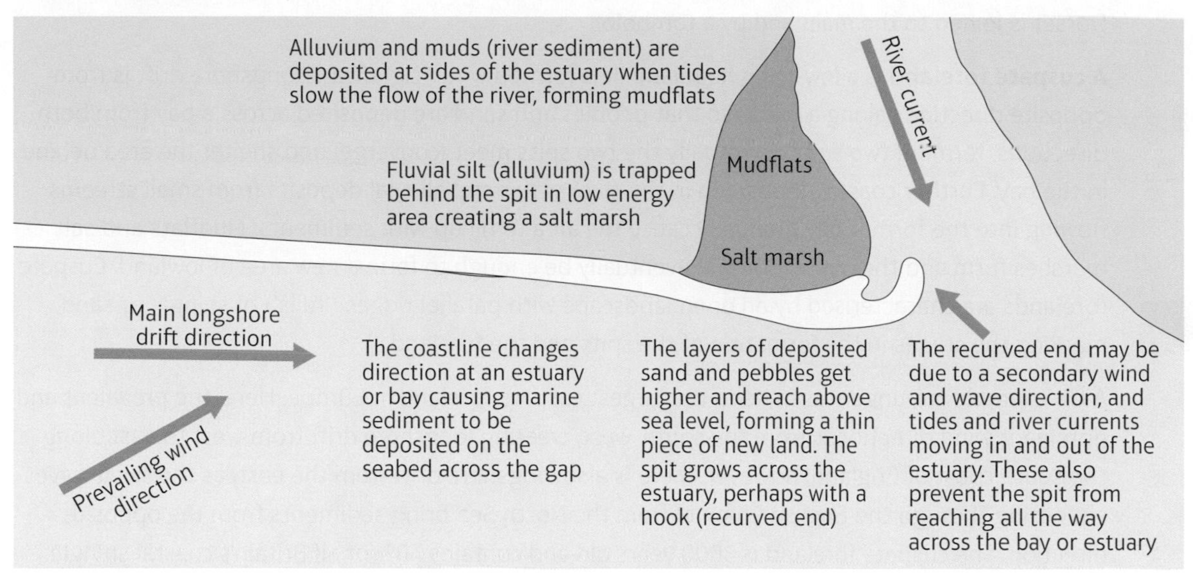

Figure 3.16: The formation of a spit

A double spit may form when there are local variations in longshore drift and a large bay with strong river currents. One example is at Poole Harbour in Dorset, comprising the Studland peninsula, formed by the dominant west-to-east longshore drift of the south coast of England, and Sandbanks, formed by a short section of east-to-west longshore drift caused by wave refraction around Durlston Head (see Figure 3.4). These two spits have not joined to form a bar or cuspate foreland, because several rivers discharge their flows into the sea through Poole Harbour and out between the spits.

Table 3.2: Pebble size data for two sites on a beach in the UK

Column 1 Site 1: pebble long axis in mm (x_1)	Column 2 ($x_1 - \bar{x}_1$)	Column 3 $(x_1 - \bar{x}_1)^2$	Column 4 Site 2: pebble long axis in mm (x_2)	Column 5 ($x_2 - \bar{x}_2$)	Column 6 $(x_2 - \bar{x}_2)^2$
83	17.67	312.23	28	−4.89	23.91
67	1.67	2.79	39		37.33
71		32.15	22		118.59
46			27	−5.89	34.69
62	−3.33	11.09	45	12.11	
49	−16.33	266.67	51	18.11	327.97
76	10.67		31	−1.89	3.57
65	−0.33	0.11	29	−3.89	15.13
69	3.67	13.47	24	−8.89	
$\Sigma x_1 = 588$		$\Sigma(x_1 - \bar{x}_1)^2 = 1126.01$	$\Sigma x_2 = 296$		$\Sigma(x_2 - \bar{x}_2)^2 = 786.87$
$n_1 = 9$		$s_1 =$	$n_2 = 9$		$s_2 = 9.35$
$\bar{x}_1 = 65.33$			$\bar{x}_2 =$		

A tombolo forms after longshore drift carries sediments across a gap between the mainland and an island, forming a narrow low ridge of sand and pebbles. It may be a spit at first, but when the deposition links the island and mainland together it is known as a tombolo. The Isle of Portland in Dorset is joined to the mainland by a tombolo.

A cuspate foreland is a low-lying headland. It is formed when significant longshore drift is from opposite directions along a coast, so that pebbles and sand are deposited across a bay from both directions, forming two spits. Eventually the two spits meet (converge) and shelter the area behind in the bay. Further coastal deposition in the slack water and alluvial deposits from small streams flowing into the former bay gradually cause the area to fill up with sediments. Mudflats and salt marshes form and the deposition will eventually be enough to form a new area of lowland. Cuspate forelands are characterised by an open landscape with parallel ridges ('fulls') of shingle or sand, marking the stages in the formation of the spits and the foreland.

A UK example is Dungeness in Kent, the largest cuspate foreland in Europe. Here, the prevalent and dominant wind direction is from the south-west, creating longshore drift from west to east along the south coast of England; however, there is also longshore drift from the east, as the swell waves squeezing through the English Channel from the North Sea bring sediments from the opposite direction. The cuspate foreland is 3000 years old and contains 40% of all Britain's coastal shingle. The flint pebbles mostly originated from the Sussex chalk cliffs. Deposition continues to take place from the west today but has reduced from the east, perhaps due to coastal management to the north. The point of greatest deposition is at the nose of the ness, where wave refraction reduces the wave energy.

ACTIVITY

1. Explain the role of sediment transport in the formation of depositional coastal features.

2. With reference to two depositional coastal landforms, explain the role of vegetation in stabilising coastal depositional features.

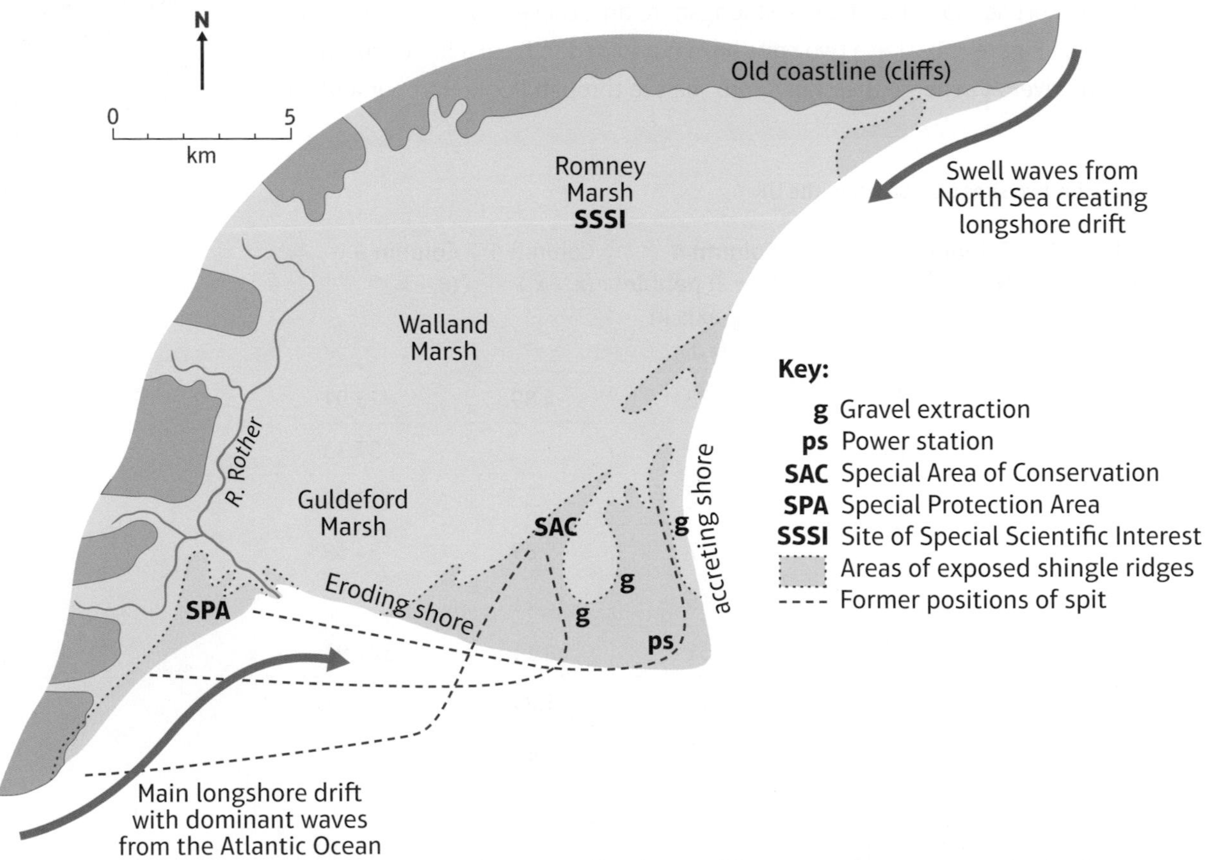

Figure 3.17: Sketch map of Dungeness, Kent, a cuspate foreland

The sediment cell concept

Along a section of coastline, erosion, transport and deposition processes operate in a linked system. A **sediment cell** (or littoral cell) has sources, transfers and sinks which are important in understanding the coast as a system (Figure 3.18). Each cell can be regarded as a closed system when there are large barriers between them, with little transfer from one cell to another. However, in reality under certain conditions transfers between the main sediment cells will take place. In theory, erosion in one place, called the source, will be balanced by deposition in another place, called a sink, and longshore drift and currents transfer sediment within the cell (Table 3.3). The amount of sediment gained from sources and lost to sinks can be quantified and a sediment budget calculated. These calculations can be useful in assessing coastal change and the effects of coastal management schemes. The coastline of England and Wales is divided into 11 primary coastal cells, with sub-cells within each primary cell.

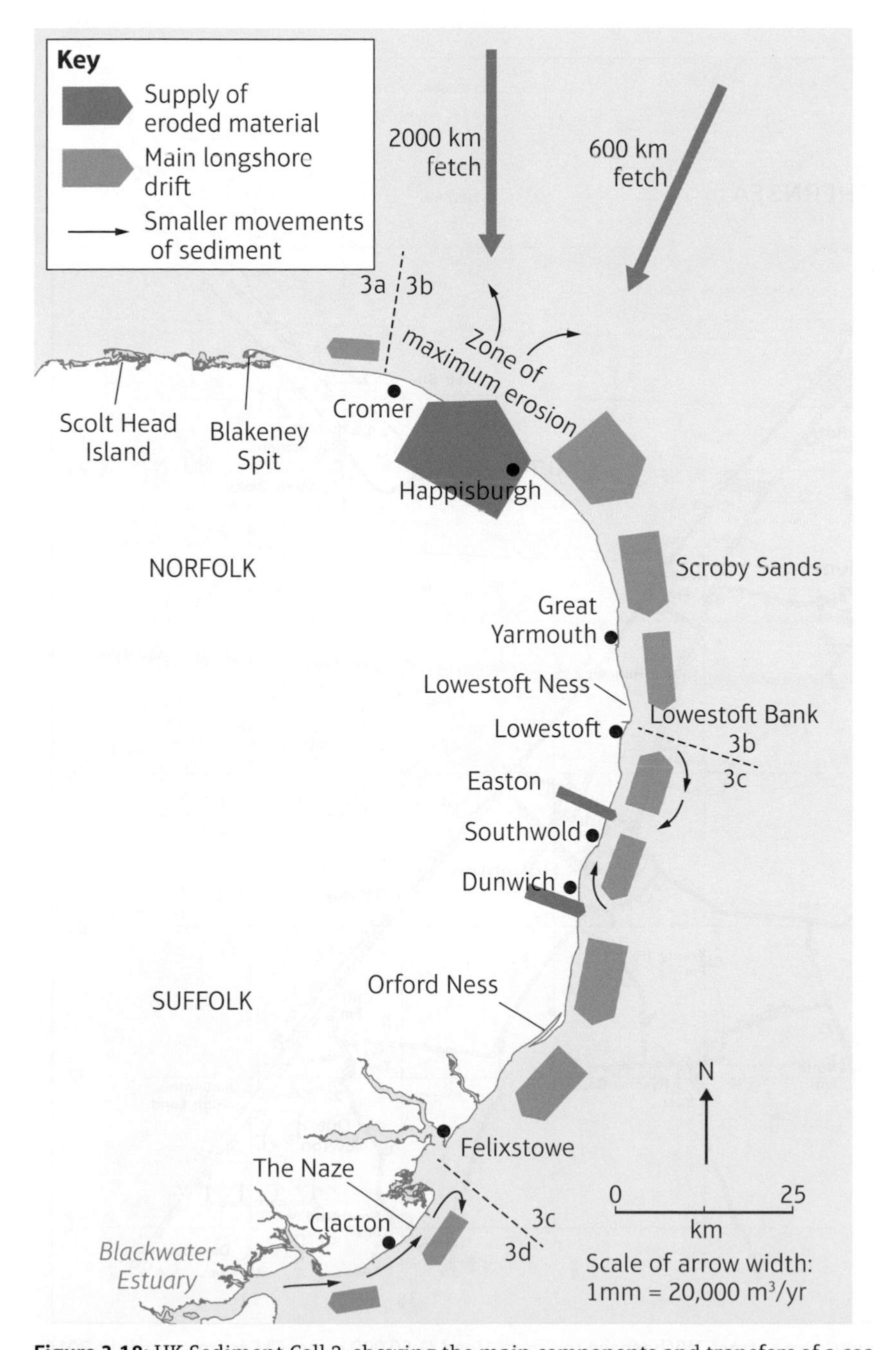

Figure 3.18: UK Sediment Cell 3, showing the main components and transfers of a coastal sediment system

Literacy tip

When using geographical terms such as erosion, transport and deposition in your answers to questions, think carefully about adding extra detail about what is actually happening – don't just make a broad general statement. For example, state the specific erosional process.

ACTIVITY

Using Figure 3.18, describe and explain what is happening along the East Anglian coast.

Table 3.3: Sediment cell sources, transfers and sinks

Sources	Transfers	Sinks
Erosion of cliffs	Longshore drift	Backshore depositional landforms (e.g. sand dunes)
Onshore currents supplying sediments to the shore	Wave transport through swash and backwash	Foreshore depositional landforms (e.g. beaches)
Land sediments eroded by rivers	Tides moving sediments in and out	Nearshore depositional landforms (e.g. bars)
Wind-blown (aeolian) sediments from land	Currents – localised (e.g. rip) or large scale	Offshore sediment deposition to deep offshore waters (e.g. through undersea canyons)
Subaerial processes (weathering and mass movement)	Wind along shore or on- and offshore	
Shells and remains of marine organisms		

ACTIVITY

CARTOGRAPHIC SKILLS

Study the OS map (Figure 3.19), showing the Withernsea to Spurn Head coastline:

1. Using names and six-figure grid references, identify and describe some of the human uses of the coastline.
2. Identify the designated natural areas along this coastline, using names and four-figure grid references.
3. Identify and describe the map evidence for erosion and deposition along this coastline; include references to processes and features.
4. Identify and locate any evidence of coastal management along this coastline.
5. Compare the map evidence with Figure 3.20, which shows the wider sediment cell. What are the implications for this coastline, given the sources, transfers and sinks?

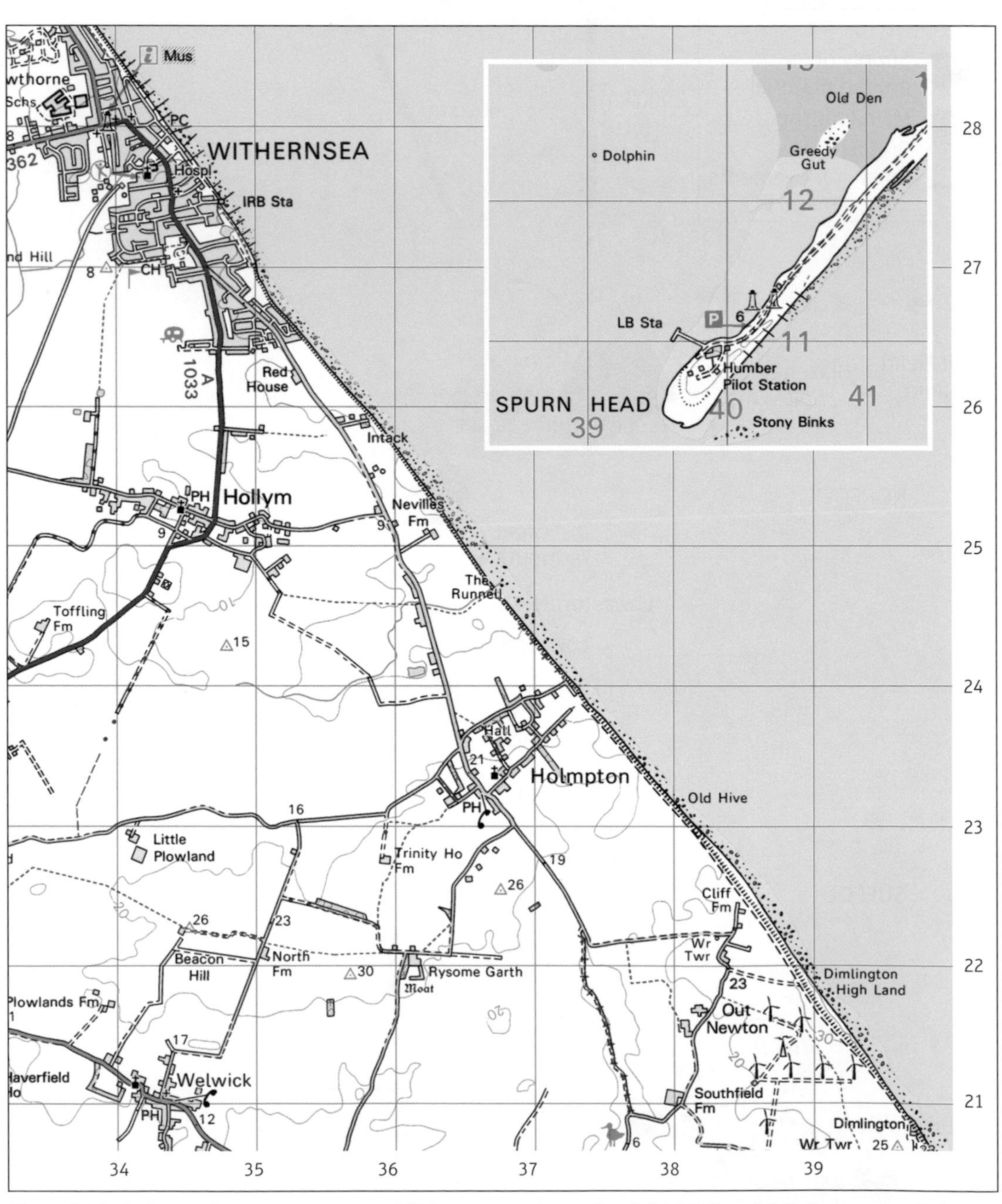

Figure 3.19: Ordnance Survey map extract, Withernsea to Spurn Head (1:50,000 scale) © Crown copyright 2016 OS 100030901

Within sediment cells there may be **negative feedback**, which usually helps to maintain a balance within the system, or **positive feedback**, which tends to change the balance until a new equilibrium is reached. An example of negative feedback is where wave erosion causes rock falls, which then protect the base of a cliff from further erosion. An example of positive feedback is damage to sand dunes during storm conditions, which may create a 'blow-out', allowing the wind to move more sand away, preventing grasses from regrowing – which allows further erosion to occur.

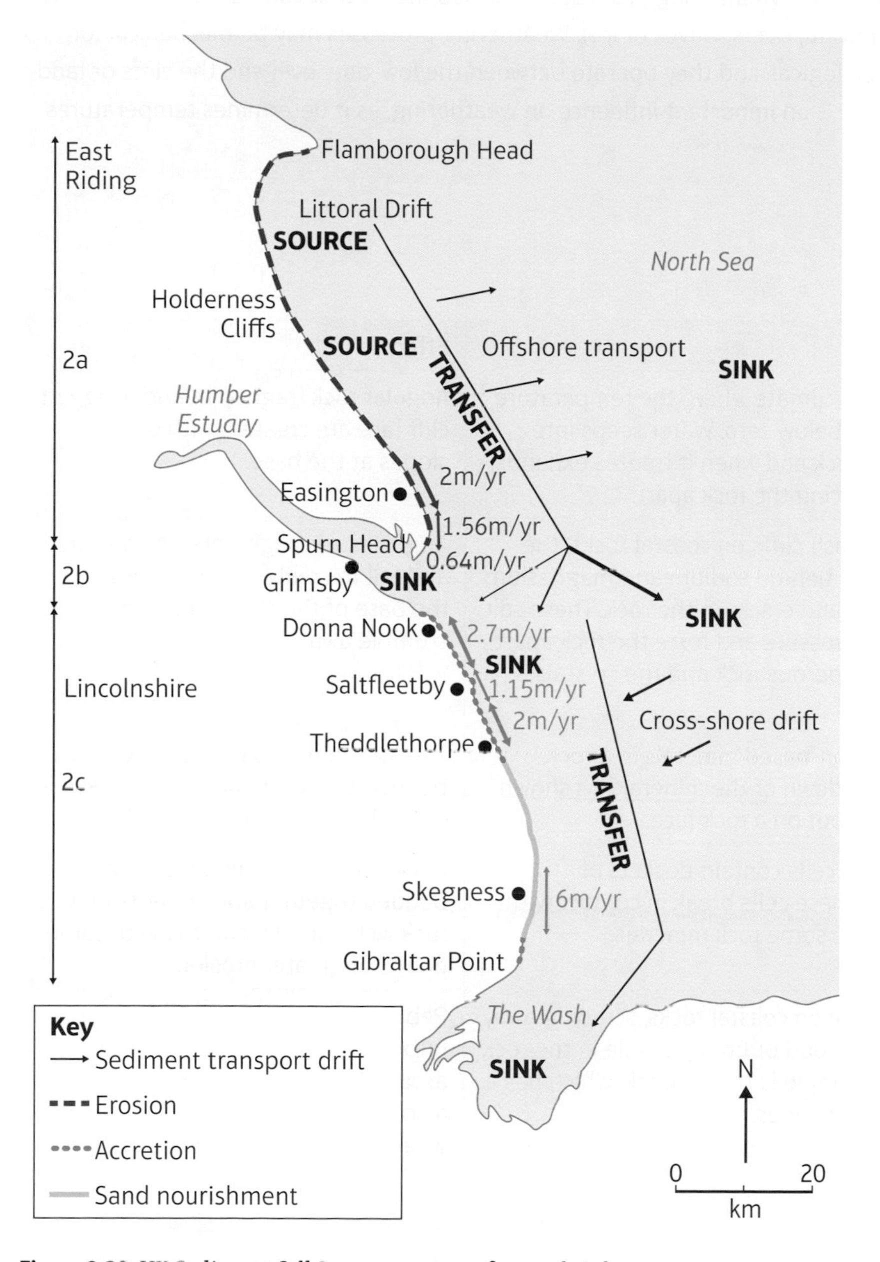

Figure 3.20: UK Sediment Cell 2 – sources, transfers and sinks

Synoptic link

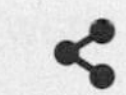

The last ice age played an important part in the geology of the British Isles. At the coast, ice sheets deposited thick layers of unconsolidated glacial till (e.g. in East Anglia, and at Holderness in Yorkshire). These deposits cannot resist coastal erosion processes and have rapid rates of retreat, to the extent that many settlements have been lost over hundreds of years. Natural climate change has also changed sea levels, creating new coastlines. Coastal communities are under threat, especially where the coast is not protected, adding to the diversity of places in the UK. (See page 96.)

Literacy tip 

Be careful not to muddle the terms erosion and weathering. Weathering is the breakdown of rock at the location where it is found (*in situ*), and does not involve any movement, while erosion is the wearing away of the rock or wider landscape, for example by moving seawater (wave action).

Subaerial processes and coastal landforms

Subaerial processes are a combination of weathering and mass movement processes that alter the shape of a coastline, as distinct from the marine erosion, transport and deposition processes.

Weathering processes at the coast

Weathering is the breakdown of rocks *in situ*, at or near the Earth's surface, exploiting weaknesses in rocks over a long timescale. Weathering produces loose sediments that can be moved by gravity or removed by erosion (wind, rivers, waves or ice). Weathering processes may be mechanical (or physical), chemical or biological, and they operate between the low-tide level and the cliffs or land of the backshore. Climate is an important influence on weathering, as it determines temperatures and moisture levels.

Table 3.4: Examples of coastal weathering processes

Type of weathering	How it works	Effects
Frost-shattering (or freeze-thaw) (mechanical type)	Only found on coasts in a climate where the temperature changes daily above and below zero. Water seeps into joints and cracks in the rock and when it freezes expands, exerting pressure and forcing the rock apart.	Angular rock fragments and a jagged cliff face are created, with scree slopes at the base.
Salt crystallisation (mechanical type)	When waves break or splash cliffs on coastal rocks, the water evaporates, leaving behind sodium and magnesium salt compounds in joints and cracks in the rock. These salt crystals grow and exert pressure and force the rock apart. Seawater may also enter porous rock and the crystals grow inside the rock itself.	Angular rock fragments are loosened and fall to create scree slopes at the base of the cliff, or rock faces crumble away.
Oxidation (chemical type)	Oxygen combines with iron-based minerals in a rock, causing a chemical breakdown of the minerals, as shown by a red-orange rusty colour on a rock face.	The rock minerals will no longer be bonded together and so the rock will crumble, making erosion easier.
Seaweed acids (biological type)	Some seaweed (e.g. kelp) cells contain pockets of sulphuric acid, so when these cells break in contact with rock, the acid will dissolve some rock minerals.	Rock minerals will no longer be bonded together and so parts of a rock will crumble, and these become points of greater erosion.
Boring molluscs (biological type)	Many marine molluscs live on coastal rocks, scraping away at the rock surface to get food or boring a hole in the rock to make a home. One example is the Piddock, which has a shell with serrated cutting edges.	Pebbles and rocks with holes bored into them are more easily moved around by the waves. The holes also provide weak points for other weathering processes to act.

ACTIVITY

1. Why is weathering important to the supply of sand and rock fragments in a sediment cell?
2. How may weathering influence rates of cliff recession?

Mass movement at the coast

Mass movement is the movement downslope of rocks, sand, clay, glacial till or soil. It is caused by gravity once a slope has become unstable, after waves have undercut resistant rocks or when rainwater enters unconsolidated rocks and forces particles apart (pore pressure) so that they no longer cohere. Water content determines the type of mass movement (Figure 3.21). Slope angle is also an important factor in determining stability, although, if not held together by roots of plants, sand and soil can move on quite shallow slopes.

Landforms of coastal mass movement

Surveys of Happisburgh in North Norfolk showed that in the early part of the 21st century, erosion rates were greater than 10 m per year in unprotected places, with 36,000 tonnes of sediment being eroded from a 200-m section of the cliffs made of glacial deposits. Landslide is a generic term for the mass movement of rock, earth or debris down a slope, and includes falls, toppling, sliding and slumping, and flows. A coast may have a combination of these, especially if different rock types are present at the same location.

Rock or block falls occur on steep slopes as a cliff face is weathered, which loosens blocks, and when wave erosion has created a wave-cut notch so that a section of the cliff is no longer supported. Rock fragments fall to the base of the slope and form talus scree slopes. These steep, fan-shaped mounds of angular material have larger boulders at their core and smaller material on top. The slope angle of the talus scree is usually between 34 and 40 degrees, depending on the size of fragments (larger fragments have a steeper angle of rest, due to friction between them). Wave processes will then work on the talus scree, gradually reducing it in size until it can be transported away.

In unconsolidated sands and clays, mass movement occurs in the form of rotational slumps (see Figure 3.5), slips or mudflows. These forms of mass movement occur along the Holderness coast, the North Norfolk coast and Lyme Bay in Dorset. Rotational slumping is where a section of a cliff remains intact as it moves down a cliff along a curved slip plane; this will leave a crescent-shaped rotational scar above it on the cliff. The vegetation layer is usually intact on top of the slump. A sequence of slumps will create benches or terraces in the cliff profile (Figure 3.21). A slip is more common where there is drier material, such as unconsolidated sands, where the cliff material tumbles to create a talus slope. Where there is plentiful water and a higher clay content, mud flows may result, which spill out over the foreshore as a lobe.

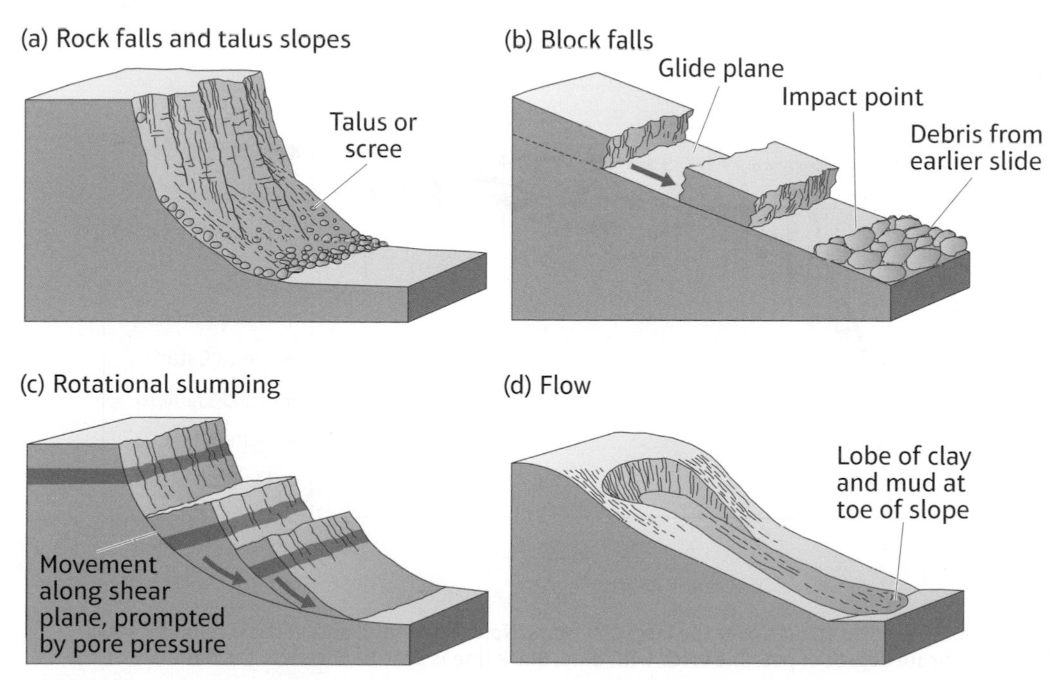

Figure 3.21: Types of mass movement – rock falls, block slides, rotational slumps and flows

ACTIVITY

TECHNOLOGY/ICT SKILLS

1. Research websites for information on the types of mass movement found at (a) Black Ven, Dorset, (b) Dunwich, Suffolk, and (c) Overstrand, Norfolk. Record your information on annotated photographs of these three coastal locations.

2. Visit the British Geological Survey website and investigate a block-fall landslide case study. Make notes to add to your study file.

AS level exam-style question

Explain why coastal processes may vary from day to day. (6 marks)

Guidance

This question is about short-term changes and so will be mainly related to wave action. Recall wave characteristics and explore how, as these change, so do the processes (erosion, transport and deposition).

A level exam-style question

Explain the formation of a cuspate foreland. (6 marks)

Guidance

Be careful not just to describe what the feature is like with factual detail of a named example. The question is asking about the processes and combination of circumstances necessary for a cuspate foreland to form.

CHAPTER 3

How do coastal erosion and sea-level change alter the physical characteristics of coastlines and increase risks?

Learning objectives

2B.7 To understand how sea-level change influences coasts on different timescales.

2B.8 To understand how rapid coastal retreat causes threats to people at the coast.

2B.9 To understand why coastal flooding is a significant and increasing risk for some coastlines.

Sea-level change

Over long periods, sea levels around the world have changed significantly in relation to the land (Figure 3.22). This is largely linked to the ice age and interglacial sequence – with lower sea levels during an ice age, because more liquid water has turned into ice, and higher sea levels when the Earth's climate is warmer, because the ice melts and water returns to the seas and oceans. The world is currently in an interglacial period and so sea levels would be expected to be higher. However, there is a range of influences that include **eustatic**, **isostatic**, tectonic and enhanced climate change.

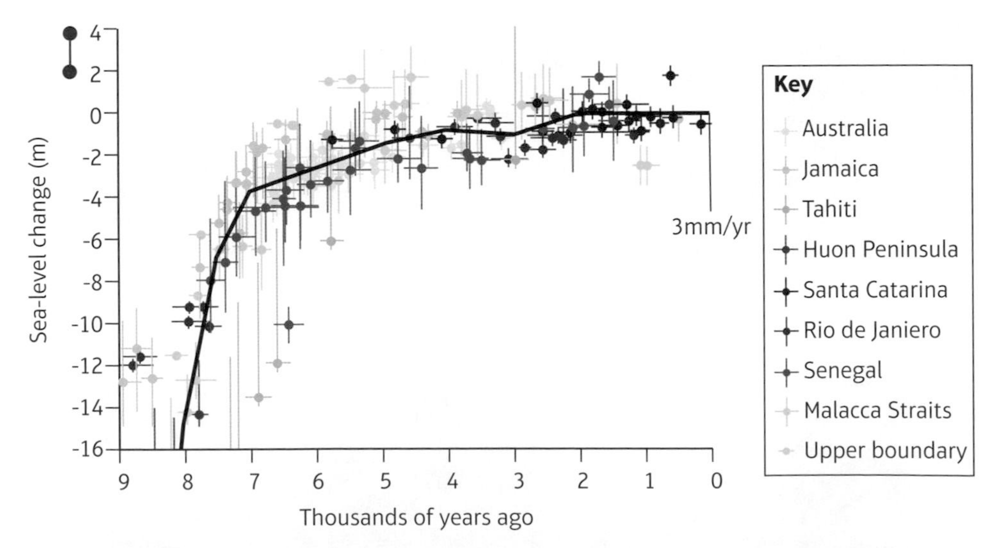

Figure 3.22: Global sea-level change over the last 9000 years. Note: Horizontal and vertical coloured lines through plots indicate degree of possible error with data. Black line is best-fit based on the data.

Eustatic, isostatic and tectonic change

Eustatic change refers to the change in sea level (up or down). During an ice age, much of the world's water is stored as ice in ice sheets, ice caps, glaciers and frozen ground. For example, the ice sheet in Antarctica is 4 km thick in places today. Consequently, sea levels fall, which is why during

Yellow = Players, Orange = Attitudes and actions, Purple = Futures and uncertainties

the last ice age there was no North Sea, and the UK was still joined to the rest of Europe. As the climate warms up, the ice on land melts and returns water to the sea. It is estimated that sea levels have risen about 120 m since the end of the last ice age 10,000 years ago; if Antarctica completely melted today, sea levels could rise a further 50 m (remember that melting sea ice does not count in this change, as it already displaces its volume).

Isostatic change refers to a change in land level, and is also a process linked to ice ages. When ice is kilometres thick, it is very heavy, and able to push the land downwards, because the upper mantle underneath the crust is a soft, viscous fluid. Land areas near or under ice sheets would thus have been depressed, but areas farther away would have been tilted upwards slightly. For example, in the UK, Scotland was pushed down by the weight of ice, while southern England, which was ice-free, rose slightly. When ice sheets melt at the end of an ice age, the land that has been pushed

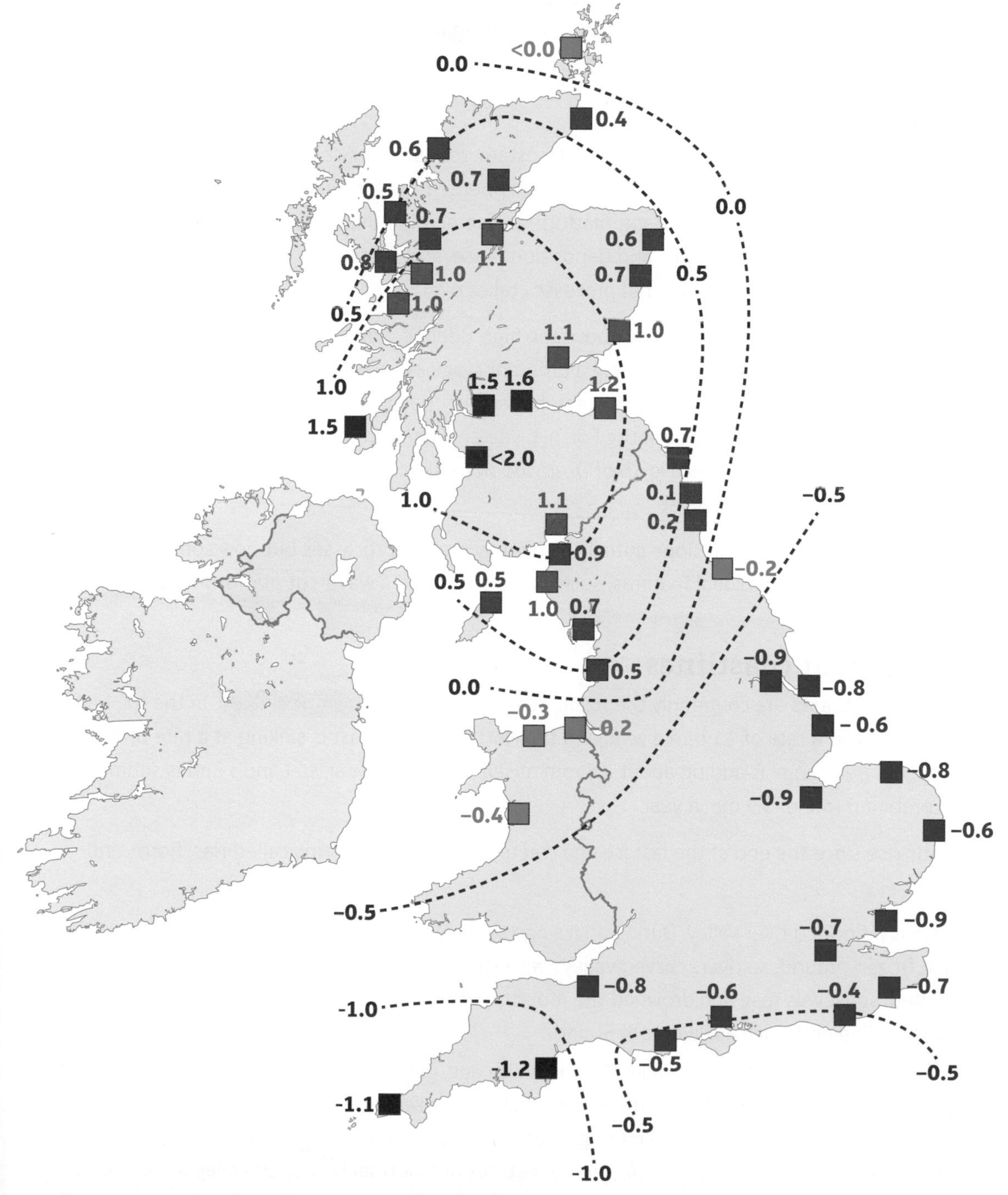

Figure 3.23: Annual isostatic change in the UK, in millimetres

Extension

Investigate Milankovitch cycles and how they would affect sea levels and coastal change.

Synoptic link

Climate change on a long timescale, as shown by Milankovitch cycles, affects the alternating pattern of ice ages and interglacials. Climate change also affects the processes found on the Earth's surface, such as moving ice and changing sea levels, which affect where waves will meet the land as well as how far rivers have to flow to meet the sea. Milankovitch cycles are also important to an understanding of present-day climate change and global warming. (See page 71.)

ACTIVITY

Consider Figures 3.22 and 3.23 and the other information in this chapter (pages 140–144). Are the majority of the world's coastlines advancing or retreating? Justify your answer.

Synoptic link

If the mantle layer did not behave as a fluid, there would be no isostatic change, because the crust could not have been pushed down by the weight of the thick ice sheets. For it to be a fluid, the mantle needs the immense heat from the Earth's core. (See page 22.)

down rebounds back upwards, while the land that has been tilted upwards slightly starts to sink (Figure 3.23). The UK, with its 500-odd miles north to south, is still showing isostatic readjustment, with Scotland rising and southern and eastern England sinking. However, those southern coasts may experience accretion of sedimentary material in the form of sand dunes, spits, bars, mudflats and salt marshes because of the creation of low-energy areas.

Tectonic change may have a direct effect on the shape of a coast and coastal processes. For example, as tectonic plates move and collide, some continental shelves and areas of land are pushed upwards, such as the Kerala coast of India. Other areas may sink; for example tectonic subsidence of up to 60 cm can be found in the Bakar-Vinodol area of the Croatian coast. Volcanic islands may also form to create new coastlines such as island arcs, or hotspot locations such as the Caribbean islands and Hawaii. There is also a great risk from tsunamis, particularly for coastal areas around the Pacific and Indian Oceans where subduction of tectonic plates causes sudden thrust or downward movements of the seabed. With increasing numbers of people living by the coast in these regions, human vulnerability as well as natural processes create risks.

Emergent coastlines

Emergent coasts are commonly the result of isostatic rebound. In Scandinavia, isostatic uplift is still 2 cm per year today, and in Scotland there is evidence of seven sets of raised beaches and fossil (relict) cliffs (some with wave-cut platforms) several miles inland. Once removed from marine processes, wave erosion, transport and deposition cease, and shaping by **terrestrial processes** such as wind, rain, rivers and subaerial processes takes over.

A **raised beach** is a former beach now above the high tideline. Some raised beaches may consist of several different levels, each indicating a different stage of uplift. Features such as rounded pebbles and boulders are likely to be present, but smaller particles have usually been removed. Current rates of isostatic rebound in the Forth, Clyde and Tay valleys in eastern Scotland are between 1.8 and 2 mm per year. North of Drumadoon on the Isle of Arran, a raised beach is now about 5 m above current sea level.

A **fossil cliff** is a near-vertical slope initially formed by marine processes but now some distance inland. Other coastal erosional features, such as sea caves and a wave-cut platform, may still be visible.

Submergent coastlines

Submergent coasts are commonly the result of sea-level rise or isostatic sinking. In the UK, Land's End is sinking at a rate of 1.1 mm a year, and the North Norfolk coast is sinking at a rate of 0.9 mm a year. Climate change is adding about 2.8 mm of sea-level rise a year, so Land's End is submerging at a combined rate of 3.9 mm a year.

Eustatic rise since the end of the last ice age created drowned landforms called rias, fjords and fjärds:

- **A ria** is a flooded river valley. During an ice age some land areas were not covered with ice but had frozen ground, so rivers carved valleys with steeper sides than normal. Then, after the ice melted, sea levels rose and drowned the mouths of these valleys. There are several in south-western England, such as Plymouth Sound.
- **A fjord** is a flooded glaciated valley. During an ice age, glaciers eroded U-shaped valleys down to the coast of the time and then, after the ice melted, the sea level rose again and flooded into the valley over a shallow threshold, creating a very deep water inlet with steep sides, such as Milford Sound in New Zealand (Figure 3.24). All the features of a normal U-shaped valley are present, such as hanging valleys and truncated spurs.

Yellow = Players, Orange = Attitudes and actions, Purple = Futures and uncertainties

- **A fjärd** is a flooded inlet with low rocky banks on either side formed by post-glacial drowning of glaciated lowland rocky terrain, such as the Gulf of Finland. The drowned glacial lowlands in western Scotland are fjärds with small islands, called skerries.

Dalmatian coasts (see page 120) feature several linked parallel flooded valleys, with long islands between them. This is a result of submergence, when sea levels rose and flooded the valleys between mountain ridges parallel to the coast.

ACTIVITY

Explain how sea-level change creates both emergent and submergent coastlines over varying time scales.

Figure 3.24: The fjord of Milford Sound, New Zealand

Climate change and sea-level rise

When the Earth is at its warmest, most ice will have melted and the oceans will absorb heat and expand (thermal expansion); these processes combine to raise sea levels. The Intergovernmental Panel on Climate Change (IPCC) has estimated that sea levels in the last interglacial period were around 5 m higher than at present. When the Earth is at its coldest, ice will have grown to a maximum and the oceans will have cooled, so sea levels will be much lower. Since the last Ice age (Pleistocene) finished about 10,000 years ago, the Earth has entered a long warming period.

There is much concern about how humans have enhanced natural warming since the 19th century through the release of greenhouse gases into the atmosphere. The impact on sea level is clear: as greater warming melts glaciers, ice caps and ice sheets and thaws permafrost, more water is added to oceans, causing a sea-level rise. However, a more significant factor in sea-level rise is thermal expansion: as seawater heats up by absorbing heat from the atmosphere, its volume expands. The ocean has stored over 90 per cent of the increased heat energy of the climate system in recent decades, and IPCC research shows that sea levels are rising faster over time, with an average rate of 3.2 mm a year between 1993 and 2010, and a projected rate by the end of the 21st century of between 8 and 16 mm per year. If the ice sheet on Greenland were to completely melt, sea levels could rise relatively quickly by 7 m.

Table 3.5: Observed and predicted (from models) causes of sea-level rise, 1993 to 2010 (IPCC data 2013)

Source of rise	Observed share	Predicted share
Thermal expansion	39.3%	53.2%
Glaciers melting	27.1%	27.9%
Greenland melting	15.4%	5.0%
Antarctica melting	9.6%	-
Others	8.6%	13.9%

Extension

Study the latest IPCC reports and discover the latest predictions of sea-level rise, and the advice on how coastal areas could prepare and adapt to the change.

ACTIVITY

Using the knowledge gained from studying this chapter and your own investigations, suggest two locations on the UK coastline – one location with slow recession rates, and one location with rapid recession rates. Explain your selections.

Low-lying islands such as the Maldives will disappear, along with coastal ecosystems such as the Sundarbans in Bangladesh and India. IPCC research shows that wave heights will increase in the Arctic Ocean as sea ice melts, increasing wave erosion on Arctic coasts. Stormier seas are also a general climate change prediction, with stronger storms such as mid-latitude depressions creating larger, more destructive waves. These, combined with higher sea levels, make it easy to predict that erosion rates, and coastal change in general, will increase in the future.

Coastal retreat

Physical and human causes of coastal erosion

As shown earlier in the chapter, coastal retreat or recession is the result of natural factors such as the geology of the coast, marine processes and subaerial processes. The most rapid rates of coastal recession will occur where there are weakly consolidated rocks, large destructive waves, submergence of the coastline, large-scale mass movement and constant weathering processes. The slowest rates of recession will be where there are resistant rocks, smaller constructive waves, little mass movement and occasional or slow weathering processes.

As well as natural influences on coastal recession, human actions also sometimes alter the natural processes operating in a sediment cell, increasing rates of retreat. For example, coastal defences built at one location will stop or limit the supply of sediment to a cell, which means that another place down drift may not receive sediment for beach-building. When the supply is cut off, the beach gets narrower and is less able to absorb wave energy, and waves hit the backshore with more force. An example of this is at Dunwich, south (down drift) of Southwold on the Suffolk coast.

Offshore dredging may also increase coastal retreat. Dredging removes sand and gravel for construction purposes (under UK government licence), to deepen entrances to ports or to supply sediments for beach nourishment. Dredging off the North Norfolk coast has been blamed for increased erosion rates, as the supply of sediments to beaches has been altered. Deeper water caused by dredging may also allow waves to maintain their circular motion and energy closer inshore, and have a more destructive impact on the coast (see page 127). Elsewhere, stronger storm activity, linked to global warming, has badly eroded beaches, such as at Brisbane (Figure 3.25).

Synoptic link 

Economic development (globalisation), population growth and urbanisation processes may increase the human use of coasts. Ports (e.g. Felixstowe), coastal retirement settlements (e.g. Bournemouth) and coastal cities (e.g. the Thames Gateway, east of London) may need to expand, and sand and gravel may be removed for construction (e.g. offshore dredging in North Norfolk and Chesil Beach in Dorset). (See page 119.)

Figure 3.25: Beach erosion at Brisbane, Australia

Yellow = Players, Orange = Attitudes and actions, Purple = Futures and uncertainties

CASE STUDY: Human activities and coastal recession

Dredging of shallow seabeds can have a negative impact on the natural environment, such as removing benthic species and communities, increasing suspended sediment levels (turbidity), which can damage coral reefs (e.g. the Great Barrier Reef, Australia), and sediments settling and covering subtidal and intertidal marine communities. However, increased deposition may raise sediment levels such as in estuaries, and artificial deposition on beaches may offset sea-level rise or isostatic sinking. Local sea currents may be altered and waves may become larger, which may affect sediment balances and erosion rates. But evidence of impacts is difficult to verify, and much research is based on computer modelling rather than measured data.

Research by Marinet has concluded that offshore dredging between 6 and 19 km out to sea between Winterton-on-Sea and Lowestoft has resulted in beach loss, undermining of sand dunes and erosion of windfarm cables and supports at Scroby Sands. However, government departments and the British Marine Aggregate Association deny that dredging is the cause of these problems.

The Nile Delta

Multiple human activities and influences can be found concentrated in some coastal locations, such as the 240-km coast of the Nile Delta in the eastern Mediterranean. Here are holiday beach resorts, coastal defences, general tourism, marine recreation, fisheries, land reclamation, agriculture, settlements, and transport and port infrastructures. The delta area is where 95 per cent of Egypt's population live. This coastline is experiencing retreat, with significant erosion on half of it, such as the Rosetta promontory, with sediments moved eastwards. There is salt intrusion into the delta due to sea-level rise, and coastal flooding will become more frequent, both due to climate change. As sea levels rise and protective offshore bars are eroded, 3.3 per cent of the delta land area will be lost. The building of the Aswan Dam in 1964 reduced terrestrial sediment supplies.

A 2015 study classified 32.4 per cent of the Nile delta coast as highly vulnerable, with only 26 per cent as low vulnerability. If sea levels rise by 1 m by the end of the century and no action is taken, 2 million hectares of fertile land will be lost and at least 6 million people displaced, including about 30 per cent of the city of Alexandria.

The United Nations Development Programme (UNDP) has implemented an Integrated Coastal Zone Management (ICZM) project (2009–16) for the Nile Delta as part of Millennium Development Goal (MDG) 7 – 'ensure environmental stability'.

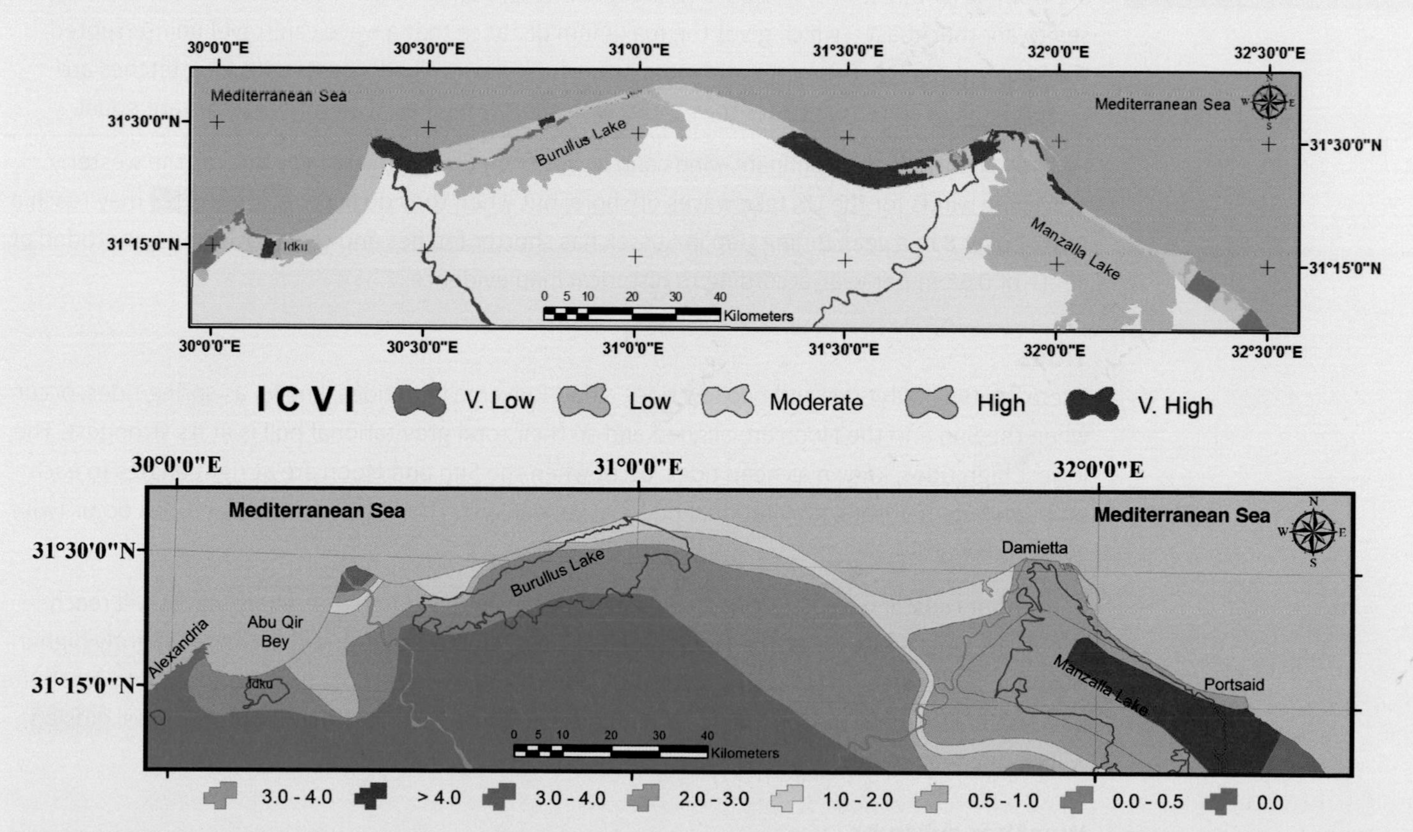

Figure 3.26: GIS images of the Nile Delta, Egypt, showing vulnerability to sea-level rise (top) and subsidence rates (bottom) (IVCI = Improved Coastal Vulnerability Index)

ACTIVITY

Outline the likely viewpoints of the following groups of people (players) towards coastal management along a stretch of coastline such as North Norfolk:

1. residents in a seaside resort with hard engineering defences
2. residents in a coastal village with old, failed defences
3. a dredging company
4. a large construction company
5. a wind energy company
6. a port authority
7. the RSPB
8. summer tourists
9. the national government (Defra and the Environment Agency).

Subaerial processes and coastal retreat

Both weathering and mass movement accelerate recession rates. Weathering weakens the rocks found at the coast and allows erosion rates to increase. Mass movement moves sediments to the base of coastal slopes where wave action and longshore drift can carry material away; this then exposes the base of coastal slopes, resulting in cliffs retreating further. Cliff failures in the chalk cliffs on the French side of the English Channel were found to be due to weathering rather than marine erosion. Rainwater has a direct impact on coastal slopes, such as by creating rills and then gullies in unconsolidated materials or at weak points in harder rock.

Mass movement is more strongly linked than wave action to seasonal climate changes. For example, at Overstrand in North Norfolk, a full range of hard engineering coastal defences is in place protecting the cliff base from marine erosion, but the cliff still retreats as a result of rain entering the rock layers, increasing pore pressure and causing slumping. This leaves clear scars and terraces (or benches) on the cliff face and lobes of clay across the promenade.

Factors affecting rates of recession

The factors affecting erosion and submergence change over time, so rates of recession (or retreat) will also change.

Wind direction

Wind direction changes daily, and since it determines wave direction, there is a complex pattern of wave activity and erosion. The dominant wind produces the largest waves. At Land's End in Cornwall it comes from the south-west, and on the North Norfolk coast it comes from the north (see Figure 3.18). Coasts will experience maximum erosion at these times, and less erosion when winds and waves are from other directions. This links to the fetch – the distance from a coast across the ocean or sea to another coast – which gives the maximum distance that a wave can travel uninterrupted; the longer the fetch, the larger and more powerful a wave will be. Coasts with long fetches are therefore likely to retreat faster than those with shorter fetches, if all other factors are equal.

For North Norfolk, the dominant winds and waves from the north are very rare, as the westerly prevailing winds for the UK take waves offshore, but when they do occur erosion rates may reach a rate of over 8 m a year. Birling Gap in Sussex has shorter fetches and chalk rock, and has eroded at a rate of 0.58 m per year, according to historical map evidence.

Tides

There are two high tides and two low tides a day. Extreme high tides, known as spring tides, occur when the Sun and the Moon are aligned and so their total gravitational pull is at its strongest. The lowest high tides, known as neap tides, occur when the Sun and Moon are at right angles to each other and so their total gravitational pull is at its weakest. These spring and neap tides occur twice every lunar cycle.

In terms of coastal retreat, tides are important because they determine where waves will reach the shore. Four times a year the Moon is at its closest to the Earth, and this creates slightly higher high tides, called 'perigean spring tides'. At high tides, waves are more likely to be able to reach the backshore and erode the land faster. If high tides coincide with destructive storm waves, erosion will be at a maximum.

Weather systems

In the northern Atlantic, and consequently the UK, there is a seasonal weather pattern based on a sequence of high-pressure areas (anticyclones) and low-pressure areas (depressions). High pressure brings calmer conditions and smaller waves. In contrast, low pressure has a mixture of air masses at different temperatures and air rises rapidly; consequently there are strong winds rotating in an anticlockwise direction, and these create larger waves. In winter, the difference between the temperature at the equator and the pole is at its greatest, which means that the depressions are at their strongest, with lower air pressure and faster wind speeds, which then create larger destructive waves and the fastest recession rates.

Global warming has added more heat to the atmosphere, which has intensified the low-pressure systems. This means that the UK now experiences even stronger winds and larger waves during these storm events (they are now named by the Met Office). Coastal erosion during winter is likely to increase, with recession being visible. The IPCC reports that a 1-cm rise in sea level will erode a beach by about 1 m horizontally as the balance of processes is altered.

Local and global influences on coastal flooding

Coastal flooding is of major concern around the world because so many people have settled by lowland coasts. By 2060, 12 per cent of the world's population will live in low elevation coastal zones (below 10 m). Flood risk increases in coastal zones that are experiencing isostatic sinking, as in Essex, and land shrinkage due to drainage and reclamation, as in the Netherlands. The Essex coast has many estuaries with mudflats and salt marshes and pasture land originally reclaimed from salt marsh and protected by sea walls; in order to manage the flood risk it was decided to retreat the coast by breaching sea walls and allowing high tides to flood areas, so returning them to a natural state. This created a buffer zone so that unexpected damage was limited. Half of the Netherlands, including cities and industrial centres, is below sea level. After the disastrous floods of 1953 the country completely protected its coast with barriers and embankments, and monitored and maintained its sand dunes and beaches.

Ecosystems such as salt marshes and mangrove forests are important for reducing flood risk, so their removal can increase risk significantly. For example, mangroves:

- reduce the height of waves by an average of 40 per cent within the first 100 m of forest, which becomes more significant at times of very high sea levels, reducing wave erosion and the distance reached inland
- stabilise sediments, trapping and adding to them, so keeping the level of coastal land higher
- reduce storm surge levels by 0.5 m for every 1 km of forest that seawater has to pass through.

In central Java, Indonesia, the coastline retreated due to flooding when mangroves were removed and soil subsidence followed.

Many large cities around the world are at or just above sea level. In developed countries there are defences against the sea, such as the Thames Barrier for London and the Maeslantkering Barrier for Rotterdam; New York is also planning barriers. But even the defences in developed countries may not be good enough, as experienced by New Orleans when Hurricane Katrina hit in 2005. Developing countries have greater difficulties building barriers because of the expense, and a sea-level rise of just 40 cm in the Bay of Bengal would submerge 11 per cent of Bangladesh's coastal land, creating 7 to 10 million environmental refugees. Island communities such as the Maldives, Tuvalu and Kiribati will cease to exist, and people will need to find new homes.

Storm surges

These events are linked to low-pressure weather systems, notably tropical cyclones and depressions (extra-tropical cyclones). When air pressure is low over the sea, the water is able to rise

ACTIVITY

Suggest which of the following factors has the strongest influence on the rate of coastal recession:

1. wind direction and fetch
2. tides
3. seasons
4. weather systems, or
5. storm activity. Produce a rank order, from strongest influence to weakest influence, and justify your suggestions.

Literacy tip

Expand your vocabulary through wider reading. Make sure you know the meaning of isostatic, eustatic, accretion, estuarine, temporal scale, morphology, dynamic equilibrium, positive feedback, negative feedback, environmental refugees, holistic management, alluvium, polder, jet stream, symbiotic algae, amenity, aesthetic and topography.

ACTIVITY

TECHNOLOGY/ICT SKILLS

Investigate sea-level rise in the coastal locations mentioned in the text, using Climate Central's 'Surging Seas Risk Zone Map'. Is there a similar pattern around the world?

upwards, forming a dome of seawater under the depression. The lower the air pressure, the higher the dome will be. This dome of water can then surge ashore when the low-pressure system moves near a coast. As tropical cyclones and depressions are also associated with strong winds, they create large storm waves that will appear on top of the dome. Collectively, this adds considerable height to sea levels and greatly increases the risk of coastal flooding and erosion. If these events coincide with an extreme high tide, there is potential for a disaster.

The shape of a coastline can also make the situation worse, if a storm surge is confined and funnelled into an area of shallow offshore water. Examples are the southern part of the North Sea, affecting countries such as the UK and the Netherlands, and the Bay of Bengal in the Indian Ocean, affecting countries such as India, Bangladesh and Myanmar.

CASE STUDY: Bangladesh and sea-level rise and storm surges

Bangladesh is the country most at risk from sea-level rise (between 6 and 20 mm a year), and from storm surges linked to tropical cyclones: 40 per cent of all recorded storm surges have occurred in Bangladesh. In 1970 a storm surge killed half a million people, and storm surges have killed about 1.3 million people in total since 1700. In the past, storm surges have reached 100 km inland, and with further sea-level rise this situation will get worse. It has been estimated that a 1.5 m rise in sea level would flood 22,000 km^2 and displace between 15 million and 17 million people.

With a higher sea level, the major rivers of Bangladesh will flow more slowly, ponding back upstream and increasing the risk of river flooding. Wetter environments by the sea and rivers will become breeding grounds for diseases such as cholera and malaria. The main port of Chittagong would be out of action, and loss of farmland due to direct flooding and salt contamination will reduce the country's GDP, as well as causing shortages of rice and vegetables leading to malnutrition and starvation. It has been estimated that 40 per cent of Bangladesh's farmland may be lost to the sea if there is a 65-cm rise in sea level by the 2080s. Around 20 million people already have their drinking water affected by salty seawater. The valuable coastal ecosystem of the Sundarbans (mangroves) would also be lost, reducing protection from coastal floods. High tide levels are also rising as the land subsides (between -8 mm and -18 mm a year) due to drainage, dredging and channelisation as well as natural compaction of the delta. In 1995, Bhola Island was submerged, displacing 500,000 people.

In 2007 Cyclone Sidr hit southwest Bangladesh as a category 4 tropical cyclone (240 km/h winds). The storm surge reached 10 m in places, and caused $1.7 billion in damage, mostly to housing. The coastal flooding affected up to 3 million households, and 2 million people lost their source of income, most of them among the poorest people in the country.

CASE STUDY: North Sea storm surges, 1953 and 2013

In January 1953 a night-time storm surge caused the deaths of 307 people in England and over 2100 in the Netherlands. Sea levels rose by more than 3 m and flooding occurred along the east coast from Yorkshire to Kent as extreme high tides coincided with the storm surge and large waves. Seawater flooded lowland areas, topped coastal defences and breached sand dunes. Almost 65,000 hectares of farmland and 20,000 homes were flooded, and 32,000 people evacuated. Damage was estimated at over £1.2 billion at today's prices. There was no flood-warning system, weather forecasts were basic, and modern communications did not exist. People went to sleep not knowing of the approaching danger.

In December 2013 the largest storm surge since 1953 took place, with sea levels higher in some places – 6.3 m at Blakeney, North Norfolk. Around 18,000 people were evacuated, but only 1400 properties were flooded. Although some flood defences were breached, the Environment Agency said 800,000 homes had been protected, and better forecasting had given people time to prepare and move to evacuation centres. Major impacts were avoided by the raising of the Thames Barrier to defend London and the closure of the floodgates on the Delta Scheme in the Netherlands. Localised flooding caused damage worth £1.7 billion, notably at Hemsby in Norfolk where seven homes and a lifeboat station were lost when sand dunes were eroded.

Extension

New York had a 'wake-up call' in October 2012 when a deep depression that had started life as a tropical cyclone (Sandy) moved onshore at the city.

- Investigate this event, focusing on the coastal flooding that took place.
- How is the city planning to ensure that the impacts of sea level rise and storm surges are reduced in the future?

CASE STUDY: Typhoon Haiyan, Philippines, 2013

Typhoon Haiyan, locally called Yolanda, was a category 5 (the highest) tropical cyclone with sustained winds recorded of up to 315 km per hour. It struck the Philippines in November 2013, leaving over 6200 dead and 28,000 injured. On Leyte and Samar islands the storm surge measured up to 7 m; Tacloban airport terminal building was destroyed by a 5.2-m storm surge and waves up to 4.6 m high. The low-lying areas on the eastern side of Tacloban city were hardest hit, flooding extended for 1 km inland and roughly 90 per cent of the city was destroyed and evacuation centres flooded.

Although wind speeds were extreme, the major cause of damage and loss of life was the storm surge. It turned the densely populated low-lying areas into wastelands of mud and debris. The World Health Organisation classed it as a category 3 disaster (the highest level): 14.1 million people were affected, with 4.1 million displaced from their homes and only 2.5 per cent of them accommodated in official evacuation centres. A year later the government had repaired only six of the 43 damaged ports, 213 out of 19,600 classrooms, and three out of 34 bridges, and in Tacloban fewer than 100 of 14,500 promised new permanent homes had been built.

Figure 3.27: Tacloban in the Philippines after Typhoon Haiyan, 2013

Extension 

Suggest why the predictions of sea-level rise by the IPCC and other reputable organisations have such a wide range of uncertainty.

Climate change and coastal flooding

When considering the effects of global warming, it is predicted that by the end of the 21st century depressions and tropical cyclones will have more energy and be stronger, with faster winds and lower air pressure (an intensity increase of up to 11 per cent by 2100). The additional heat energy causing this change is from the atmosphere and warmer oceans. It is unlikely that the number of tropical cyclones will increase, because of the combination of factors needed to form them, but it is still unclear how a faster polar front jet stream in temperate latitudes may affect depressions. Lower air pressures will cause higher doming of seawater, and higher wind speeds will create larger storm waves.

Although protection and preparation are improving in many areas of the world, the potential risk to people is increasing as more and more people live in and use lowland coasts – especially in southern, eastern and south-east Asia. In 2013 the IPCC reported that mean sea levels rose by 1.5 mm a year between 1901 and 1990, and by 3.2 mm a year between 1993 and 2010. IPCC prediction models suggest that mean sea levels will rise by between 0.45 m and 0.74 m by 2100, and will continue to rise for centuries. A government study in Australia concluded that a sea level that is higher by 10 cm will increase the frequency of extreme sea-level events by up to six times, and a 50-cm rise will increase frequency by between 100 and 10,000 times.

CASE STUDY: Kiribati and climate change

Situated in the middle of the Pacific Ocean, Kiribati is composed of 33 coral atolls. Most of the population live on Tarawa, where the maximum height above sea level is just 3 m. Sea-level rise and storm surges due to climate change have increased flooding, beach erosion and contamination of the limited supplies of fresh water.

In addition, the rise in sea temperatures due to climate change is causing 'bleaching' of coral reefs, the health of which is vital for the growth of the islands which are coral atolls. Bleaching is a process where the additional heat of the ocean produces stress in the coral polyps, causing them to expel their symbiotic (algae) zooxanthellae, so losing their food supply and colour (hence the term 'bleaching'). Bleaching has occurred in 1998, 2002, 2006, 2010 and 2015 in particular.

Not only are oceans warming on the surface – up to 3°C by 2100 – but also down to a considerable depth. Scientists estimate that the risk of flooding in Kiribati will increase by 200 times between 2000 and 2080. However, other scientists point out that if sea-level rise is slow enough, then coral growth may keep pace with the rise, depending on the level of bleaching and any 'tipping points' such as a sudden increase in sea-level rise due to the melting of the Greenland ice sheet. The Kiribati government has looked at buying 23 km² on Fiji's Vanua Levu Island, ready to relocate its population, who may become environmental refugees.

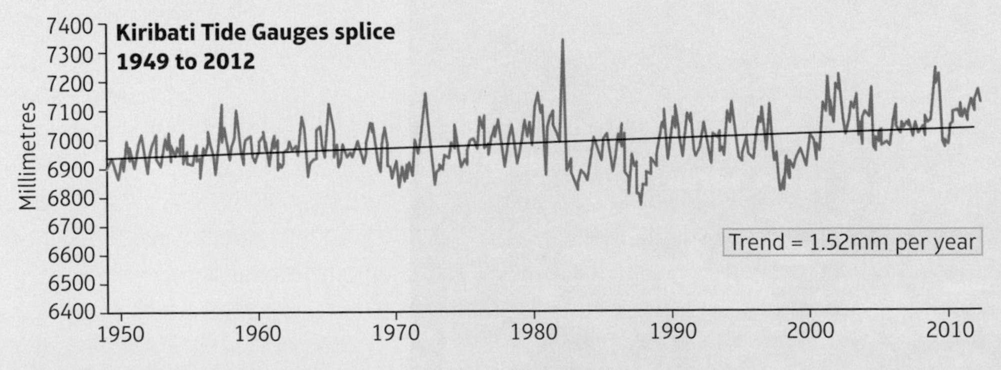

Figure 3.28: Graph showing tide levels in Kiribati, 1949 to 2012

Yellow = Players, Orange = Attitudes and actions, Purple = Futures and uncertainties

Figure 3.29: Flooding at high tide in Kiribati

AS level exam-style question

Explain how local factors may increase the risk of coastal flooding. (6 marks)

Guidance

The question states 'local factors', so be careful to stick to these, and not drift into global factors. Things to consider include topography (height of land), subsidence, removal of natural vegetation, and shape of coastline. Remember to assess – rating the importance of the local factors in a location.

A level exam-style question

Explain the impacts of storm surges on lowland coastal areas. (8 marks)

Guidance

In order to evaluate the impacts, you will need to use located examples. But do not spend most of your answer explaining what storm surges are; concentrate on the impacts, for example in the North Sea and the Bay of Bengal, and suggest which impacts are greatest and the reasons why.

CHAPTER 3

How can coastlines be managed to meet the needs of all players?

Learning objectives

2B.10 To understand that the increasing risk of coastal recession and flooding has serious consequences for affected communities.

2B.11 To understand the different approaches to managing the risks associated with coastal recession and flooding.

2B.12 To understand why coastlines are now increasingly managed by holistic integrated coastal zone management (ICZM).

Erosion and flood risk for coastal communities

Economic and social costs

The impacts of coastal flooding and coastal erosion can be measured in terms of economic costs – such as the costs of repairs and rebuilding (including insurance), loss of jobs and income, and the money and resources needed in the form of aid. Some of this burden falls on governments and international aid organisations (non-governmental organisations – NGOs), but there is still a considerable burden for individuals and families, especially if there is no insurance or compensation.

In the UK the government does not provide compensation to people who lose their homes to coastal erosion: at Happisburgh in North Norfolk, a house on Beach Road was valued at just £1 in 2006. In developing countries, poverty means that rebuilding homes may cost people a large proportion of their income.

Social costs are those that directly affect people themselves, such as losing family members and friends, having to relocate temporarily or permanently, loss of livelihood with destruction of farms or places of work, or losses of **amenity** areas linked to tourism and recreation or aesthetic value.

In 2014 *The Guardian* reported that the UK Environment Agency had estimated that 7000 properties in England and Wales, worth £1 billion, would be lost to sea-level rise this century, and over 800 properties lost to coastal erosion by 2035. Without coastal defences the number of properties lost would reach 74,000 by 2100. Without adaptation, by 2050, global costs of sea-level rise could reach $1 trillion, with a global loss of GDP of between 0.3 per cent and 9.3 per cent per year (a figure near the upper end would prompt a global recession). Some developing countries, such as Mozambique and Small Island Developing States (SIDS) such as the Maldives, are especially vulnerable economically, and could lose everything.

Yellow = Players, Orange = Attitudes and actions, Purple = Futures and uncertainties

Consequences in developing and developed countries

CASE STUDY: The consequences of coastal flooding for Australia

More than half of Australia's coasts are vulnerable to erosion and retreat due to sea-level rise, especially in the state of Victoria where 80 per cent of the coastline is at risk. It is expected that with just a 10-cm rise in sea level, the risk of coastal flooding will triple, affecting the cities of Sydney and Fremantle and much infrastructure not designed to accommodate sea-level rise. Coastal flooding in Sydney, on the scale of what is now considered a 1-in-100-year coastal flood, may become a daily occurrence by 2100. A 1-m rise in sea level will expose more than US$162 billion of industrial, commercial, transport and housing infrastructure to coastal flooding and erosion. Groundwater supplies may be affected by salt intrusion.

Economic costs increase greatly with every small increase in sea level. For example, it has been estimated for south-eastern Queensland that a 0.2-m rise in sea level would cause damage of US$1.4 billion, and a 0.5-m rise US$2.8 billion. In terms of amenity value, many of Australia's famous natural areas, such as the Great Barrier Reef and ecosystems such as mangrove forests, salt marshes and sea grass, will be trapped between rising sea levels and coastal protection measures, such as higher sea walls. Sea-level rise will also push salt water further up estuaries, affecting natural ecosystems such as the freshwater habitats in Kakadu National Park, Northern Territory. Coral reefs may be unable to grow fast enough to keep up with sea-level rise; those more than 50 m deep will die. Fisheries may be affected, and the tourism industry, which contributed US$30 billion to Australia's GDP in 2013 and employed 8 per cent of the workforce, uses sandy beaches such as on the Gold Coast that could be eroded (Figure 3.25). Expenditure on beach nourishment could increase by as much as US$39 million per year (Figure 3.30).

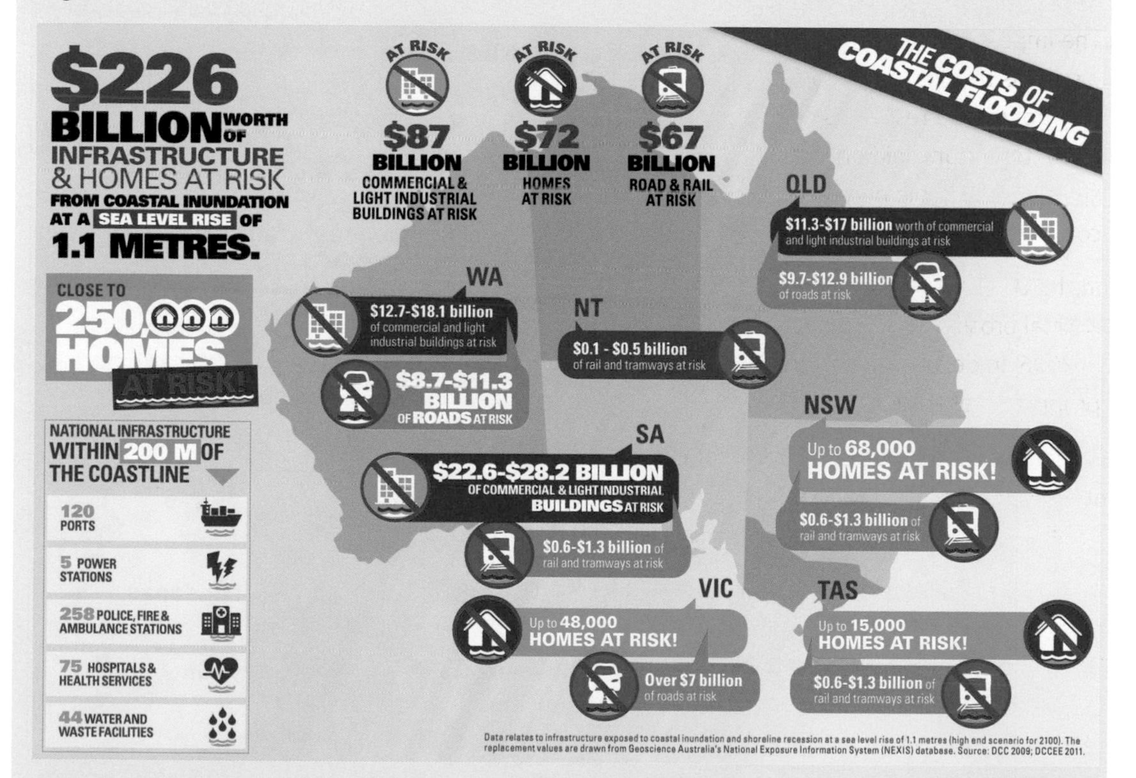

Figure 3.30: The costs of coastal flooding in Australia (values in Australian dollars)

Extension

Investigate NOAA's Digital Coast, which shows the impacts of sea-level rise on US coasts through GIS mapping. Explore this GIS resource:

- View the impact of various sea-level rise scenarios on the city of New York and on the state of Florida.
- View the impacts of sea-level rise on those classified as having high social vulnerability. Make notes on what you discover.

ACTIVITY

Compare the social, economic and environmental impacts of coastal flooding in Australia with the Philippines.

CASE STUDY: The consequences of coastal flooding for the Philippines

Out of the 'coral triangle' countries in the tropical western Pacific Ocean, the costs of damage due to sea-level rise are predicted to be greatest in the Philippines, with losses of $6.5 billion a year, without adaptation costs. The rate of sea-level rise is around 5.8 mm a year, higher than the global average, because the Pacific Ocean currents and trade winds move water towards the islands. The country's high level of poverty (over 90 per cent of the country's wealth is controlled by about 15 per cent of the population) makes the country economically vulnerable.

There are many coastal cities and communities in the Philippines, including the capital Manila, with growing populations. In Manila Bay, the natural ecosystems of mangroves, corals and seagrass have been damaged by pollution, over-exploitation and siltation, which greatly reduces their ability to protect the backshore from sea-level rise. An analysis has shown that parts of the urban area around Manila Bay, such as Cavite City and Las Pinas, would be flooded by 2100 with a 1-m sea-level rise, and other areas would be frequently flooded with storm surges or high tides. Up to 2.3 million people could be affected, 62 per cent of them in metro Manila.

San Fernando is also threatened: a 2012 study estimated that by 2100 the city will lose 300 buildings, over 283,000 m^2 of land and over 123,000 m^2 of beach. Property losses are estimated at about $2.5 million and land loss at $21 million; the recreational value of the beach is estimated at $95,000 a year. In addition, there will be social amenity losses of schools, churches and beach. The bay on which the city is located is also used for mooring fishing boats and other vessels; 130 fishermen earn about $12 a week. Due to high unemployment, alternative jobs are difficult to find, so when sea levels rise and moorings are lost, there will be an estimated welfare loss to the local community of $168,000 per year. Small breakwaters of rip-rap have been constructed to protect government buildings and infrastructure, for example at San Agustin, at a cost of $21,000.

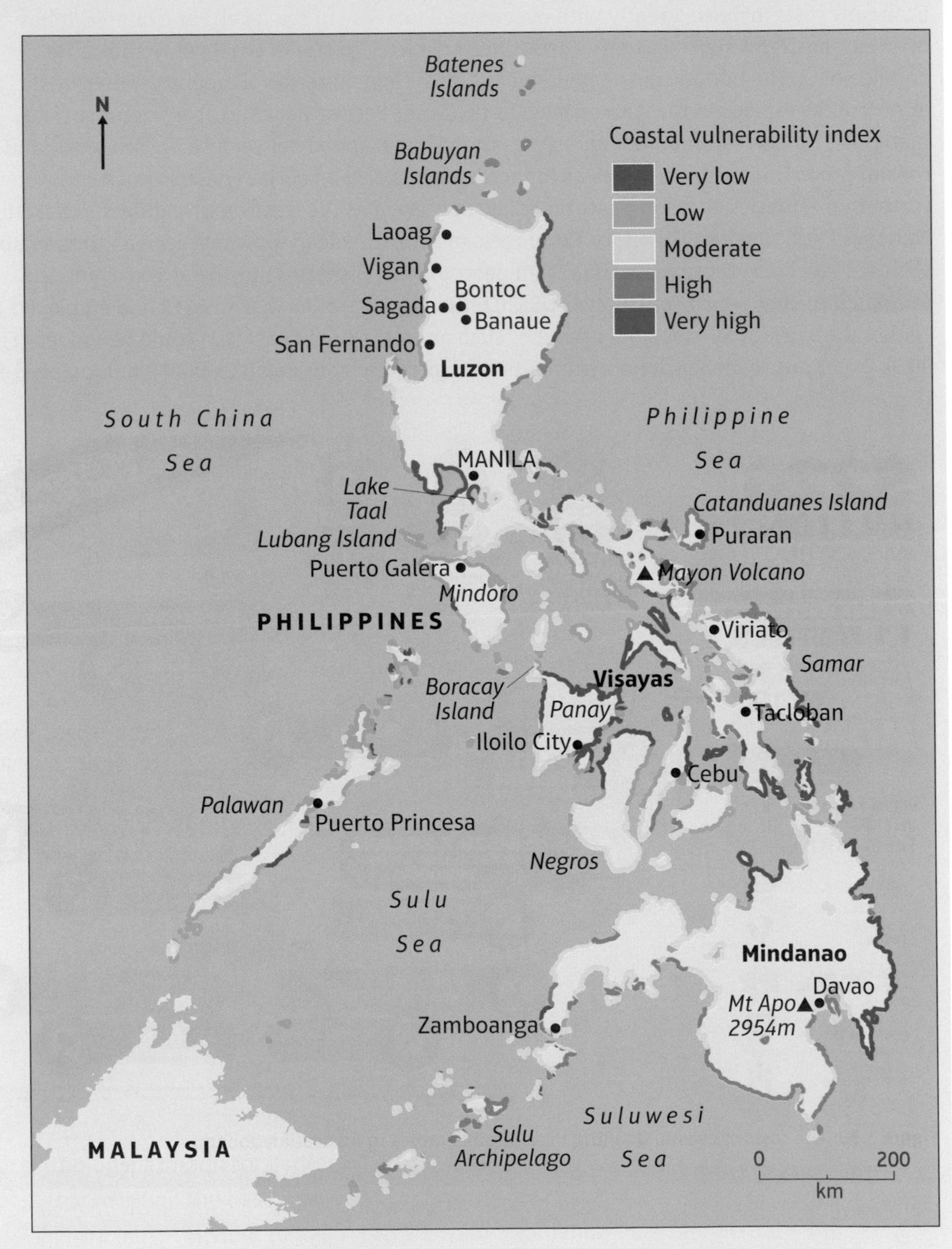

Figure 3.31: GIS map showing changing sea levels in the Philippines

Yellow = Players, Orange = Attitudes and actions, Purple = Futures and uncertainties

Climate change and environmental refugees

Officially a refugee is someone seeking safety in another country due to fear of persecution in their own country, but not all climate change migrations will be across international borders: many take place within the country affected. For example, between 2008 and 2013 the Philippines had the third-largest number of internally displaced people due to natural hazards, with 94 per cent of these displacements in 2013 – totalling 6.8 million people – due to 'storms'.

This is not just a feature of developing countries. For example, in 2005 over 1 million people evacuated from New Orleans and other nearby coastal communities to escape Hurricane Katrina, so avoiding the 7.6- to 8.5-m storm surge that flooded large areas of the city and coast. The evacuees went to many places across the USA, but 30 per cent did not return, and became environmental refugees, leaving New Orleans a partially abandoned city.

The United Nations does not include those fleeing from climate change within its definition of refugee, but most of them are likely to be people escaping rising sea levels by the end of the 21st century. Major cities such as London, Washington DC, Venice, Shanghai and Kolkata will be affected, as will the rice farmers in Asia's delta areas, such as Bangladesh, and island states such as Tuvalu and Kiribati.

Synoptic link

One effect of globalisation is increased migration, especially of refugees. While environmental refugees may only be a small part of this process at the moment, the global link between countries is likely to become much more important in the future, as more people are displaced by sea-level rise and coastal storms. This will also affect sovereignty, for example, if whole island communities move to another country; what happens to territorial waters if the land is flooded? (See page 207.)

CASE STUDY: Environmental refugee actions

The Maldives

After the Indian Ocean tsunami of 2004, the Maldives government started moving people from its lowest 200 islands to the few higher ones. The average ground-level elevation is 1.5 m above sea level and, with an economy based on tourism and fishing, jobs are likely to disappear, encouraging people to move to other countries. With a population of nearly 400,000, it may prove difficult to find mainland countries willing to take so many people, but the Maldives government is looking at the possibility of buying land elsewhere to move to and is in negotiation with Australia, India and Sri Lanka.

'It's a slow onset, little impacts,' said Thilmeeza Hussain, a former UN representative from the Maldives. 'For example, people living closest to the beach, their houses may get washed away by a storm but nobody's talking about it because it's just one or two houses. But what do they do? It's a small country and we don't have homeowner's insurance and things like that. People spend their life savings to build these homes and they're left with nothing.'

New Zealand

In New Zealand, the Pacific Access Category Ballot allows 75 citizens of Kiribati, 75 citizens of Tuvalu and 250 citizens of Tonga to be granted residency status in the country each year, and the seasonal worker programme allows employers in farming and accommodation industries to use labour from the Pacific Islands. Already 3000 of Tuvalu's 10,800 people have moved to New Zealand as part of a labour migration plan, which will eventually allow all of its citizens to move from their eight coral atolls to New Zealand.

The New Zealand courts have considered applications for permanent residence and citizenship on the basis of climate refugee status, and in 2014 they granted residency to a family from Tuvalu on 'exceptional humanitarian grounds'. The Alesana family had argued that climate change had made it too difficult to live in Tuvalu because of frequent coastal flooding, coastal erosion, limited fresh water and loss of crops. The legal ruling mentioned that the children were particularly 'vulnerable to natural disasters and the adverse impact of climate change'.

ACTIVITY

Why might countries such as India and Australia be worried about the numbers of environmental refugees?

Managing the risks of erosion and flooding

Hard engineering

Table 3.6: Hard engineering at the coast

Type of hard engineering	How it works	Advantages	Disadvantages
Groynes	These are commonly wooden walls on a beach at right angles to the coast, to slow down longshore drift movement. They may be open or closed, depending on how much beach sediment engineers wish to trap. They make a beach wider and higher so that waves expend their energy on it rather than on the backshore.	• Maintain the size of a beach, which protects the coast at that point • Enhance the beach for recreational amenity, assisting tourism • Less expensive than sea walls	• Expensive (about £1000 per linear m) due to the difficulties in getting firm foundations in a beach • May be an obstacle to people moving freely along a beach, because they may be high, or create a large drop on the down drift side • Are clearly not natural and may be considered unsightly • Greatly reduced longshore drift transport may cause narrower beaches and increased erosion down drift
Sea walls	These appear in different shapes parallel to the backshore. Recurved sea walls use the shape of a wave to direct the water into the following wave, so reducing wave energy. Stepped sea walls use the pointed edges to break up a wave as it hits the wall, so dissipating wave energy.	• Made of long-lasting concrete and able to absorb wave energy • Give people confidence and a sense of security • The tops can be used as promenades, providing access along the coast • Prevent high water levels from moving inland	• The most expensive of all coastal defences (about £5000 per linear m) • Do not fit in with the natural surroundings and may be considered unsightly • May make accessing a beach difficult due to their height • Scouring by waves makes it necessary to have deep pilings at their base to prevent it being undercut • Greatly reduce the supply of sediments, which may affect other coastal areas nearby
Rip-rap	These are boulders (usually granite) or specially designed concrete shapes (tetrahedrons). These are resistant to erosion and with a large surface area break up waves, so dissipating their energy. They may also be used to hold back mass movement on an unstable cliff.	• Long lasting and flexible in use • Can be placed at susceptible points on the backshore, to protect the base of a sea wall or to be used as a breakwater or groyne	• Cost about £50 per m^3 • May look unsightly and, even if natural rock, may contrast with the local geology • Can create access difficulties as they are dangerous to climb over • Seawater may still move through it, so some weathering and erosion may still occur on the backshore.
Revetments	These are sloped walls, often made of wood, placed parallel to the backshore but a short distance in front of it. They are able to take the force of breaking waves, so weakening their erosive strength and protecting the backshore.	• Absorb wave energy • Trap beach sediments behind them, reducing its removal by backwash or longshore drift • Longshore drift able to continue • Allow linear access along a beach. • A cheaper alternative to sea walls	• Cost about £1500 per linear m • Reduce access up and down the width of a beach • Look unsightly, especially if they stretch several kilometres along a coast. • May need constant maintenance as the wood is abraded by powerful waves

Type of hard engineering	How it works	Advantages	Disadvantages
Offshore breakwaters (reefs)	Rock boulders (usually granite) may be dropped and aligned in short lengths in shallow nearshore waters parallel to the shore. Their purpose is to absorb wave energy and dissipate waves before they have a chance to damage the foreshore or backshore. Being offshore, they allow longshore drift to continue behind them.	• Have proved effective in protecting vulnerable sections of coast • Can create sheltered water areas behind them for water sports, as well as keeping a beach in place for recreational and tourist use • Can be used to protect the entrance to harbours, creating calmer waters for safe entry or exit	• Costly, between £1 million and £2 million • May look unsightly at low tide as the geology of the boulders usually does not match the local geology • May need other coastal engineering to complement them, such as sea walls in the gaps between the reefs • May create increased deposition on the landward side, reducing longshore drift
Gabions	These are pebbles in wire baskets, which when tied together can make a wall where the great surface area absorbs wave energy and breaks up waves.	• Very flexible in terms of placement on the backshore as an additional defence above a sea wall, or to help hold back mass movement on a cliff • Relatively cheap and easy to maintain	• Not very strong, so not suitable for high-energy locations • Need frequent repair • May be considered unsightly

Figure 3.32: Examples of hard engineering – revetments and groynes at Overstrand, Norfolk

ACTIVITY

For each type of engineering shown in Tables 3.6 and 3.7, locate an example on the UK coastline and a photograph of this example. Annotate each photograph with the points from the tables that are also visible in the photograph.

Soft engineering

Table 3.7: Soft engineering at the coast

Type of soft engineering	How it works	Advantages	Disadvantages
Beach nourishment	Replaces beach sediments that may have been eroded or transported by longshore drift. A large beach will absorb wave energy and protect the backshore from erosion.	• Uses natural sediments, so the beach looks natural • Provides an amenity for recreation and so supports the local tourist industry	• Cost about £10 per m^3 • Does not last long, especially under winter storm conditions, and so may have to be repeated frequently • Sediments may have been dredged from offshore, so changing sediment cell balances and deepening the water, so that waves may approach the shore with more destructive energy
Cliff regrading	The lithology of a cliff may be unstable and prone to sudden collapses, so to remove this uncertainty engineers calculate a stable slope angle based on the rock characteristics. The cliff slope is then artificially cut back to the stable angle.	• Creates a natural-looking slope • Brings some certainty and confidence to property owners close to the cliff edge and reduces risk of sudden loss of property • Should remain stable if the base of the slope is protected from marine erosion	• Cost about £1 million • Some land and property will be lost when the slope angle is changed • Engineers may get their calculations wrong, or extreme natural conditions may overwhelm the changes • Stabilisation measures such as cliff drainage and vegetation planting also needed
Cliff drainage	Where the cliff lithology consists of a permeable layer above an impermeable layer, it may be unstable after rainfall, due to pore pressure. To reduce this pressure and reduce mass movement, drains with gravel can be inserted into the cliff to drain water out quickly.	• Looks natural once completed, as the engineering is not generally visible • Reduces mass movement, bringing some confidence to those owning land and property near the cliff edge	• Difficult to implement along the whole of a cliff without disturbing the cohesion of the rock layers • Will not prevent mass movement; only reduces it to some extent
Dune stabilisation	Dunes can provide a natural barrier to sea-level rise and storm waves, but are fragile. Monitoring their condition and repairing them with a geofabric or replanting of grasses (e.g. marram grass), together with infilling of slacks, will help keep them in place.	• Looks natural and is an effective barrier to higher sea levels and tides • Provides a natural ecosystem and recreational area	• May need to be fenced off during works, which reduces the amenity value in the short term • Powerful storms may ensure that this approach only works for a short while.
Managed retreat	Where it is accepted that there is little that can be done to stop high sea levels changing the coast, areas can be set aside for the sea to flood or erode. For example, former salt marshes that are now farmland may be allowed to flood again, or buffer zones next to eroding cliffs can be created so that no valuable property is lost.	• Allows natural processes to take place uninterrupted • Can extend current ecosystems	• Needs agreement from land and property owners to create these areas • Does not prevent land being lost, and may only be a medium-term solution – depending on the rate of sea-level rise • Currently no compensation in the UK for land or property loss • Possible loss of archaeological evidence

Yellow = Players, Orange = Attitudes and actions, Purple = Futures and uncertainties

Sustainable management

Sustainable management of a coast is a long-term approach that considers future threats such as sea-level rise and storm surges. In some cases this may mean abandoning coastlines, such as whole island states, or lowland areas or those where coastal erosion may increase. Such plans may conflict with the wishes of local people, who may lose their land, home, job and sense of community and feel that there is no social justice.

Adaptation in the UK has involved a policy change away from hard engineering defences towards 'more sustainable approaches' that aim to:

- use the latest scientific understanding
- evaluate new coastal developments
- ensure that any necessary developments provide social and economic benefits
- ensure the overall long-term sustainability of coastal areas.

Adaptation has included the second phase of Shoreline Management Plans (see page 161), which reflect predictions of sea-level rise by 2100. The Thames Estuary 2100 Project is intended to evaluate the risks to London and the Thames estuary, to ensure protection from future sea-level rise and storm surges (including building a larger Thames Barrier by 2070). The high costs of building and maintaining a coastal defence against large-scale natural processes mean that some locations will not be supported by government funding. The potential loss of property and cultural heritage has led to conflict between governments and communities. The UK government, through the Environment Agency (coastal flooding and erosion) and Defra (coastal erosion), encourages local communities and councils to adapt to coastal change themselves, such as through the 'Pathfinder Initiative' (see page 295).

CASE STUDY: Coastal realignment in Essex, UK

The UK has adopted a 'Finding Space for Water' philosophy which involves managing coastal retreat. The Essex Wildlife Trust (EWT) and other landowners, especially farmers, have discussed giving up land to the sea. EWT purchased the Abbot's Hall estate in Essex, on the Blackwater Estuary, and converted more than 84 hectares of farmland into salt marsh and grassland by breaching old embankments in four places in 2002. A **cost-benefit analysis** concluded that, since the soil quality and land value were low, it was not worth maintaining the embankments or building them higher.

This is the largest coastal realignment project in Europe, costing £645,000. It provides sustainable sea defences for the future, because the 49 ha of additional mudflats and salt marshes absorb incoming higher seas and storm surges without any local damage, and also reduce the volume of water travelling up the Essex estuaries and creeks, which could flood settlements. Wildlife also benefits, as the mudflats and salt marsh create new intertidal habitats (see page 124).

Two stakeholder groups were concerned that the plans would change the local water environments. The West Mersea fishermen who carry out oyster fishing in the estuary were concerned that sediments could choke and kill oysters and water level changes could remove oyster habitats. The RSPB was concerned about increased erosion and flooding of their site opposite the farm, which would change bird habitats.

CASE STUDY: Namibia's coastal strategy

Namibia established a national policy on climate change in 2011, emphasising adaptation measures to reduce vulnerability, and enhancing adaptive capacity and sustainable development. In 2013 a development policy was launched as part of the country's Vision 2030. Part of this aimed to strengthen the governance of the country's 1570-km coastline, and improve the quality of life of coastal communities while maintaining coastal ecosystems.

Namibia identified three categories of coastal adaptation:

- Normal actions: these are not directly related to sea-level rise, but are sensible low-cost coastal management options. They include preventing future development within a coastal buffer zone, conserving wetland and estuary environments, stabilising and reducing degradation of sand dunes and bars, integrating sea-level rise predictions into future planning and risk management, and reducing poverty so that people can better prepare themselves (e.g. build stronger homes).
- Additional actions: these are actions for managing the risk of sea-level rise at moderate cost. They include soft engineering such as beach nourishment, e.g. at Henties Bay, protecting and enhancing kelp beds, planting grasses and restoring dunes, and rehabilitating estuary and wetland environments. Institutional support is provided through mapping vulnerable areas, communicating risks to people, ensuring that insurance covers risks, establishing laws to enforce coastal-zone management, research and monitoring, and early-warning systems based on tide levels and meteorology.
- Expensive actions: these are more effective actions for protecting people and property from sea-level rise. They including hard engineering such as constructing or raising sea walls in key locations, e.g. Walvis Bay, relocating and resettling communities from vulnerable areas, raising infrastructure such as ports and roads, and building barrages and barriers.

ACTIVITY

Consider the actions suggested in Namibia's coastal strategy. Evaluate the plans from the point of view of

1. cost-effectiveness
2. adaptation to sea-level rise
3. impacts on people living and working in the coastal zone.

Holistic coastal management strategies

The physical processes operating in a littoral (sediment) cell are interlinked for long stretches of a coast. Human activities also have an impact on land, on the shore and in the sea. To manage a coastline effectively, it is therefore necessary to consider a long length as well as all stakeholders using the coastal environment. It is also important to consider a long timescale in coastal management as the physical processes take hundreds if not thousands of years to complete, and climate change has introduced an urgent need for planning over the next 100 years.

Integrated coastal zone management

The idea of Integrated Coastal Zone Management (ICZM) was introduced at the 1992 Earth Summit meeting and aims to bring together economic decision-makers such as in tourism, fishing and ports, and different government levels (including international). Other aims include the conservation of coastal ecosystems, ensuring the ability of future generations to use the coast, and the involvement of all relevant stakeholders.

ICZM emphasises cooperation between all stakeholders so that everyone benefits. It is a dynamic process because of changing demands at the coast, and decisions are likely to be increasingly concerned with vulnerability and sustainability. The EU adopted the Recommendation on Integrated Coastal Zone Management in 2002, and the Marine Strategy Framework Directive of 2008 also emphasised a comprehensive and integrated approach to protect all European coasts and seas. ICZMs can now be found worldwide, at varying states of sophistication and with various levels of action. Parts of the Philippines have ICZMs, such as Lingayen Gulf and Batangas Bay, with an emphasis

Yellow = Players, Orange = Attitudes and actions, Purple = Futures and uncertainties

on environmental management and use of coastal resources with some consideration of disaster preparedness and response, but little consideration of climate change, tsunamis or El Niño events.

Shoreline management policies

In 1993 the Ministry of Agriculture, Fisheries and Food (Defra from 2001) set out a national Strategy for Flood and Coastal Defence in England and Wales, based on 11 natural sediment (littoral) cells and sub-cells. This led to the development of Shoreline Management Plans (SMPs), produced through cooperation between all the relevant Coast Protection Authorities, Defra and the Environment Agency, but criticised for not 'listening' to local people. SMP1 covered 1997 to 2003. New SMPs (SMP2) were proposed in 2004 with the aim of planning the management of the UK coastline for 100 years, to take into account medium-term changes such as sea-level rise.

Shoreline Management Plan 2

Coastal defences for each coastal management unit are judged on the following criteria: technically sound coastal engineering appropriate to natural processes; environmentally acceptable protection measures; economically viable coastal defences (considering land uses and property losses); long-term sustainability of natural coastal processes and ecosystems; and maintenance and repair costs. Each management unit is then allocated one of four management policies:

1. **Hold the existing shoreline**: intervention with maintenance of existing defences and building of new defences.
2. **Advance the existing shoreline:** intervention with the building of new defences on the seaward side.
3. **Managed realignment of the shoreline:** no planned intervention, with natural processes operating without interference, but monitoring and managing the processes in certain places where necessary.
4. **No active intervention:** doing nothing, with no investment in defences or maintenance of any existing defences; these coastlines will normally retreat.

SMP2, from 2004, was controversial in a few places because the policy changed from holding the line to no active intervention. This affected property values and the security of communities. In both periods the emphasis was on the **cost-benefit analysis** (CBA), which compares the cost of coastal defences with the value of land to be protected. So places with low land value, such as farmland and recreational land, had difficulty qualifying for government money for sea defences. Although local governments can build coastal defences, they cannot afford to build them without money from national government. For example, in SMP1 and SMP2, Happisburgh, North Norfolk, failed to secure funding because of a poor CBA, the land uses were not valuable enough to justify spending money on defences. The UK government did investigate alternative community approaches to shoreline management through Pathfinder pilot projects, and in 2012 provided £11 million for 15 small areas to investigate ways of managing the coast.

CASE STUDY: Happisburgh, North Norfolk – SMP and Pathfinder

Since Happisburgh repeatedly failed to qualify for government grants for coastal defences, a soft engineering approach of 'managed retreat' has operated, much to the dismay of locals and especially the residents of Beach Road who, one by one, have lost their homes to the sea. By 2105, the shoreline may recede by 200 m, with a loss of a further 50 homes, caravan site, and property losses totalling £6 m. In 2003 the lifeboat and beach access ramp were lost, prompting concerns that tourism would be affected.

The 1996 SMP1 was Managed Retreat and this did not change in the 2004 SMP2, just renamed to No Active Intervention. By 2055 a further 35 homes could be lost with 250 m of land, and

ACTIVITY

TECHNOLOGY/ICT SKILLS

Investigate the aims and actions of a UK coastal ICZM, perhaps an area in which you have undertaken fieldwork. Assess how successful the ICZM is for managing the coastal zone sustainably.

Extension

The Shoreline Management Plans for England and Wales are on the Environment Agency website, and often on local council websites. Investigate the SMP for your fieldwork area and analyse the local coastal processes, land use, defences and plans, and the reasons behind the SMP.

Synoptic link

Coastal settlements are one type of 'diverse place' within the UK, and the people living in these communities have strong feelings about them and the issues affecting them. Several players influence the vitality and future of these places, including the national government through policies such as SMPs, local governments such as North Norfolk District Council, coastal engineers who produce reports for different government levels, and pressure groups such as the Coastal Concern Action Group. (See page 295.)

Literacy tip

Official reports can be very long and often repetitive. When doing research, the best approach is to read the abstract or summary first, to understand the key points and to identify whether any information in the main report will be relevant to the set task. The contents list and any tables of data may also be useful. Look over the diagrams, too, as these may be useful in several ways, not just in the way they are used in the report.

ACTIVITY

1. Consider the attitudes and views of different players towards coastal policies, plans and schemes, for example:
   a. Defra
   b. the Bangladeshi government
   c. a Happisburgh resident living close to the cliff edge
   d. a poor farmer in the coastal lowlands of Chittagong
   e. environmental pressure groups such as Greenpeace
   f. the IFAD and the IPCC
   g. coastal tourist businesses
   h. local fishermen.
2. Outline the conflicts that might arise between these different players.

the main village and the church by 2105. House values are very low and people cannot afford to move elsewhere, having invested in buying a home that at the time would have been some distance from the eroding cliff. There is no national system of compensation for these homeowners, although in 2009 the government did start providing grant aid of £5000 to assist with demolition costs and £1000 for relocation costs.

Campaigning by the CCAG (Coastal Concern Action Group) raised awareness of the coastal erosion issue, not only locally (with the Buy a Rock campaign) but throughout the UK. The campaign culminated in the launch of a Pathfinder pilot project in 2009. The North Norfolk District Council Pathfinder bid covered six settlements, including Happisburgh, and obtained government allocation of £3 million, partly because of its emphasis on integrated coastal management and stakeholder engagement.

Key parts of the plans were to be completed by June 2011. £1.4 m was set aside for 'purchase and lease-back' of 11 houses on Beach Road (part of a 'rollback' strategy to create a buffer zone between the eroding cliff and the main village); owners were offered half the 'non-blighted' value of their homes and all but one person took the offer. This gave people the chance to relocate (but 'compensation' was a word avoided and there are no similar national plans). Beach Road houses have been demolished and a buffer zone between the village and the cliff established. £0.45 m was allocated for relocating Manor Farm Caravan Park, £165,000 for new public toilets, and £100,000 for removing derelict coastal defences. By the end of 2013, only the caravan park had not been moved, due to difficulties finding an alternative site. The planning rights for the Beach Road properties were bought and sold to the local council, which funded the purchase of farmland on the landward side of the village to replace the homes. This allowed the community to remain the same size, and profits from the sale of the new houses will be used in the future to buy (at 40/50 per cent of non-risk market value) properties on the seaward side that are under threat from coastal erosion. Further lobbying by the CCAG and others has managed to change the SMP2 policy for Happisburgh to 'managed realignment', which does allow for some coastal defences in the future, if funding can be found.

Figure 3.33: Failed defences and coastal erosion at Happisburgh, Norfolk

Environmental impact assessment

Environmental impact assessment (EIA) is a key part of the decision-making process. EIAs aim to identify the positives and negatives of a development or scheme before it is implemented. The feedback then informs the process so that improvements and modifications can be made to any plans before they are started. In the UK, shoreline management plans are judged by a strategic environmental assessment (SEA), to ensure that any proposals are central to SMP policy. There is

Yellow = Players, Orange = Attitudes and actions, Purple = Futures and uncertainties

a rating scale from 'significant positive effects' to 'significant negative effects'. The SEA for North Norfolk SMP2 (largely 'no active intervention') found the following:

Significant positives:

- Protects local tourist assets, such as at Blakeney and Wells.
- Provides sand dune stability, such as at Holkham.
- Allows natural processes and natural evolution of the coastline.
- Retains critical infrastructure and enhances navigable channels, as at Thornham.
- Allows landward migration of intertidal habitats, as at Morston.
- Increases biodiversity, as at Holme.

Significant negatives:

- Increases 'coastal squeeze', leading to a loss of national and international intertidal habitats (Ramsar sites).
- Increases loss of reed beds and habitat for bitterns and geese species.

Extension

Investigate the views of an international environmental pressure group such as Greenpeace, regarding coastal management and impact assessments, especially in relation to sea-level rise and coastal erosion.

CASE STUDY: Coastal management in Chittagong, Bangladesh

In Chittagong, Bangladesh, a Coastal Climate Resilient Infrastructure Project (2012) supported by the Asian Development Bank (ADB) aims to 'climate-proof' the area. The project involves:

- improving road connections (for farmers and markets) while raising embankments to 60 cm above normal flood levels and making them resistant to coastal erosion
- creating new market areas with sheds raised on platforms above the expected 2050 sea level
- constructing, improving or extending 25 tropical cyclone shelters, taking account of sea-level rise and higher wind speeds
- training in climate resilience and adaptation measures.

As this is a collection of small schemes, the ADB did not require a full environmental impact assessment (EIA); instead, an Environmental Assessment and Review Framework was used. Positives were seen as helping to alleviate poverty (by 10 per cent) by generating income opportunities, adapting to climate change and reducing disaster risk, for example road flooding five days a year instead of 20, and environmental enhancement such as the planting of trees. Negatives were seen as disturbance of people and natural habitats, especially during construction phases, permanent removal of natural vegetation, and relocation of some households (200 people) by road realignment. In May 2015 the International Fund for Agricultural Development (IFAD), a UN agency, which loaned the project US$60 million, rated the progress of the project as satisfactory overall, but with slow progress on road embankments.

Figure 3.34: A multipurpose cyclone shelter at Chittagong, Bangladesh, with solar power, rainwater harvesting and first-aid room – built with US government funding to withstand strong earthquakes and 260 km/h winds

AS level exam-style question

Explain the possible social impacts of coastal recession on coastal communities. (6 marks)

Guidance

Note that the question only asks for social impacts, so consider impacts on people – individually, as families or as communities. Be careful not to drift into economic impacts, although there is some overlap. It may be best to refer to an example to help put points into context.

A level exam-style question

Explain why UK government coastal management policies vary from place to place. (6 marks)

Guidance

There are four categories, from 'hold the line' to 'no active intervention'. These are national government policies, so your answer needs to consider why the government has these policies and how they use them. This may reflect costs, sea-level rise or natural processes, for example.

Fieldwork exemplar: Coastal landscapes and change

Enquiry question

How do coastal sediments and profiles change?

Fieldwork hypotheses

Null hypothesis (H_0)

The distribution of marram grass at Oxwich has a random pattern.

Alternative hypotheses (H_1)

The proportion of plant cover by marram grass on sand dunes at Oxwich decreases from the yellow to the grey dunes.

Location of the study

Oxwich Bay is the second-largest beach on the Gower Peninsula and is visited by around 250,000 visitors each year. Behind the wide, 4-km long sandy beach is an extensive system of sand dunes, a salt marsh behind the dunes known as Oxwich Burrows, woodland and cliffs. A small stream, Nicholaston Pill, crosses the dunes and the salt marsh. The limestone promontory of Oxwich Point shelters the bay to the south; the dominant waves are from the south-west and sediment is transported to and deposited in the bay. It is thought that a spit initially formed here and gradually extended across the bay to create a bar and lagoon, trapping the river sediment to form the salt marsh.

Great Tor, a smaller headland, encloses the north-eastern end of the bay, although at low tide the beach connects with Three Cliffs Bay beyond. The area is a National Nature Reserve and the Countryside Council for Wales manages the dune system by planting marram grass to fix the dunes and maintaining footpaths to reduce trampling.

Methodology

Three linear transects were chosen from the strand line (the high-water mark) to the salt marsh, a distance of about 270 m, with each end marked by a ranging pole. This provided a stratified sample of the whole sand dune ecosystem and three sets of data to compare. Care was taken to avoid areas of excessive trampling, as this would have made it difficult to observe and measure natural patterns. Tape measures were laid in a straight line from one ranging pole to the distant one to keep transects straight. A pre-prepared record sheet was completed using a systematic sample every 5 m along the tape measure, to avoid a biased selection of points. A quadrat (0.25m^2) was placed on the ground exactly every 5 m and all measurements were then completed within the quadrat.

The main data was the type of vegetation present, which was identified using picture diagrams and descriptions. The amount of each type of plant, or bare sand, was recorded in the table as a percentage using the 100 squares of the quadrat to make an accurate estimate. Each table was numbered in sequence and the distance from the start noted, so that later analysis could be completed efficiently. In addition, other measurements such as slope angle, height of species, evidence of trampling, microclimate and soil conditions (such as temperature and acidity) were measured to correlate with the vegetation pattern and help explain it. Photographs were taken at each visual change in vegetation, and photographs of unidentified species were also taken to allow further research later. Questionnaires were conducted among visitors to determine opinions about the condition and management of the dunes.

Care was taken on the dunes to avoid health risks, including wearing sturdy walking shoes, light clothing suitable for the warm, sunny weather conditions, including hats, and everyone applied sun cream. Students were briefed on the safe use of equipment and teachers controlled the use of ranging poles.

Data presentation

Results for vegetation frequency were summarised in tables for each of the three transects. Data was compared to reveal any anomalous measurements that needed checking. Kite diagrams were then drawn to show the vegetation pattern for bare sand and the most abundant species. These were drawn to scale parallel to each other to give an accurate cross-section of the sand dune ecosystem. These diagrams were then annotated at the appropriate distance with a summary of the other data collected, and photographs added to the display. To further present the data, a Chi^2 statistical analysis was carried out to help analyse the pattern found.

Analysis and conclusion

There was a clear visual change in vegetation type with distance from the strand line, but also between the yellow and grey dunes and the slacks. Marram and red fescue grasses were co-dominant on the fore and fixed dunes and creeping willow was the main species in the dune slack between them. The tortula moss was more common in the lower sections of the dunes where soil moisture levels were higher. Bracken was not found on the fore dune but became more common on the fixed and 'wasting' dunes, where brambles and dewberry were also found. Other species present were saltwort, sea rocket, sand wort and sea couch on the main dune ridge and sand sedge in the sheltered slacks. As well as soil moisture, the pH changed from 8 on the fore dune to 6 on the grey dunes. There was little evidence of human trampling along transects, even though it was present elsewhere, so a natural pattern was recorded.

The alternative hypothesis was accepted: there is definitely a change in the pattern of marram grass from the younger dunes nearer the sea to the older dunes nearer the land, and most is found on the yellow dunes and hardly any in slacks or the grey dunes. This was proved by the Chi^2 statistical test result of 127.25 (well above the critical value of 12.59 at the 0.05 probability level).

Distance from transect start (m)	Av. number of marram grass plants per quadrat (*O*)	Expected number of marram grass plants per quadrat (*E*)	$O - E$	$O - E^2$	$\frac{O - E^2}{E}$
0-40	37	30	7	49	1.63
40-80	63	30	33	1089	36.30
80-120	56	30	26	676	22.53
120-160	38	30	8	64	2.13
160-200	2	30	−28	784	26.13
200-240	14	30	−16	256	8.53
240-280	0	30	−30	900	30.00
df = 6 Totals:	210	210			Σ127.25

Evaluation

Data collection could be extended with further transects in other parts of the dune system, or studies of different UK sand dune ecosystems in different parts of the country (or even in other countries), to see if there are similar patterns. The quadrat survey required observers to make a judgement on percentages, and while there was some discussion in the small groups, there is always the possibility of human error at this point. Three transects did provide comparable data and the same observers recorded the data, so there should have been consistency. The measurements were completed in the least disturbed area, ensuring that the most natural pattern present was recorded. It was difficult to identify all the plants and some had to be recorded just as 'plant A', so inviting a local expert – perhaps one of those managing the dunes – might have been useful.

Summary: Knowledge check

Through reading this chapter and by completing the tasks and activities, as well as your wider reading, you should have learned the following and be able to demonstrate your knowledge and understanding of coastal landscapes and change (Topic 2B).

a. What are the distinctive features and landscapes of the coast and the littoral zone?

b. How does geological structure influence the development of coastal landscapes at different scales?

c. Why do rates of coastal recession vary?

d. How does marine erosion create distinctive coastal landforms and landscapes?

e. How does sediment transport create distinctive coastal landforms and landscapes?

f. How does coastal deposition create distinctive coastal landforms and landscapes?

g. What is the influence of subaerial processes on coastal landforms?

h. How does mass movement create coastal landforms?

i. Explain the causes of sea-level change?

j. How have sea-level changes affected the coastline in the past?

k. How does sea-level change affect the coastal landscape?

l. What causes coastal retreat and how does this threaten people?

m. Explain the factors that are increasing the risk of coastal flooding?

n. How are communities affected by the increasing risk of coastal recession and coastal flooding?

o. How can coastal areas be managed and how sustainable are these approaches?

p. Explain the different views of management policies?

As well as checking your knowledge and understanding through these questions, you also need to be able to make links and explain ideas, such as:

- Littoral or sediment cells consist of sources, transfers and sinks, showing that coastal processes are linked.
- Variations in geological characteristics create a wide variety of small-scale and large-scale coastal features.
- Coastal retreat is the result of long-term natural processes and some short-term human processes.
- Subaerial processes work with marine processes to shape a coastline.
- Large numbers of people live and work in lowland coastal areas and are at risk from contemporary coastal changes.
- There are similarities and differences between developed and developing countries in terms of impacts of coastal flooding and storm surge events.
- The designation of environmental refugees linked to climate change at the coast is not straightforward.
- There is much debate about the sustainability of hard and soft engineering approaches.
- Integrated coastal zone management has only been partially introduced in many parts of the world, raising questions about the existence of holistic coastal management.
- Shoreline management plans are controversial and bring into conflict economic and social objectives.

Preparing for your AS level exams

Sample answer with comments

Assess the benefits of soft engineering approaches when managing threatened coasts. (12 marks)

Soft engineering approaches include beach nourishment and dune stabilisation as well as changing the shape of the cliff. Soft engineering approaches are a benefit because they work with nature instead of against it to protect threatened coasts. This is a key consideration for places down drift of coastal engineering.

Soft engineering techniques have been used at Mappleton on the Holderness coast in Yorkshire. They have protected the coastline with cliff stabilisation, dune regeneration, creating marshland and beach nourishment. They have also done managed retreat. These all look natural and work with natural processes, keeping a balance within the sediment cell.

In the 1700s Mappleton was 3½ km from the sea but now it's on the coast and houses are literally falling into the sea! The locals wanted to save the village so they got funding for a scheme costing over £2 million in 1991. They used soft and hard engineering with rock groynes and rip-rap to protect the cliffs. They also regraded the cliff to stop slumping as well as beach nourishment. But there are problems because the cliffs further south are being eroded faster now.

Beach nourishment involves adding sand and/or shingle to beaches that are being eroded to make them wider. It is quite expensive to do because it needs to be done all the time or it will be washed away. The advantages are that the coastline is protected and there is more beach for holidaymakers, so there's an economic benefit. Also, if the waves have to travel further to reach the cliff, they will lose energy before they get there. Marram grass can be planted to encourage sand dunes to grow, which help protect the coast. This is called dune stabilisation.

Cliff regrading means changing the slope angle of the cliff so it won't collapse so easily. It can be helped by putting drainage pipes in the cliff to remove water and by planting grass to bind the soil. They can also put in piles to hold the cliff in place. Dead trees can be used to do this. The cliff foot can be protected with beach nourishment and sand dunes. These approaches can be used as well as hard engineering to get holistic management. Soft engineering may not work as well as hard engineering but it is cheaper and more natural, so it is preferable.

A further soft engineering method is mentioned but 'doing nothing' is not. If it had been it would have allowed a wider range of objectives to be considered. A one-sentence conclusion is insufficient.

No need for these definitions. Instead, show that you understand the question and what it wants you to think about. The command word 'assess' is asking you to consider how successful the approaches are.

Useful to mention a located place but it must be more than just 'name dropping': the candidate needs to give more details that fit the question. What are the objectives behind the management of the coast?

The candidate has fallen into the trap of just repeating facts rather than making them fit the question set. There is a hint of a cost rather than a benefit, but does it relate to soft engineering?

This paragraph is more in line with providing an answer to the question. Benefits and problems are mentioned and so marks are now being gained. However, there is a lack of depth, such as how sand dunes and beaches protect the coast and how much income is earned from tourism.

Verdict

This is an average answer, perhaps more at GCSE than AS level, and does not directly answer the question or give enough detail. The command word 'assess' requires a wide range of criteria to be used to judge soft engineering, including environmental, social and economic considerations. The views of players such as local people and the government could have been included. At least two threatened coastal locations should have been used.

Preparing for your A level exams

Sample answer with comments

Evaluate the threats for lowland coastal areas arising from future sea-level rise. (20 marks)

Sea levels have risen significantly since the end of the last ice age some 12,000 years ago but the sea level is now thought to be rising as a result of global warming caused by human activities; that is, the sea level is rising due to melting glaciers and ice sheets and the thermal expansion of the seawater. This is threatening low-lying coastal regions around the world. The sea level rose by about 1.8 mm a year in the 20th century but now it is rising faster – by over 3 mm a year. More than 200 million people are already at risk of coastal flooding and this number will rise due to population growth and migration to megacities on the coast like London and Shanghai. Some places such as the Netherlands can afford to protect themselves by building coastal defences. But poorer places such as Bangladesh cannot afford to do this. Ecosystems are also threatened, such as in the Sundarbans in India and Bangladesh. Here, the mangrove swamps are being inundated by the sea and people are having to move away.

> Good understanding of some of the causes of sea-level rise, but no mention of subsidence and isostatic change. The threat is put into context with mention of population numbers, but small island states could have been mentioned and the case studies developed further.

Coral islands such as the Maldives, Kiribati and Tuvalu are under threat from sea-level rise as they are all low-lying, being less than 2 m above sea level. Coral reefs are also threatened as the sea level may rise faster than they can grow. They also need sea temperatures between 23 and 29 degrees C and cannot survive if the sea becomes too warm. A rise in sea level by the end of this century of between 18 and 59 cm has been forecast by the IPCC. But if the Greenland ice sheet totally melts, sea levels could rise by 6 m! And if Antarctica melts, the sea level could rise by 60 m!! If this happens, the Maldives, Kiribati and Tuvalu will have to be abandoned to the waves and the locals will have to migrate.

> Island states now mentioned but the candidate has strayed from the question – the issue to explore is how 'bleaching' of corals increases flooding. The general threat is covered, but it is time for specifics.

The southern Mediterranean, West and East Africa, and South and South-East Asia are the regions most threatened by rising sea levels. Places with low-lying deltas and rapid population growth are unable to adapt to rising sea levels as the financial costs will be too high for poorer countries. For example, in Bangladesh a quarter of the land is less than 2 m above sea level and by 2050 rising sea levels will inundate some 17% of the land and displace about 18 million people. People will lose their land, which is mostly used for agriculture, and this will have an impact on food supplies.

> Specific threats are covered here – the displacement of people and the impact on food production – but these needed to be explored further. The Nile Delta would make an interesting contrast to Bangladesh.

The rise in sea level could mean major cities such as London ending up under water, and destroying a massive amount of infrastructure and economy. The Thames Barrier exists to help protect the city and there are already plans to build a larger barrier further east on the Thames estuary by 2070. The Netherlands has already totally protected its coast and cities and ports. Low-lying regions are under threat from sea-level rise, especially in poorer parts of the world where people can't afford proper sea defences. As sea levels are rising faster, the situation will only get worse. About a half of the Netherlands is below sea level but they can afford to build defences.

> The threat to cities is made clear and how the threats are different in different countries. This could have been explored further, perhaps.

The threat to lowland coastal areas is still a matter of some guessing, as it is not known by how much sea levels will rise. They have risen by 120 m since the end of the last ice age and are predicted to rise perhaps as much as a metre by the end of the century. Countries will have to prepare themselves but what can they do against the power of the sea?

> This conclusion appears rushed; perhaps the candidate was under time pressure. What are the main threats and what can be done about them?

Verdict

This is an average answer, and perhaps a little short – although it is quality of thinking that is judged and not quantity of information. Some good understanding of the threats is shown, as asked for in the question, but they have not all been covered. Other points that could have been made are threats from high tides and storm surges; natural areas under threat such as the Sundarbans and corals; salinity of coastal soils; internal displacement of people and environmental refugees; loss of land and jobs; risk of death and disease; political unrest – is anyone responsible? Alternative scenarios are briefly suggested in the conclusion, but more judgements on the seriousness of the threats was possible.

THINKING SYNOPTICALLY

Read the following extract carefully and study Figure A. Think about how all the geographical ideas link together or overlap. Answer the questions posed at the end of the article. This article first appeared in Carbon Brief: Clear on Climate on 25 September 2014 by Robert McSweeney.

CHINA TOPS NEW LIST OF COUNTRIES MOST AT RISK FROM COASTAL FLOODING

Over 50 million people in China will be at risk from coastal flooding by the end of the century if greenhouse gas emissions continue to stay high, a new study finds.

The research shows Asian countries dominate a top-20 ranking of most vulnerable nations from rising sea levels, with China topping the list.

A team of researchers from the US climate news website Climate Central mapped sea levels around the world using a global database of tide gauge measurements. They then combined these measurements with projections of how much scientists expect sea levels to rise with climate change. The researchers also present their results as a 'top-20' list, ranking countries by the total population at risk in 2100.

China has the largest population at risk, more than twice as many people as Vietnam, which sits at number two on the list.

European countries are quite prominent, with the Netherlands, UK, Germany, France and Italy in the top 20. The UK sits twelfth with over 2.5 million people predicted to be at risk of flooding by the end of the century.

Flood defences

While the study shows how many people could be at risk from coastal flooding in the future, it doesn't consider the impact of measures that reduce flood risk, such as building flood defences.

Almost half of the population of the Netherlands could be at risk from flooding, the study notes. But the country has one of the most sophisticated flood protection systems in the world. As a result, the population exposed to increased flood risk is likely to be smaller.

But flood defences are expensive, with large upfront costs and continuous maintenance. London's Thames Barrier, for example, cost over £500 million to build (about £1.6 billion in today's money) and will require around £1.5 billion to maintain over the next 20 years.

Not all countries can afford to take similar measures to protect themselves, particularly those with large coastal populations to defend. When a similar study ranked port cities around the world according to population exposed to flood risk in 2070, they found 17 of the top 20 cities were in developing countries.

Projections

In addition to the business-as-usual scenario, the new research looks at two more optimistic scenarios, requiring 'sharp' and 'extremely sharp' reductions in carbon emissions.

They give an idea of how the numbers might change. The population at risk in China could be as low as 36 million by the end of the century. That's if emissions are cut rapidly, and sea levels rise relatively slowly. Taking the most pessimistic view, the number of people at risk jumps to 62 million.

The study doesn't take into account the likely increase in global population. Recent projections have suggested a 95 per cent likelihood of the world's population rising to between nine and 13.2 billion by 2100.

Predicting the actual number of people at risk is hard and depends on population, and flood protection. But it illustrates this is a problem that's only going to get bigger as sea levels rise.

Table A: Global estimates of population number and proportion at risk from coastal flooding by 2100 by country. Assumes current emissions trends continue, and a central estimate of sea level rise.

Country	Population number at risk (millions)	Proportion of population at risk (%)	Country	Population number at risk (millions)	Proportion of population at risk (%)
China	50.47	4	Taiwan	1.03	4
Vietnam	23.41	26	South Korea	1.03	2
Japan	12.75	10	Nigeria	0.85	1
India	12.64	1	Italy	0.84	1
Bangladesh	10.23	7	Iraq	0.63	2
Indonesia	10.16	4	North Korea	0.63	2.5
Thailand	8.18	12	Belgium	0.62	6
Netherlands	7.79	47	UAE	0.57	7
Philippines	6.21	7	Mozambique	0.54	2
Myanmar	4.74	9	Cambodia	0.45	3
USA	3.09	1	Ecuador	0.34	2
UK	2.57	4	Hong Kong	0.24	3
Brazil	1.74	1	Denmark	0.23	4
Germany	1.67	2	Oman	0.15	5
France	1.26	2	Ireland	0.13	3
Malaysia	1.17	4	Bahrain	0.08	6

ACADEMIC SKILLS

1. How accurate is data from tide gauges when judging sea level rise?

2. Why are the estimates quoted in this article only until the end of the 21st century?

3. a. What are the weaknesses of the assumptions made by Climate Central in their risk projections?

b. To what extent do these weaknesses affect the reliability of their findings?

4. Climate Central is a player in the climate change debate. To what extent do you think that they add a balanced viewpoint to the debate? Give two reasons for your answer.

ACADEMIC QUESTIONS

5. Study Table A carefully.

a. Suggest a suitable cartographic method of displaying the two pieces of information (number and proportion) on the same world map. Explain why your choice would be a suitable method.

b. To what extent do these two pieces of information (number and proportion) have a positive correlation?

c. Analyse the patterns of coastal flooding risk with reference to selected countries, continents and levels of economic development.

6. Which aspects of physical and human geography have combined to place China at the top of the risk list?

7. What actions can people living and working by the coast, urban managers and national governments take to reduce these risks?

8. How significant will the future pattern of population growth, as predicted by the UN, be to the risk pattern predicted and shown in Table A?

9. How significant could further socio-economic globalisation be to reducing the future risk of coastal flooding in developing countries?

Dynamic places

Globalisation

Introduction

Globalisation is a process. It doesn't have a beginning or, necessarily, an end. It is a process by which the world is becoming increasingly interconnected as a result of massively increased trade and cultural exchange across borders. Growing interdependence within the global economy is the result of the movement of goods, services, capital, technology, people and ideas. This process is facilitated by governments and international institutions, built by transnational organisations, held together by transport and communication networks and underpinned by the choices and dreams of consumers. It is a powerful process that has led to a rapid and profound increase in wealth and improved quality of life for the Western world since the end of the Second World War, for emerging economies since the 1970s, and for the former communist world since the 1990s. The impact of globalisation has also been felt in popular culture, with successful music artists and film stars becoming globally famous.

However, the economic gains of globalisation for some are accompanied by extreme inequalities. There are definite patterns of winners and losers, and the process is disruptive to some places and at particular times. Areas that were winners at one point in the globalisation process can become losers at a later stage. Areas that were disadvantaged can start to experience rapid increases in economic activity and quality of life as technologies and patterns of investment change.

The consumption generated by globalisation drives an improved quality of life for many, but threatens environmental unsustainability for all. Some argue that the process is leading to the emergence of a global culture based on Western ideas and consumption. Others argue that ethical and environmental concerns about unsustainability are leading to increased localism and a growing awareness of the impacts of the consumer society. Most agree, however, that the future of globalisation is uncertain because it depends on the collective engagement and choices of people everywhere (except for a few isolated indigenous peoples in far-flung corners of the world).

In this topic

After studying this chapter, you will be able to discuss and explain the ideas and concepts contained within the following enquiry questions, and provide information on relevant located examples:

- What are the causes of globalisation, and why has it accelerated in recent decades?
- What are the impacts of globalisation for countries, different groups of people and cultures?
- What are the consequences of globalisation for global development, and how should different players respond to its challenges?

Figure 4.1: The City of London skyline. The influence of globalisation on our lives is profound, but some of the processes may be hidden from view. What global processes do you think lie behind this 'finance-scape' of the City of London?

Synoptic links

Globalisation shapes patterns of trade, development, technology, urbanisation, culture and power and so overlaps with many topics. Players include national governments and blocs such as the EU, non-governmental organisations such as the World Social Forum, technology providers such as Google, transnational corporations such as Toyota and HSBC and consumers. Their attitudes towards trade and actions on factory locations for example, shape the nature, extent and speed of connections within and between economies. The history of globalisation has been one of increasing wealth but also rising inequality, affected by economic booms and busts (Kondratiev cycles). On the one hand, billions of people have been taken out of poverty but, on the other hand, increased consumption threatens cultural identities and environmental stability. Therefore, the future is hopeful but highly uncertain as globalisation is an ongoing process.

Useful knowledge and understanding

During your previous studies of Geography (KS3 and GCSE), you may have learned about some of the ideas and concepts covered in this chapter, such as:

- the process of globalisation
- transnational corporations as key players in the globalisation process
- how technology and connectivity support economic development
- ways of measuring development
- how measures of economic development reveal global patterns of inequality
- the consequences of uneven global development, including deindustrialisation and migration
- urbanisation as a result of rural–urban migration
- fair trade, ethical consumption and recycling.

This chapter will reinforce this learning and also modify and extend your knowledge and understanding of globalisation. Remember that the material in this chapter features in both the AS and A level examinations.

Skills covered within this topic

- Use of proportional flow lines showing global networks.
- Ranking and scaling data to create indices.
- Analysis of human and physical features on maps to understand lack of connectedness.
- Use of population, deprivation and land-use datasets to quantify the impacts of deindustrialisation.
- Use of proportional flow arrows to show global movement of migrants from source to host areas.
- Analysis of global TNCs and brand-value datasets to quantify the influence of Western brands.
- Critical use of World Bank and United Nations (UN) data sets to analyse trends in human and economic development, including the use of line graphs, bar charts and trend lines.
- Plotting Lorenz curves and calculating the Gini coefficient.

CHAPTER 4

What are the causes of globalisation and its accelerating pace?

Learning objectives

4.1 To understand that globalisation has accelerated in recent decades because of rapid developments in transport, communications and business

4.2 To understand the importance of political and economic decision making in the acceleration of globalisation

4.3 To understand how globalisation has affected some places and organisations more than others

Transport, communications and business

The International Monetary Fund (IMF) defines **globalisation** as 'the growing economic interdependence of countries worldwide through increasing volume and variety of cross-border transactions in **goods** and **services**, freer international **capital** flows, and more rapid and widespread diffusion of **technology**'. However, the idea of growing **interdependence** can also be applied to people, **culture** and our impact on the natural environment.

As transport and communications technologies have improved, the same market forces that govern national economies have been able to extend their reach internationally. Free markets promote competition and the division of labour (with people specialising in what they do best). This improves efficiency, increases profits and allows wealth creation, sometimes known as capital formation. As markets expand in size due to globalisation, so the potential of these economic factors increases. It is this potential for economic growth that persuades governments, businesses and individuals to make the necessary compromises to join the global economy. But markets do not ensure that everyone shares in the increased benefits, and critics of globalisation argue that the process reinforces inequality between nations. However, supporters of globalisation argue that all countries that establish good governance, education and training and openness to foreign investment and technology transfer stand to benefit.

A core element of globalisation is the expansion of world trade through the elimination or reduction of trade barriers. Greater imports of goods offer consumers more choice at lower prices, while providing strong incentives for domestic industries to be competitive. Exports, often a source of economic growth for developing nations, can stimulate job creation as well as earning valuable foreign currency. More and better global connections can encourage foreign investment, which creates employment for the local workforce and brings new technologies, which promote higher **productivity**.

Table 4.1: Timeline of selected global trade agreements

1948	**The General Agreement on Tariffs and Trade (GATT)** Initially, 23 countries agreed 45,000 tariff concessions affecting US$10 billion of trade, rising by 1973 to 102 countries agreeing tariff reductions worth more than US$300 billion.
1994	**The World Trade Organisation (WTO)** GATT became the WTO, with 123 countries agreeing to major reductions in tariffs (of about 40%) and agricultural subsidies, allowing full access for textiles and clothing from developing countries and an extension of intellectual property rights.
2001 to present	**The Doha Development Round** So far, the WTO's attempts to negotiate further agreements reducing barriers to trade have been unsuccessful. The main disagreement between developed economies (EU, US, Japan) and emerging economies (led by Brazil, India and China) is over agricultural subsidies, which are viewed as a type of trade barrier.

ACTIVITY

1. Complete a survey of the products that you own and where they were made by studying the labels. Cover a range of types of product such as clothing, electrical goods, foods, and sports equipment.

2. Merge your results with others and calculate the percentage of the total products originating in each country.

3. Which were the dominant countries? Explain why.

4. Which country is furthest away from the UK? How did the products from this country reach you?

The shrinking world: developments in transport and trade

Five factors have accelerated the process of globalisation.

1. Transnational corporations (TNCs): TNCs are the architects of globalisation. They invest abroad (foreign direct investment) and build the links between the places that make products and the places that consume goods and services.
2. Lower transport costs: In the 19th century, railways, the telegraph and steamships reduced the costs of moving goods internationally. In the 20th century, containerisation further reduced transport and handling costs, and international aviation made fast travel and airfreight more common and affordable.
3. Computer and internet technology: Manufacturing in diverse locations can be coordinated using computer software (CAD/CAM) for the transfer of ideas. Global communication, including social media, has enabled the creation of recognisable global brands.
4. International organisations: Trade relies on trust and cooperation, and the global economy has rules and referees. For example, the World Trade Organisation agrees trade rules, the United Nations promotes peace and cooperation, the World Bank provides loans for development, and the International Monetary Fund provides loans to governments.
5. New markets: Companies invest in new markets order to make profits. More consumers create more potential sales, and thus higher potential profits. The opening up to global trade of populous new markets such as India and China has encouraged businesses to invest money in the hope of generating more sales. The success of investment and sales is reflected in global stock markets and share prices.

Table 4.2: Timeline of selected transport innovations

1712	Invention of the steam engine, leading to steam road locomotives in 1784, steam boats in 1802 and steam rail locomotives in 1804
1886	First automobiles in commercial production
1903	First controllable aeroplane demonstrated
1929	Jet engine designed, first prototype in 1937
1955	Development of the shipping container
1960s	Commercial development of jet aircraft

CASE STUDY: Containerisation's contribution to a 'shrinking world'

Containerisation is a system of transporting goods in strong, standard-sized steel containers that can each carry up to 25,000 kg in weight of goods and can easily be transferred from ships to lorries and trains. The introduction of container ships in the 1960s kick-started a revolution in world trade. Transporting goods had previously been slow and expensive, because ships had to be loaded and unloaded manually by longshoremen/stevedores.

The container idea was taken up across the world when the patent for them was given to the industry. Businesses embraced container services, because they reduced the unit cost of international transport from 30 per cent to 1 per cent, as well as increasing speed and cutting theft and losses due to breakages. Retailers and manufacturers can also use containers as mobile storerooms, reducing the cost of warehousing. With developments in information technology, companies can predict demand in factories or shops, and so deliver just the right amount of goods at just the right time (JIT = just in time). This reduces the cost of storing stock.

Container ships have rapidly increased in size (Figure 4.2), enabling the efficient movement of huge quantities of cargo. The value of trade (goods and services) as a percentage of world GDP increased from 42.1 per cent in 1980 to 60 per cent in 2013. Foreign direct investment increased to more than US$1.4 trillion in 2012.

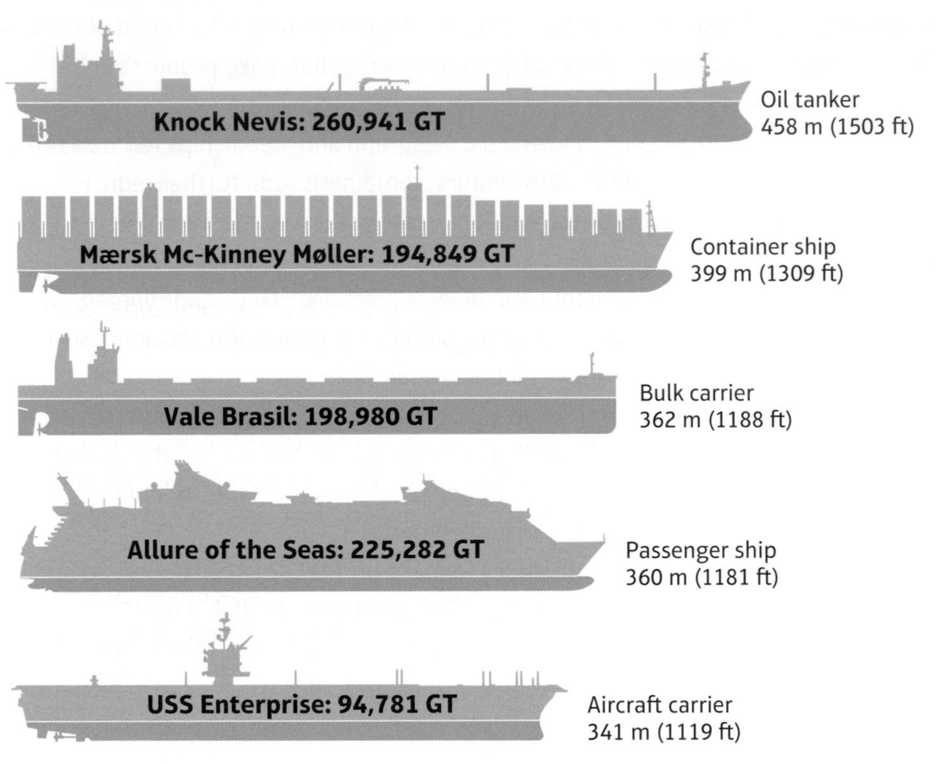

Figure 4.2: The world's largest ships in 2013 (GT = gross tonnes)

Developments in ICT and mobile communication

The idea of the 'digital economy' arose in 1995 in response to the development of the internet. Since then, the supporting infrastructure (hardware, software, telecommunications and networks) has developed rapidly, strengthening the digital economy. Initially the focus was on e-business (reshaping business organisations and processes via computer networks) and e-commerce (the sale of goods via the internet). However, the advent of mobile technologies and social media has extended the scope of the digital economy. It has fundamentally changed the way people interact and conduct every aspect of their lives, not just in business processes and the exchange of goods, but also culturally.

Yellow = Players, Orange = Attitudes and actions, Purple = Futures and uncertainties

Table 4.3: Timeline of selected ICT and mobile communication developments

1832	First demonstration of wireless telegraphy (radio)
1837	First commercial electric telegraph
1925	First demonstration of television
1940	Teletype (early form of communication between networked computers)
1957	First satellite, leading to modern mass communication, electronic navigation and global positioning and data collection systems
1969	ARPANET, becoming the basic internet in 1981
1989	World Wide Web invented

This rapid and fundamental change has implications for patterns and processes within the global economy. The scale of the digital economy (estimated to be worth US$1.5 trillion in 2015) has caused the demise of some established businesses:

- First, when online retailers like Amazon started selling media such as books, films and music, traditional retailers like Woolworths in the UK and Blockbuster in the US closed because they were unable to compete.
- Second, more general goods went online, such as consumer electronics and clothing. Manufacturers of such goods have adapted by building their own online shopping sites, such as Nike Official Site.
- The next development was the establishment of online marketplaces that gave small businesses access to a global customer base. In 2011, Amazon had a 12 per cent online share in North America, Taobao had 80 per cent in China, and Rakuten had 30 per cent in Japan. These platforms are estimated to take more than US$1 trillion in annual sales.
- The traditional supermarket model is also now being challenged, with online food and grocery shopping and the 'click and collect' model of home shopping.

Meeting the needs of shoppers online has caused businesses to change the way they operate. The volume of sales and innovations in warehousing have brought down costs for companies and consumers, and many online services offer free delivery to build customer loyalty. High sales volumes also enable the new digital marketplaces to put pressure on other suppliers to reduce their costs and speed up delivery of stock. Ownership and control of widely recognised **global brands** has become increasingly important in these vast markets, because these products are attractive to consumers with disposable income. Businesses have taken full advantage of flexible employment laws to build supply chains of goods and services from smaller, sometimes self-employed businesses and workers on zero-hours contracts. These recent changes suggest that there will be further changes to patterns of employment, contracts and pay in the future.

Table 4.4: UK international trade in goods (September 2015)

Imports (£ billion)	Exports (£ billion)
Germany (5.3)	USA (4.1)
China (3.4)	Germany (2.6)
USA (3.0)	Switzerland (1.6)
Netherlands (2.5)	France (1.6)
France (2.1)	China (1.6)

Extension

David Harvey, in his book *The Condition of Postmodernity* (1989), was the first person to articulate the idea of time–space compression. He described how the relationship between space and time is affected by technological change. As a result, our cultures and communities are exposed to rapid growth and change, challenging our identities and sense of place.

Make a mind map of the technologies you are using in your A level studies. Consider the impact of each on your ability to learn and the impact that this may have on your own identity and sense of place.

ACTIVITY

CARTOGRAPHIC SKILLS

Maps show spatial data. Flow line maps show how things move from one place to another, and the size of movement. The best flow line maps consist of arrows showing the direction of movement, where the width of the arrow is drawn to a scale to represent the volume of movement from one place to another.

1. Investigate the Office for National Statistics' (ONS) website and find the latest interactive flow line map to showing internal migration within the UK. Review your home region, and discover the detail of the movements into, and out of, the region.
2. Draw a flow line map on an outline map of the world using the data in Table 4.4.
   a. Draw a proportional arrow between the UK and each of the other countries in the appropriate direction. Choose a scale, such as 1 mm per £1 billion of value. Colour export arrows in one colour, and import arrows in another. Be careful to fit the arrows so that they do not overlap.
   b. Investigate additional UK import and export data and add these to the map.
   c. Describe and explain the pattern shown by your completed map.

Political and economic decision making

Trade requires goods that meet the specified or expected standard of quality, together with the prompt payment of bills according to the contract or trading agreement. Societies have developed laws to regulate trade, guaranteeing trust and promoting greater amounts of trade. Governments recognise the benefits of increasing trade flows for economic growth, employment and taxation revenue. As a result, governments make agreements to regulate cross-border trade so as to encourage greater flows.

However, there are potential disadvantages from cross-border trade: domestic companies may become over-specialised, and if the specialism suffers a decline in demand there can then be a large fall in employment and tax revenue. An example of this is the specialisation in financial services in the UK, which suffered a steep decline in economic output as a result of the financial crisis of 2008. Also, competition from abroad may lead to some domestic firms reducing production or going out of business altogether, such as the UK steel industry. Therefore, governments can face political opposition when making trade deals. Some industries and communities may call for governments to impose taxes and **tariffs** on imported goods, to protect their trade from competition. This is called **protectionism**.

International political and economic organisations

After the Second World War, the Western allies viewed the resumption and growth of international trade as crucial in rebuilding shattered economies, and also as a way of helping to prevent future conflicts. The Bretton Woods Agreement that came into force in 1946 was a system of rules, institutions and procedures to regulate the international monetary system for the USA, Canada, Western Europe, Australia, New Zealand and Japan. All these countries had linked their currencies to gold and the US dollar. The agreement also established the International Monetary Fund (IMF) and the International Bank for Reconstruction and Development (IBRD).

These international organisations provide the confidence required for countries and businesses to trade internationally. They increase the flow of trade by reducing taxes and tariffs, standardise production so that consumers can buy compatible and reliable products, and outlaw damaging competition, such as flooding a market with subsidised products so as to put competitors out of business.

The International Monetary Fund

The purpose of the IMF was to help governments balance their payments within this system, in case of economic difficulties. The IMF grants loans to member countries if they cannot pay their debts. The idea is that if countries can receive mutual support in times of **asymmetric economic shock**, they are less likely to resort to the protectionism seen in the run-up to the Second World War. IMF members are assigned financial ratings reflecting their relative economic power, and, as a sort of credit deposit, they pay a 'subscription' amount matching their rating. The main purpose of the fund today remains to ensure the stability of the international monetary system – the system of exchange rates and international payments that enables countries (and their citizens) to make transactions with each other.

The International Bank for Reconstruction and Development

The aim of the IBRD was to provide loans to help rebuild economies after the Second World War, and to alleviate poverty. (The Soviet Union, as a communist state, did not participate in the IBRD. It was not until 1992, after the collapse of the USSR, that the Russian Federation and other successor states of the USSR became involved.) The IBRD is now part of the World Bank Group and, because the Western world has long since recovered from the Second World War, the World Bank has changed its focus to tackling extreme poverty. It aims to reduce the percentage of people living on less than US$1.25 a day to no more than 3 per cent of the world's population by 2030, and to reduce inequalities by encouraging income growth for the bottom 40 per cent of every country. It does this by providing low-interest loans, grants and technical assistance to developing countries. The bank often works in partnership with governments and other **multilateral** institutions as well as commercial banks to deliver projects.

GATT and the World Trade Organisation

In 1947, 23 countries signed the General Agreement on Tariffs and Trade (GATT) to begin the global process of facilitating trade, which has continued through negotiations, or 'trade rounds'. Over time, more countries signed up, and the participants agreed to further reductions in tariffs and more standardisation of products. In 1995 this process led to the establishment of the World Trade Organisation (WTO). This is still the forum where, according to the GATT preamble, countries seek a 'substantial reduction of tariffs and other trade barriers and the elimination of preferences, on a reciprocal and mutually advantageous basis'.

ACTIVITY

Describe and explain the role of international organisations in the promotion of globalisation processes.

Foreign direct investment (FDI)

A **foreign direct investment** (FDI) is a controlling ownership in a business enterprise in one country by a company or organisation based in another country. International agreements to promote trade have enabled a significant increase in FDI since the Second World War. Governments wishing to receive investment in their economies from abroad must make it legal for foreigners to own and control businesses and property in their country. FDI can be 'inorganic', where a foreign firm buys a company in another country; for example, the UK received £12 billion of FDI from the USA when Kraft Foods bought Cadburys in 2010. FDI can also be 'organic', where foreign investment expands the operations of an existing business in a foreign country; for example, Nissan's decision in 2015 to invest £100 million in its existing car plant in Sunderland to produce the next generation of its Juke car. In this latter example the FDI will guarantee 6700 jobs at the plant and support a further 27,000 positions in the supply chain.

The role of national governments in FDI

National governments encourage FDI because the influx of capital can boost economic growth, employment opportunities and tax revenues for the host country. FDI often leads to improvements in infrastructure that can also benefit domestic firms. Greater competition from new companies can lead to productivity gains and greater efficiency in the host country. This is because of the transfer of skills through additional training and job creation, and the availability of more advanced technology and access to research and development (R&D) resources.

Government mechanisms to encourage FDI include:

- low corporation tax and low individual income tax rates
- tax 'holidays' or other types of tax concessions
- preferential tariffs (sometimes with **special economic zones**)
- **export processing zones** (special economic zones established to produce goods for export)
- free or subsidised land
- visas and approval for the relocation and expatriation of key staff
- subsidies for infrastructure spending
- subsidies for R&D.

Privatisation and market liberalisation

Some national governments view the **privatisation** of companies and the **liberalisation** of markets as a means of increasing economic growth. Privatisation means transferring ownership of a business, agency, service or property from the public (government-controlled) sector to the private sector. It can also mean government outsourcing of services to private firms, such as G4S managing prisons in the UK. Liberalisation means to reduce and remove rules restricting economic activity and companies. Governments do this because they think that rules make businesses inefficient, and reduce motivation and innovation. There are also concerns that regulators may become more concerned with the industries they are regulating than with the interests of consumers and the general public.

CASE STUDY: Privatisation and market liberalisation in the UK

The Conservative government of Margaret Thatcher started a programme of deregulation and privatisation after the general election of 1979. This included British Telecom (1984), local bus services (the Transport Act 1985), British Gas (1986) and British Rail (1993). In 1997 a Labour government took the Bank of England out of direct government control and removed its power to control the financial activities of banks in the UK. The Legislative and Regulatory Reform Act 2006 was introduced to enable ministers to make Regulatory Reform Orders (RROs) to deal with older laws that they deemed to be out of date, obscure or irrelevant.

Trade blocs

A **trade bloc** is a type of intergovernmental agreement, where barriers to trade in a world region (tariffs and non-tariff barriers) are reduced or eliminated among the participating states. They can be stand-alone agreements between several states (such as the Association of Southeast Asian Nations) or part of a regional bloc (such as the European Union).

Governments within trade blocs recognise that innovation and branding add value to secondary and tertiary products over time. However, primary products tend to go through boom and bust cycles, in terms of their value. This generic relationship is shown in Figure 4.3.

Therefore governments seek to create better trade terms for domestic companies that produce secondary and tertiary products. They hope that, if they can increase their trade, they will be encouraged to invest in their workforce and their products. This will benefit the country by increasing the tax base and creating high skill levels and high wage employment. This helps to explain why governments choose to join trade blocs. They are also attracted by an increase in FDI, the majority of which will come from members of the trade bloc, but also countries outside the trade bloc will try to invest inside it to avoid tariffs and to access the larger market.

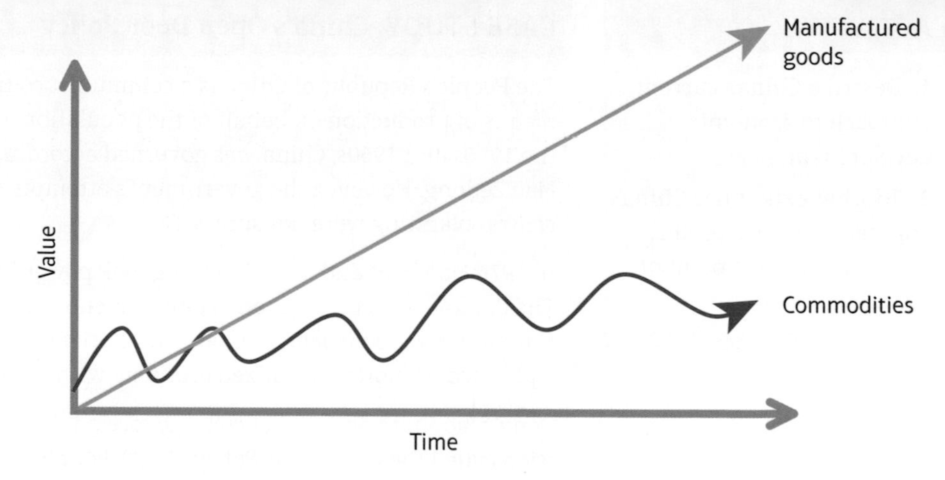

Figure 4.3: Generic change over time in the value of manufactured goods and commodities

Table 4.5: Advantages and disadvantages of trade bloc membership

Advantages of trade bloc membership	Disadvantages of trade bloc membership
Bigger markets (but no extra taxes): The UK has a population of 65 million and the EU a population of 508 million. UK companies like Tesco have benefited by expanding into other EU countries and sourcing their goods at the best price from within the 28 member states.	**Loss of sovereignty:** A trading bloc (and political union) is likely to lead to some loss of sovereignty. For example, the EU deals not only with trade matters but also with human rights, consumer protection, greenhouse gas emissions and other issues only marginally related to trade.
National firms can merge to form transnational companies: TNCs can compete globally, but they need large domestic markets to generate economies of scale. Increased sales lead to lower relative production costs and hence higher profits and consequent investment; for example, Vodafone became the world's largest mobile telecommunications company in 2000 by merging with Germany's Mannesmann.	**Interdependence:** Because trading blocs increase trade among participating countries, the countries become increasingly dependent on one another. A disruption of trade within a trading bloc may have severe consequences for the economies of all participating countries. The current challenges facing the banking sector of all eurozone countries are a good example of this.
Protection from foreign competitors and political stability: For example, in 2007 the EU blocked £50 million of Chinese-made clothes from entering the UK because the annual quota had already been filled. (This was called the 'bra wars' in the tabloid newspapers.) The idea is to limit the import of cheap goods to protect domestic manufacturers. By limiting such confrontations, trade blocs are said to bring political stability.	**Compromise and concession:** Countries entering into a trade bloc must allow foreign firms to gain domestic market share, sometimes at the expense of local companies. They do this in the expectation that their consumers will benefit from better products and keener prices, as well as in the hope that their firms will also expand abroad.

Special economic zones and government subsidies

Governments can also build infrastructure to attract foreign direct investment and increase trade flows. Special economic zones (SEZs), sometimes known as export processing zones, are a good example. These are often large areas of land set aside by government in locations well placed for international trade, such as seaports. Companies can import raw materials and export finished manufactures from these zones without incurring domestic taxes, so encouraging TNCs to set up branch plants (subsidiaries). This provides employment for locals, payment of payroll taxes, and technology transfer. Domestic firms can link in to supply the branch plant. In time, the strengthening of the domestic economy in and around the zone may enable these domestic firms to expand abroad themselves.

ACTIVITY

Using Table 4.5 and other information:

Evaluate the political importance of trade blocs in helping countries compete in a globalised world.

ACTIVITY

1. Describe China's current approach to economic development.
2. To what extent has China's approach been successful from an economic point of view.

CASE STUDY: China's Open Door Policy

The People's Republic of China is a communist country. This means that the state controls the means of production on behalf of the population, to ensure that everyone benefits equally. In the 1950s and 1960s, China was governed according to Marxist–Leninist ideology as adapted by Mao Zedong. However, the government's attempts to improve agriculture and industry through central planning were not successful.

In 1976 Mao died and Deng Xiaoping took power. He carried out significant economic reforms. The Communist Party loosened governmental control over citizens' personal lives, and the communes were disbanded in favour of private land leases. This marked China's transition from a planned economy to a mixed economy with an increasingly open market environment.

Deng Xiaoping's desire to encourage foreign direct investment to help secure his reforms has been called the Open Door Policy (1978). Four SEZs were set up in 1980: Shenzhen, Zhuhai and Shantou in Guangdong, and Xiamen in Fujian. These are near Hong Kong, Macau and Taiwan, with the aim of attracting capital and business from these external Chinese communities. Economic growth was rapid: in 1978 China's exports were negligible but by 2013 China had overtaken the USA to become the world's biggest trading nation in goods. But, as is often the case with rapid economic growth, it was not sustained and in 2015/16 there was a significant slowdown in China's economy.

The unequal impacts of globalisation

The global economy is increasingly connected, but the degree of globalisation varies by country. Since the pattern of connection is not uniform across all countries, the benefits of globalisation are not distributed uniformly. For example, **global hubs** are 'cores' that demonstrate a number of intense connections to the rest of the world, because others wish to connect to them. Many hubs host major TNCs and have increasingly diverse populations. **Demographic** flows, as well as flows of finance, trade and ideas, move towards them.

Transport and communications technologies help networks operate efficiently, and air travel and containerisation enable the cheap and efficient movement of people and goods between these hubs. The transport and communications **networks** are focused on the core areas, and this increases the incentives for people and businesses to invest in the core regions. Telecommunications allow long-distance links between producers and consumers. The internet in particular has increased the capacity of the telecommunications network to handle data, with GIS and GPS enabling accurate monitoring of flows between nodes. The opportunities for creative and productive collaboration between well-connected places have increased, so the best-connected places are also best placed to specialise and produce goods. This is a **cumulative causation** process.

The global core–periphery model (Figure 4.4) shows this process.

- Clusters of globally significant economic activity are focused in core regions.
- Upward transition areas benefit from FDI.
- Offshoring and outsourcing come from the core (see page 184).
- Downward transition areas lose out due to loss of FDI and the negative impacts of offshoring and outsourcing.
- Resource frontiers experience economic boom and bust as the value of their commodities responds to changing demand from the upward transition areas and the **core regions**.

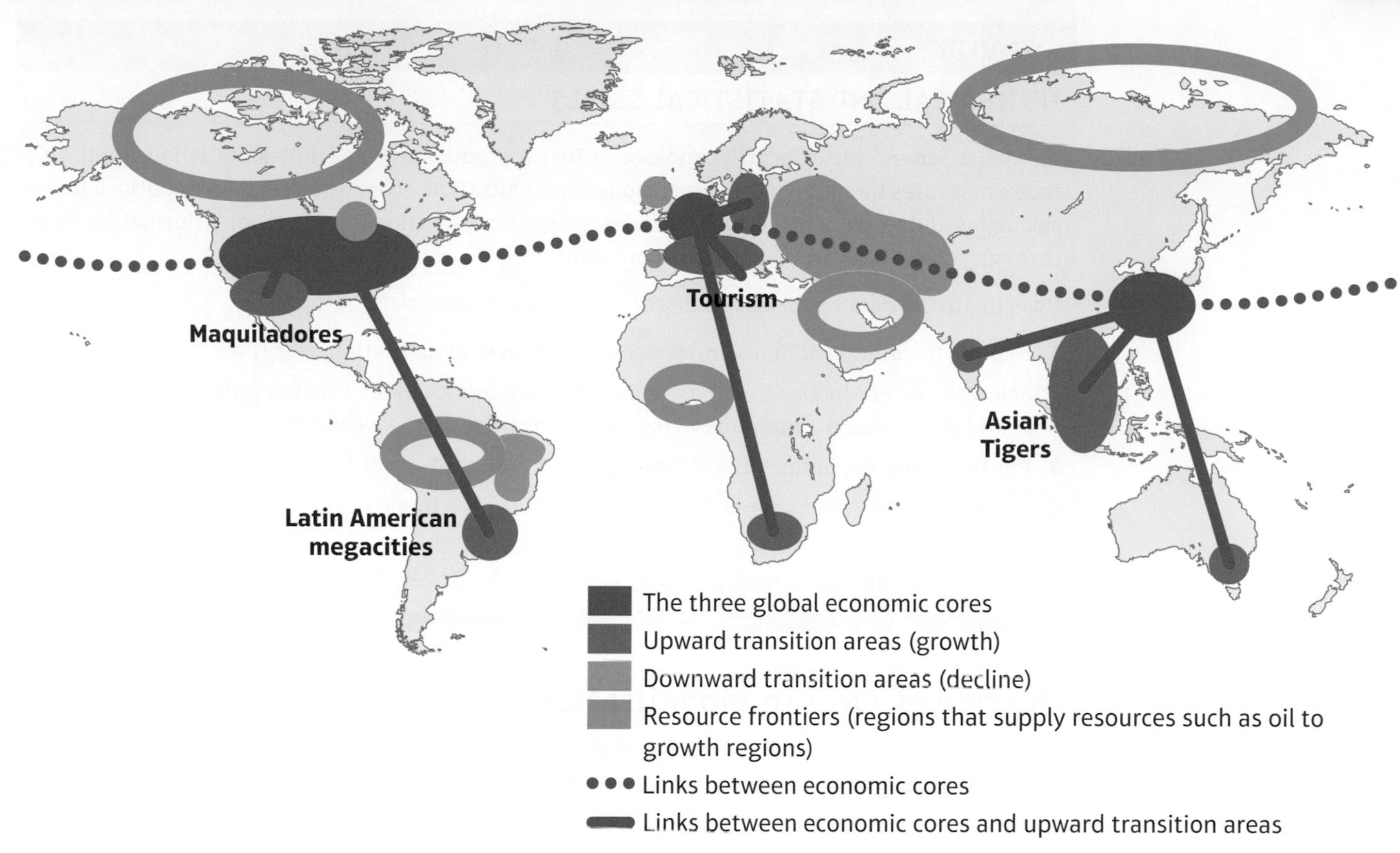

Figure 4.4: The global core–periphery model

Indices and indicators for measuring globalisation

The patterns and extent of globalisation can be measured in several ways:

- The KOF Globalisation Index (from 2002) measures the extent to which countries are socially, politically and economically linked to others. The more globalised a country is, the more links it will have in terms of tourism, communication, trade, foreign direct investment (FDI) and socio-political processes. It is one of the few measures that consider the political element, but its major weakness is that data for some countries is not available. In 2015, Ireland was ranked as the most globalised country overall, and the Solomon Islands as least: The top country in economic globalisation was Singapore, social globalisation was Austria and political globalisation was Italy.
- The US management consultancy AT Kearney produces a Global Cities Index. This index uses measures of business activity, human capital, information exchange, cultural experience and political engagement to rank cities in terms of the quantity and quality of their global connections. In 2014 New York and London were unchanged in the top spots in the index, but emerging economy cities such as Jakarta and Manila were increasing their global connections at a faster rate.
- Another index to measure globalisation is the IMF's Annual Report on Exchange Arrangements and Exchange Restrictions (AREAER). This records the existence of restrictions to trade in different countries.
- A further way to measure globalisation is the KAOPEN Index of openness to capital. This index was established by Chinn and Ito in 2005 and measures how easy it is to invest and withdraw investment in different countries.

ACTIVITY

NUMERICAL AND STATISTICAL SKILLS

The Swiss Federal Institute of Technology in Zurich produces the KOF Index of Globalisation. This index measures the three main dimensions of globalisation: economic, social and political. It also calculates and ranks actual economic flows, economic restrictions, and data on information flows through personal contact and cultural proximity.

Investigate the data on the KOF Index of Globalisation website:

1. Explore the animated maps and watch the spread of globalisation over time.
2. Select the Query Index tab, choose table format, select an index (such as economic), choose the year and then select a region (following the instructions on the website).
3. Put the countries, in the table of data obtained, into rank order from highest to lowest.
4. Repeat this for a later year and note the changes in the impact of globalisation for the countries in your rank order.
5. Explain the changes in rank positions using key ideas from this chapter.

The role of TNCs in globalisation

TNCs can be described as the 'architects' of globalisation. They bolt together different national markets with their supply chains and marketing strategies. It is often difficult to map the geography of a TNC because the largest firms have branch plants (factories or offices) in almost every country in the world, each having business links with local partners.

TNCs import and export goods and services, and in the process they make significant investments in foreign countries, sometimes by buying and selling licences for production or sale in foreign markets. Many engage in contract manufacturing, permitting a local manufacturer in a foreign country to produce their products, although some also open their own manufacturing facilities or assembly operations in those countries.

The prospect of increased sales and potential profit encourages TNCs to expand internationally. **Outsourcing** is where a business makes a contract with another company to complete some of the work, rather than doing it within the company. This may involve making products or providing a service; the aim is to reduce costs because other companies can do the work at a lower cost. **Offshoring** is where a company moves part of its operations to another country, often because labour costs are lower or because the economic situation is more favourable for profit making, such as lower taxes and tariffs.

CASE STUDY: Jaguar Land Rover – outsourcing and offshoring

Jaguar Land Rover (JLR) is the UK's largest automotive manufacturer. It has been owned by the Indian transnational corporation Tata since 2008. In 2014, JLR sold over 460,000 vehicles, with 26% of vehicles sold in China and 21% overseas (outside of UK, Europe and North America, not including China). JLR has 3 UK manufacturing sites, Solihull, Castle Bromwich and Halewood, employing over 17,000 employees.

In order to gain better access to emerging markets such as China and India, and to reduce its costs, the company has invested in international manufacturing sites. JLR established an assembly plant in India for several of its models in 2011.

In 2012, JLR established a joint partnership in China with Chinese company Chery Automobile Company, to produce vehicles and engines for several models. An assembly plant was built in

Yellow = Players, Orange = Attitudes and actions, Purple = Futures and uncertainties

Changshu, North of Shanghai, and production on the first model began in 2014. The investment in this plant was around £1 billion.

In 2014, JLR confirmed it was to build a new manufacturing facility in Rio de Janeiro Brazil, becoming the first UK manufacturer to invest in vehicle manufacture in the country and create around 400 jobs in the country.

In 2015, the company announced plans to build a new manufacturing plant in Slovakia and also to manufacture vehicles in Austria, with Austrian company Magna Steyr.

Glocalisation and the development of new markets

Glocalisation means adapting the goods or services of a business to increase consumer appeal in different local markets. It is a marketing technique for TNCs that combines cultural respect and self-interest. Glocalisation may be necessary because tastes differ from place to place, and clearly there is no advantage in offering a uniform product if it is not popular in a particular location. Glocalisation is commonly used to address religious or cultural objections, or regulatory or design restrictions. For example:

- BMW makes right-hand drive cars for the UK market, but left-hand drive cars for the German market because the countries drive on different sides of the road.
- Coca-Cola hosts different websites in different countries, so that each site is culturally appropriate and engaging, while still maintaining a strong brand identity.
- Tesco doesn't wrap fruit and vegetables in plastic in Thailand, because of the country's 'wet market' tradition where customers trust only produce they select by hand.
- Disney in the USA does not serve alcohol in its parks, but you can buy wine in Disneyland Paris. McDonalds sell McVeggie Burgers in India because the majority population is Hindu and vegetarian, especially avoiding beef.

Table 4.6: Impacts of TNCs on host countries

Positive impacts	Negative impacts
Raised living standards: FDI increases the productivity of the labour force in developing economies. This leads to higher wages and rising living standards.	**Tax avoidance:** TNCs pay tax in the lowest tax regimes they can. For example, in 2012 Amazon paid only £2.4 million corporation tax on £4.2 billion of sales in the UK.
Technology transfer: TNCs transfer technology from their parent companies to their branch plants. This can help accelerate economic development in emerging economies. An example is computer-aided design (CAD), which has enabled small manufacturing firms in developing countries to become suppliers to OECD manufacturers.	**Growing global inequalities:** TNCs cluster in selected economies concentrating FDI in favoured regions, such as eastern rather than western China. This can increase regional inequalities.
Political stability: In eastern Europe and China, investment from TNCs has contributed to economic growth. In providing work for the 'floating population', TNCs have helped reduce conflict between rural and urban populations.	**Environmental degradation:** TNCs can export the negative externalities of their activities to the less developed countries where they operate. For example, they may move manufacturing production out of the EU to avoid carbon taxes.
Higher environmental standards: TNCs have international brands to maintain and can set high environmental standards as a result. They bring good practice into countries that sometimes do not have good records of environmental protection. For example, Unilever launched its 'sustainable living' plan in 2010.	**Unemployment:** Outsourcing and offshoring can lead to unemployment in developed economies and higher social security spending.

CASE STUDY: Cargill – opening up new markets

TNCs have the scale to build new markets in countries that previously were weakly connected to the global economy. Cargill, the largest grain trader in the world, lends money to wheat farmers, owns and runs a vast grain transportation and storage business, and has a joint venture with Monsanto to provide grain seeds. Its scale enables huge investment in the infrastructure of food production. For example, Cargill has built nine animal-feed mills in Vietnam in the last 17 years. This creates huge demand for shrimp farmers and other input producers in Vietnam, as well as investment in transport infrastructure to connect agricultural areas to the mills. Such investments spur further agricultural and economic development.

ACTIVITY

Why has globalisation been slower to reach some regions of the world than others?

AS level exam-style question

Explain how changes in communication and ICT have accelerated globalisation. (6 marks)

Guidance

Explain means to give reasons. Each sentence in your answer should therefore have two parts: a named communication and ICT technology, and a reason why each one has accelerated globalisation.

A level exam-style question

Explain how national governments encourage foreign direct investment in their economies. (4 marks)

Guidance

Name at least three techniques and, for each, explain how it may lead to an increase in FDI.

Regions isolated from globalisation

Some peripheral locations remain largely 'switched off' from globalisation, as shown in Figure 4.4. There are few, if any, connections between these regions and the core global economy. Table 4.7 shows some of the physical/environmental, political and economic factors that help explain this isolation.

Table 4.7: Reasons for global isolation

Physical/environmental	Political	Economic
Distance from market discouraging FDI	Corruption and presence of organised crime/terrorist groups	High level of government debt
Wilderness (desert, tundra)	Weak commitment of government to development	Weak education levels and poor workforce skills
Low agricultural potential (such as a short growing season)	Civil or tribal conflict (active or legacy), possibly a relic of colonialism	Poor transport and telecommunications infrastructure
Lack (or poor quality) of energy and mineral deposits	Exclusion from trade blocs or disadvantaged by trade rules	Dependence on particular industries (vulnerable to commodity cycle)

CASE STUDY: North Korea – an isolated country

Following the Korean War (1950–53), Korea was split in half to form two countries, North Korea and South Korea. North Korea came under Soviet-sponsored communist control and, under its founder President Kim Il Sung, it adopted a policy of diplomatic and economic 'self-reliance' as a way of stopping outside influence.

After decades of economic mismanagement and resource misallocation, the country now relies heavily on Chinese aid to feed its population. The Korean War ended with an armistice and no 'final peaceful settlement' has yet been achieved. The Korean Demilitarised Zone (de facto a new border between North and South Korea) put into force a ceasefire, which continues to this day. Political factors therefore best explain North Korea's lack of connection to the global economy.

ACTIVITY

CARTOGRAPHIC SKILLS

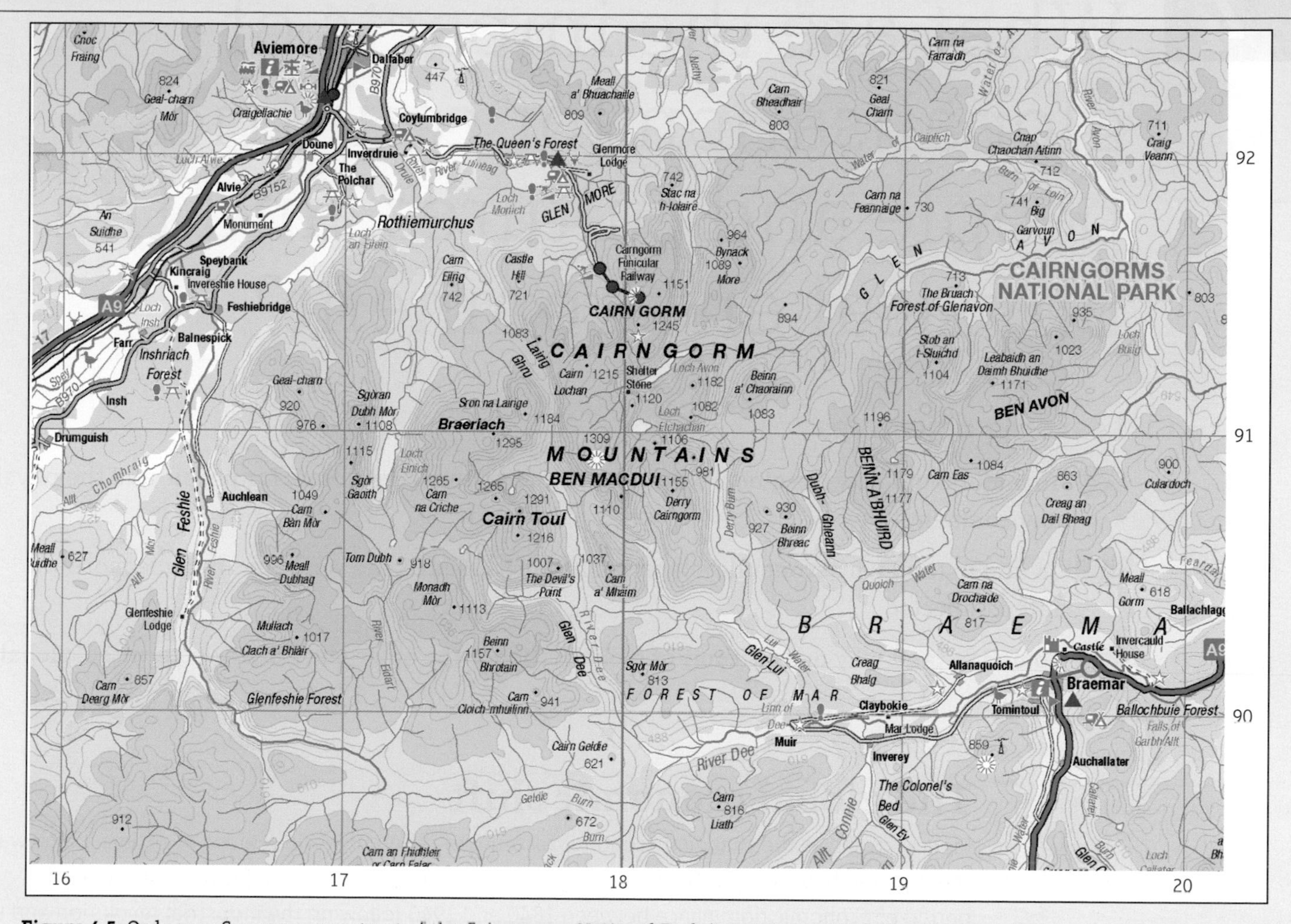

Figure 4.5: Ordnance Survey map extract of the Cairngorms National Park (1:250,000 scale) © Crown copyright 2016 OS 100030901

1. Using the information in Table 4.7, identify the human and physical features, using names, distances and six figure grid references from Figure 4.5 that may explain why the Cairngorms in Scotland is less well connected than many other regions of the UK.

2. Suggest what changes may be necessary to the human and physical characteristics of the Cairngorms, shown in Figure 4.5, to enable the area to become more connected to the rest of the UK and to the world.

CHAPTER 4

What are the impacts of globalisation for countries, people and cultures, and the physical environment?

Learning objectives

4.4 To understand how the global shift has created winners and losers for people and the physical environment

4.5 To understand why economic migration has increased as the world has become more interconnected

4.6 To understand why one outcome of globalisation is the emergence of a 'Westernised' global culture

The global economic shift

Figure 4.6 shows the changing economic fortunes of different world regions over time. China was the dominant economy at the beginning of the 19th century, and then Europe experienced a period of economic dominance between 1840 and 1940 following the industrial revolution and the maturing of industry. However, the steady rise of the economic power of the USA was also evident from the end of the 19th century, and since the end of the Second World War the USA has been the dominant global economy.

The global economic shift to Asia

The late 20th century saw China and India increase their share of global GDP. The IMF predicts that on current trends, China will regain the largest share of global GDP. This trend can be explained by the shift of manufacturing to east, southeast and south Asia and the outsourcing of services from developed to emerging economies. A greater share of economic activity brings with it an increasing share of global influence. The changing location of production is termed the '**global shift**'.

China is an emerging economy with a huge supply of cheap labour. Its arrival as a global economic power has been due to the shifting pattern of global trade and production; in particular, it has been able to lower the prices of labour-intensive manufacturing and increase the prices of goods (commodities). The shift began in the 1950s and accelerated in the 1960s as low-tech production, for example of textiles and toys, moved to the emerging 'Asian Tiger' economies such as Taiwan. It accelerated again in the 1980s when consumer electronics flourished in Japan, and again in the 1990s when electronics production was outsourced to China and other low-cost locations. The 2008 global financial crisis reinforced this shift, because trade balances between the emerging economies increased while those of developed countries experienced relative decline.

Extension

Assess the role of China in enabling the shift of the global economic centre of gravity towards south and east Asia. Investigate sources such as *The Economist* and *Financial Times*.

Yellow = Players, Orange = Attitudes and actions, Purple = Futures and uncertainties

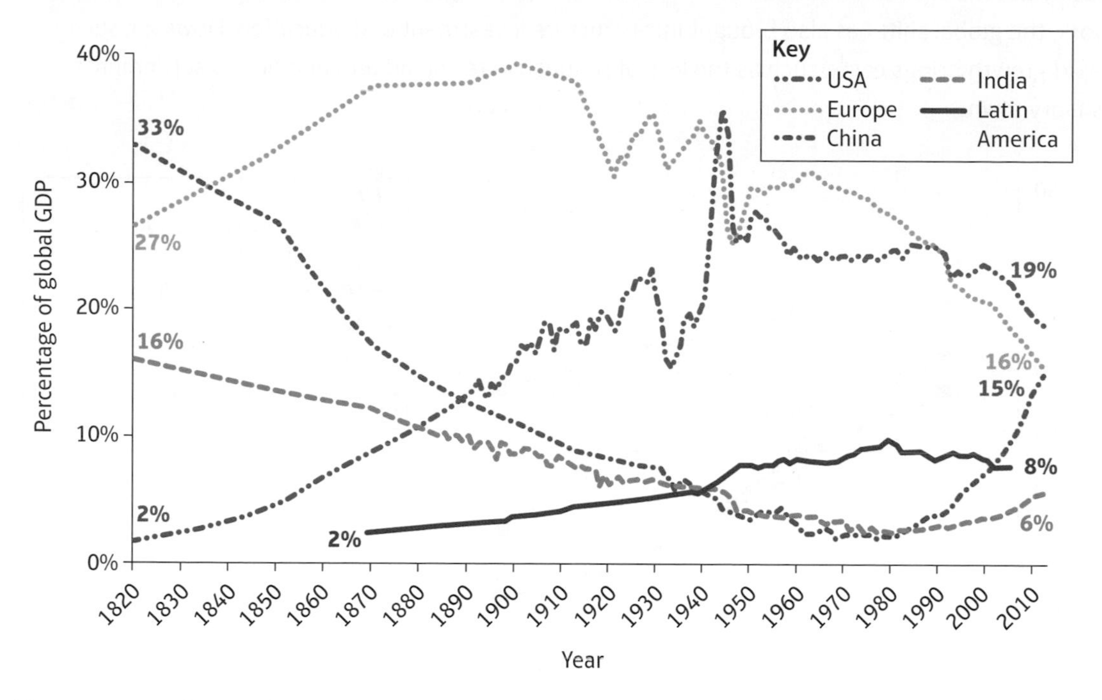

Figure 4.6: Distribution of share of global GDP over time

ACTIVITY

GRAPHICAL SKILLS

Study Figure 4.6. Describe and explain the changes to global GDP between 1820 and 2012.

CASE STUDY: China's role in the global shift in manufacturing

The USA and the EU dominated global trade for decades because their markets were far bigger than those of other economic powers, but this is no longer true. In 1990 China's share of global trade in manufactures (the sum of exports and imports) was just 2 per cent. By 2010 it had reached 10 per cent.

In 2002 the US market bought 26 per cent of Brazil's exports and the EU bought a further 25 per cent, so the US and the EU had significant influence in Brazil as a result. In 1990 the share of the Chinese market in Brazil's merchandise exports was 2 per cent; it was 5 per cent in the middle of the last decade and 15 per cent in 2010. At the same time the US market share declined to 10 per cent. China is now a more important market for Brazil than the USA, and China's influence in Brazil has consequently increased. This is why geographers believe that the global economic 'centre of gravity' has shifted to east, south-east and south Asia.

The global shift in services shows a different pattern. For example, financial services have started to disperse to world cities, largely to provide around-the-clock operation and to reflect the growing Far Eastern business market. For back-office and customer care call service operations, the English language is important. English is widely spoken in India, which helps to explain why the country accounts for 50 per cent of the global business outsourcing market. Bangalore was recently named the world's largest outsourcing city.

Winners and losers in a globalising world

The movement of the global economic centre of gravity to Asia has brought benefits. Millions have migrated from rural areas to higher-productivity secondary industry employment in urban areas. Figure 4.7 shows how the number of people living on less than $1.25 a day has declined since 1990. One of the United Nation's Millennium Development Goals was to halve, between 1990 and 2015, the number of people living on less than $1.25 a day: in 1990, 1.9 bn lived on less than this amount but by 2015, the number was 836 nm. Further to the poverty reduction associated with waged work, the global shift has also brought infrastructure investment and education. However, such rapid growth brings costs, such as the loss of productive farmland and unplanned settlements (shanty towns).

ACTIVITY

ICT SKILLS

1. Using Figure 4.7, describe what happened to the number of people in the world living on less than US$1 a day in the 1980s and 1990s.
2. Research United Nations website information to discover the success of the Millennium Development Goals (2000 to 2015) in further reducing the number of people living on less than US$1 a day.

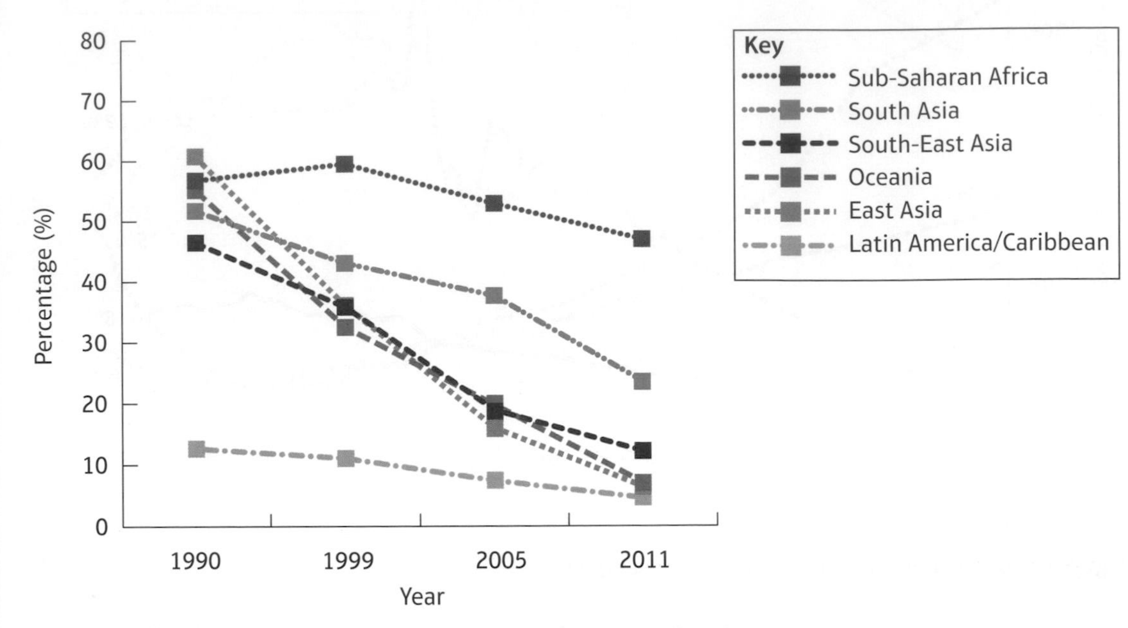

Figure 4.7: Number of people in the world living on less than US$1 a day

Synoptic link

Unplanned settlements in emerging economies are less able to cope with tectonic events than settlements in developed economies. The 2015 Nepal earthquake had a death toll of 9000, of which the majority were in unplanned settlements within the cities of Kathmandu and Bharatpur and in the village of Langtang. Many areas are also in need of regeneration, having decayed following the closure of industries as global shifts take place. (See page 44.)

CASE STUDY: The benefits and costs of the global shift in Mumbai, India

Mumbai is the largest city in India and one of the 10th largest in the world. It has benefited from the global shift and is home to clusters of key industries – in particular finance, nuclear power generation, music, film and textiles. A huge infrastructure investment has been required to support this growth.

- The Chhatrapati Shivaji International Airport is the main aviation hub for the city, and the second-busiest airport in India in terms of passenger traffic. The capacity of the airport is being increased to handle up to 40 million passengers annually. A new international airport has been sanctioned by the Indian government and will help relieve the increasing burden on the existing airport.
- The Jawaharlal Nehru Port, which currently handles 55–60 per cent of India's containerised cargo, is a hub port for the city, for India and for the Arabian Sea.
- Mumbai is also the headquarters of two of the Indian Railways' zones, the Central Railway and the Western Railway.

Rapid rates of **urbanisation** have led to poverty and unemployment, poor public health care, and poor civic and educational standards for a large section of the population. With overcrowding and shortages of land, Mumbai residents often live in cramped, relatively expensive housing. The average price of 95 m^2 of accommodation in the city is US$250,000, or 90 times GDP per capita. With flats out of reach, the proportion of people in slums has risen to around 60 per cent, compared with 20 per cent in Rio de Janeiro and Delhi. Of the rest, about half live in dilapidated, rent-controlled houses or flats for public-sector employees.

Environmental problems in developing countries

Rapid industrialisation and urbanisation resulting from the global shift have caused pollution, overexploitation of resources and the dumping of waste, which impact negatively on human health and wellbeing. The United Nations settlements programme UN Habitat estimates that 1 billion people live in urban slums. Their life expectancy, state of health, employment opportunities and standard of education are no higher than for people living in the countryside. Megacities generate a 'poison cocktail' of pollutants arising from traffic congestion and increased energy **consumption** – cities are responsible for 75 per cent of CO_2 emissions. The American management consultant firm Booz Allen Hamilton estimates that US$40 trillion needs to be invested by 2030 in energy and water networks to make cities sustainable.

CASE STUDY: Environmental damage in China

The Chinese government issues fairly strict environmental regulations, but monitoring and enforcement are the responsibility of local governments which lack the necessary expertise and are more interested in economic growth. Corruption is a further hindrance to effective enforcement. As a result, 60 per cent of groundwater is of poor or extremely poor quality, and 36 per cent of forests are facing pressure from urban expansion. The UN has identified the Yellow Sea and South China Sea as the most degraded marine areas on Earth, and 67 km^2 of land is lost to desertification annually. In January 2013, fine airborne particulates, which pose the largest health risks, were at levels as high as 993 $\mu g/m^3$ in Beijing, compared with World Health Organisation guidelines of no more than 25. The World Bank estimates that 5 of the world's most polluted cities are in China.

Problems of deindustrialisation in developed countries

Developed economies have gained from the reduced cost of manufacturing, redeployment of capital and labour to new industries, and the development of tertiary industries associated with the global shift. However, certain regions in developed countries have experienced considerable social and environmental problems associated with **deindustrialisation**. Industrial estates may be left derelict unless they can be repurposed for new industries or residential developments. Workers with skills and expectations associated with secondary industries may struggle to find employment in tertiary and high-technology industries. As a result, deindustrialised regions face above-average unemployment rates and have suffered **depopulation**, as people seek employment opportunities elsewhere, away from the high crime rates and dereliction often found in areas and cities around the world associated with traditional mining and manufacturing economies.

Synoptic link 

The 2012 London Olympics site was built on the derelict Lea Valley as part of a regeneration strategy for East London. As a result of deindustrialisation, the area had been the second-most deprived borough in London. The London Docklands area nearby has also seen massive regeneration and redevelopment, after suffering deindustrialisation when it was unable to compete with new container ports. (See pages 254–255.)

CASE STUDY: Deindustrialisation in Detroit, Michigan

During the 1920s Detroit was the fourth-largest city in the USA, with a thriving car industry (including Ford). However, industrial restructuring resulting from global shift, such as car manufacturing moving to Japan, resulted in dereliction, depopulation, crime and high unemployment. The city lost a quarter of its population and is now only the USA's 18th largest. The resulting reduction in tax revenue led to the city being declared bankrupt in 2013, with debts of £18.5 billion. Over one-third of all families in the city have income below the US federal government's official poverty level. In 2014, Detroit's murder rate was 45 per 100,000, the highest of any US city. City authorities estimate that two-thirds of murders are drugs-related.

ACTIVITY

NUMERICAL AND STATISTICAL SKILLS

The UK government collects data on population, deprivation and land use. This information is available from the Office for National Statistics and Neighbourhood Statistics websites. Middlesbrough is a traditional industrial city in the north-east of England that has suffered dereliction, depopulation, crime and high unemployment due to the decline of the chemical industry, partly as a result of global shift. You can access data showing the impact of deindustrialisation by using postcodes, such as TS6 6AZ for part of Middlesbrough.

1. Review the data from the 2011 census. Identify all indicators where Middlesbrough scores less well than England as a whole. In which measure is the difference between the local score and the national score greatest? How does this link to economic restructuring?

2. Review the seven measures of deprivation. Middlesbrough performs well on only two measures. Which ones are they, and how are they linked to economic restructuring?

3. Compare the data from your own postcode area with TS6 6AZ. What do the results tell you about the impact of the global shift and economic restructuring on your area?

Synoptic link

The Harris–Todaro model of rural–urban migration assumes that the migration decision is based on expected income differentials between rural and urban areas, rather than actual wage differentials. Therefore, stories of successful migration outcomes that filter back to the countryside can influence many more migration decisions, despite the fact that many migrants are less successful in the city. (See page 194.)

Economic migration in an interconnected world

The rapid growth of **megacities** is driven by rural to urban migration. Fast population growth in rural areas results in depressed wages (underemployment) and scarcity of goods. Loss or degradation of farmland and pastureland due to development, pollution, land grabs or conflict accelerate this process. Also, the growth of mass transportation and improved communication within developing countries means that there are fewer obstacles to movement. Higher productivity in cities leads to higher wages and an improving quality of life in urban areas. Moreover, public services are easier to fund in densely populated areas. Cities therefore have better health and education services, which tend to increase productivity and income. The bigger the perceived contrast in life chances between rural and urban areas, the greater will be the rate of migration.

Natural rates of population increase can also be high in urban areas, because many migrants are young adults – women of childbearing age. Half the world's urban population today is below the age of 25. As a result, the UN has predicted that Asia's urban population will rise from 1.4 billion now to 2.6 billion in 2030; Africa's will rise from 300 to 750 million; Latin America and the Caribbean's from 400 to 600 million.

Social and environmental challenges in megacities

People benefit from moving to a megacity only if their productivity gains exceed the increased cost of living. A rapidly growing megacity faces many challenges:

- **Overcrowding:** when cities become overcrowded, the resultant competition for resources drives up prices and can lead to a lower quality of life for many.
- **Poor housing:** because migration from the countryside is fast and on a huge scale, the city authorities cannot plan for it. Moreover, new migrants cannot afford high rents for decent housing, and are forced to live in poor conditions.
- **Traffic congestion:** rapid growth means that cities grow faster than the rate at which the authorities can build new routes. Also, informal building creates a densely populated city where it is hard to build new transport infrastructure (or proper housing).

Yellow = Players, Orange = Attitudes and actions, Purple = Futures and uncertainties

- **Air pollution:** rapid growth causes serious air pollution due to the concentration of solid fuel stoves, industrial fumes and waste, and emissions from personal and commercial vehicles serving the strong urban economy.

ACTIVITY

Describe and explain why megacities in developing countries experience significant social and environmental problems.

CASE STUDY: Social and environmental challenges in Karachi, Pakistan

Karachi has a population of over 24 million and the population is estimated to be growing at rate of 5 per cent annually. Most migrants are drawn from rural areas of Pakistan, but some come from neighbouring countries such as Afghanistan. UN Habitat estimates that half of Karachi's population lives in slums and, since around 75 per cent of these slum dwellers work in the informal sector, the tax base of the city is low. Services are provided on the basis of neighbourhoods pooling their resources to lay sewerage lines to the nearest disposal points, which are often natural storm drainage channels, rather than a proper system. The efficiency of the city's waste management is severely compromised as a result. Poverty causes the political and social alienation of young people, which in turn leads to ethnic violence and crime.

However, even worse conditions exist in ad hoc settlements consisting of homes made out of bamboo and industrial waste material in the dry riverbeds within the city. Similar settlements also exist on landfill sites and informal garbage sorting yards within and on the periphery of the city. In almost all such cases, the residents pay rent to criminal gangs supported by the police or officials of local government agencies. These settlements have no legal electricity or water connections, no schools, health centres or open spaces. These peripheral settlements have expanded greatly because of the city government's failure to manage the structure of the city and the great difficulty of finding a place to live in a more established slum near the city centre.

Increased international migration

When nations, regions and cities are strongly connected to similar places globally through the production and consumption of goods and services, they experience flows of **elite migrants**. These may be affluent individuals, whose investment is sought by countries through tax breaks and other inducements. Otherwise, they might be highly skilled or influential people within industries or TNCs. Such individuals face few barriers because their immigration is considered by governments to be beneficial to economic growth and competitiveness in a globalised economy. For example, in the UK the Immigration, Asylum and Nationality Act 2006 created a five-tier points system for awarding entry visas. Tier 1 is for high-value migrants (investing at least £1 million in a UK-based institution and earning the 75 points required for a visa). However, the countries that lose these 'elites' may experience a development disadvantage through lost investment or skills.

CASE STUDY: Russian oligarchs in 'Londongrad'

In the 1990s the Soviet Union was dissolved and became the Commonwealth of Independent States. During the process of dissolution, businesspeople acquired many Soviet-era industries. Many of these businesses proceeded to make their new owners incredibly rich, in particular those selling commodities to China and other countries growing as a result of the global shift. The Greek term 'oligarch' is often used to describe them, partly because of their great wealth but also because of the political power they have, linked to their wealth.

A significant number of Russian oligarchs have bought homes in the wealthiest boroughs of London, or, according to some journalists, 'Londongrad'. Some, like Len Blavatnik, Eugene Shvidler, Alexander Knaster, Konstantin Kagalovsky and Abram Reznikov, are expatriates, having taken permanent residence in London. All own property in both countries, and some have acquired controlling interests in major European companies. For example, Roman Abramovich bought Chelsea Football Club in 2003.

This migration allows the Russian elite access to global financial markets and in return the UK benefits from their investment: The combined total of bonds and loans raised by Russian businesses in London between 2004 and 2013 was over £250 billion; UK banks charge fees of up to 3 per cent of the amount borrowed.

Some countries rely on low-wage foreign labour as they globalise their economies. The Gulf states rely heavily on such migration in order to make the most of their wealth gained from oil and natural gas sales, especially to build cities, infrastructure and to deliver services to their citizens. Foreign nationals make up 30 per cent of the workforce in Saudi Arabia, rising to 80 per cent in Qatar and the United Arab Emirates. The low labour participation rates for females and the small native populations help explain this reliance on foreign workers. Most male migrants come from India, Pakistan, Egypt and Bangladesh, while female migrants come to the Gulf states from Sri Lanka, the Philippines and Bangladesh to work in domestic service. The emerging economies that are the source of these migrants receive **remittances** that the migrants send back home. For example, the World Bank estimated that in 2010 the Philippines received US$21.3 billion in remittances, accounting for 8.9 per cent of its GDP.

ACTIVITY

Suggest how economic globalisation may cause social challenges in emerging countries.

CASE STUDY: Low-wage economic migration to Dubai

Demographically, Dubai is not really an Arab city state, because fewer than one in eight of its residents are citizens of the UAE. Many educated Indians live a comfortable life in Dubai, and a few have become rich. For example, in 2005 Indian industrialist Bavaguthu Raghuram Shetty paid US$12.25 million to buy the 100th floor of the Burj Khalifa, a Dubai skyscraper twice as tall as New York's Empire State Building. However, South Asian guest workers, who make up more than 60 per cent of the population, work 12-hour shifts in high temperatures for about $5 a day. Many workers are trapped in the UAE by debts to agents in their source country, to whom they will have paid high fees for their work visas. Their motivation in migrating was to send remittances to their families, leaving them little to live on in the foreign country.

The costs and benefits of migration

Migration has costs as well as benefits for both source countries (where the migrants are leaving from) and host countries (where they are moving to). These impacts can be categorised as economic, political, social and environmental. In host countries, skilled migrants raise the GDP per capita because they have higher pay and employment rates than locals, and the resident population benefits from the dynamic effects of skilled immigration on productivity and **innovation**.

In source countries, emigration can relieve pressure on the labour market, and remittances can make a significant contribution to GDP. The social impact of migrants ought to be the same as an increase in the native-born population, as greater numbers of people increase the demand for service provision. Migrants also pay taxes and often work in essential public services, but services may be overused, there may be overcrowding, congestion and some crime. However, if the characteristics of migrants vary greatly from the local population, then their impact may be greater. Environmental impacts in host countries are associated with congestion and pressure on marginal land. In source countries the reverse may be true, with depopulation leading to dereliction and decline.

Yellow = Players, Orange = Attitudes and actions, Purple = Futures and uncertainties

The rise of UKIP in the UK and *Front National* in France shows the high level of popular concern regarding migration. In some countries, however, migration is seen as a development tool; for example, 10 per cent of the population of the Philippines are currently working abroad.

Table 4.8: The costs and benefits of migration for source and host countries

Benefits for source country	**Benefits for host country**
Reduced pressure on welfare spending Reduction of workforce balanced by remittances Returning migrants bring professional, social and political experience	Filling of skills gaps and labour shortages Migrants contribute to economy as consumers More taxpaying workers to support and offset an ageing population
Costs for source country Skills shortages in key areas of the economy Demographic imbalance – implications for reduced birth rate and higher dependency ratio in future Depopulation leading to dereliction	**Costs for host country** Extra community policing and translation costs Need for extra school places and health services in areas of concentrated immigration Pressure on rented sector of housing market

ACTIVITY

TECHNOLOGY/ICT SKILLS

Research the real and perceived costs and benefits of migration for the UK as both a host nation for immigrants and a source nation for emigrants. The Migration Observatory website at Oxford University is an excellent source of information on this topic.

Global culture

Culture is a system of shared meaning, as illustrated in Figure 4.8. It is collective, which means it is shared by groups of people and will have a common influence on the way they live their lives and interpret the world around them. Moreover, culture is dynamic; it evolves over time as it is passed from generation to generation. This process is influenced by contact with other cultures, which is increasing with globalisation.

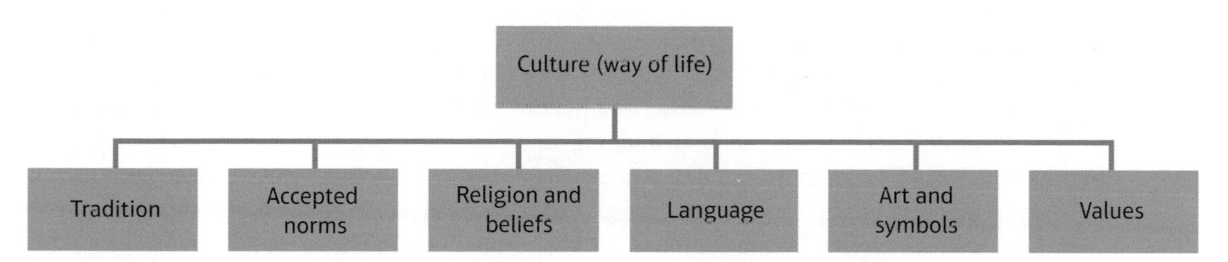

Figure 4.8: Components of culture

The spread of 'Westernised' culture

Globalisation presents increased opportunities for cultural exchange. In an increasingly interdependent global economy, with increasing cross-border movement of goods, services, capital, technology, people and ideas, cultures will influence others and in turn be influenced themselves. This is known as **cultural diffusion**. However, we have already seen that certain countries and regions have accrued greater benefits than others from globalisation. Therefore, players from the most successful countries and regions – the Western countries – are more able to project their culture through globalisation processes.

Cultures that perceive globalisation as an opportunity may be more open to external influences and more likely to change culturally, while cultures that perceive globalisation as a threat are more likely to adopt policies that seek to defend their cultural integrity. Table 4.9 shows contrasting views of the likely impact of cultural globalisation.

Table 4.9: Contrasting views of the potential impact of cultural globalisation

Group	Hyper-globalisers	Transformationalists	Sceptics
View	Globalisation is a successful process. Cultures will become ever more integrated as economies become more integrated. The world will move towards homogeneous cultures as a result.	Cultures are dynamic in their response to globalisation. It is not inevitable that the world will move to a homogeneous culture. All cultures will change, but in different ways, and new hybrid cultures may evolve.	Globalisation is profound in the core global economies and reflects their interdependence. Beyond this core there is a marginalisation, not destruction, of poorer groups (and their cultures).
Evidence	TNC marketing strategies create similar consumer demand across cultures, leading to uniformity in the components of culture and therefore a decline in local and national identity.	Rap music emerged in the cities of the USA and has spread globally. However, it has developed culturally distinct variants in different countries, such as France and Japan.	The rise of China, India and Iran will limit the dominance of 'Western' cultures and ensure continuing cultural heterogeneity at the global scale.

CASE STUDY: The Disney Corporation and Americanisation

Thomas L. Friedman, writing in the *New York Times* in 1999, suggested that American's need for bigger and better, their desire for fast food and high tech sold in a free market, and their promotion of this globally, was *de facto* promoting Americanisation. The largest TNCs are American (US) and therefore in promoting their brands globally, they are pushing American culture and values.

For example, the Disney Corporation is one of the world's largest media companies. It produces films, animations, television, theme parks and holidays. It builds its brand globally, redubbing films and animations for most markets. In doing so, it promotes American values, including family values, traditional morals and consumerism. A particular cultural meme that regularly appears in Disney films is the white wedding. The term originates from the white colour of the wedding dress, which first became popular with Victorian era elites after Queen Victoria wore a white lace dress at her wedding. However, the term now also encapsulates the entire Western wedding routine, especially in the Christian religious tradition. The global influence of Disney's promotion of this is clear, as the 'white wedding' is now the dominant global wedding style, despite most cultures having different wedding traditions.

New opportunities for disadvantaged groups

The emerging global culture has also created opportunities for disadvantaged groups. Globalisation has enabled mass migrations of people from different cultures. The British Nationality Act 1948 granted the subjects of the British Empire the right to live and work in the UK. The dissolution of the British Empire and associated social and economic dislocation created push factors encouraging migration from former colonies to the UK. Moreover, demand for labour in the UK in the aftermath of the Second World War provided a powerful pull factor. Migration rules have been progressively tightened since the 1960s and 1970s, but the result has been the establishment of many cultures within the UK.

Yellow = Players, Orange = Attitudes and actions, Purple = Futures and uncertainties

The Race Relations Act 1968 created rights for all races with respect to employment, housing, commercial and other services. This was extended again with the Race Relations Act 1976. Attitudes have developed over time to become known as the diversity agenda: governments have responsibility for legislating to prevent discrimination against all disadvantaged groups. The activity of global players, such as IGOs and TNCs, is also important in conveying these values. For example, football's governing body FIFA is a global player, representing football federations around the world. It promotes the sport as well as selling valuable television rights and marketing for competitions such as the World Cup. However, in doing so, it also promotes anti-discrimination policies and attitudes: an example is FIFA's #SayNoToRacism campaign.

CASE STUDY: The Paralympic movement

The first organised event for disabled athletes that coincided with the Olympic Games took place on the opening day of the 1948 Summer Olympics in London. Dr Ludwig Guttmann of Stoke Mandeville Hospital hosted a sports competition for British Second World War veteran patients with spinal cord injuries. The aim was to create an elite sports competition for people with disabilities that would be equivalent to the Olympic Games.

The first official Paralympic Games, no longer open solely to war veterans, was held in Rome in 1960. At Seoul in 1988, the Paralympic Summer Games were held directly after the Olympic Summer Games, in the same host city, using the same facilities. This set a precedent that has now been formalised. The success of the Paralympic movement alongside the globalisation of sports by key players such as the International Olympic Committee shows how globalisation can create new opportunities for disadvantaged groups.

ACTIVITY

Assess the extent to which globalisation has improved the quality of life for disadvantaged groups of people.

Cultural erosion

The ease and frequency with which people move around the world, and improvements in communications and the global marketing of styles, places and images can lead to a cultural supermarket effect. People are no longer confined to developing an identity based upon the place in which they live, but can choose from a wide range of different identities. They now adopt clothes, ways of speaking, values and lifestyles from any group of their choice. In some places this can lead to **cultural erosion,** including the loss of language, tradition and social relations. Cultural erosion can also change the built and natural environments.

Landscapes are shaped by our culture: they may be historic (the remains of an ancient culture's landscape) like Stonehenge, modern (a 'new' landscape reflecting the culture of today) like London's Docklands, or mixed (a fusion of ancient and modern) like the Louvre and its Pyramid in Paris. As a result, most cultural landscapes are mixed and complex, with traces of past cultures intermingled with those of today.

Developed countries protect their cultural landscapes. For example, the UK has 400,000 listed buildings, 20,000 scheduled ancient monuments and over 40 registered historic battlefields. Emerging economies may have a limited capacity to directly protect their cultural landscapes, in particular their ethnographic landscapes, but UNESCO aims to help preserve and promote the common heritage of humanity, protecting nearly 1000 natural, cultural and mixed sites worldwide. However, the least developed countries remain highly vulnerable to cultural erosion.

CASE STUDY: Loss of tribal lifestyles in Papua New Guinea

It is estimated that more than 7000 different cultural groups exist in Papua New Guinea, and most of them have their own language. Because of this diversity, there are many different cultural forms of art, dance, weaponry, costumes, singing, music and architecture. People typically live in villages or dispersed hamlets and rely on the subsistence farming of yams and taro. The principal livestock in traditional Papua New Guinea is the oceanic pig. To balance their diet, people hunt, collect wild plants and fish. People who become skilled at farming, hunting or fishing (and are generous) earn a great deal of respect in Papua New Guinea.

The island became a partly British and partly German colony in 1884. By 1905 British New Guinea had come under the control of Australia and the country gained its independence in 1975. Under colonial rule, Papua New Guineans experienced political, social and economic integration. Missionaries and administrators suppressed 'tribal' warfare to allow freedom of movement, and integrated villagers into the colonial economy as plantation workers and mission helpers. Missionary activities also led to the spread of Christianity and Western education. Class differences emerged as educated parents with good jobs provided for their children's future. Increasing intermarriage between different cultural groups meant that many couples failed to pass on their native language to their children, alienating their village kin, and in places direct cultural conflict has taken place.

In 1964 the discovery of copper in Bougainville resulted in the construction of a giant copper mine. It was argued that the profits from the mine would benefit all of Papua New Guinea. Bougainvilleans were suspicious of the motives of the Australians and Rio Tinto (the British TNC that established the mine). They were also resentful of the mainland Papua New Guineans who were brought in to work the mine. In November 1988 a guerrilla operation began that became the Bougainville Revolutionary Army (BRA). The conflict continued throughout the 1990s and it has been difficult for the police who have been fighting against their fellow citizens. The nation's 'law and order' problem is multifaceted, but attacks by youthful gangs, outbreaks of rioting and looting, and the resurgence of tribal warfare are major sources of disorder and misery.

Cultural erosion and opposition to globalisation

Concern about cultural impacts and economic and environmental exploitation has led to opposition to globalisation from some groups. These concerns are similar to those expressed during the independence movements and struggles during the process of **decolonisation**, when poverty was regarded as a product of colonial history. These groups are known as **structuralists**, as they explain the inequalities arising from globalisation with structures such as 'capital vs labour', 'men vs women', or 'one race against another'. These groups often oppose globalisation and argue that inequality in the global economy will only be resolved by structural change.

Extension

Compare the viewpoints of globalisation of the WSF and WEF shown in Table 4.10. What evidence exists to support each viewpoint?

However, other groups regard inequality in a globalised world as the product of winners and losers in global competition, and promote free trade and free markets as a means of eradicating inequality. They support globalisation because they believe that all countries will eventually receive the same benefits as Western economies and cultures. Table 4.10 compares the views of the World Economic Forum with those of the World Social Forum to illustrate the attitudes of pro- and anti-globalisation groups.

Yellow = Players, Orange = Attitudes and actions, Purple = Futures and uncertainties

Table 4.10: Contrasting attitudes to globalisation

Organisation	World Social Forum	World Economic Forum, Davos
Description	An open meeting place where social movements, networks, NGOs and other civil society organisations opposed to neo-liberalism and a world dominated by capital or by any form of imperialism come together to pursue their thinking, to debate ideas democratically, to formulate proposals, share their experiences freely and network for effective action.	Exclusive Swiss mountaintop resort (with very high security) where 2500 business leaders (and heads of state) meet to make deals and discuss global issues. Cooperation occurs in such forums. This is where the big deals on FDI and trade are made.
Viewpoint	Winnie Byanyima, the Oxfam executive director, has said: 'It is staggering that, in the 21st century, half of the world's population – that's three-and-a-half billion people – own no more than a tiny elite whose numbers could all fit comfortably on a double-decker bus.' She argues that this is no accident either, saying that growing inequality has been driven by a 'power grab' by wealthy elites, who have co-opted the political process to rig the rules of the economic system in their favour.	The WEF's mission is cited as 'committing to improving the state of the world by engaging business, political, academic, and other leaders of society to shape global, regional, and industry agendas'. They see globalisation as the means to drive economic growth and cooperation as the means to ensure that all countries and companies can benefit.

Loss of ecosystems can increase opposition to globalisation

Increased consumption associated with globalisation may have significant negative consequences for biodiversity. Some areas have experienced destruction of whole ecosystems, for example in the hinterlands of expanding megacities. Other areas have seen degradation of ecosystems, due to pollution and resource exploitation.

CASE STUDY: Environmental concerns and the anti-globalisation movement in India

According to a World Health Organisation study, of India's 3119 towns and cities, just 209 have partial sewage treatment facilities, and only eight have full wastewater treatment facilities. Over 100 Indian cities dump untreated sewage directly into the Ganges River. Environmentalists and others suggest that globalisation promotes negative externalities in the pursuit of economic growth. Evidence for this includes agricultural run-off, and small-scale factories along the rivers and lakes of India; fertilisers and pesticides used in intensive agriculture in north-west India are found in rivers, lakes and ground water; and flooding during monsoons worsen India's water pollution problem, as it washes contaminated soils into rivers and wetlands. Some economists argue that globalisation creates wealth which can fund better environmental management; they would argue that investment is needed to bridge the gap between the 29,000 million litres per day of sewage that India generates, and a treatment capacity of a mere 6000 million litres per day.

Environmental concerns and anti-globalisation concerns are often connected. India has seen campaigns against US soft drinks TNCs, due to their practices of drawing too much ground water so that the local area has suffered. India has also seen many campaigns against high-profile dam developments. Dam creation has been thought of as a way for India to catch up with the West, by providing cheap hydro-electric power. On a smaller scale, Jhola Aandolan is a popular movement, fighting against polyethylene bag use. They promote cloth, jute and paper bags to protect the natural environment. This also recognises the important role of recycling in the informal job sector of many Indian cities which have developed as a result of globalisation and rural-to-urban migration.

AS level exam-style question

Explain why some emerging economies have experienced major environmental problems as a result of the global shift. (6 marks)

Guidance

Brief, topical examples from the news would enhance your answer. Distinguish between problems due to the actions of TNCs and poor enforcement of regulations by governments.

A level exam-style question

Explain why globalisation has caused change to built environments in economically emerging countries. (4 marks)

Guidance

Remember that changes can be positive as well as negative. Consider both the formal and informal changes required to accommodate larger urban populations.

CHAPTER 4

What are the consequences of globalisation for global development and the physical environment?

Learning objectives

- **4.7** To understand why globalisation has led to a better quality of life for some, and widening inequality for others
- **4.8** To understand how globalisation has led to social and political and environmental tensions
- **4.9** To understand ethical and environmental concerns about and responses to unsustainability and consumerism

Measuring quality of life and inequality

Quality of life or standard of living is not easy to measure. It is possible to use standardised, and therefore comparable, measures of economic activity. For example, gross domestic product per capita measures the total value of goods and services produced in a country over one year per person. It is measured in the same way for all countries and therefore enables meaningful comparison of quality of life, as shown in Figure 4.9.

ACTIVITY

Describe the pattern shown in Figure 4.9.

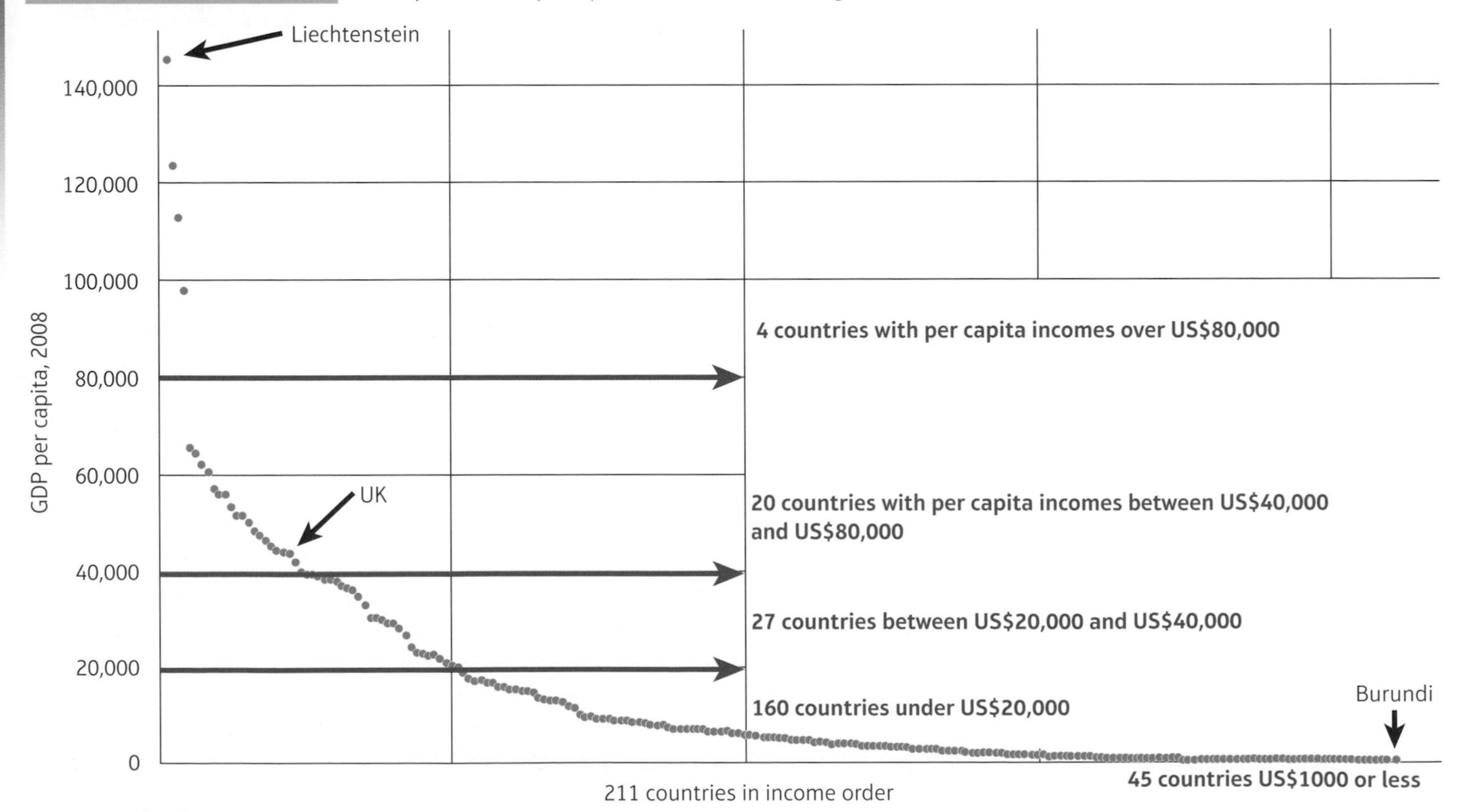

Figure 4.9: GDP per capita for countries in rank order, 2011

Single or **composite indicators** can hide wide variations within a given population. The total economic output of a country may be rising and, given a stable population, GDP per capita will rise as a result. However, if the rewards of this growth are distributed unequally, then the indicator will underestimate both the wealth of the wealthiest and the poverty of the poorest. This would then not accurately reflect the quality of life of the average citizen.

Indicators of development

There is more to quality of life than just economic activity. It is also necessary to measure social and environmental factors. For example, the UN's Human Development Index (HDI) measures average achievement in three basic dimensions of human development: a long and healthy life, knowledge, and a decent standard of living. This is known as a composite indicator because it measures more than one characteristic of a country's standard of living. It allows a comparison so that countries may consider their different human development outcomes. For example, Malaysia has a higher GNP per capita than Chile, but in Malaysia life expectancy at birth is about seven years shorter and expected years of schooling are 2.5 years shorter than in Chile. Chile therefore has a higher HDI value than Malaysia. This contrast can stimulate debate about the impact of government policies on quality of life.

There is also more to quality of life than that which is easily measured. Below is a list of significant factors:

- freedom of speech and political voice
- the impact of conflict or corruption
- equality of opportunity, such as level of gender and racial discrimination
- individual perception, such as aspirations and hope for the future
- quality of the environment, including housing
- social, economic and environmental **sustainability**
- religion and respect for tradition.

The traditional economic and social indicators of development omit these factors because they require qualitative measurements. As a result, the UN General Assembly placed 'happiness' on the global development agenda in July 2011. The Happiness Index goes beyond economic concerns and measures sustainable development, preservation and promotion of cultural values, conservation of the natural environment and the establishment of good governance. In 2015 the UN also established its Sustainable Development Goals.

ACTIVITY

1. Study the list of significant factors that may affect quality of life. For each, suggest how it could improve or reduce the quality of life of a person.

2. Compare the GII with other indicators of globalisation. Is there a link between the level of development and GII or the amount of globalisation and GII? Suggest reasons for your answers.

CASE STUDY: Gender Inequality Index (GII)

The Gender Inequality Index (GII) measures gender disparity. It has been created by the United Nations Development Programme (UNDP). According to the UNDP, this index is a composite measure which captures the loss of achievement within a country due to gender inequality. It uses three dimensions to do so: reproductive health, empowerment and labour market participation. Reproductive health measures maternal mortality and adolescent fertility rates. Women's health during pregnancy and childbearing is considered by the UN to be a sign of women's status in society. High rates of adolescent pregnancy are considered to reflect low educational opportunities for women and weak life choices. Empowerment measures female membership of parliament and participation in higher education. Both measures are considered to expand women's freedom by increasing their ability to question. They also measure access to information and public involvement. Labour market participation accounts for paid work, unpaid work, and actively looking for work, and is considered to measure economic aspects of gender inequality. However, it does not measure family work, a significant limitation. In 2014, Slovenia ranked highest, followed by Switzerland; the UK was ranked 39th out of 155 countries, USA 55th, Qatar 116th. The greatest inequality levels were measured in Niger and Yemen.

Measuring income inequality

The Lorenz curve

The Lorenz curve (Figure 4.10) can be used to graphically represent the distribution of income. The percentage of households is plotted on the x-axis, the percentage of income on the y-axis. The curve therefore represents income distribution. The line of equality shows what an equal distribution of wealth would look like, whereas the Lorenz curve shows the actual distribution.

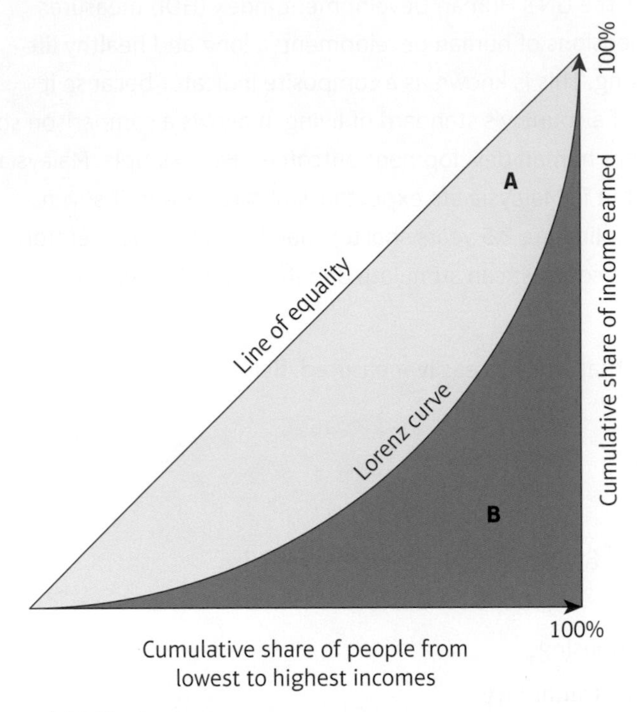

Figure 4.10: The Lorenz curve

The Gini coefficient

The Gini coefficient is a number between zero and one that measures the degree of inequality in the distribution of income or wealth. The coefficient would be 0 for a society in which each member received exactly the same amount. A coefficient of 1 would mean that one member got everything and the rest got nothing.

The Gini index is defined as a ratio of the areas on the Lorenz curve diagram. If the area between the Line of Equality and the Lorenz curve is A and the area under the Lorenz curve is B, then the Gini index = A ÷ (A + B). The higher the coefficient, the more unequal the distribution will be.

ACTIVITY

NUMERICAL AND STATISTICAL SKILLS

Figure 4.11 shows the Lorenz curve for the world distribution of wealth in 1988.

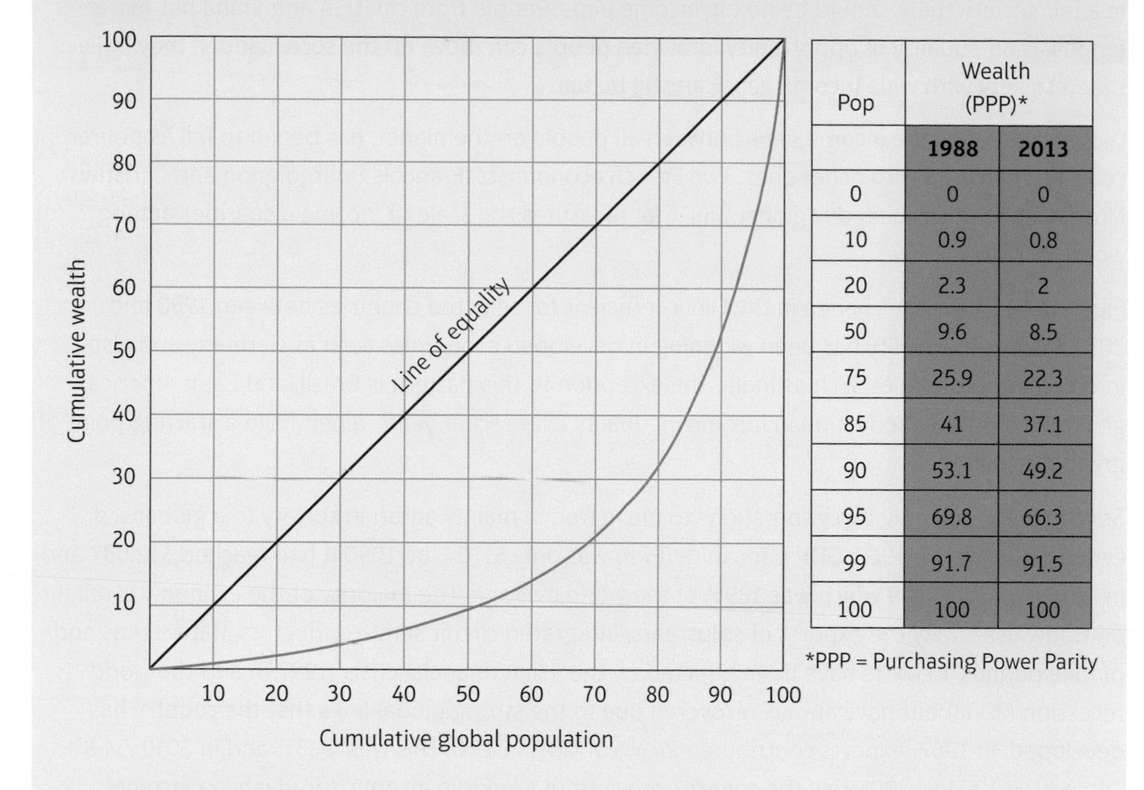

Pop	Wealth (PPP)*	
	1988	2013
0	0	0
10	0.9	0.8
20	2.3	2
50	9.6	8.5
75	25.9	22.3
85	41	37.1
90	53.1	49.2
95	69.8	66.3
99	91.7	91.5
100	100	100

*PPP = Purchasing Power Parity

Figure 4.11: The Lorenz curve for the world distribution of wealth, 1988

1. Calculate the percentage of global wealth for the poorest 30 per cent of the world's population in 1988.

2. Calculate the percentage of global wealth for the richest 30 per cent of the world population in 1988.

3. Using data from the table in Figure 4.11 and a piece of graph paper, plot a Lorenz curve for the world distribution of wealth in 2013.

4. Calculate the percentage of global wealth for the poorest 10 per cent of the world population in 2013.

5. Calculate the percentage of global wealth for the richest 10 per cent of the world population in 2013.

6. Describe and explain the change in the percentage global wealth of the poorest 10 per cent and richest 10 per cent of the world population between 1988 and 2013.

7. Review Figure 4.10. *A* equals 30.5. *B* equals 50. 30.5 divided by 50 equals 0.61. Therefore the Gini coefficient for the world distribution of wealth in 1988 was 0.61. Calculate the Gini coefficient for the world distribution of wealth in 2013 from your Lorenz curve.

8. Describe and explain the change in the Gini coefficient for the world population between 1988 and 2013.

Maths tip

To find a good estimate of the area on the graph paper area under the Lorenz curve (*A*), first count the complete squares. Then count the bisected squares. Add the number of complete squares to the half the number of bisected squares to estimate *A*. (Furthermore, *A* + *B* is equal to half the total area of the graph.)

Globalisation's winners and losers

Globalisation has created winners and losers between and within developed, emerging and the least developed economies. Some societies are more concerned about equality of opportunity, others more about equality of outcome. Europeans tend to be more egalitarian, believing that in a fair society there should be no big income gaps. People from the USA and China put more emphasis on equality of opportunity; provided people can move up the social ladder, they believe that a society with wide income gaps can still be fair.

Global inequality, the income gaps between all people on the planet, has begun to fall as poorer countries catch up with richer ones. Two French economists, François Bourguignon and Christian Morrisson, have calculated a 'global Gini' that measures the scale of income disparities among world citizens.

Extension

Using data to support your answer (e.g. World Bank, national government statistical websites), explain why globalisation may widen the development gap within countries while narrowing it between countries.

Figure 4.12 shows the change in the Gini coefficient for selected countries between 1980 and 2010. Income inequality has been widening in developed economies such as Germany and also in emerging economies such as India. The exception to this pattern is Brazil, and Latin America generally has had a reduction in income inequality over recent years, albeit from a starting point of great inequality.

South Korea took only two generations to move from a mainly agrarian society to a globalised economic leader. In 1962, GDP per capita (PPP) was only $1704, by 1990 it had reached $12,087 and in 2014 it was $33,629 which was 189% of the world average. The majority of the economy is reliant on trade, especially the exports of ships, cars, integrated circuit semi-conductors, flat screens and mobile phones. Exports have been affected by the Asian financial crisis (1997/8) and the world recession (2008) but have always recovered due to the strong global links that the country has developed. In 1962, exports contributed 24.4% of GDP, in 1998 this was 44.3% and in 2010, 54.8%. The period 1994 to 2004 saw the country move from a 'middle income' to 'advanced income' country. Global links have helped South Korea, starting with the USA in the 1950s and 1960s,and then Japan and the EU, and more recently China: in 2014, 26% of exports went to China and 13% to the USA. South Korean culture has also had a measure of global impact, for example, the musician Psy, and manhwa.

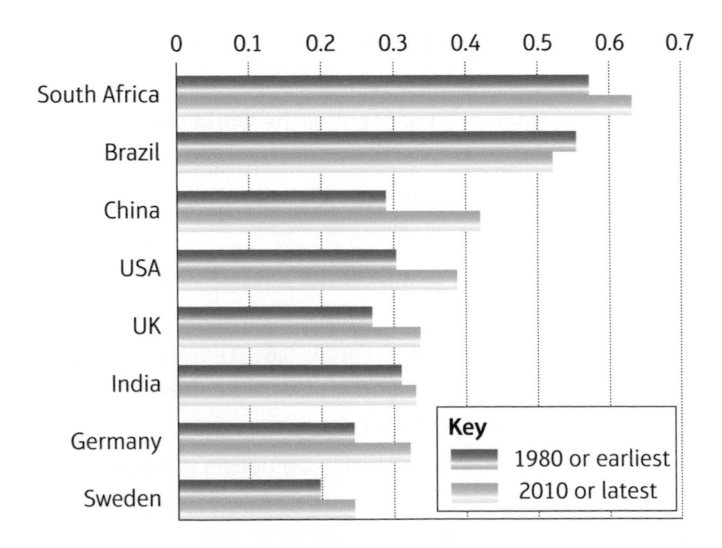

Figure 4.12: Change in income inequality over time for selected countries

Yellow = Players, Orange = Attitudes and actions, Purple = Futures and uncertainties

CASE STUDY: Air Pollution Indices

Winners and losers from globalisation can also be measured in terms of environmental quality. The eastern zones of UK cities were often dominated by poorer residents during the period of the industrial revolution, as the prevailing westerly winds blew air pollution from coal fires and factories over these areas. Similar issues regarding air quality can be seen in emerging economies today. In Malaysia, air quality is reported as the API (Air Pollutant Index). This measures sulphur dioxide (SO_2), nitrogen dioxide (NO_2), suspended particulates (PM10), carbon monoxide (CO) and ozone (O_3). If the API exceeds 500, a state of emergency is declared. This means that non-essential government services are suspended, and all ports in the affected area are closed. Private sector commercial and industrial activities may also be forced to suspend activities. The highest API value ever recorded was 1986, in Palangkaraya on 22 September 2015 during the 2015 South-East Asian haze. In Hong Kong, they use an Air Quality Health Index. This informs residents of the short-term health risk of air pollution, so they can take precautionary measures. The drive to industrialise and economically develop as part of the globalisation process can be seen as responsible for the increased energy use and air pollution emissions.

Trends in economic development

The world as a whole has not become more unequal as a result of globalisation. However, trends in economic and environmental development since 1970 do show that some regions are more unequal than others. After the Second World War, Western nations experienced increased economic wealth, while their former colonies and the communist world did not. This happened because the industrial economies of the 'West' were more advanced, with more productive workers and political policies that ensured a relatively equal distribution of the economic benefits. By the 1970s, average income per person in the ten richest countries was around 40 times higher than that in the ten poorest.

From around 1980 poorer countries began to catch up with richer ones, and within those countries richer people gained wealth faster than poorer people. The surge in emerging markets began with China's Open Door Policy (see page 182). By the 2000s the majority of emerging economies were growing faster than rich countries, to such an extent that global inequality started to fall, even as the gaps within many countries increased. Figure 4.13 shows the Port of Ningbo-Zhoushan, located on the coast of the East China Sea. It provides evidence of China's rapid growth; this port is now the busiest in the world in terms of cargo tonnage.

The huge changes that have swept the world economy since 1980 – globalisation, **deregulation**, the information technology revolution and the associated expansion of trade, capital flows and global supply chains – all narrowed the income gaps between countries, yet widened them within countries at the same time. The global reach of the modern economy hugely increased the size of markets and the rewards to the most successful. New technologies pushed up demand for well-educated people, boosting the incomes of skilled workers. The integration of some 1.5 billion emerging economy workers into the global market economy boosted economic returns, ensuring that the 'haves' would have more. It also exposed low-skilled workers in developed economies to competition from low-skilled workers in emerging economies. The relatively high wages of low-skilled workers in the developed world put them at a severe disadvantage and as a result these groups have experienced unemployment and wage uncertainties due to globalisation. As shown by the closure of steel works in the UK at the beginning of 2016, when Indian company Tata decided that it was no longer economical to run them.

Synoptic link

Globalisation has affected the success of industries in developed countries as competition has increased, for example, car manufacturing in the USA met competition from Japan, and the steel industry in the UK met competition from east Asian countries. These impacts shaped places in the UK for example and prompted the need for regeneration (see page 225).

CASE STUDY: RANA PLAZA

In 2013, a eight-storey factory building housing the Rana Plaza clothing company collapsed, killing over 1000 people. the building was typical of many that had been built quickly to meet the rising demands of the world clothing industry; where Bangladesh found itself able to fill a gap in the world market for cheap clothing. Prior to this disaster, extra floors had been added to the building, and cracks appeared in columns leading to a closure due to safety concerns. However, the building was then declared safe and workers had reluctantly returned to work, as other buildings had been known to collapse. Bangladesh has low wages and so is able to meet the needs of suppliers on high streets of developed countries, and the growing 'fast fashion' business due to global communication and IT links which allow it to respond quickly to consumer demands. The country's clothing exports increased from $6 bn in 2004 to $21 bn in 2013 and the country is regarded by some, such as Goldman Sachs, as having the potential for economic 'take-off'. There have been global campaigns to improve working conditions (health and safety), pay and child labour, but prior to 2013, this did not include the construction of buildings. Typical working data in these factories are: 10 hours a day, 6 days a week, 300 days a year for $90 a month. An estimated 20% of the Bangladesh population depend on the clothing industry.

The United Nations, clothing brands, unions and the Bangladesh government have combined to try and ensure that all factory buildings are properly surveyed. The Rana Plaza Arrangement Fund was set up to provide compensation for survivors and families of those killed, with contributions from 29 global brands who had their clothing made in the factory. 'Who is to blame for the disaster?' is a difficult question to answer due to the complexities of globalisation.

Figure 4.13: The port of Ningbo-Zhoushan, China

ACTIVITY

GRAPHICAL SKILLS

Look for the Human Development Reports in the Research and Publications section of the United Nations Development Programme (UNDP) website. The data can be viewed using the Public Data Explorer.

1. Download the data for trends in the Human Development Index, 1990–2015.
2. Choose one country from the very high human development category, one from high, one from medium and one from low.
   a. Plot their HDI scores against time on a line graph.
   b. Describe the trend in HDI over time for each country.
   c. Compare the trend for each country. Remember that these represent the broad groups (very high, high, medium and low) within the HDI.
   d. Explain any contrasting trends using the key ideas from this chapter.

Maths tip

Consider the range of data you will be plotting before you start drawing a graph. This will enable you to design the scale for the HDI axis so that you can distinguish between the trend lines clearly. You should also plot the countries in different colours to make them easier to compare.

Yellow = Players, Orange = Attitudes and actions, Purple = Futures and uncertainties

Globalisation and social and political tensions

Open borders, deregulation and encouragement of foreign direct investment have created culturally mixed societies. As a result, there are many migrant **diasporas**; for example, the UK is a **multicultural** society resulting from globalisation.

CASE STUDY: Immigration and the changing face of the UK

In the years immediately following the Second World War, the UK received over 500,000 migrants from the Caribbean. The pull factor was work due to the rebuilding of industry in the UK following the war. There were no intervening obstacles because visa restrictions did not apply at this time to countries in the British Commonwealth. The push factors were poverty in the Caribbean associated with decolonisation. These migrants settled mainly in London.

During the 1960s, UK textile industries in the Midlands and North West were booming and needed labour. About 750,000 Pakistanis and a million Indians were attracted by these pull factors. Again, few visa restrictions and cheaper air transport help to explain this migration.

Conflicts in other former British colonies also led to refugee migration to the UK. Two notable refugee migrations were of Ugandan Asians (1972) and Vietnamese boat people from Hong Kong (1975), who settled in cities around the UK. This was a much smaller migration, with about 30,000 Ugandan Asians and 20,000 Vietnamese involved.

As migration became more controversial, UK governments increased the restrictions on migrants from overseas. The one exception is EU migrants, because all EU citizens have the right of free movement within the EU. In 2004, poorer countries in central and eastern Europe joined the EU and since then it is estimated that over a million migrants from these countries have come to the UK. Unlike previous migrations, many more of these migrants have settled in provincial towns and rural areas.

Figure 4.14 shows the distribution of foreign-born diasporas in UK regions, based on the 2011 census. It shows that the number of foreign-born residents in England and Wales has increased by nearly 3 million since 2001 to 7.5 million people, which means that 13 per cent of UK residents were born outside the country. The most common countries of birth outside the UK are India, Poland and Pakistan. The number of ethnic white British people is down to 80 per cent, and London has become the first region where white British people have become a minority. In 2011 45 per cent (3.7 million) of people in the capital described themselves as white British, down from 58 per cent (4.3 million) in 2001.

Rapid social change and the rise of social tensions

Viewed from one perspective, globalisation has eroded the traditional power of the state and players such as unions to protect citizens from rapid social change. Decisions on the best location for investment, and therefore employment, are taken by TNCs and international financial institutions. At a national level, social security systems are being reshaped and the ability of governments to control migration flows is being reduced. This leads to insecurity at work, greater inequality within communities, and a loss of confidence in elected governments to manage the negative impacts of globalisation.

It is no surprise that social tensions are increasing as a result of these trends. As a result, popular right-wing political parties that argue for controls on globalisation are gaining support. In Europe, the *Front National* in France, Swedish Democrats and the Austrian Freedom Party are good examples of this. Figure 4.15 shows Marine Le Pen, President of the Front National in France, speaking at a rally in Paris on 1 May 2015.

Region – foreign-born population					
North East – foreign-born population **134**	India 14	Germany 12	Poland 11	China 7	Pakistan 7
North West – 515	Pakistan 74	Poland 49	India 47	Ireland 42	Germany 19
Yorkshire and the Humber – 403	Pakistan 68	India 40	Poland 37	Germany 21	Ireland 14
East Midlands – 410	India 62	Poland 46	Germany 23	Ireland 22	Pakistan 17
West Midlands – 538	Pakistan 84	India 83	Poland 47	Ireland 37	Jamaica 27
East of England – 540	India 49	Poland 40	Ireland 36	Germany 29	USA 27
London – 2674	India 254	Poland 123	Bangladesh 116	Ireland 108	Nigeria 93
South East – 910	India 72	South Africa 63	Ireland 52	Poland 51	Germany 50
South West – 347	Poland 36	Germany 31	Ireland 23	India 23	South Africa 18
Wales – 147	Poland 14	India 12	Germany 11	Ireland 9	Philippines 6
Scotland – 331	Poland 53	India 27	Ireland 23	Germany 20	Pakistan 16
Northern Ireland – 101	Ireland 26	Poland 15	Lithuania 6	Canada 4	Slovakia 4

Figure 4.14: Distribution of foreign-born UK residents by region, 2011 (in thousands)

ACTIVITY

1. Study Figure 4.14. Describe and explain the distribution of the foreign-born population in the UK.
2. Outline the costs and benefits of migration for host and source areas.
3. Suggest why nationalist political parties, such as Front National in France, have attracted increasing support.

Figure 4.15: Marine Le Pen, President of the *Front National* in France, speaking to a rally in the Place de l'Opéra, Paris, 1 May 2015

Yellow = Players, Orange = Attitudes and actions, Purple = Futures and uncertainties

According to a 2015 report 'The Rise of Populist Extremism in Europe' by Matthew Goodwin for Chatham House, support for these parties is mainly from specific social groups: economically insecure lower middle-class citizens, and skilled and unskilled manual workers. These groups have a profound hostility towards immigration, multiculturalism and cultural and ethnic diversity. They are less motivated by feelings of economic competition from immigrants and minority groups, than by the feeling that immigration and rising diversity threaten their national culture, the unity of the nation and the national way of life. Popular extremist parties suggest that minority groups (with an increasing focus on Muslims) pose an economic and major cultural threat to European societies. They also claim that mainstream parties are unable or unwilling to respond to this threat.

Environmental tensions from globalisation

The same transnational processes which have reshaped the world of work have significant impacts on the natural environment. The apparent weakness of government and non-government organisations to manage these changes and impacts also applies to transboundary environmental resources. An example is transboundary water sources. A water conflict may arise in any location that experiences rising demand, diminishing supply or conflicting user needs. The UN has established the Helsinki Rules to encourage management of transboundary resources on an equitable basis, but the pressures of development and population growth may prevent cooperation.

CASE STUDY: Transboundary water conflict in the Nile Basin

There is significant rising demand for water in the Nile Basin. For example, Ethiopia wants to use the Nile River for HEP plants and industrial development. All countries with access to the river have invested in irrigation following the 2008 food crisis. Moreover, high population growth rates are increasing demand – for example, Egypt's population is forecast to almost double to 150 million by 2050. There is a diminishing supply of water, due to desertification, salinisation and increased evaporation linked to climate change. Deteriorating water quality has resulted in the increased prevalence of waterborne disease. Furthermore, inefficient dams worsen problems of supply, due to siltation. Finally, there are many conflicting user needs. The Nile is the world's longest river, flowing through ten countries, with 360 million people depending on it for survival. About 85 per cent of its water originates in Eritrea and Ethiopia, but 94 per cent is used by Sudan and Egypt. The waters are split 75/25 between Egypt and Sudan by the Nile Waters Treaty of 1959.

However, Ethiopia and Kenya need increased water allocation for their development (HEP, irrigation of commercial farms, tourism facilities). Egypt fears that storing water behind the proposed Ethiopian Renaissance dam will reduce the capacity of its own Lake Nasser (thereby reducing the power-generating capacity of Egypt's giant hydroelectric plant at Aswan). Ethiopian officials have sought to allay fears by pointing out that storing water in their cooler climate will ensure much less water is lost to evaporation, but Egyptians are unconvinced. The Nile Basin Initiative is an IGO with the potential to allow for equitable use of this transboundary resource along the lines of the Helsinki rules. All countries have been participating in its meetings since 2001. The secretariat deals with technical matters and holds ministerial gatherings. It agrees on irrigation and HEP projects, many with World Bank support (agreements on equitable use are often a criterion for loans). However, Egypt is the dominant economic and military power in the region and is not prepared to negotiate away its advantages under the 1959 treaty. Moreover, many countries within the river basin are politically unstable and suffer from corruption, which hinders transnational cooperation.

ACTIVITY

Suggest how globalisation has increased environmental tension between some countries.

Attempts to control the spread of globalisation

Some countries, such as China, have taken more radical steps to control the impact of globalisation, through censorship.

CASE STUDY: The Great Firewall of China

Amnesty International reported in 2014 that China has the largest recorded number of imprisoned journalists and cyber-dissidents in the world. These people were imprisoned for communicating with groups abroad, signing online petitions, or calling for reform and an end to corruption. The government is concerned that online tools such as instant messaging services, chat rooms and text messages will help to organise or publicise anti-pollution and anti-corruption protests and ethnic riots.

The government of China has made 60 internet regulations, to be implemented by provincial branches of state-owned internet service providers. The BBC has estimated that more than 2 million people are directly or indirectly monitoring the internet for the Chinese government. This has the effect of blocking website content and restricting the internet access of individuals.

More broadly, the motivation for censorship is shown by a favourite saying of Deng Xiaoping (the paramount leader of China from 1978 until his retirement in 1992) in the early 1980s: 'If you open the window for fresh air, you have to expect some flies to blow in.' The saying is related to a period of economic reform in China that became known as the 'socialist market economy'. Superseding the political ideologies of the Cultural Revolution, these reforms led China towards a market economy, open to foreign investors. Nonetheless, the Communist Party of China has moved to protect its values and political ideas by 'swatting the flies' of other ideologies.

Other countries have sought to place limits on migration. For example, despite facing an imminent labour shortage due to its ageing population, Japan has done little to open itself up to immigration. Restrictive immigration laws bar the country's farms and factories from employing foreign labour, and stringent qualification requirements shut out skilled foreign professionals. There is a web of complex rules and procedures that discourage entrepreneurs from setting up in Japan. The Liberal Democratic Party unveiled a plan in 2008 calling for Japan to accept at least 10 million immigrants. However, opinion polls showed that most Japanese were opposed to this, and the Liberal Democrat Party went on to lose the election. A survey by the newspaper *Asahi Shimbun* showed that 65 per cent of respondents opposed a more open immigration policy.

Attitudes towards migration in the UK

Ipsos MORI conducts a monthly poll asking respondents to name the most important issue facing the UK, and after they reply they are then asked to name any 'other important issues'. Respondents are not prompted with particular topics; rather, they reply with whatever comes to mind. Immigration ranks consistently among the top five issues. As of June 2015, it was the issue picked most often by respondents (45 per cent). The 2013 British Social Attitudes Survey found that 77 per cent of respondents endorsed reducing immigration either 'a lot' or 'a little'. However, a 2010 survey by Transatlantic Trends found that 72 per cent of the UK public supported admitting more doctors and nurses from other countries to cope with increasing health care demands, and 51 per cent supported admitting more care workers to help manage the burdens of an ageing population.

Yellow = Players, Orange = Attitudes and actions, Purple = Futures and uncertainties

Trade protectionism

Protectionism is the economic policy of limiting trade between countries through tariffs on imported goods, restrictive quotas, and regulations that disadvantage foreign companies compared to domestic ones. Protectionists believe that there is a need for restrictions on trade in order to protect the economy, the standard of living of domestic workers and the dominant culture. Moreover, many argue that it is impossible for young businesses and industries at an early stage of development to become established unless they are protected from the full rigours of free trade for a period of time.

CASE STUDY: *L'exception culturelle*

L'exception culturelle is a political concept introduced by France in the General Agreement on Tariffs and Trade (GATT) negotiations in 1993. It allows for culture to be treated differently from other commercial products. The aim is to ensure that countries can restrict imports of foreign films and media in order to limit cultural erosion. It also allows countries to subsidise cultural activities, such as film production, in a way that would be banned for the manufacture of cars, for example.

France has taken advantage of *l'exception culturelle*. For example, the National Centre of Cinematography taxes cinema ticket sales and uses the revenue to aid the production and distribution of French cinema. Another example of protectionist measures is the *Loi sur l'audiovisuel*, which specifies that 'radio has to broadcast 40 per cent French songs'. The policy has had some limited success; for example, between 2005 and 2011, only 50 per cent of films shown in French cinemas were American imports, compared to 60 to 90 per cent in other European countries.

Protecting cultural identity

Some minority groups seek to retain their cultural identity in the face of a globalising world, while embracing its economic advantages. Figure 4.16 shows members of the Haida Nation standing with elders during a 1985 protest over logging on South Moresby Island in British Columbia, Canada. Following the 1993 Gwaii Haanas Agreement between the Canadian government and the Council of Haida Nation, the entire archipelago and surrounding waters became protected and were renamed the Gwaii Haanas National Park Reserve and Haida Heritage Site.

ACTIVITY

Evaluate the ways in which cultural identity can be protected from change due to globalisation forces.

Figure 4.16: Members of the Haida Nation, a Canadian First Nation, protesting against logging, November 1985

CASE STUDY: Canada's First Nations

The First Nations are the various aboriginal peoples of Canada. There are currently 634 recognised First Nations governments, which are represented at the Assembly of First Nations, which protects their rights and culture. They have cultures spanning thousands of years, often recorded in oral histories. Self-government has given First Nations powers that combine those of a province, school board, health board and municipality. For example, the Fort McKay First Nation in Alberta has worked with the Canadian oil sands industry to protect their land and culture, but also to benefit economically by providing services to the industry. This First Nation agreed a 20-km exclusion zone between an oil sands development, which produces 250,000 barrels of bitumen a day, and its Moose Lake reserve lands. Second, they negotiated contracts to provide services for the oil sands industry from their own First Nations companies worth more than US$100 million annually.

Concerns about the consumer society

Globalisation has resulted in masses of cheaply produced food and goods crossing between continents. Low production costs have driven global levels of **consumption** higher as prices have fallen relative to earnings. There are also more consumers: the global population was 3.7 billion in 1970 but is predicted by the UN to reach 10 billion by 2050.

Here are some examples of how higher incomes have changed consumer patterns:

- Global meat production has more than quadrupled in the last half century, to over 308 million tonnes in 2013, bringing with it considerable environmental and health costs due to the large-scale use of water, feed grains, antibiotics and grazing land required.
- Coffee production has doubled since the early 1960s – although an estimated 25 million coffee growers worldwide are at the mercy of extreme price volatility.
- Since 1960 global plastic production has continued to rise, with 270 million tonnes of plastics produced in 2013 alone. **Recycling** rates remain low and most plastics end up in landfill and the oceans – polluting ecosystems, entangling wildlife and blighting communities.
- The world's fleet of cars is now over 1 billion, with each vehicle contributing greenhouse gases and reducing air quality.

The rise of localism

Owing to globalisation, goods that were once sourced locally are now imported. The negative effects of our consumption are often experienced in places distant from where consumers live. Consumers often do not know the full footprint of the products they are buying, such as the embedded water in a T-shirt or steak, the pesticide exposure of cotton farmers, or the local devastation caused by timber companies cutting down softwood forests to produce paper.

However, some local groups and NGOs promote local sourcing as a response to globalisation, with the aim of increasing sustainability. Localism describes a range of political philosophies that prioritise local over regional and global. Generally, localism supports local production and consumption of goods, local control of government, and promotion of local history, local culture and local identity.

Yellow = Players, Orange = Attitudes and actions, Purple = Futures and uncertainties

CASE STUDY: Winchester Action on Climate Change (WinACC)

WinACC is an example of a local group that is promoting local actions to mitigate the negative effects of globalisation, in particular the negative environmental impact of a **consumer society**. WinACC helps people save energy in their homes and businesses, and save money and keep warm in winter. WinACC presses for better public transport, supports sustainable renewable energy schemes, large and small, and works closely with local councils to support the emerging low-carbon economy.

An example is the rolling out of the Cool Communities programme in Winchester. This creates EcoGroups, comprising five or six individuals from different households in a local community. They form neighbourhood groups that meet in each other's homes to discuss ways of reducing their carbon footprint in a fun and sociable environment. WinACC is also involved in the SAVE programme, a two-year project funded by OFGEM, the energy industry regulator, to research the best method of communicating energy reduction messages to the public.

The Bristol Pound is a local and community currency that was created to improve Bristol's local economy. Its primary aim is to support independent traders, in order to maintain diversity in business around the city. The scheme is a joint not-for-profit enterprise between the Bristol Pound Community Interest Company and Bristol Credit Union (Table 4.11).

Table 4.11: The economic, social and environmental costs and benefits of the Bristol Pound

	Costs	Benefits
Economic	Consumers can miss out on the price benefits of competition in national and regional markets. For example, the supermarkets Aldi and Lidl often charge lower prices than local traders, as a result of their European scale and buying power.	Money keeps on circulating locally to benefit local independent businesses in the area; this is called the local multiplier effect. If the money is spent at a supermarket chain instead, more than 80 per cent will leave the area almost immediately.
Social	Like all localism schemes, the Bristol Pound is very inclusive locally, but excludes outsiders. The mirror image of the social capital being built locally is the disincentive created to build wider connections.	Using a local currency creates stronger bonds between local consumers and businesses. This increases social capital (the value of social networking in spreading trust and cooperation so that everyone benefits).
Environmental	Global trade allows for commodities to be produced in the most resource-efficient location. For example, flowers can be grown with fewer energy inputs in Kenya than in European glasshouses. The costs of transport do not necessarily outweigh these gains.	Local trade reduces dependence on international trade and enhances self-sufficiency. This should decrease emissions through reduced transportation costs.

"It is not just about saving energy and reducing carbon emissions, it helps people feel more a part of the community"

"The Transition Network doesn't aim to frighten people about the future, it wants to find opportunities where there are threats."

"I love our community orchard that supplies local produce and reduces 'food miles'."

CASE STUDY: Transition towns and sustainability

A transition town is a community project that seeks to build resilience in response to the issues of peak oil, climate change and economic instability, by creating local groups that uphold the values of the transition network. The idea began in Totnes in the UK in 2006, and has now spread to more than 1000 communities worldwide, such as Albuquerque in the USA. The main purpose of transition towns is to raise awareness of sustainable living and to build local ecological resilience in the near future. Ecological resilience is the ability of a natural area to maintain its normal patterns and processes without being damaged by a disturbance. A key aim is build local resilience to the challenge of peak oil, by reducing dependence on fossil fuels. Communities are encouraged to seek out methods for reducing energy usage, as well as reducing their reliance on long supply chains that depend on fossil fuels. Food is a key area for transition; sometimes the slogan 'Food feet, not food miles' is used.

Initiatives so far have included creating community gardens and replacing ornamental tree plantings with fruit or nut trees for food. Communities seek to match the waste of one industry with another industry that uses that waste material, a scheme sometimes referred to as 'industrial symbiosis'. Repairing old items rather than throwing them away is encouraged. Central to the transition town movement is the idea that a life without oil could in fact be far more enjoyable and fulfilling than the current dependent pattern. Towns aspire to be somewhere much better to live in than places built on a consumer culture, which is perceived as wasteful and greedy. Therefore, the philosophy is to rebuild relationships within the community and with the natural world. Finally, transition initiatives generally include the global financial crisis as a further threat to local communities, and promote local complementary currencies as a way to create sustainable economies.

Fair trade and ethical consumption

Ethical consumption is practised through 'positive buying'. This means deliberately choosing a purchase because of the product's ethical nature, or alternatively it could mean a 'moral boycott', such as not buying a product because of concerns over its ethical nature. The aim of ethical consumption is to reduce the inequalities of global trade and improve the working conditions for disadvantaged groups. This approach was popularised by the UK magazine *Ethical Consumer*, which produces 'ratings tables' for different products and services based on ethical criteria such as animal rights, human rights and pollution and toxics. This approach empowers consumers to make ethically informed consumption choices and provides campaigners with reliable information on corporate behaviour.

ACTIVITY

Compare the success of 'transition towns' and 'fair trade' in helping ordinary people reduce the possible negative impacts of globalisation.

Such criteria-based ethical and environmental ratings have subsequently become commonplace for business-to-business corporate social responsibility and sustainability ratings, such as those provided by Innovest, Calvert and Domini. Businesses have become aware of the importance of ethical considerations, and increasingly present themselves to their consumers as morally and environmentally aware. For example, Marks & Spencer announced their 'Plan A' in 2007. This set out 100 commitments to source responsibly, reduce waste and help communities over five years. To support their goal of becoming the world's most sustainable retailer Marks & Spencer have launched their Plan A 2020. The plan combines 100 existing commitments with revised and new ones.

Yellow = Players, Orange = Attitudes and actions, Purple = Futures and uncertainties

CASE STUDY: The Fairtrade system

The international Fairtrade system seeks greater equity in international trade by promoting fair terms of trade to benefit farmers and workers. It is a trading partnership of NGOs, based on dialogue, transparency and respect that includes three producer networks, 25 Fairtrade organisations, Fairtrade International, and FLOCERT, the independent certification body. The Fairtrade mark on a product guarantees that its ingredients have been produced by small-scale farmer organisations or plantations that meet Fairtrade social, economic and environmental standards. The Fairtrade minimum price and an additional Fairtrade premium are paid to the producer to invest in business or community projects.

For certain products, such as coffee, cocoa, cotton and rice, farmers are encouraged to build small-scale democratic organisations. These offer rural families a stable income, which enables them to plan for the future. For some products, such as bananas, tea and flowers, Fairtrade also certifies plantations, the companies that employ large numbers of workers on estates. These large-scale producers must protect workers' basic rights, keep them safe and healthy, allow them freedom of association and collective bargaining, prevent discrimination and guarantee that they use no bonded or illegal child labour. Fairtrade also requires employers to pay wages that progress towards living wage benchmarks. The producers themselves decide how the Fairtrade premium should be invested to improve schools, transport, health care, sanitation, the natural environment and business equipment and practices.

Resource recycling

Recycling has a role in managing resource consumption and ecological footprints. EU countries are aiming to recycle at least half of their municipal waste by 2020. By doing so, waste will be diverted from landfill and raw materials reused for new consumption, offering lower environmental costs as well as employment. For example, recycling an aluminium can saves 95 per cent of the energy required to make a new one from scratch.

From a lifecycle perspective, increased recycling in the EU between 2001 and 2010 successfully cut greenhouse gas emissions from municipal waste by 56 per cent. However, this EU-wide figure hides great variation within the EU: Germany has a recycling rate of 62 per cent whereas Romania's rate is only 1 per cent. Moreover, some resources – such as paper, metals and glass – are commonly recycled, while other materials – such as cling film, medicine packaging, razor blades and crockery – are rarely recycled.

CASE STUDY: Keep Britain Tidy

In 1954 the Keep Britain Tidy organisation was formed by the National Federation of Women's Institutes. They wanted to address the rising problem of litter in what they saw as an emerging 'throwaway society'. By 1969, the iconic 'Tidyman' logo began appearing on bins and packaging, to encourage people not to drop litter. The organisation is much larger now and engaged in a range of environmental programmes, including projects to reduce litter on beaches, in parks and in schools. The organisation works with business, government and local communities to reduce and manage waste in environmentally friendly ways.

AS level exam-style question

Explain why it might be considered unethical to buy consumer products made in developing countries. (6 marks)

Guidance

You could categorise your reasons: social concerns and environmental concerns, for example.

A level exam-style question

Explain how globalisation may change cultural identity. (4 marks)

Guidance

Give specific case-study examples rather than making generic comments on cultural identity.

Summary: Knowledge check

Through reading this chapter and by completing the tasks and activities, as well as your wider reading, you should have learned the following and be able to demonstrate your knowledge and understanding of globalisation (Topic 4).

- **a.** What is the definition of globalisation?
- **b.** Explain five factors that have accelerated globalisation?
- **c.** What is time–space compression?
- **d.** Name the main intergovernmental institutions of the globalised world and describe their functions.
- **e.** How and why do national governments encourage foreign direct investment?
- **f.** How can the extent of globalisation be measured using indicators and indices?
- **g.** List the positive and negative impacts of TNCs on host countries.
- **h.** Give reasons why some regions are 'switched off' to globalisation.
- **i.** Describe the 'global shift'.
- **j.** What environmental problems have resulted from the global shift?
- **k.** Why have some regions in Western economies declined as a result of the global shift?
- **l.** Why have megacities grown rapidly over recent decades?
- **m.** List the costs and benefits of migration for host and source countries.
- **n.** Describe the three contrasting views of the potential impact of globalisation on cultural diversity.
- **o.** In what ways do nations and groups seek to protect themselves against the negative impacts of globalisation?

As well as checking your knowledge and understanding through these questions, you also need to be able to make links and explain ideas, such as:

- Globalisation is a long-standing process that has accelerated because of rapid developments in transport and communications.
- Political and economic decision making is an important factor in the acceleration of globalisation.
- Globalisation has affected some places and organisations more than others.
- The global shift has created winners and losers in a globalising world.
- The scale and pace of economic migration has increased as the world has become more interconnected.
- The emergence of a global culture, based on Western ideas and consumption, is one outcome of globalisation.
- Globalisation has led to dramatic increases in development for some countries, but also to widening development inequalities.
- Social and political tensions have resulted from the rapidity of global change caused by globalisation.
- Ethical and environmental concerns about unsustainability have led to increased localism and awareness of the impacts of a consumer society.

Preparing for your AS level exams:

Sample answer with comments

Assess the social and economic problems caused by the rapid growth of megacities. (12 marks)

The rapid growth of megacities is driven by rural to urban migration. Natural rates of population increase can also be high in urban areas that receive high numbers of migrants of childbearing age. It is the speed of this growth, widespread poverty and weak government that cause major social and environmental problems.

The introduction should refer to rates of megacity growth and the types of social and environmental problems that result, and not just in developing countries but worldwide.

For example, over a million people live in a slum called Dharavi in Mumbai, India. It is estimated that 60 per cent of Mumbai's population live in slums. Housing is a challenge for megacities because migration from the countryside is so fast. The city authorities cannot plan for it. Moreover, new migrants cannot afford high rents for better housing and are forced to live in poor conditions.

Several points mentioned but the links between them are unclear – how megacity growth leads to problems. It is good to see facts in terms of a place name and percentage.

Transport congestion is a challenge caused by rapid growth. Traffic congestion has been caused by rapid growth because the city is growing faster than the authorities can build new routes. Also, informal building has created a very densely populated city where it is hard to build new transport infrastructure.

Case study knowledge needed here, such as a brief example of congestion in a named megacity such as Mumbai and its plans for improving its transport system. There was also an opportunity here to refer to congestion in megacities in developed countries (e.g. London).

Air pollution is a challenge caused by rapid growth. The World Health Organisation ranks New Delhi as the world's most polluted city. New Delhi has the highest number of registered cars in India and taxis are replacing rickshaws at a rapid rate. Rapid growth causes such air pollution due to the proximity of industrial production to the city centre. Weak governments may struggle to enforce environmental regulations. Also, rapid growth concentrates diesel personal and commercial vehicles in the urban economy. The large numbers of people are the major problem facing megacities because they move around a lot, need somewhere to live and work, and need services and food.

This paragraph reads like a list and therefore few definite links are made, although the theme is sound. The concluding sentence is a valid judgement, but it has not really been supported in the rest of the answer.

Verdict

This is an average answer, which would have been improved by recognising that megacities exist in developed, emerging, and developing countries and that their problems can be slightly different. While each paragraph does focus on a different social or environmental problem caused by rapid megacity growth, there needed to be a more detailed explanation of how the issues arose. Some paragraphs have factual content but one megacity could have been a focus with facts, and others mentioned in context. The conclusion should have a strong assessment of the significance of the problems in order to answer the question.

Preparing for your A level exams

Sample answer with comments

Assess the role of technology and trade blocs in accelerating globalisation. (12 marks)

The world is now often described as being globalised, and this has taken place over the last 60 years, at an increasingly fast rate. Trading links between countries have existed since the European empires, but there is a growing economic interdependence of countries in a wider variety of goods and services today. There are many reasons for this, such as faster transport and communications and international capital flows between TNCs.

This is a sound start; the candidate avoids stating the rote-learned definition and links the definition of globalisation nicely into the question theme, thereby demonstrating understanding of it. Better would have been to include the idea of time–space compression.

New technologies such as the shipping container have contributed to globalisation. In the 1960s shipping costs accounted for 30 per cent of total cost. Today shipping costs make up less than 1 per cent. There has been a consequent boom in world trade. The global shift in manufacturing from high-cost west to low-cost east is the result. For example, Datang in Zhejiang Province, China, has become known as Sock City. The town currently produces 8 billion pairs of socks each year, a third of the world's sock production.

Too many short sentences and ideas are not linked together. Detailed explanations are essential at this level. There are examples of a technology and an impact, but the assessment is only implied. It needed to say how significant a contribution to the acceleration of globalisation it is.

New technologies such as satellite TV and the Internet have contributed to globalisation. Manufacturing in diverse locations can be coordinated using CAD/CAM. Internet communication technologies and social media have enabled the advertising industry to create global brands. These enable TNCs to build business worldwide. This is a highly significant contribution to the acceleration of globalisation. TNCs can scale up their cross-border connections as rapidly as the scale and capacity of the technology allow.

Begins to assess the significance of the contribution to globalisation made by the technology but lacks supporting evidence, for example when the internet developed and how many people and businesses are now linked to it.

Trade blocs such as the EU contribute to globalisation by increasing the size of markets for TNCs. The UK has a population of 65 million. The EU has a population of 505 million. UK companies like Tesco have benefited by expanding into other EU countries and sourcing their goods at the best price from across the 28 member states.

Trade blocs such as the EU contribute to globalisation by encouraging the creation of TNCs. TNCs can compete globally, but they need large domestic markets to generate economies of scale. Increased sales lead to lower production costs and therefore the potential for higher profits (which should encourage greater investment). For example, Unilever was a result of the merger of British soap maker Lever Brothers and Dutch margarine producer Unie.

The link to the question is unclear here. Transnational corporations may be one of the architects of globalisation. The example is old and is perhaps not a good example of acceleration.

Overall, therefore, technology makes a very significant contribution to the acceleration of globalisation. The extent to which countries have become interdependent is determined by the capacity of transport and communications technologies to sustain cross-border exchange. Trade blocs contribute to the acceleration of globalisation by encouraging the growth of TNCs that are themselves the architects of the cross-border links that characterise globalisation. However, this is not as significant a contribution as technology.

Begins to assess the contribution of technologies and trade blocs to the acceleration of globalisation. Better if these were put in context of the other factors that have accelerated globalisation, such as the WTO.

'Assess' means comparing the two factors mentioned to others that have enabled globalisation. This answer only explains how technology and trade blocs *do* accelerate globalisation. For example, trade blocs promote regional trade and investment, which may create world regions that are more isolated from globalisation, there may be other important factors.

Exam tip

Expectations step up at each level of study – i.e. from GCSE to AS to A level. Questions at A level may appear to be straightforward, but have a higher level of complexity behind them requiring considerable expansion of ideas with supporting evidence. Be prepared to do this!

Verdict

An average answer that begins to assess the role of technology and trade blocs but lacks supporting evidence. More assessment of the contributions to the acceleration of globalisation is needed. Case study evidence is present but it should have been in greater depth and up to date. The conclusion assesses the contributions to the acceleration of globalisation, which demonstrates understanding but again is largely unsupported. Planning answers to A level questions will hope you cope with the complexity and gain a high mark.

THINKING SYNOPTICALLY

Read the following extract carefully. Think about how all the geographical ideas link together or overlap. Answer the questions posed at the end of the article. This article first appeared in the *Financial Times* (online) on 19 May 2014 by Edwin Heathcote.

ARE CREATIVE PEOPLE THE KEY TO CITY REGENERATION?

The idea that the creative quarter is the key to the regeneration of any city has become so entrenched that it has become almost a cliché. The orthodoxy is that it is the cultural pioneers who are best able to turn around decaying districts and transform them from neglected and economically stagnant sites into thriving, hipsterish hotspots. Richard Florida's 2002 book *The Rise of the Creative Class* became the cornerstone of this notion and one that was adopted by planners, sociologists and politicians as a kind of default position. Creative quarters, what's not to like? But, perhaps, now it is time to reassess the results of this almost obsessive drive to attract creatives, to better understand how this process has worked, and whether it is always positive.

London, with its almost maniacal churn, a city irredeemably in thrall to property prices and with real estate as investment rather than home, is the ideal place to start. The city's bohemian centre has been shifting around for centuries, from Chelsea to Bloomsbury, from Soho to Shoreditch and now on to Hackney and Dalston. Each of these areas became artistic, literary and design centres and each was, in turn, gentrified as the creative classes made once unattractive areas edgy and seductive, a process that attracted younger, affluent middle classes who wanted to be associated with hipness. While this kind of regeneration can seem an unalloyed good thing to city boosters and economists, it has its downsides. The speed and intensity of change in London's property market has, in recent years, highlighted those problems. Creative quarters need time to grow. They need to build an infrastructure of the different trades, venues, office and workshop spaces and, most importantly, people, who are then able to embed themselves into the fabric of the city, establishing the kind of network that builds into a specific urban character something strong enough to attract others. There are no fixed rules for the kinds of infrastructure needed to foster a creative community but there are some features that have consistently helped. Among these is a particular and fine balance between cost and centrality. All the areas above, along with Clerkenwell, Stoke Newington, Peckham, Bethnal Green, Bermondsey and others were blessed with proximity to the city centre and an abundance of cheap space. That kind of loose-fit space, whether it was once industrial or warehouse, dockside or commercial, does not dictate how it should be used. A factory or a printing works, an office block or a warehouse can accommodate big studios or small incubator offices alongside apartments and cafés. Also the grain of historic fabric, even if it only 50 years old, adds an air of authenticity that always seems lacking in the new.

But there should not be too much heritage. Where the architecture is over-protected, rapid change is difficult. Where its use is too prescribed or zoned, again, change and adaptation are stymied. It is precisely in the blend and the flexibility of that particular cocktail of typology, age, disuse and adaptability to changing trends that a quarter's creative resilience can lie. London's booming property market, however, ensures that even the cheapest areas are no longer truly cheap and the kinds of spaces that were once attractive only to artists and designers – lofts and converted industrial spaces – have become among the most desirable residential spaces, to the extent that developers now build new domestic buildings to resemble industrial interiors. London's lofts are now, as they are in New York where the trend kicked off, out of bounds to creatives. Yet their successors are not being built. There is, understandably, no looseness in new development. Uses are ruthlessly prescribed as commercial, residential, retail or cultural – that's it. And the retail streets, once the city's rich incubator of everything from workshops to markets, are being built only to attract the big chains. There are no adaptable spaces, none of the big-scale industrial-type infrastructure that has proved so enduring. Developers and architects should build more anonymously, creating boxes with less defined uses. It is, of course, difficult to convince a bank of the value in this as-yet-undefined future. Regeneration in Britain is almost always conceived in terms of shops and shopping and apartments with balconies. It is extremely two dimensional.

Is anywhere in Europe doing it better?

Milan, another expensive metropolis, has done it well. The city might be known for fashion but design occupies an equally prominent role, notably with the Salone del Mobile, the world's biggest design fair, by far. The fair is on the unremarkable Fiera site but the real action goes on in events dotted around the city.

First it spread to the Zona Tortona, the residential and industrial area around the Via Tortona, and then on to Ventura Lambrate, a gritty industrial district on the city's edge. In both areas the design events have seeded workshops, cafés, studios and new cultural buildings, often accommodated in former industrial structures – exactly the kind of framework needed for a creative district. It can manage the difficult but critical shift from high fashion to artsy bohemianism within a single block.

Barcelona is often held up as the most visionary city and it is difficult to disagree. It is also instructive to see the parallels with London. Both are big port cities with rich historic centres, both are cosmopolitan and tourist centres and both are post-Olympic cities. UK politicians have enthusiastically picked up on Old Street's Silicon Roundabout (a place as unattractive as its name suggests) but Barcelona was in the forefront of developing a digital and innovation district with its 22@, in the former industrial district of Poblenou. The success of this huge chunk of creative city (equivalent to 115 historic city blocks) has been down to visionary politicians (notably former mayor Joan Clos, now head of UN Habitat), sophisticated urban planning and a clever use of zoning. This mixes residential with commercial, and historic industrial with fine contemporary architecture, so that the blend in types of space is maintained and the kind of gentrification that is so apparent and seemingly unstoppable in London has been halted or decelerated. It should not be forgotten that the city's infrastructure is almost impeccable: a fantastic metro system ensures one is never more than a few minutes away from a beach or a major station. It is a connected city in every way.

Berlin had an experience that was different again. As Germany reunited, its new capital found itself with a glut of empty commercial space as state and municipal bureaucracies that had once been duplicated were rationalised. In part, the freeing up of the massive accommodation of the Stasi, East Germany's overbearing secret police, ironically created the space that now houses the city's creatives. East Berlin's now-defunct industry, propped up by sales to other communist bloc economies, also left its legacy of generous space. A relative lack of speculation in the property markets helps Berlin sustain its creativity. Housing is mostly owned by pension funds and big organisations that are keen to secure long-term, hassle-free returns so rent is cheap and the young are able to stay in city centre accommodation as long as they like – although rental and purchase prices have accelerated recently. After the fall of the Wall it was Mitte and Kreuzberg that took on the creative mantle, followed more recently by one-time workers' district Friedrichshain. Yet even Berlin's coolest districts are not immune to gentrification – though here it tends to be bigger bars pushing out smaller ones and squatters being forced out of blocks that no one previously cared about.

These different narratives show there is no single rule, nor even a set of rules, that guarantees the seeding or the survival of a creative city. Yet, ironically, one of the critical factors may well be failure. Creative economies depend on slack and the kind of redundant space that is the result of economic crisis, political upheaval, the collapse of industry or some other massive change. The potential for revival is there, in the infrastructure, but people need other reasons to come.

ACADEMIC SKILLS

1. Is there a strong expression of opinion in this article? Whose opinion is presented?
2. To what extent does this article provide qualitative evidence and quantitative evidence?
3. From an academic geography point of view, how would you specifically improve this article? (i.e. what else would you have included).

ACADEMIC QUESTIONS

4. a. Who are the cultural pioneers?
   b. How can 'cultural pioneers' help to regenerate a range of different places?
5. Evaluate the success of using 'creative quarters' to regenerate urban areas?
6. a. Explain the role of globalisation processes in helping to create change in the European cities featured in this article.
   b. Identify the possible obstacles linked to globalisation processes that may restrict creative regeneration.
7. Who have been the important players in the regeneration of Barcelona?
8. How does globalisation combine with diversity to create a vibrant future for cities such as Milan?

Dynamic places

Regenerating places

Introduction

The announcements by Caparo Industries in October 2015 and, more recently, by Tata Steel that they were closing steel mills in Redcar, Scunthorpe and Port Talbot were a reminder that deindustrialisation is an inevitable by-product of globalisation. The world is a dynamic place, and global changes in trade (in this case, rising exports of low-cost steel from China) and global increase in demand have meant that long-established factories in places like Redcar and Scunthorpe began operating at a loss, until production was no longer viable.

As factories shut down, their locations and the identities of people living there change, often in quite unexpected ways. Across the UK as a whole, deindustrialisation has led to tremendous growth in a post-industrial economy, but at the same time it has created and deepened economic and social inequalities. For example, while Cambridge has made the most of its famous university to increase wealth across the region, and London has become a financial hub for transnational companies and the associated 24-hour-economy of West End shows, hotels and restaurants, other areas of the country have declined. Towns that were dependent on a particular industry, such as Ipswich, which grew up surrounded by agricultural production, and Liverpool, with its major port, are especially vulnerable to deindustrialisation, possible unemployment and associated decline and deprivation.

These issues mean that regeneration projects are sometimes needed to help places adjust to a post-industrial economy. The projects themselves can be controversial, and the impact they will have on people's identity needs to be explored and understood. Urban and rural regeneration programmes aim to improve the environment and quality of life for stakeholders, and also to make them more attractive to potential investors. However, these initiatives can cause conflict among various groups in the affected communities, which may have different views about the priorities and best strategies for regeneration.

In this topic

After studying this chapter, you will be able to discuss and explain the ideas and concepts contained within the following enquiry questions, and provide information on relevant located examples:

- What is the place where you live or study like, and how does it compare to another place?
- Why might regeneration be needed?
- How is regeneration managed?
- How successful is regeneration?

Figure 5.1: The new visitor centre at the Giant's Causeway, Northern Ireland's only World Heritage site and the only UNESCO site with a bus stop. Has the visitor centre altered the reality of living on the North Antrim coast?

Synoptic links

Changes in the UK and around the world have directly affected local places, in both urban and rural areas. This creates different problems, depending on where you live, and some of these problems are difficult to solve. Players include national and local government, transnational companies (TNCs) and workers in a variety of sectors. At one level, every individual is included, because everyone interacts and engages directly or indirectly with problems in their community. The attitudes of these players determine whether people are empowered to make changes in the places where they live and work, and this may determine whether communities and places succeed or permanently struggle.

Useful knowledge and understanding

During your previous studies of Geography (KS3 and KS4), you may have learned about some of the ideas and concepts covered in this chapter, such as:

- Economic sectors in the UK (the Clark-Fisher model)
- Changes to the UK industrial structure (globalisation, deindustrialisation, the post-industrial economy)
- Impacts of the decline (the post-industrial economy, the spiral of decline)
- Contrasts between different parts of the UK (the N-S divide)
- The need for regeneration (rebranding, re-imaging)
- The development of the digital economy (tertiary, quaternary, quinary, green sectors)
- The changing workforce (home working, self-employment)
- Demographic change (migration, services, the rural–urban continuum)
- Strategies to improve urban areas (comprehensive redevelopment, the New Deal for Communities).

This chapter will reinforce this learning, but also modify and extend your knowledge and understanding of regenerating places. Remember that the material in this chapter features in both the AS and A level examinations.

Skills covered within this topic

- Use of GIS to represent data about place characteristics
- Interpretation of oral accounts of the values and lived experiences of places from different interest groups and ethnic communities
- Use of the Index of Multiple Deprivation (IMD) database to understand variations in levels and types of deprivation
- Investigation of social media to understand how people relate to the places where they live
- Testing of the strength of relationships through the use of scatter graphs and Spearman's rank correlation
- Use of different newspaper sources to understand conflicting views about plans for regeneration
- Evaluation of different sources (music, photography, film, art, literature) and appreciation of why they create different representations and image of a local place
- Exploration of discursive/creative media sources to find out how place identity has been used as part of rebranding
- The interpretation of photographic and map evidence showing 'before and after' cross-sections of regenerated urban and rural places
- Interrogation of blog entries and other social media to understand different views of the success of regeneration projects

Fieldwork

This topic includes a fieldwork element, examining one aspect of a relevant urban or rural area. Students could investigate questions about local regeneration strategies, devising an appropriate methodology.

CHAPTER 5

How and why do places vary?

Learning objectives

4A.1 To understand the different ways in which economies can be classified and how they vary from place to place

4A.2 To understand why places have changed their function and characteristics over time

4A.3 To understand how past and present connections have shaped the economic and social characteristics of your chosen places

Economic activity and employment

When Colin Clark and Alan Fisher devised their three-sector theory of economic activity in the 1930s (Figure 5.2), they envisaged a positive model of change, in which countries moved from a focus on the primary to the secondary to the tertiary sector as they developed. Improved education and cultural change led to higher qualifications and the ability to obtain higher paid employment, with prospects of promotion. Later, the tertiary sector was supplemented by the quaternary and quinary sectors. Although the majority of workers in the UK today work in the tertiary sector, some industrial towns are still reliant on the secondary sector.

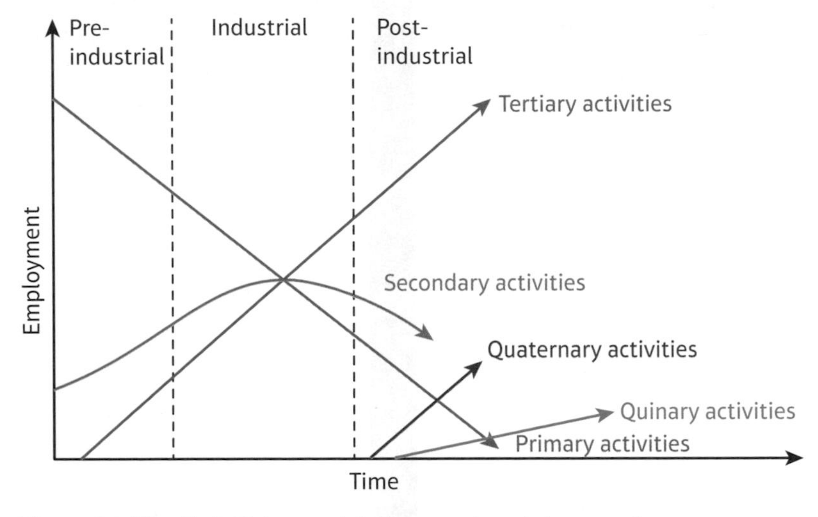

Figure 5.2: The Clark-Fisher model of economic activity over time

Table 5.1: The five economic sectors, their description and examples of places characterised by these jobs

Primary	Secondary	Tertiary	Quaternary	Quinary
Extraction of raw resources Mining Farming	Manufacturing and processing Iron and steel Car manufacturing	Service sector, tourism and banking	High-tech research and design	Knowledge management, consultancy Leadership/CEOs
Plympton, Dartmoor, SW England	Scunthorpe Sunderland, NE England	Aylesbury, SE England	Cambridge, E England	London, SE England

Yellow = Players, Orange = Attitudes and actions, Purple = Futures and uncertainties

Types of employment

In some UK towns, deindustrialisation has altered the types of employment that are available. With an increasing focus on the tertiary sector, the ability to be flexible and respond to global market changes makes employing people on a **temporary** basis increasingly desirable; businesses are starting to avoid employing workers on **permanent** contracts. In some rural areas, temporary migrant workers are vital to fill seasonal jobs but, for some workers, rising pressure on their household budget means that self-employment (for example taxi-drivers, and farmers selling food in high-street markets) can be a risky choice. In some towns there are also concerns about the exploitation of temporary and **self-employed** workers: since the discovery that nearly 3000 migrant workers were living illegally in Slough in 2013, the local council has used planes equipped with thermal imaging cameras to identify so-called 'sheds with beds'.

Port Talbot in South Wales is a good example of a town where heavy industry and chemical plants operate side-by-side with what might become the world's largest biomass power station, industrial and energy park. However, the older traditional industries remain economically marginal, as shown by the withdrawal of Tata steel from Port Talbot in 2016. Distinct patterns of employment have developed to reflect the location of major industries and key infrastructure, or economic sectors. Port Talbot has large numbers of part-time and full-time employees dependent on the success of its secondary industries, while nearby Swansea has experienced regeneration of its central area and waterfront which has attracted many university students and created tertiary jobs linked to shops and leisure activities.

In Windsor there is a divide between east and west, with the majority of **full-time** managers and directors living in the east, closer to London. In more rural parts of the country, for example in Oxfordshire, there are a many small clusters of **part-time** self-employed workers, including university students working in temporary jobs and graduates establishing start-up hubs (see Oxford Science Park, page 252).

ACTIVITY

TECHNOLOGY/ICT SKILLS

1. Explore the Datashine website to discover employment and other economic data about Swansea (using post code SA1 4AN) and Port Talbot (using post code SA13 1JB). Make notes on the patterns that you find.
2. Use the Datashine website to explore the pattern of economic activity in your local place.

Employment and social factors

Transition into a post-industrial economy has not been even across the UK. There are tremendous inequalities in the levels of pay and type of work. Hartlepool, a former shipbuilding and steel town in Teesside, has had the highest local unemployment rate in the country for some time, at almost twice the national average. In 2014 there was nobody in work in 30 per cent of the town's working-age households.

Unemployment has a clear impact on **health**, **life expectancy** and **education**. In Camden, one of North London's most deprived wards, 34 per cent of children live in poverty, compared to 21 per cent nationally. The unemployment rate for young people is higher than in the rest of the UK, and, although youth unemployment in Camden has been falling since 2013, in that year 7 per cent of young people were not in employment, education or training (NEETs). Research suggests that growing up in poverty, and the likelihood of dropping out of further education, are associated with poorer health in later life. Similarly, ill-health and disability have an impact on people's ability to work. While life expectancy has been increasing in Camden, to a point where the figures for both men and women in 2010 were above the national average. However, 43% of deaths were considered to be premature, especially amongst deprived communities.

'The causes of deaths that are disproportionately affecting those more from the deprived communities compared to the least deprived, and contributing to the life expectancy gap are cardiovascular conditions, lung cancer, chronic cirrhosis of the liver, respiratory diseases and suicides.'

Literacy tip

When describing patterns, tell the reader what is happening in the north, south, east and west of an area or region.

Look also for features of a map that could be described as 'surrounding' or 'clustered', as opposed to 'scattered' or 'surrounded by'.

Social deprivation does not always have to mean that businesses struggle. Although Sparkbrook in Birmingham ranks high on the Index of Multiple Deprivation, the British Pakistani community living there is highly entrepreneurial. Empty factories have become boutique fashion stores. Other small businesses have been established, such as bakeries, restaurants and halal steak houses, women-only gyms and wedding dress shops, and the trickle-down effects from these businesses all help diversify the market and create yet more businesses. These self-employed ventures were set up during a recession and, although there is always a risk of businesses collapsing, as the UK economy begins to grow again, they are likely to thrive as sustainable, permanent employers.

Synoptic link 

While income inequality between countries has been reducing, inequality within countries is on the rise (for example, in the USA where the percentage of income going to the richest 1 per cent doubled from 10 to 20 per cent between 1980 and 2010 (see page 205 and Figure 4.12).

Inequalities in pay levels

There are stark differences in rates of pay across the UK. The highest median earnings in 2011 (Figure 5.3) were found in south-west London, with London by far the most prosperous region. Older industrial cities, still suffering from deindustrialisation, tend to have lower average pay than elsewhere; however, the lowest rates are in north-west Wales. Interactive maps such as those found on the *Guardian*'s datablog (Figure 5.4) show how this inequality of pay correlates with both distribution of jobs in different economic sectors and **quality of life** across the country. With the poorest health occurring in post-industrial South Wales, north-west England and western Scotland, it is clear that communities in these areas have suffered the most from the changes brought about by globalisation. The healthiest area in the UK to live is Richmond upon Thames, in south-west London, within the most prosperous area.

London's economic structure is also one of great inequality. Although the living wage in the capital city has increased by 3 per cent (2014–15), there are still many people earning below the London living wage (estimated by the Living Wage Foundation to be £9.40 at the start of 2016). In April 2016, the government introduced a minimum National Living Wage for over 25s of £7.20 an hour, inevitably other organisations representing workers believe that this should be higher. Bankers and doctors are the highest earners, while those working in construction, education and other public sector jobs earn the least. In 2014 the Annual Survey of Hours and Earnings suggested that the fastest-growing rate of pay was for directors and senior managers (particularly insurance brokers, with a rise of 17 per cent), while pay growth for low earners was much slower, if not negative (for example waiters and waitresses working part-time saw their pay fall by 11 per cent).

How places change over time

Over time, the places where people choose to live change as the inhabitants reshape and reconfigure them to meet their shifting needs and priorities. The **function**, shape, land-use pattern and details of the surrounding social, economic and natural environments are determined by the types of people that live there and the many different influences on their lives.

Far from being tranquil natural landscapes, many rural areas are sites of intensive food production and they sustain the lives of people in urban as well as rural environments. Goods produced here are taken to market towns and cities, and in these settlements various services, such as banks and lawyers, have evolved to facilitate the commercial transactions that are involved.

In the mid-1800s mechanisation reduced the need for manual labour in fields, so workers moved into the **industrial** towns and cities. These workers provided cheap labour for factories producing iron, steel, textiles and many other commodities. Some towns took advantage of their location (next to a river or coast, near valuable natural resources or at bridging points or centrally located nodal points) and grew as commercial centres. The emerging middle class needed ways to organise day-to-day life, and county towns began to host key **administrative** functions, such as

Yellow = Players, Orange = Attitudes and actions, Purple = Futures and uncertainties

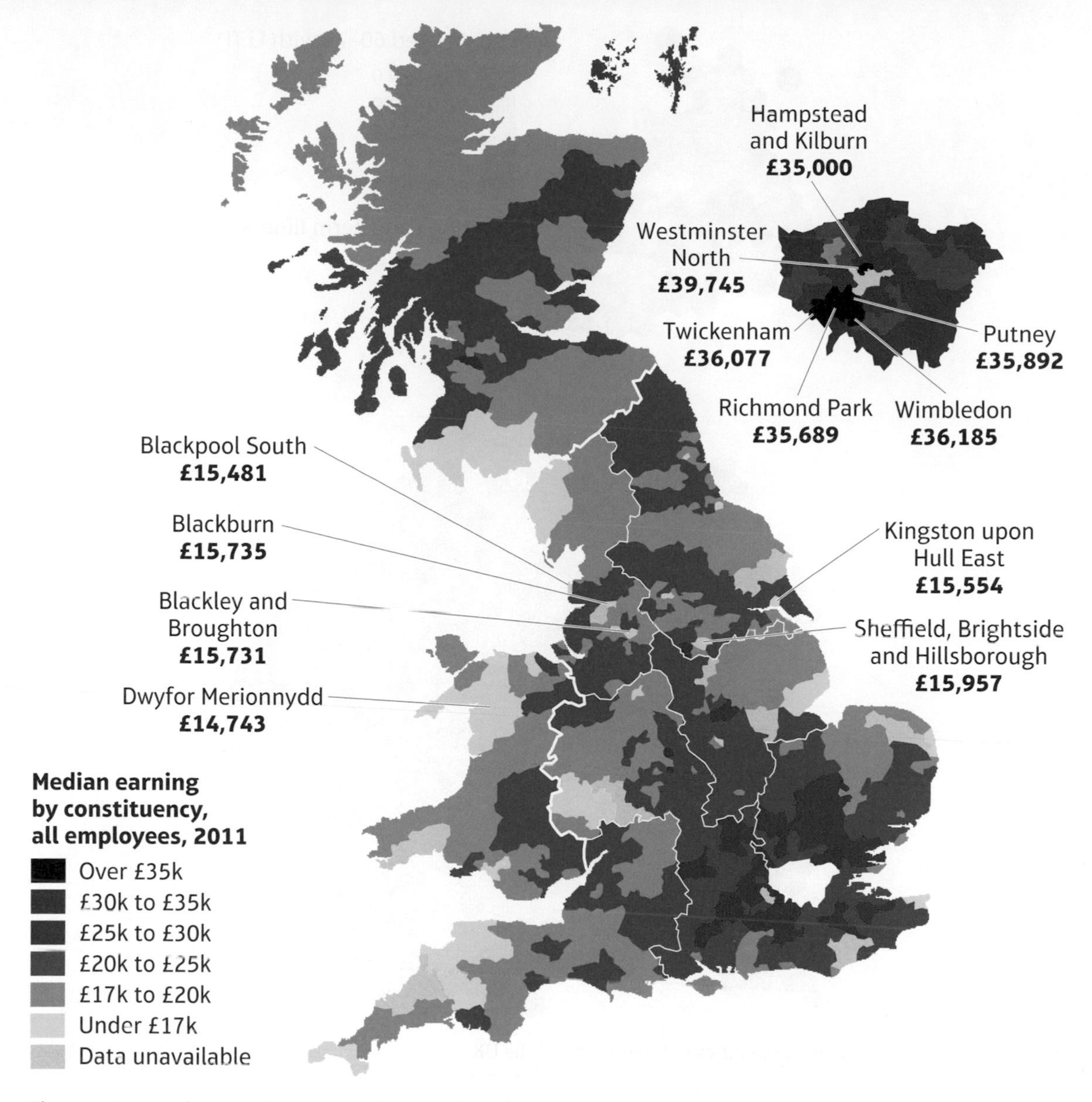

Figure 5.3: Map showing the pattern of unequal median wages around Britain (2011)

county courts, banks and corn exchanges, and, more recently, regional headquarters of various **commercial** enterprises.

Some of these functions have been superseded in today's post-industrial economy, as goods are now transported from all over the world and distributed around the country. Towns compete to become important **retail** destinations, with shopping often a leisure activity. US-style shopping malls encourage retailers to co-locate with restaurants, bowling alleys and cinemas, to create an 'ultimate retail experience'.

Alongside these functional changes, there have been **demographic changes** in the UK – in both **ethnic composition** and **age structure**. Migrants arriving from the Caribbean, West Africa and India after the Second World War clustered together in major cities:

- Jamaican migrants arriving on the MV *Empire Windrush* settled near Brixton in south London.
- Indian workers settled around Neasden and Southall in west London.
- A large Pakistani community settled in the industrial heartlands of Bradford, Huddersfield and Birmingham.

ACTIVITY

CARTOGRAPHIC SKILLS

1. In statistical terms, what is meant by the term median?

2. Describe the general pattern of median earnings found in Britain shown in Figure 5.3.

3. Compare the patterns shown in Figures 5.3 and 5.4 (i.e. look for similarities and differences). Explain what you notice.

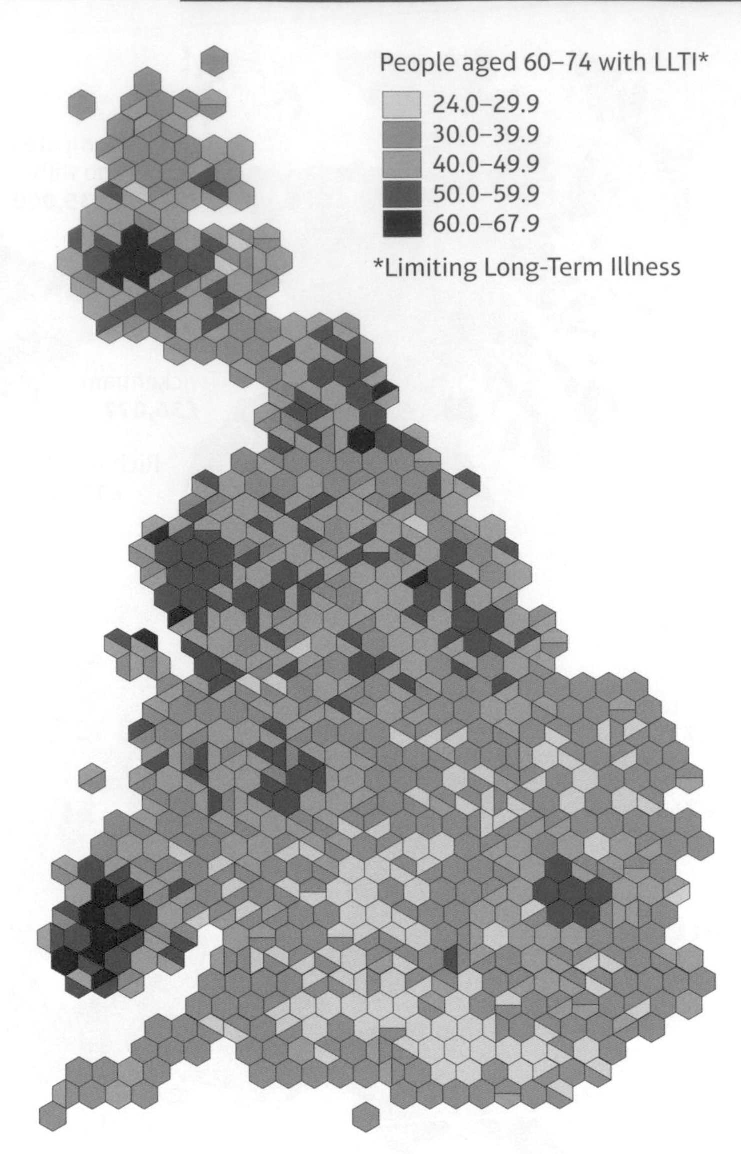

Figure 5.4: Map showing inequalities in health around the UK

Figure 5.5: The food court at the Trafford shopping centre, Manchester, showing the integration of leisure and retail

Yellow = Players, Orange = Attitudes and actions, Purple = Futures and uncertainties

As the UK struggled with racism and prejudice, areas of some towns associated with ethnic minority communities saw collapsing property prices and spirals of decline. These conditions were made worse by poor housing, out-migration and deindustrialisation, and parts of some towns, such as Middlesbrough in the North East, still suffer extreme deprivation today (see page 239). However, some areas, because of their relative affordability in comparison to exceptionally expensive parts of major cities, have been transformed into Bohemian hotspots of thriving alternative music and artists. Whitechapel, Dalston and Brixton are good examples of rapidly changing places in London that are attracting young graduates, who are reshaping and rebranding these areas into 'cultural villages'; if this process also involves an influx of wealth it may be called **gentrification**.

Evening Standard

FINAL NIGHT EXTRA

MONDAY, JUNE 21, 1948

ONE PENNY

MEAT: LAST RESERVES

WELCOME HOME!

Evening Standard 'plane greets the 400 sons of Empire

Smithfield emptying fast

£100 may now be spent on a building

FIRST BACK-TO-WORK MOVE AT THE DOCKS

Evening Standard Industrial Reporter

London markets are now drawing on their last reserves of meat following the gravest turn yet in the dock strike.

Exports up £3,500,000, imports cut £8,900,000

NOW THE GAP NARROWS TO £40,400,000

Owner's labour

MEAT, CLOTHES ARE FREED

In Australia

100 go back

Ivan and the MP

His advice cost 20s.

Page Five, Column Three

WEEK'S WEATHER

£2000 mink goes

To-morrow's weather

Quiet Wimbledon

—And the rains came

Evening Standard Reporter

Wife deserts war-blinded husband

Berlin, by air

Freight fleet taking over

"My home is Hoover-Cleaned LIKE 'GROSVENOR HOUSE'"

'MAD PARSON'

HOOVER LIMITED

Figure 5.6: Newspaper report of the arrival of MV *Empire Windrush* from Jamaica in 1948 (*Evening Standard*)

The heritage and cultural history of a place may be of particular significance for its residents. There might be pride in particular tourist attractions, sports clubs, schools or universities, and many towns have different functions that interact. This results in unique places whose identity is interwoven into what becomes a vital part of success in places, as they attempt to rebrand, re-image and regenerate.

Why do places change?

Five major factors have shaped how places have changed in the UK:

1. Physical factors

Dynamic changing landscapes present management challenges. Despite coastal defences, sea-level rise and climate change are causing rapid coastal erosion in some places, and threatening the livelihood of agricultural workers. At Happisburgh in Norfolk, 250 m of land has been lost to the sea over 250 years (see pages 161–162). Towns like Tewkesbury, Shrewsbury and Carlisle are facing up to the reality of more frequent floods and the need to invest in flood defences.

ACTIVITY

ICT SKILLS

1. Investigate the pattern of gentrification across London, by looking for house prices on websites such as Zoopla, whose 'heat maps' can give you an idea of one area compared to another, and then of house price change over time.
2. Search the ONS census database for the changes in number of people working in media, culture and sport. Estate agents use this information to predict the next area of gentrification.
3. Search for news articles that investigate how local people are being priced out of areas by rising house prices.

Extension

Decide whether gentrification is a curse or a blessing for local residents. Thinking of places near where you live, where do you think the next area of gentrification might be?

Figure 5.7: Floods in Carlisle, Cumbria, December 2015

Figure 5.8: The eco-village at Greenwich Millennium Village, London

Concern about climate change is starting to shape policy, architecture and land-use decisions. For example, increasing amounts of farmland are being used to create large solar farms, and renewable resources and zero-emission buildings are integrated into urban planning. Buildings at the Greenwich Millennium Village (2000) in south-east London (Figure 5.8) showcase innovative design, with integration of combined heat and power, careful use of glazing and technology, together with planned use of space for an ecology park and prioritisation of public transport. In Blackpool, newly built flood defences (2014–16) protect against flooding, but have also regenerated the seafront with six new 'headlands' to create more open space for tourist attractions and better access to the beach.

2. Accessibility and connectedness

The development of the UK's motorway and rail networks has changed the importance of different towns and villages around the UK. Former major railway towns like Crewe and Nuneaton have seen relative decline since the development of the motorways after the 1960s. With the completion of the new High Speed 1 railway line (2007) between London and the Channel coast, and as London's Thames Gateway redevelopment extends eastwards along the River Thames, villages in Essex and Kent are becoming increasingly popular rural alternatives for those who cannot afford London property prices. The government is also intending to develop a brownfield 'Garden City' in Ebbsfleet Valley, to take advantage of the fast rail connections to London.

Yellow = Players, Orange = Attitudes and actions, Purple = Futures and uncertainties

A by-product of improved transport infrastructure has been a steady flow of migration both within and into the UK. The growth of regional airports, such as East Midlands and Newcastle, has facilitated immigration from Eastern Europe into rural areas of the UK, reinforcing 'bridgeheader' communities that have been established since the early 2000s.

Communications infrastructure has also been upgraded. Two-thirds of the UK has access to fibre-optic broadband. The government has provided £530 million of funding for Broadband Delivery UK to extend broadband into rural areas, with subsequent upgrades to superfast fibre broadband.

3. Historical development

Some places have changed slowly over time, and their current layout and characteristics still reflect their history. One example of this is Totnes, which has deliberately introduced 'Transition Town' projects (see page 214) to protect its local culture and history. The town grew as a traditional bridging point and then developed on the valley side of the River Dart. The establishment of a local currency, the Totnes Pound, has helped local businesses along 'Buttermilk Walk' to thrive, and St Mary's Church and Totnes Castle still exist as major cultural attractions in the town.

Figure 5.9: Advertising the Totnes Pound

4. Local and national planning

The UK government has struggled for some time to tackle a chronic shortage of housing stock. The National Infrastructure Plan (2010) has designated towns like Bicester as new 'garden cities', with up to 13,000 new homes and a new railway station. Bicester Village, a retail outlet on the edge of the town, has already expanded. Sometimes rural villages risk being overrun by the expansion of urban areas. For example, the growth of Milton Keynes has seen the incorporation of villages like Middleton and Great Linford, as well as smaller towns like Bletchley and Stony Stratford. As the government contemplates building an east–west rail link between Oxford and Cambridge, villages in North Buckinghamshire and Bedfordshire can expect increasing house prices and congestion, as they become more desirable and accessible places to live.

5. Other factors

Globalisation, with its developments in transport technology and communications infrastructure, has made it more cost-effective for manufacturing companies to transfer operations to other parts of the world, particularly Asia. The closure of factories in the UK, or deindustrialisation, has triggered major changes in towns and cities, for example, Redcar steel plant (2015), Goodyear in Wolverhampton (2015), and LG Phillips in Newport (2003). Equally, migration into the UK has changed the character of some towns and cities (see page 303).

AS level exam-style question

Explain two reasons why the function of a place might change over time. (4 marks)

Guidance

This is a simple question that asks you to explain any two of the reasons listed in this chapter, including physical factors, accessibility and connectedness, historical development and the role of local and national planning.

How can we measure these changes?

As geographers look at changes to places over time, judgements can be made about their stage of economic development, their employment structure and demographic characteristics. The consequences of these changes for local residents and employers can be explored, as well as the associated impacts of how land is used and managed, and how the culture is managed.

A level exam-style question

Explain why employment factors may create variations in quality of life between places. (6 marks)

Guidance

This question requires accurate and relevant geographical information. Make sure you mention a range of reasons, linking jobs to money, and how wealth can be used to improve living conditions.

How past and present connections shape places

The economic and social characteristics of a place are shaped by past and present connections to other places, both real and imagined. The consequences of these changes are not felt evenly.

Regional and national influences

There is increasing evidence of a new north–south divide between UK cities. The towns and cities that have grown the most are in the south, particularly those designated as areas for new growth, such as Milton Keynes and Swindon. The cities that have shrunk are all in the north of the UK,

ACTIVITY

NUMERICAL AND STATISTICAL SKILLS/ TECHNOLOGY/ICT SKILLS

1. Investigate how the place where you live has changed and decide whether its function has changed.

a. Use the Neighbourhood Statistics (from the ONS) to interrogate the census to find out how key economic and demographic indicators changed between 2001 and 2011.

b. Find qualitative evidence – e.g. oral accounts and stories about the experience of your place – from the perspective of different interest groups and ethnic communities.

2. Investigate another, contrasting place. How do the changes in one place compare to the changes in the other?

Synoptic link 

A by-product of the UK being within the EU trade bloc is the free movement of goods, services, people, money and ideas. This widens opportunities for workers able to move to the UK, and TNCs investing in some towns and cities create employment and wealth. EU membership also creates challenges for the identity of some places around the UK, as job opportunities in local industries are lost to newcomers. This links to globalisation processes. (See page 300.)

Table 5.2: Typical changes to places over time

	Agricultural	Industrial	Post-industrial decline	Post-industrial regeneration
Employment trends	Primary Slowly downwards	Secondary Under threat	High unemployment and rising	Tertiary/ quaternary Quickly rising
Demographic changes	Counter-urbanisation Increasingly elderly High fertility and mortality	Urbanisation	Suburbanisation Young move outwards	Young graduates move in
Political allegiance	Conservative and resistant to change	Socialist	Liberal and more radical elements	Variable depending on who is credited with the improvements
Land-use changes	Farming gradually more mechanised Maybe suburbanised villages	Heavy/light industry and associated infrastructure	Some abandoned derelict land	Brownfield regeneration, e.g. flats, offices, leisure
Income deprivation	Relatively high and probably increasing	Beginning to decrease	High	Falling for some who are appropriately skilled
Employment deprivation	Mixed – high for low-skilled, low for high-skilled	Improving	High	Falling for some who are appropriately skilled
Health deprivation	Probably high	Deteriorating because of pollution	High	Improving
Crime	Probably low	Increasing	High	Falling
Quality of environment	Pristine scenery	Increasingly poor	Poor	Improving

particularly those once famous for their heavy industry. While some UK cities still , for example, Middlesbrough rely on manufacturing, globalisation and technology mean that fewer workers are needed to make different products, and the products themselves have changed (for example, the growing aerospace industry, for example in Bristol).

The government's 2015 plan to develop a 'northern powerhouse' is a good example of how national policy is used to enable good trade and industry connections to be made across the country. But there are some deep rooted problems which may be difficult for the government to resolve, especially during an economic recession. A similar focus on 'enterprise' in the 1980s by the then Conservative government led to free-market privatisation and the tremendous growth of the financial sector in the City of London.

Yellow = Players, Orange = Attitudes and actions, Purple = Futures and uncertainties

International and global influences

The UK's global connections, often developed during the time of the British Empire, have resulted in increasing diversity in the UK. The majority of Londoners would now describe themselves as being from an ethnic minority. There has been a dramatic increase in the number of Eastern European workers in rural areas as well. For example, in Boston, Lincolnshire, 10 per cent of the town's population comes from one of the 'Accession 8' countries that joined the EU in 2004. This can put pressure on school places, and affect the attitude of local students towards diversity, particularly those with connections to unemployed people, where the perception may be that their unemployment has resulted from increased competition for jobs.

Some ethnic groups have suffered from prejudiced attacks, but many towns have benefited from immigration because, instead of having an ageing population with declining skills, they now have a younger population with greater aspirations. As migrant groups become more settled, entrepreneurs set up restaurants and construction firms, marketing agencies and bakeries. New migrants, in turn, fill their previous jobs.

Impacts on identity

For any country undergoing rapid change, identity becomes increasingly important. Cities naturally bring together identity politics and minority influences. In London, high immigration levels have caused what UK sociologist and geographer Stephen Vertovec has termed 'superdiversity'. Southall in west London is characterised by its increasingly mixed Indian and Somali community. Within both communities there are newly arrived migrants, refugees, and long-established migrants. Many of the Somali diaspora work to provide financial support for families and communities back in Somaliland. Socio-economic conflict mixes with racial tension, together with various civil rights issues and LGBT social movements. Identity politics is sometimes expressed through art or music genres, and in parts of London any combination of these shape and define local places. In Manchester, Canal Street became the centre of a 'gay village' in the 1980s, when various laws attacking civil liberties forced the gay community to cluster together for support. The shops and services on this pedestrianised street now cater for this community.

Since online shopping has to some extent superseded the weekly trip to the town centre, or even the suburban superstore, some residents may now be bypassing their nearest small town altogether, in favour of a nearby city or specialist services in other local towns. Some towns are experiencing the suburbanisation of ethnicity, and occasionally a clash of cultures. For example, in Beaconsfield (south Buckinghamshire), there have been protests against a Sikh free school opening to educate children spilling over from nearby Slough. Pluralistic societies are often complex; the local Sikh community in Beaconsfield itself is actually opposed to the school because of increased traffic as people travel from Slough to the school.

Agricultural decline means that rural areas are having to become increasingly multifunctional. Digital communication such as Skype and rural broadband helps bridge the physical gap (distance) between communities. Now that teenagers in isolated villages are able to communicate with school friends 24 hours a day, virtual identity is increasingly important. Despite the illusion of a rural 'idyll', there is a new risk of a growing gap between highly connected, relatively affluent, online residents, and digitally excluded, less affluent residents with restricted mobility, who remain reliant on services in the nearest small towns – for which there is dwindling demand. Older people (retired), unable to access public or private transport or use electronic communication, risk becoming examples of 'hidden poverty'.

ACTIVITY

1. How have local, national and global economic changes influenced people's identity and sense of belonging in Berbera?
2. How have local, national and global social changes influenced people's identity and sense of belonging in London?
3. How have economic and social changes influenced people's identity and sense of belonging in the place where you live?

Synoptic link

Some groups of people are positive about change. Others are not. The internet can revive local and lost cultures, such as an appreciation of local architecture or music. However, new ideas coming into societies and influencing younger people can set up tensions between different age groups. This relates to concerns about the cultural impact of globalisation and its role in changing places both in the UK and around the world. (See pages 299–300.)

ACTIVITY

TECHNOLOGY/ICT SKILLS

Investigate how different types of people relate to, and interact with, the places where they live. Social media are good sources to use.

1. Based on their comments, what do people think about where they live?
2. Based on the hashtags in their comments, do they seem to attribute events and change to anything in particular?
3. Is there evidence that particular types of people follow the social media feeds of particular organisations or other key cultural or economic players?

CASE STUDY: Berbera, a contrasting urban area

Berbera is a city in north-western Somaliland, a self-declared autonomous region of Somalia. Somaliland was a British Protectorate until 1960, and until 1941 Berbera was its capital city. During a brief period of good relations with the former USSR, it became a significant naval and missile base. Later, in agreement with the USA, the very long runway at Berbera Airport was used as an emergency landing strip for the Space Shuttle. The country lacks international recognition, however, and this keeps international investors away.

Berbera's deep sheltered seaport, strategic location along an oil route and proximity to Aden (in nearby Yemen) made the city a natural administrative and commercial hub in the country. The port continues to export goats and cattle, mainly to Yemen but also Ethiopia. Very few shipping companies call at the port; it can take months for factory parts to arrive for companies like Coca-Cola, which has a $17 million bottling plant a two-hour drive from the port. The road is in a state of disrepair, and despite an extensive crowdsourcing attempt to raise money for a 480-km (300-mile) road linking Berbera with Hargeisa (Somaliland's present-day capital), only enough money for 30 km (20 miles) was raised. Many of the older Ottoman and merchant buildings have been abandoned, with squatters occupying many of the floors in those that remain.

Approximately 60,000 people live in the city, which is a close-knit community. Small-scale economic activity sustains some workers (frying fish, selling ghat), although youth (15–30) unemployment in 2015 was 67 per cent. Many rely on money sent back from the diaspora living overseas. But the long beautiful beaches are starting to become a site for watersports activities. The water in the Sea of Aden is warm enough to swim in all year round, while plentiful fish means it is both a suitable base for shore diving and a reliable income for fisherman.

Exports through Berbera's port generate 50 per cent of Somaliland's GDP and the government is investing in the port to make it safe and attractive in order to compete with nearby Djibouti. Since the UAE cut off diplomatic ties with Djibouti in 2015, the availability of capital funds for investment in the port looks increasingly possible. Across the country as a whole, companies like Dahabshiil Group (a fund transfer company founded in Somalia but with its headquarters in Dubai) are funding 'green microfinancing' projects, such as solar-powered water pumps, to help farmers and small-business owners cope with the extreme and variable climate and create jobs within the country to limit outmigration.

Extension

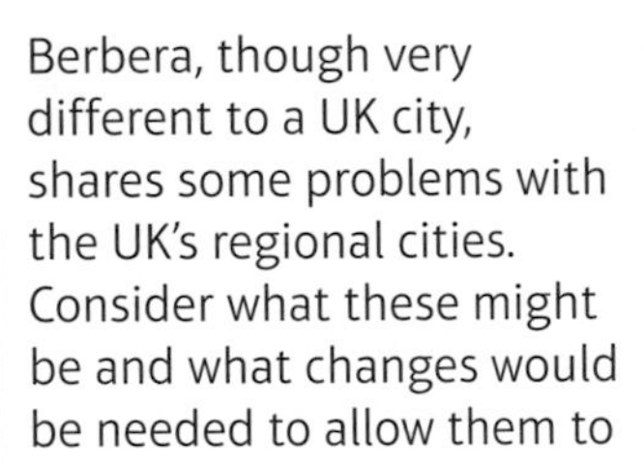

Berbera, though very different to a UK city, shares some problems with the UK's regional cities. Consider what these might be and what changes would be needed to allow them to regenerate.

Yellow = Players, Orange = Attitudes and actions, Purple = Futures and uncertainties

CHAPTER 5

Why might regeneration be needed?

Learning objectives

4A.4 To understand how economic and social inequalities change people's perceptions of an area

4A.5 To understand why there are significant variations in the lived experience of place and engagement with them

4A.6 To understand that there is a range of ways to evaluate the need for regeneration

What makes a region successful?

Some parts of a country are highly desirable and attract **inward migration**, sometimes internationally but more typically from elsewhere in the country. Figure 5.10 shows how a process of 'cumulative causation' triggered by new industry, often a TNC, attracts employees and a host of supporting companies such as those involved with supplies, infrastructure and leisure.

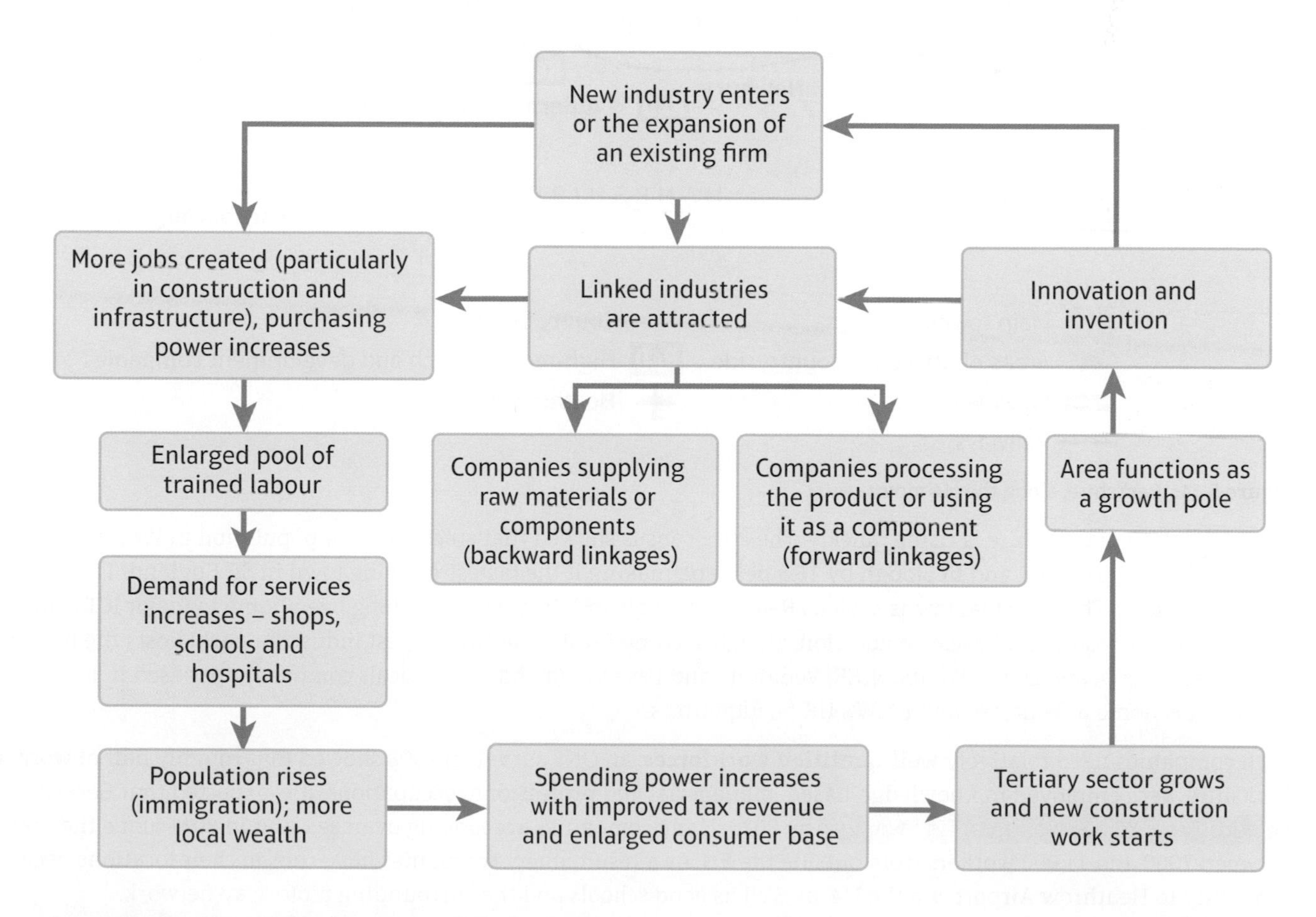

Figure 5.10: Based on Myrdal's cumulative causation model

Over time, the variety of jobs created can result in a **two-tier economy**, with the risk that many workers in less well-paid jobs will be out-priced by the housing market, when high demand leads to **high property prices**. Equally, there may be a **skills shortage,** with insufficient trained workers to do the quaternary and quinary jobs required by the new growing economy. Solving these issues requires either investment in training or recruitment of workers from overseas and, to alleviate high prices, building more affordable housing. However, the many benefits of these changes often mean that these places typically have low levels of **multiple deprivation,** and benefit from the constant renewal and improvement of infrastructure and the living environment.

CASE STUDY: Berkshire, SE England – a successful region

The M4 motorway runs along the county of Berkshire from east to west. Much of the county is influenced by Heathrow Airport and the M25 in the east. Towards the west, the rivers Thames and Kennet have created a wide, flat floodplain, which is a great location for high-quality urban living and work.

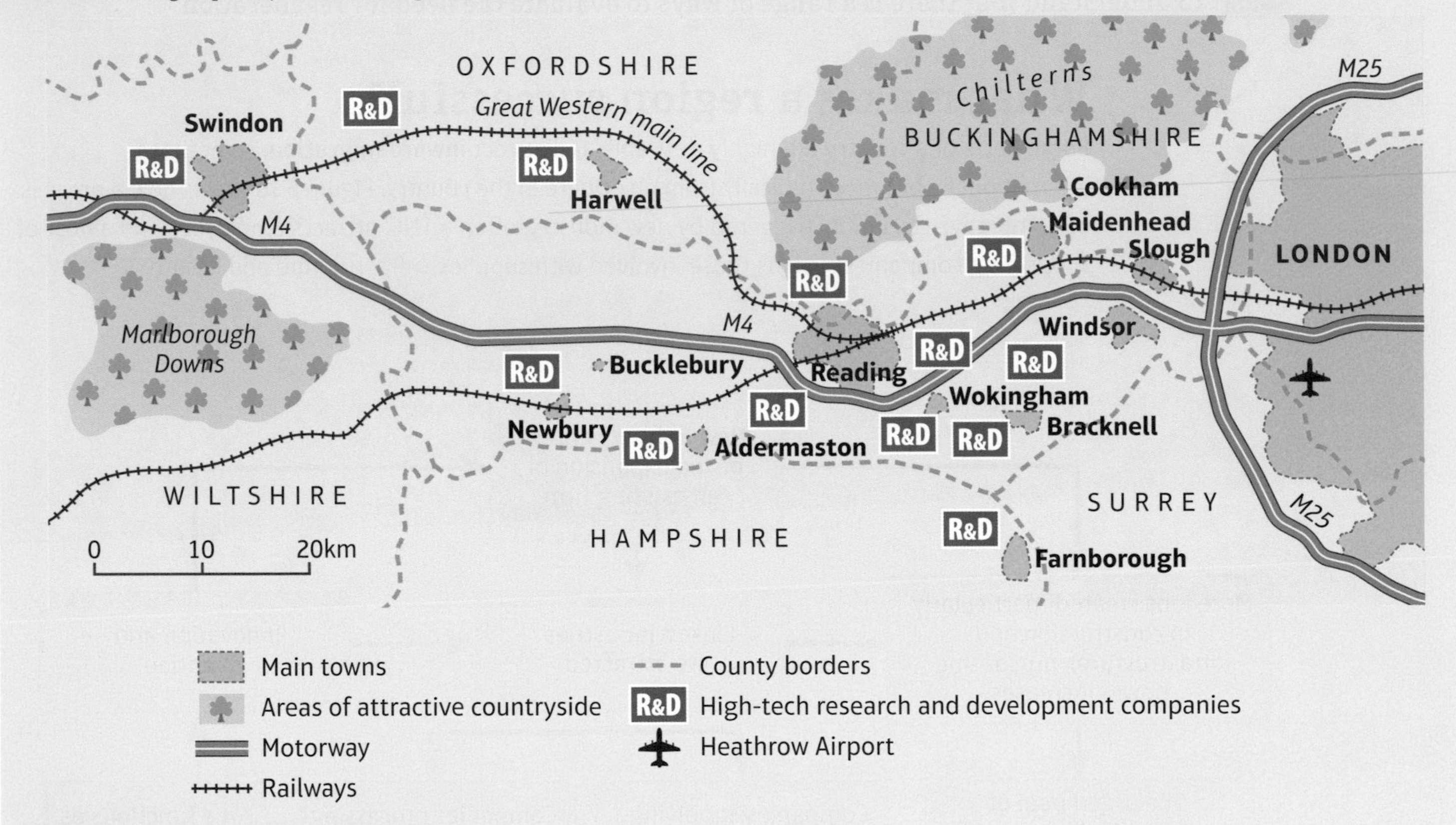

Figure 5.11: Berkshire, along the M4 motorway

The population of Berkshire is rising quickly. The 2011 census showed that since 2001 the population in West Berkshire had increased by 6.4 per cent, and in Slough by 16.3 per cent, making it the fastest-growing town in SE England. There are high **rates of employment**. The principal towns such as Reading, Slough and Bracknell have long been home to major ICT companies, particularly Microsoft and Oracle. In addition, Slough Trading Estate (the UK's largest industrial park) hosts the headquarters of TNCs such as O_2 and Dulux Paints. NfER, Vodafone and BayerAG (a pharmaceuticals company) are based in Newbury, while Bracknell is home to Waitrose and BMW's UK headquarters.

Such companies need relatively **well-qualified workforces**; an ONS survey in 2008 showed that roughly half of workers in Berkshire were employed in knowledge-based, managerial and professional occupations. It is expected that Berkshire will need an additional 70,000 well-qualified workers by 2020. Many companies are looking overseas, and in 2011 alone they recruited between 7000 and 11,000 workers from outside the EU. As a result, many companies have chosen their locations because of the proximity to Heathrow Airport, via the M4, as well as good schools and the surrounding motorway network.

The demand for living space has meant that Berkshire has some of the most expensive villages in the UK. Historical villages sometimes have royal connections (e.g. Bucklebury was the home of the Middleton family), or are the setting for TV dramas (such

Yellow = Players, Orange = Attitudes and actions, Purple = Futures and uncertainties

as Midsomer Murders, filmed in the so-called 'millionaire villages' of Cookham and Cookham Dene). The wealth of the residents has helped support the rural economy. There are many dairy-based farms, where the farm shops sometimes thrive on Royal Commission. The 2000-hectare Windsor Great Park is an historic royal hunting ground. Historical sites (not least Windsor Castle) and National Trust properties bring tourists to the area, and the racehorse industry is active around Ascot Racecourse.

Figure 5.12: Cookham in Berkshire, along the River Thames

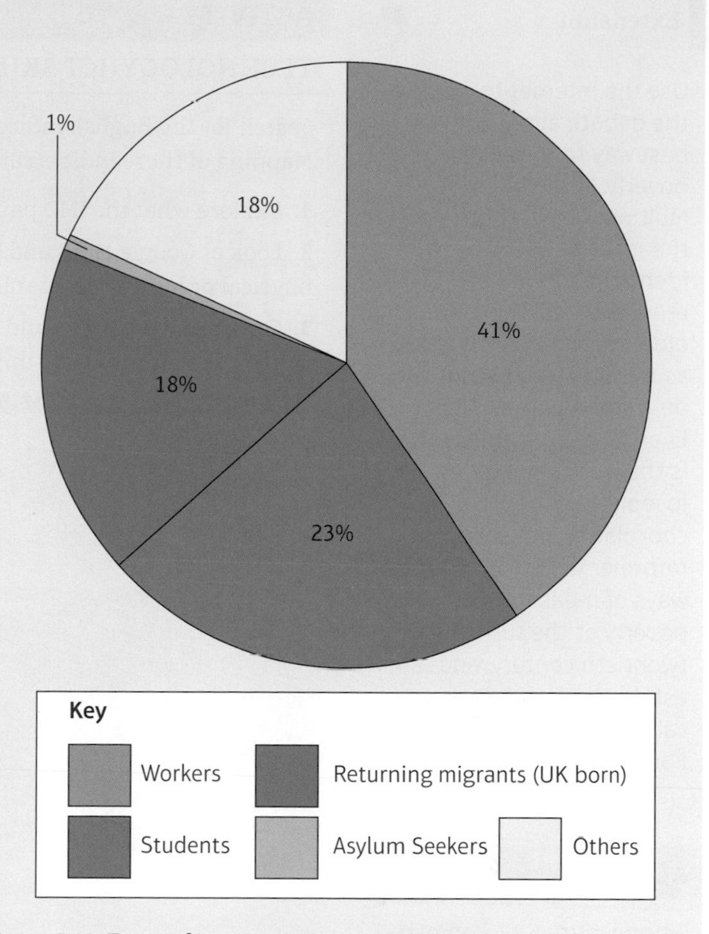

Figure 5.13: Types of immigration into Berkshire

Property prices, however, are a concern, with prices rising 40–50 per cent in the ten years from 2005, according to Zoopla. In September 2015, prices for some six-bedroom properties were in excess of £2 million. Property prices are out of reach for 20 per cent of the working age population. Thames Valley Berkshire Local Enterprise Partnership (LEP) had only 0.4% of its neighbourhoods in the most deprived 10% of national areas according to the 2015 Index of Multiple Deprivation.. Lambourn and Aldermaston villages both have increasingly elderly populations for whom the cost of energy is too high to use cars and keep homes heated. However, the populations are too small to commercially justify a bus service.

What makes a region decline?

In places that have suffered from deindustrialisation, unless workers are able to utilise a different set of skills, unemployment can trigger a downward spiral of economic decay, as illustrated in Figure 5.15. For some towns this spiral of decline can become almost impossible to reverse. Over time, the quality of life in areas within such towns is reflected by a high index of multiple deprivation. This is a statistic used by the government to identify relative deprivation, ranking each area on a scale from 1 (most deprived) to 32,844 (least deprived). It reflects seven domains: income, employment, education, health, crime, barriers to housing and services and living environment. The IMD Report published in 2015 revealed that the area of highest deprivation was Middlesbrough, followed by Knowsley, Kingston upon Hull, Liverpool and Manchester – all areas affected by deindustrialisation.

ACTIVITY

1. What do you think is the main factor that makes the Berkshire region successful? Why this factor? Give evidence to support your answer.
2. What evidence shows that this factor is shaping change in the place where you live?
3. Assess the challenges faced by growing regions.

Extension 

Use the internet to explore the debate about the best way to measure poverty. Is the Index of Multiple Deprivation the most helpful way to identify areas in need of regeneration? There are older measurements, such as the Townsend score, developed in the 1980s. Even longer-established is the work done by Joseph Rowntree, the chocolatier and social reformer who pioneered ways of measuring poverty at the turn of the twentieth century, and established what is now called the Joseph Rowntree Foundation.

ACTIVITY

Using Figure 5.14, compare multiple deprivation levels west of London along the M4 corridor (which includes Berkshire) with those in the north-east of England (which includes Middlesbrough).

ACTIVITY

TECHNOLOGY/ICT SKILLS

Search for the English indices of multiple deprivation, which are on the UK Government website. Mapping of these indices can be completed on websites like OpenDataCommunities.org.

1. Explore what the IMD pattern is for the place where you live.

2. Look at Google Maps and work out whether there are any reasons for the pattern, e.g. the physical or human geography.

3. Compare the pattern and level of IMD of the place where you live to another contrasting place. Suggest reasons for any similarities and differences.

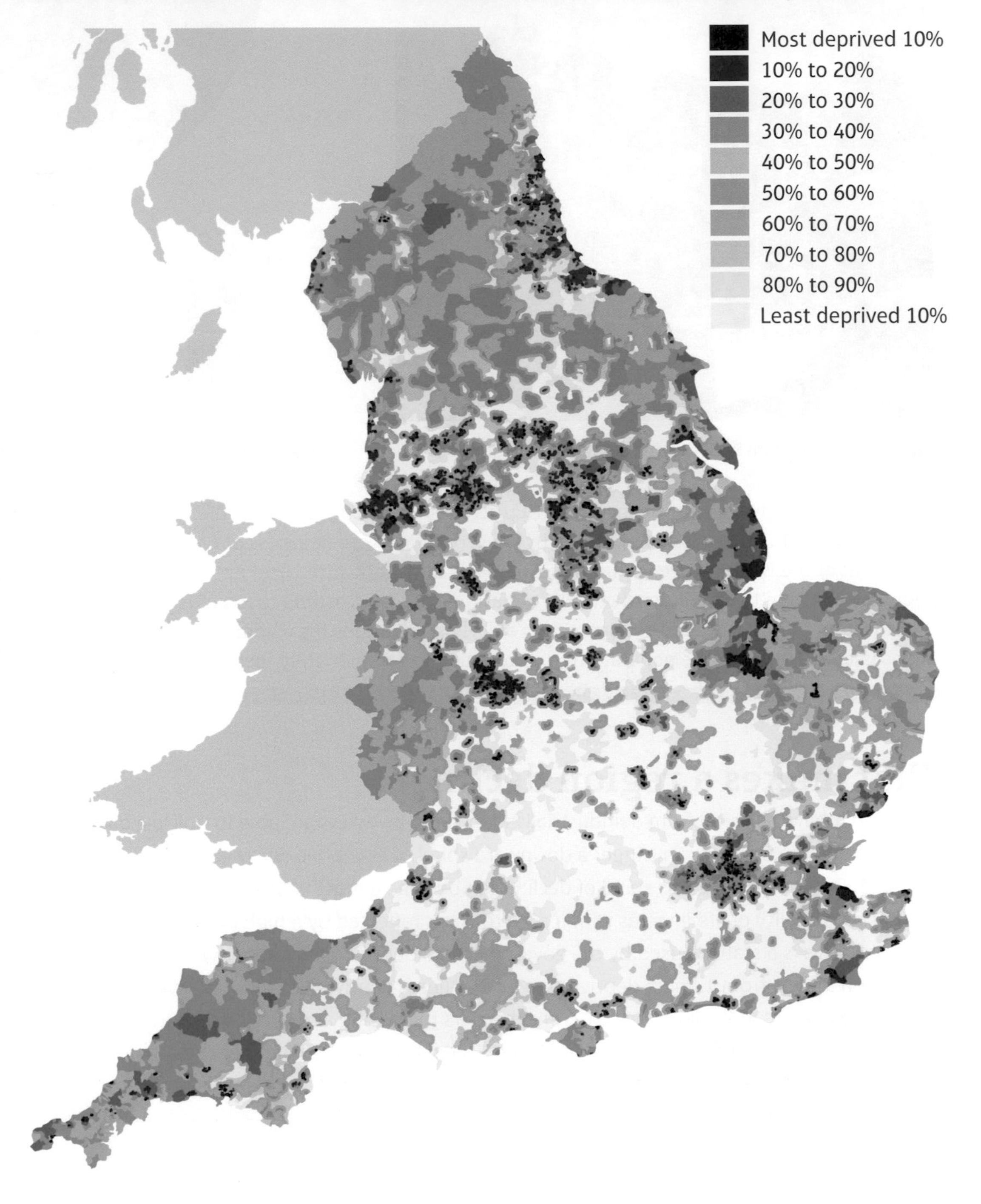

Figure 5.14: The Index of Multiple Deprivation (IMD) across England, 2015

Yellow = Players, Orange = Attitudes and actions, Purple = Futures and uncertainties

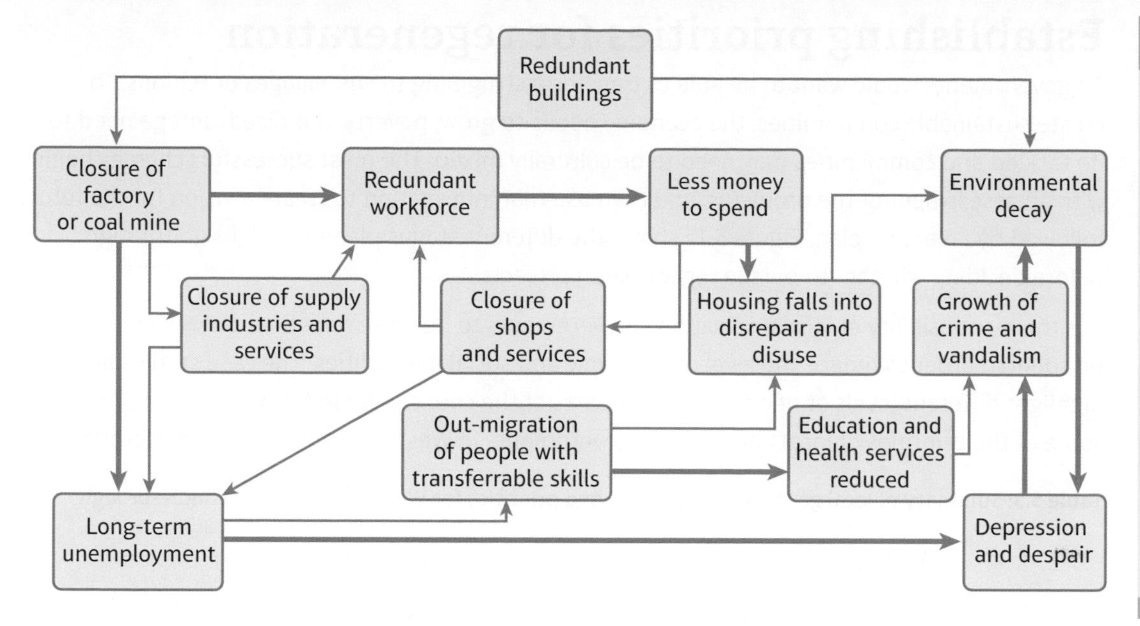

Figure 5.15: The negative multiplier effect, or spiral of decline

ACTIVITY

1. On a copy of Figure 5.15, add factual information and extension points to summarise the Middlesbrough case study.
2. What do you think should be priority for Middlesbrough, an area suffering from a spiral of decline? Explain your choice.
3. Is there evidence of a potential spiral of decline happening in your local area or contrasting place?

CASE STUDY: Middlesbrough – a declining town

Along with its neighbouring Teesside town Hartlepool, Middlesbrough has struggled with the consequences of deindustrialisation since the 1980s. Around 20,000 people have left the town since 1990. The 2008 global recession worsened the decline: many remaining small-scale businesses and services were forced to close after the large chain stores closed down as the number of customers decreased. The 2015 closure of the SSI Steelworks in Redcar has caused further decline and presents another obstacle to successful regeneration.

The local unemployment rate is almost twice the national average (13 per cent), and in 30 per cent of the town's working-age households there is nobody in paid employment. The **income** of local residents has plummeted, removing their ability to change jobs or move house. Middlesbrough contains some of the most deprived wards in the UK, with 10 per cent in the bottom 1 per cent of **deprived areas** of England. Average house prices in the North Ormesby area dropped to £57,000 in March 2015 (compared to a UK average of £180,000). Life expectancy is lower than in the rest of the UK. Landlords buy up properties cheaply and rent them out to tenants on housing benefit, but may not maintain them properly.

The effect on the **built environment** has been extreme. Whole streets of terraced houses are boarded up. There are high levels of **anti-social behaviour** and the cost of tackling flytipping has been rising by 10 per cent per year. Dumping of waste affects quality of life and creates health and safety problems. The lack of opportunities has lowered motivation for academic achievement: an Ofsted report in 2014 found that one-third of pupils (and half of secondary school students) attend schools that are 'requires improvement'. All these changes put off owner-occupiers, who choose to look elsewhere for a place to live.

Figure 5.16: Contrasting images of Middlesbrough

The council's regeneration team is working on a number of projects to improve the area and the local standard of housing. However, in times of financial difficulty and reductions in welfare spending, a continued spiral of social and economic decline (Figure 5.15) may be likely unless private investment responds to government initiatives. Regeneration may involve Middlesbrough's industrial heritage or another theme.

Establishing priorities for regeneration

All governments would want to be able to transform struggling towns, villages or regions. To create sustainable communities, the economy needs to grow, poverty and disadvantage need to be tackled and communities may need to be culturally mixed. The most successful schemes begin with an assessment of the problems, and then use that information to create a vision for the future followed by an action plan. Figure 5.17 shows the different stages planners might go through before deciding whether to pursue regeneration projects.

It is the responsibility of local and national governments to decide where financial resources should be spent in order to reduce the level of economic and social inequalities. The case studies above highlight different levels of need in different parts of the country. Table 5.3 below summarises some of the priorities of local and national governments in West Berkshire and Middlesbrough.

Table 5.3: Summary of local government concerns and priorities for West Berkshire and Middlesbrough

	West Berkshire	Middlesbrough
Major concerns	• Skills shortages (lack of labour) • Lack of higher education provision • Small pockets of deprivation, e.g. Aldermaston, together with poor bus services • Over-reliance on key business, e.g. ICT in Reading • Lack of affordable housing in **commuter villages** – e.g. Cookham and Windsor	• **Sink estates**, e.g. East Middlesbrough and Grove Hill, in comparison to relative affluence in **gated community,** owner-occupied housing estates, e.g. Coulby Newham • Gap in educational attainment – e.g. half of all students in schools 'require improvement'
Main priorities	• Deliver 10,500 new homes, mainly on brownfield land • Encourage a mixture of house sizes, types and tenures • Invest in sustainable public transport • The East Kennet Valley Plan – protected employment areas for Aldermaston, planning for new homes, improved access to rail connections • Superfast Broadband Deployment Plan	• Protect and enhance sports facilities, e.g. Middlesbrough College • Regenerate Greater Middlehaven – invest £215 million to create 1500 jobs, 100 new homes, a new police headquarters and waterside office developments • Build new high-quality housing, e.g. 11,500 homes in Grove Hill

ACTIVITY

Complete a diamond ranking exercise for either Berkshire or Middlesbrough following the steps below.

1. Find nine possible priorities for regeneration from Table 5.3 and the case studies on page 239 and page 244.
2. Put them in order: 1 is the most important; 2 and 3 rank equally important but are slightly less important than 1; 4, 5 and 6 again rank equally important but are even less important; 7 and 8 rank equally important but are less important again; and 9 is the least important.
3. Justify your decisions using information from the case studies.

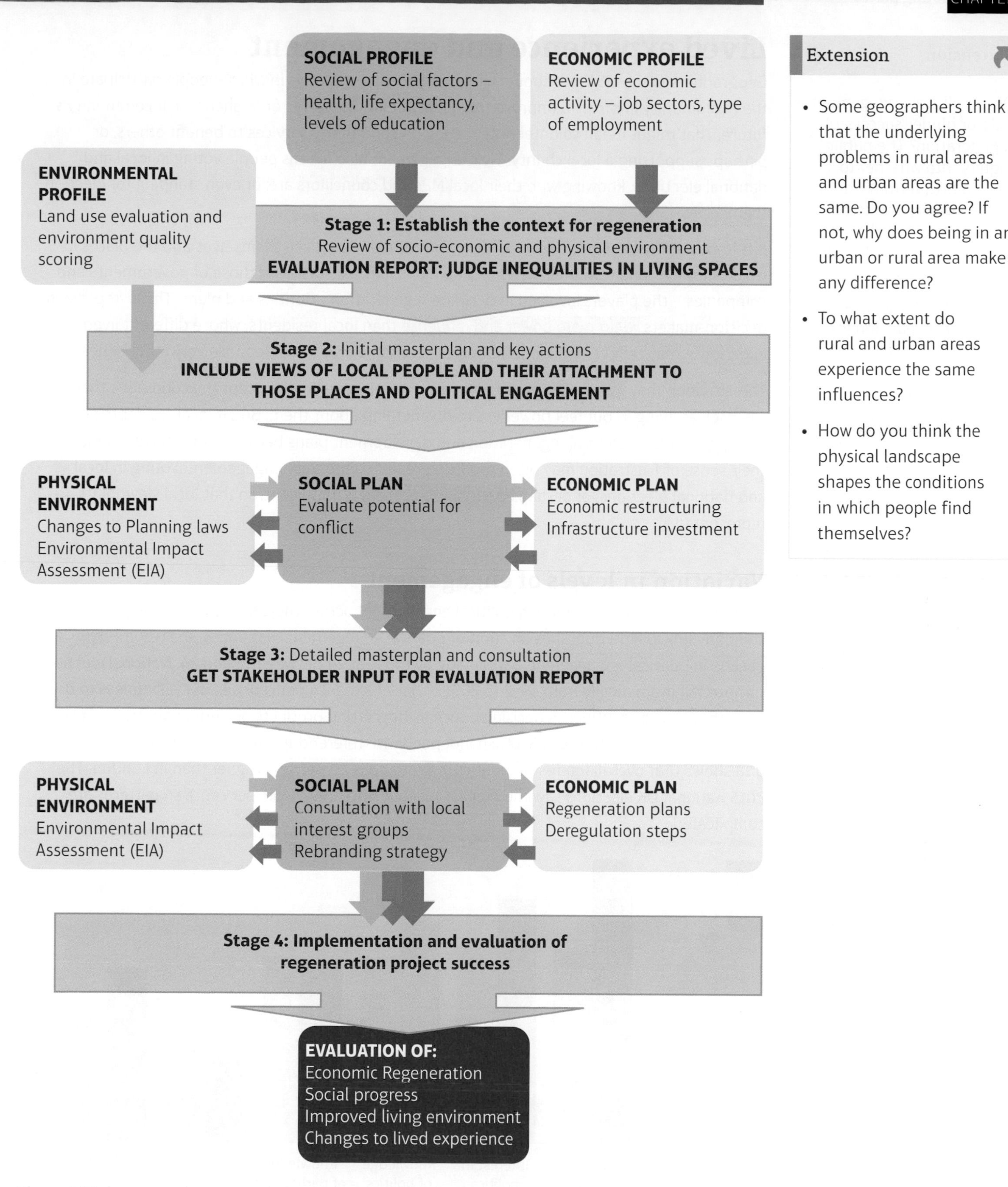

Figure 5.17: A regeneration master plan

Extension

- Some geographers think that the underlying problems in rural areas and urban areas are the same. Do you agree? If not, why does being in an urban or rural area make any difference?
- To what extent do rural and urban areas experience the same influences?
- How do you think the physical landscape shapes the conditions in which people find themselves?

Extension

Does political engagement help or hinder successful regeneration? The public inquiry into whether to build a fifth terminal at Heathrow Airport lasted four years, concluding in 1999, and the terminal did not open until 2008.

Investigate the debate on airport expansion for London.

ACTIVITY

GRAPHICAL SKILLS

Conduct an audit of political engagement in your class based on the six categories shown in Figure 5.18. Plot the results on a graph and annotate it to explain your findings.

Synoptic link

Think carefully about how the level of engagement reflects political attitudes towards change and regeneration of a place. For example, not all groups of people feel the same way about changes brought about by the globalisation of culture or industry, because some may gain or lose more than others. In some countries around the world there is corruption in regeneration and globalisation decision-making which impacts on human rights and causes some people or groups to think about making interventions (e.g. protests, donations). (See page 253.)

Lived experience and engagement

Geographers in the USA talk about civic engagement – the ways in which people participate in their community in order to improve the quality of life for others or to shape their community's future. That might mean volunteering, setting up community services to benefit others, or perhaps supporting a local charity. Civic engagement also means people voting in local and national elections, knowing who their local MPs and councillors are, or even standing for election themselves.

It is important for geographers to understand how residents see themselves and the places in which they live, and how their views and opinions may contrast with those of governments and companies – the players involved in deciding regeneration priorities and plans. These 'top-down' decision-makers wield more power and influence than local residents, whose different lived experience may reflect inequalities, but their decisions may increase those very inequalities.

This situation may cause political apathy, when people unwillingly accept the conditions they find themselves living in but feel powerless to do anything about them. But, if local residents object strongly to proposed local regeneration and development plans because of perceived inequalities, their sense of frustration may often produce greater community engagement. Voting in local and national elections, or perhaps deciding to protest, is important, so that local viewpoints are represented.

Variation in levels of engagement

There is considerable variation in political engagement across the UK. Every year the polling company IPSOS Mori publishes an audit of political engagement. Nationally, the results show a decline in political engagement, particularly by those aged between 18 and 24. National election turnout fell dramatically from 1992 to 2001 (from 77.7 to 59.4 per cent), as did willingness to do something more to influence decisions, such as boycotting products or signing petitions. Despite this, voter turnout in the 2014 Scottish independence referendum was nearly 85 per cent. Figure 5.18 shows that overall interest in politics in Scotland is 20 per cent higher than in London. The 2015 national UK elections saw the highest turnout since 1992 (66.1 per cent), so patterns are complicated.

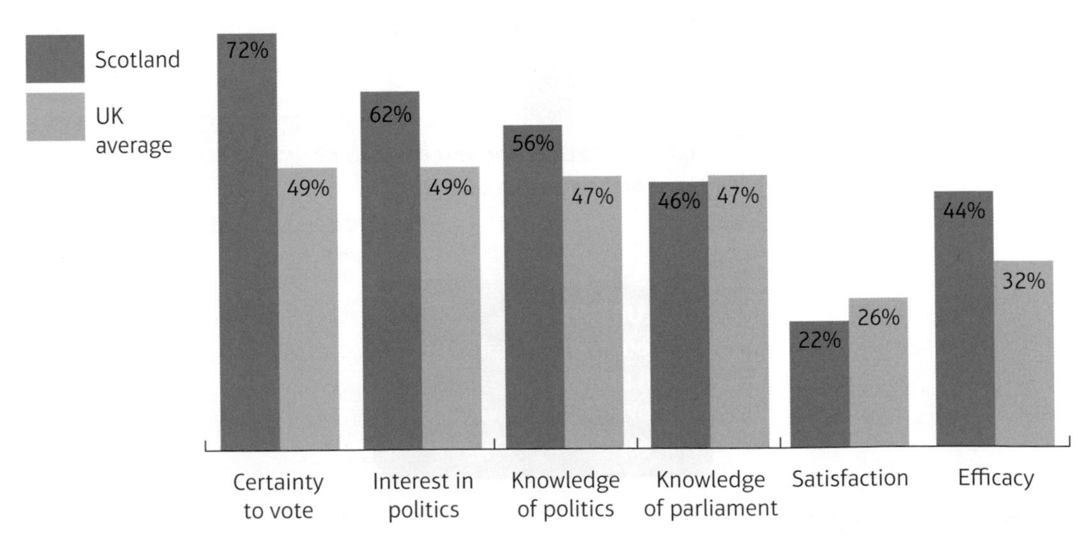

Figure 5.18: Audit of political engagement in Scotland compared to the rest of the UK, 2015

Yellow = Players, Orange = Attitudes and actions, Purple = Futures and uncertainties

Lived experience, deprivation and engagement

Increased political engagement around the UK often reflects increased mobilisation of minority groups in different communities. Where there is an inequality or issue to be addressed, participation is often higher. The decision not to fly the British flag outside Belfast City Hall (December 2012) prompted protests from loyalist communities. Similarly, after the legalisation of same-sex marriage in 2013, political engagement in areas of London such as Vauxhall (home to the largest gay community in London) began to reflect wider issues (Conservative Party), rather than just a desire for progressive social policy (Labour and Liberal Democrat parties). At the same time, research by the Electoral Commission suggests that those who experience social deprivation tend to be the most politically excluded. While the causes are difficult to determine, the danger is that political exclusion and deprivation tend to reinforce each other over time.

Figure 5.19: Amir Khan carrying the 2012 Olympic Torch in Bolton

Since their post-war arrival, many non-white communities have dispersed to different towns and cities throughout the UK. Loyalty to a particular city or region can be more important to people in the UK than their own religion or ethnicity. In Bolton, people of all religions and races supported local boxer Amir Khan (Figure 5.19) in his boxing successes. Muslim political participation has increased at all levels in British society. Clearly, this change is concentrated in certain parts of the UK.

Political engagement often reflects the need to protect the past and the present – a need to respond to imposed changes from 'outside'. George Galloway's Respect Party took advantage of the Arab Spring and Iraq War to unexpectedly win a by-election in Bradford West in 2012 with 52 per cent of the vote, only to lose again to Labour in the 2015 election (which took 63 per cent of the vote). In some areas there has been an increase in violent political extremism, and groups such as the English Defence League (EDL) have organised protests and encouraged antipathy towards Muslim communities. The only MP for UKIP represents Clacton, where just 4 per cent of the population is foreign born, compared to a national average of 13 per cent. Meanwhile in York, EDL protests in 2013 were defused by the collective response of individuals and the opening up of dialogue between many different age and cultural groups at the protest site.

Extension

Geographers sometimes analyse landscapes and places by looking at the artefacts (objects), mentefacts (shared meanings and understandings about those objects or the place) and sociofacts (the behaviours you get as a result from local people). The combination of all these things makes up the urban or rural 'fabric', and it is worth researching a particular place or landscape using more qualitative skills to understand the symbolism and shared meanings about that place.

Search for Professor Doreen Massey's fieldwork description of life in Kilburn High Road, in north-west London. This could indicate a useful approach for your independent investigation.

ACTIVITY

1. Evaluate the causes of the 2011 London riots.
2. Should these causes affect the development priorities for that part of London?

ACTIVITY

Use the DWAGES acronym (**D**isability, **W**ealth, **A**ge, **G**ender, **E**thnicity, **S**exuality) to consider what it is like to live in your local place from the perspective of each group that may have been marginalised.

Although public action in individual communities is variable, across the country the rise of 'clicktivism' (organising protest through and on social media, perhaps just by 'liking' a viewpoint, to express support) and online campaigning is starting to change the ways in which political engagement occurs. There is increasing evidence of social campaign groups acting on media reports of injustice and motivating others to take action online. The All Out campaign group has successfully mobilised support on a number of LGBT (Lesbian Gay Bisexual Transgender) human rights issues around the world.

Conflicts about priorities for redevelopment

Different groups within a city often have contrasting experiences of day-to-day issues, which frequently results in radically different views about the need for regeneration. Depending on where people live and their perspective, explanations for the August 2011 London riots reflect a broad spectrum of initial causes, summarised in Figure 5.20.

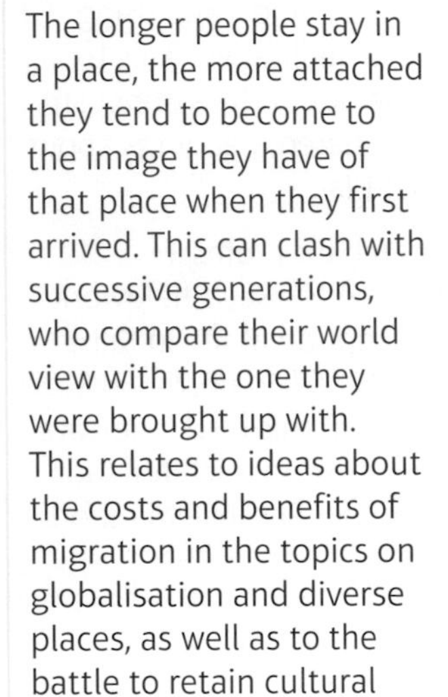

Synoptic link

The longer people stay in a place, the more attached they tend to become to the image they have of that place when they first arrived. This can clash with successive generations, who compare their world view with the one they were brought up with. This relates to ideas about the costs and benefits of migration in the topics on globalisation and diverse places, as well as to the battle to retain cultural identity. (See page 211.)

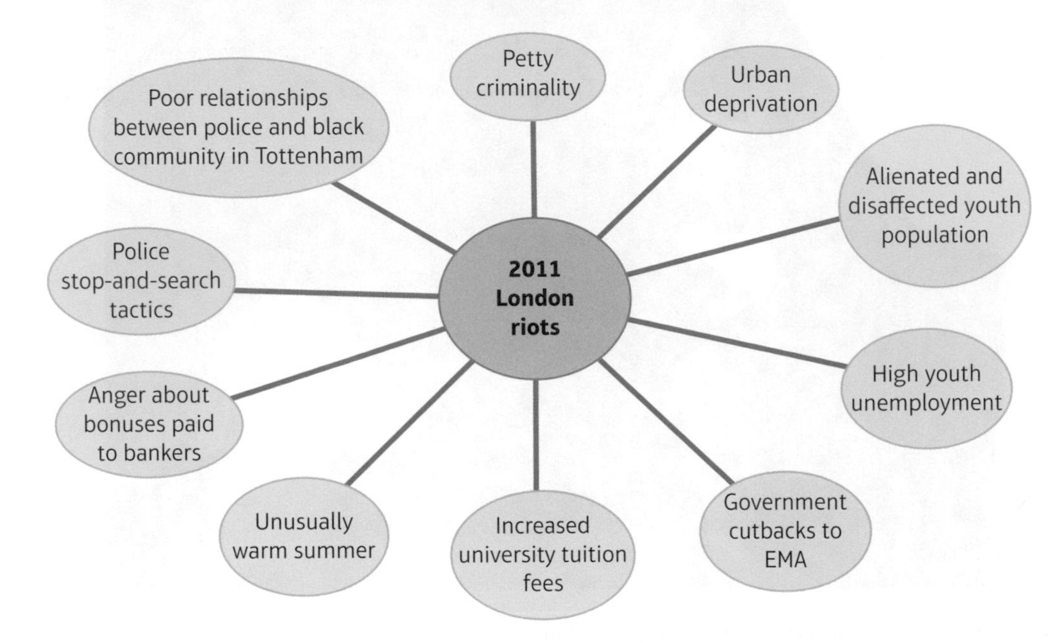

Figure 5.20: Multiple explanations for the 2011 London riots (EMA = Education Maintenance Allowance)

CASE STUDY: Regeneration in Tottenham, London

The north London borough of Haringey has seen continual economic decline since the 1970s, with some of the highest unemployment in the city. Illegal landlords pack families into overcrowded flats and houses, and gang culture makes the integration of communities difficult. The area of Tottenham was the scene of riots in the 1980s, concentrated on Broadwater Farm, a large post-war housing estate.

Different groups of people have contrasting ideas about whether plans to regenerate the area will be successful. The Mayor of London has invested £28 million in Tottenham, but often a scheme will benefit some yet disenfranchise others. By rebuilding the Peacock Industrial Estate, 5000 new jobs were created. However, this required the temporary or permanent closure of small businesses, such as the small-scale furniture, clothing and chemical companies that are unique in London's economy. On the other hand, design companies (such as N17) took on apprentices from local schools to help plan the redevelopment.

Haringey Council has engaged Arup, a planning and design consultancy, to build two housing zones with 10,000 new homes around Tottenham Hale station. Tottenham Hotspur football club have long had ambitions to rebuild their stadium. However, aside from losing funding to

Yellow = Players, Orange = Attitudes and actions, Purple = Futures and uncertainties

West Ham for redevelopment of the Olympic Stadium, the requirement to include affordable housing in a project removes some of the profit necessary for further regeneration, even though it has the potential to benefit local residents. Some are worried that the religious diversity of churches created by the African-Caribbean community will be lost. However, since new-build homes often come with a high price, most local residents are more worried about losing out to gentrification, particularly if the prices are lower than the city average, as workers struggling to afford London's sky-high property prices will be attracted. The latter group of people see regeneration as a positive, while the former see it as negative. In response to what some newspapers have described as 'social cleansing', a network called 'Our Tottenham' has grown up, supported by companies like N17, to help voice opposition to regeneration plans.

Evaluating the need for regeneration

Qualitative and quantitative data

Decision-makers will access data from both qualitative and quantitative sources during a research phase before devising plans for regeneration. Local community groups will provide opinions and lived experiences (qualitative), while government sources such as the national census or Index of Multiple Deprivation will provide measured facts (quantitative). Census data is collected every ten years, the most recent being 2011. This enables comparisons to be made with the earlier censuses, such as the 2001 census. Websites such as Neighbourhood Statistics and other census products such as the Labour Force Survey make this data available for a local area. Many datasets of social and economic indicators provide information about places, both on variations within and on larger spatial scales.

Qualitative sources explore and develop an appreciation and understanding of the views of different stakeholders on the priorities for regeneration. Different newspapers write from contrasting political slants and they will tend to showcase one particular set of views and ideas. For example, these newspaper headlines reflect different perceptions about Middlesbrough:

- 'Is Middlesbrough the worst place to live in the UK? Ofcom has backed up condemnation of Middlesbrough on *Location, Location, Location*. But do the residents agree?' (*The Guardian*)
- 'Middlesbrough really *is* the worst place to live: Damning verdict of *Location, Location, Location*' (*Daily Mail*)
- 'What can Middlesbrough teach Britain? For lessons on how to rebalance the economy, the government should look north' (*The Telegraph*)

The article from *The Guardian* is more sympathetic towards minority groups and confronting the causes of social injustice. The *Daily Mail* article is more 'sensationalist' and tends to highlight the more negative aspects of quality of life. The third, from the *Telegraph*, has a more right-wing economic focus to show that the priorities for regeneration should be around micro-businesses. Regardless of this, the Office for National Statistics (ONS) Government Report released in 2015 confirmed more objectively that the worst cases of deprivation are in Middlesbrough.

ACTIVITY

TECHNOLOGY/ICT SKILLS

Read through the posts on Twitter from organisations like 'Our Tottenham'.

1. How many groups are connected to this organisation, and in how many ways are they presenting their own perspective on their place?

2. Explain why these perspectives may create difficulties within the local communities.

Synoptic link

Different players have different motivations and act differently when faced with an opportunity for redevelopment. Smaller-scale stakeholders are often more community-based and cultivate a perception of more ethical activity. Large stakeholders have to reconcile competing viewpoints with the need to maintain profit margins. In many aspects of geography, decision-making has to consider different viewpoints, including coastal defences, location of TNCs, use of 'greenfield sites' for house building, immigration quotas, and mitigation of climate change effects. (See page 252.)

ACTIVITY

TECHNOLOGY/ICT SKILLS

1. Investigate the population structure and characteristics of two contrasting places.

a. Use the census to find out about age structure, ethnicity, deprivation measurements and see whether they have changed over the past ten years.

b. Once you have provided a postcode, you can interrogate census data on various geographic scales: by ward, for part of a town; by super output area, which is almost street by street; and by local authority, for the majority of a town or urban area. These datasets can be downloaded into Excel.

c. By collecting different datasets for the same set of postcodes, you might be able to find relationships between different measures. For example, there might be a strong correlation between areas with high unemployment and low educational qualification.

d. You can test the strength of these relationships using Spearman's Rank Correlation (explained on page 48).

e. You could also find the DfE performance tables and look up the latest performance tables for all the schools in your local authority.

f. The census data can also be mapped in a browser, or by using GIS software like ESRI-online. Mapping the different datasets might help you establish whether there are similar patterns between social class, unemployment or ethnicity. But beware of making assumptions about incorrect causal correlations.

2. Investigate your two contrasting places using Google Scholar – which provides access to university-level research into different places and often distinguishes between 'space' and 'place', particularly concerning 'representation' – and Google News search, which provides many different publications and reports, often with short interviews with local people and key decision-makers.

a. Which of the two places is more 'successful', and which is more at risk of getting into a 'spiral of decline'?

b. For each place, what are the priorities for regeneration in order to successfully reduce inequality?

c. For each place, assess the level of political engagement by the communities.

d. Evaluate how local people in the two places view their area and their attachment to it.

e. For each place, outline any regeneration plans and suggest whether these are likely to result in conflict between different groups with different needs,

AS level exam-style question

Explain two reasons why levels of engagement in local communities vary. (4 marks)

Guidance

For this 'explain' question, you need to write about any two of the following reasons: political, environmental, economic, social or technological.

A level exam-style question

Using a named place, explain why some regions are regarded as 'successful'. (6 marks)

Guidance

This question is based on part of the specification that is marked by a globe icon. The question will assess your knowledge and understanding of a located example. You will be expected to structure that knowledge as a 'broad range of ideas'.

Yellow = Players, Orange = Attitudes and actions, Purple = Futures and uncertainties

Representation through media

Different media – whether books, films, music or art – incorporate both a representation and an interpreted meaning of a place. Deciding to use this kind of evidence requires an assumption and acknowledgement that – in addition to an objective, 'scientific', and data-led analysis of places – qualitative viewpoints are valuable. An author's experience of a reality may be just as valid as a measurement. In order to evaluate the needs for regeneration, decision-makers, planners, architects and geographers must try to understand the perspective of the people who live there, how others influence their lives and the power relationships that are created as a result.

CASE STUDY: New York City through the media

Consider these representations of New York City.

- *Gossip Girl*

This TV series is set in the Upper East Side, just east of Central Park. It is portrayed as a cheaper area of Manhattan, filling up with young professionals and graduates, who don't step below 59th Street. In reality, Lower Manhattan is the more trendy part of NYC. But the point of this portrayal is to show how uptown residents look down on places like Brooklyn but eventually have to give in and consider living in less pricey lower Manhattan.

- *How I Met Your Mother*

A TV sitcom that characterised the singles lifestyle in a large city; where the characters work all day but still have enough energy to spend the evening in bars meeting strangers who are smart, beautiful and interesting. The programme points towards the reality of constant loneliness experienced by recently graduated successful twenty-somethings in major cities, as well as the perceived characteristics of different types of people from different 'zip codes' (postcodes).

- *Empire State of Mind* sung by Alicia Keys

The writers of this song were inspired during a visit to London. Feeling homesick despite the same busy streets and crowds, they noticed the unique characteristics of New York.

- David Harvey, Marxist geographer

A renowned Professor of Geography who famously wrote, 'There is nothing unnatural about New York City'. The point was to highlight how capitalism has led to a reformulation of nature to meet the need for profit at the expense of other social groups. Even Central Park is an artificial product of the decision to rebuild nature in the city, at the expense of ethnic minority groups who lived there and were moved to Harlem.

- *Humans of New York*

A photographic census of New York by photographer Brandon Stanton, in which 10,000 photos of residents were plotted on a map, together with random quotes. The intention was to get a daily glimpse into the lives of people who were otherwise just strangers in the city.

CHAPTER 5

How is regeneration managed?

Learning objectives

4A.7 To understand how national governments play a key role in regeneration

4A.8 To understand how local government policies aim to make areas attractive for inward investment

4A.9 To understand rebranding attempts to make areas more attractive by changing public perception of them

The role of national government

The UK government is responsible for considering the level of inequality across the country, as disparity between places and regions does not aid the overall development of the country. There is a widely perceived and evident gap between the north and south of the UK, in which the south is dominated by the metropolitan region of London. Lack of investment in rail infrastructure in the North East has led to some large inequalities and lack of labour mobility. Whereas a 65-km journey from Middlesbrough to Newcastle takes 90 minutes, a journey of the same distance from Chelmsford in Essex to London takes only 36 minutes. The main arterial road between Leeds and Manchester (the M62) is often jammed, and the main road from Newcastle to Scotland is still a single carriageway. Spending on infrastructure is £2595 per person per year in London, compared to just £5 in the North East. The UK government has tried to address this imbalance by developing a so-called 'northern powerhouse' (2015), where infrastructure investment will connect major industrial towns and cities in the North East and North West.

Synoptic link 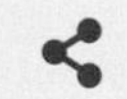

The national government often works with local and regional organisations and governance to ensure that infrastructure projects become reality. Property developers need to work in the right economic conditions, and charities often understand the needs of local people, However, both lack capital funds to reliably generate a return on the initial investment required to start large projects like these. Large, top-down projects can help diverse places across the UK and other countries, but there are always concerns about their effect on local people and the natural environment. (See page 242.)

Infrastructure investment

Some decisions about **infrastructure** centre on expanding capacity at London's main airports (Heathrow and Gatwick) and building Crossrail to ease congestion in Central London. Other schemes outside the capital region include High Speed 2 (HS2), a rail link planned between London and Birmingham and then to Leeds and Manchester. High Speed 1 has already been constructed between London St Pancras and Kent.

One project nearing completion is the Northern Hub project, a series of improvements to the rail network around Manchester. In 2011 ten local councils united to form the Greater Manchester Combined Authority (GMCA) and took control of the city's transport policy, among other things. A short curved section of track (the 'Ordsall Chord') has been built to connect Manchester's two main railway stations (Piccadilly and Victoria) and the airport station south of the city. The main stations have been expanded, and journey times to Liverpool, Sheffield and Newcastle have decreased. In addition, the government is funding electrification projects in Manchester, Liverpool and Blackpool and across to Leeds, to allow for cheaper, faster and longer electric trains. Improving access for a larger pool of labour, business is expected to increase. It is hoped the £600 million investment will eventually yield economic growth of £2.4 billion.

Yellow = Players, Orange = Attitudes and actions, Purple = Futures and uncertainties

The positive impact of these decisions cannot be taken for granted. Nearly a decade after the completion of High Speed 1, new stations such as Ebbsfleet are still surrounded by undeveloped land, although London itself has continued to see tremendous economic growth. But with air connections from regional airports such as Newcastle to New York and Dubai improving, and control over housing, education and health also in the hands of super local authorities in the North West, the overall aim of the national government in 2015 was to ensure that growth is not concentrated in any one place in the UK.

Synoptic link 

When it comes to matters of national interest, the government may have to act in the best interests of the country, rather than considering the viewpoints of local stakeholders. This may concern immigration and the development of military installations in key places. The EU and large countries such as the USA may act as superpowers with global considerations. (See page 207.)

Planning decisions

Plans for development can potentially have a significant impact on the natural environment and people's lives, but the decision may be deemed as being in the national interest. The UK government can make decisions that affect the rate and type of development (planning laws, house-building targets, housing affordability, permission for 'fracking'), which, in turn, affect the economic regeneration of both urban and rural regions.

Many planning laws are in place to limit the negative impact of development and regeneration on the social, economic and natural environment. Equally, because of the rise in single owner-occupancy, longer life expectancy and a lack of housing supply, house prices have rapidly increased in the UK. The government is under pressure to get builders to build more homes. Although targets have not been specified, the *Independent* newspaper carried a report in 2015 in which the Housing Minister claimed that 1 million new homes would be needed by 2020.

Decisions to permit **hydraulic fracturing ('fracking')** are particularly controversial. Since the UK will be forced to import nearly 70 per cent of its gas by 2020, government regulations are being loosened to make it easier for local councils to win approval for fracking. The British Geological Society estimates that there are 37 million m^3 of shale gas in the north of England alone. In 2015 £300 billion (8.6 billion barrels) of oil and gas were discovered close to Gatwick Airport in Sussex. Although it might not be possible to extract any more than a fraction of the oil there, there is a national interest in investing in this energy source. Potential mining sites are often found beneath large cities, valuable farmland and national parks. Since the national parks are mainly owned by the Crown, compensation for those using the land is very unlikely. In 2011 initial drilling in Blackpool led to minor earthquakes, and fracking was suspended, but in 2012, the government lifted this moratorium. In 2015 Lancashire County Council was expected to approve a plan to start drilling in the North West but, faced with public opposition (green groups, local landowners), this plan was rejected because of the noise impact and 'adverse urbanisation effect on the landscape'.

Figure 5.21: A fracking exploration site in West Sussex, and consequent protests

ACTIVITY

CARTOGRAPHIC SKILLS

1. Using the map in Figure 5.22, describe the distribution of prospective shale resources in the Britain.
2. Which types of area do you think will be affected the most?

Extension

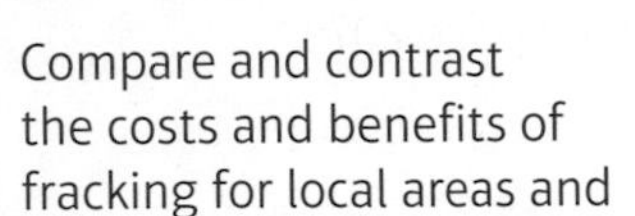

Compare and contrast the costs and benefits of fracking for local areas and for the country as a whole.

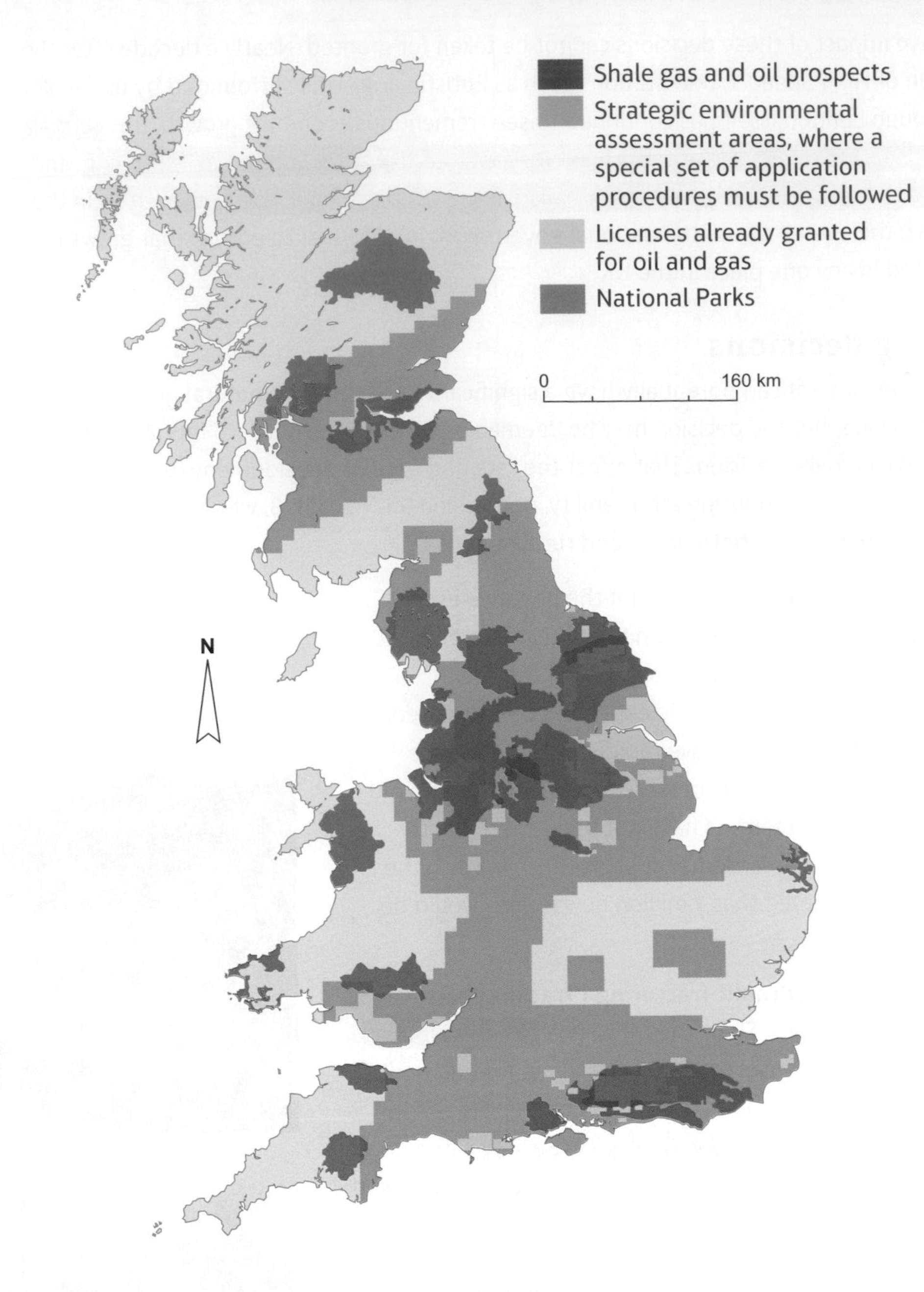

Figure 5.22: A map of prospective shale resources in Britain

International deregulation

The deregulation of capital markets since the 1970s has made it much easier for companies to locate to the UK and for foreign companies to invest in the UK's infrastructure. The UK's common law legal system, where cases are decided on the basis of previous judicial outcomes, also encourages companies to trade in London; the London Stock Exchange is one of the largest in the world. The City of London has lawyers, accountants and consultants of every description, creating an overall industry worth £95 billion. Although financial regulation is now tighter, making it harder for banks to make risky business investments, when some banks like HSBC threatened to leave London for Asia, the government decided in 2015 to halve a proposed banking levy. Some of this wealth has created regeneration in the former docklands area of east London, which has been developed as a financial centre (see Figure 5.23).

Yellow = Players, Orange = Attitudes and actions, Purple = Futures and uncertainties

Figure 5.23: The skyline of Canary Wharf, one of London's two main financial centres

ACTIVITY

Which decision has had more impact on the UK: international deregulation or open-door migration? Why do you think this is?

Extension

The UK has gone through fluctuating periods of border restrictions and open borders. What is the evidence for or against what the government should do to limit immigration?

Migration policy

Opening the UK's borders to migration is another major government decision that has helped the regeneration of local areas. In 2004 eight eastern European countries joined the EU, followed by two others in 2007. Many people from those countries then migrated to the UK, including Poles – now the second largest ethnic minority group in the UK after the Indian population. Although some Poles have returned home, 1.2 million have settled in the UK, giving birth to over 21,000 children in 2012 alone. Many have settled in towns like Slough and Corby and other rural market towns and small cities. These towns have become some of the fastest-growing parts of the UK.

Corby was previously a town with an ageing population and an increasingly poorly skilled workforce with few opportunities. East European entrepreneurs have since 2006 set up many businesses there, including restaurants, bakeries, construction firms, building design firms and marketing agencies. Property crime and antisocial behaviour in Corby have halved since 2006, clear indicators of the success of the much-needed regeneration of the town.

AS level exam-style question

Explain two ways in which local national government can affect decision-making about regeneration. (4 marks)

Guidance

Make sure you can name two completely different ways.

Local government policies

Local governments are keen to encourage innovation and investment, both from within the UK and by overseas companies. Local councils, often in partnership with major stakeholders in the area, may use a range of incentives to attract and keep companies that might improve the reputation of different towns.

Since the 1970s, cities and towns have worked with major universities, for example Cambridge, to focus on the commercialisation of research, by ensuring that the right infrastructure is in place, not only for established technology companies but also for innovative start-ups seeking to pioneer new research. The right support is important for these new companies, whether telecommunications and power supply, expertise in intellectual property law (to protect new ideas and inventions) or shared banking services. Investment has focused on innovation, high-tech and medical technology companies, which have thrived despite the 2008 recession, for example Claritest and Spectral Edge on the Norwich Research Park.

A level exam-style question

Explain how the deregulation of capital markets affects growth and investment. (6 marks)

Guidance

Make sure you can link these factors together as you develop two to three distinct reasons.

Synoptic link

Part of globalisation is the spread of tertiary and quaternary TNCs, which link places around the world together. Oxford and Cambridge are examples of where multiple stakeholders in a local area need to be governed by local councils to ensure that regeneration meets the needs of local businesses, while taking advantage of existing local talent. (See page 231.)

Competing for investment: science parks

Since the 1970s, Cambridge University has made a concerted effort to utilise its scientific expertise to increase wealth across the region. Cambridge Science Park and the St John's Innovation Centre opened with both start-up companies and large TNCs such as AstraZeneca, Toshiba and Microsoft, as well as the Royal Society of Chemistry all basing themselves in the park, not least so they can benefit from the technological expertise at the university.

Cambridge Science Park's continued growth has left Oxford in its wake. Although both cities are wealthy, Oxford has only just started to build large numbers of new houses, and the price of housing (11 times average local earnings) is discouraging potentially world-leading academics from moving to Oxford University. In response, the University has begun to take a much larger role in local decision-making. Not only are Victorian and post-war buildings being replaced by new buildings, also Oxford Science Park and Begbroke Science Park, both linked to Oxford University, have been built in the city's outskirts. In 2014, funding was granted to Begbroke to host the Innovation Accelerator to encourage high tech, aerospace and medical engineering as well as superfast computers and robotics. Oxford Science Park hosts many start up bioscience and computer hard- and software companies. The return on this investment has justified change in the town centre; the 1960s Westgate car park is being replaced by a £500 million shopping centre, and a second railway line to London has just opened. Continued growth is reflected in the first commercial flights between Oxford Airport and Edinburgh, which started in spring 2016.

Figure 5.24: Landscaping at the Oxford Science Park

Local decision-making

Major regeneration projects need the cooperation of many local interest groups. It is normally the job of the local council, such as a county council or a district council, to take the lead in ensuring that projects are successful. They must reconcile the many different interests and stakeholder groups. For example, local businesses, sometimes represented by local chambers of commerce, want economic growth even if this means demolishing old buildings, while local people often have a nostalgic attachment to historic buildings, and campaign to save them.

A Parliamentary Select Committee in 2004 concluded that many successful regeneration schemes should use historic buildings as a foundation for projects because they 'reinforce a sense of

Yellow = Players, Orange = Attitudes and actions, Purple = Futures and uncertainties

community, make an important contribution to the local economy and act as a catalyst for improvements to the wider area. They should not be retained as artefacts, relics of a bygone age. New uses should be allowed in the buildings and sensitive adaptations facilitated, when the original use of a historic building is no longer relevant or viable.'

CASE STUDY: Regeneration in Aylesbury

Aylesbury is the county town of Buckinghamshire. The Buckinghamshire Chamber of Commerce was formed in 2010, from Thames Valley, to facilitate business growth in the Aylesbury area.

Led by Aylesbury Vale District Council, the Waterside redevelopment project began in 2003 with a canal basin regeneration scheme to replace the Bucks Herald Printing Press, former police stations and old county offices. The development was delayed by the Environment Agency because of contamination of the land, some of which had been the site of a coal-fired power station and an oil depot in the past.

Figure 5.25: Aylesbury's Waterside Theatre

The first phase of development saw the demolition of a 1960s car park and construction of the £47 million Waterside Theatre and more recently the Aylesbury campus of Bucks New University, the first tertiary education provider in the town. A Waitrose supermarket and restaurants such as Nando's and Gourmet Burger Kitchen have moved to the development. Since the changes, there has been a 2.2 per cent increase in footfall since 2013 (compared to a national decrease over the same period). Plans for more restaurants and shops to open have, however, met resistance from established restaurants in the town centre.

The most recent phase of the redevelopment proposed a controversial demolition of some of the town's historic buildings, including the old County Hall and offices, police station and Judge's Lodgings. Following a campaign by local organisations (notably the Aylesbury Society) – and since the Court Buildings and Judge's Lodgings are Grade II Listed – the local council agreed to preserve some of the buildings and convert them into hotels and restaurants. One restaurant would be themed around Ronnie Biggs and the Great Train Robbery, which happened near to Aylesbury. The plan for regeneration is now focused more on the surrounding landscape, with other buildings now demolished to facilitate better pedestrian and transport access. A statue of the comedian Ronnie Barker, who began his theatre career in Aylesbury, has been added to the site.

ACTIVITY

1. Complete a conflict matrix for all the players in Aylesbury's Waterside regeneration scheme.
2. For each player, decide whether they would be in favour of the plans to regenerate Aylesbury's canal basin, (colour the square green), in disagreement (colour the square red) or might/might not be in conflict (colour the square orange).

Contemporary regeneration strategies

The eventual hope of regeneration strategies is that they will attract business investment and workers from many different contexts to create vibrant new places in urban and rural areas. The two following case studies highlight the changes that have happened in one area of London and in a rural area as a result of public–private diversification.

Figure 5.26: Queen Elizabeth Olympic Park

CASE STUDY: Queen Elizabeth Olympic Park, London

Following the Olympic and Paralympic Games of 2012, the former Olympic Delivery Authority (ODA) began a comprehensive redevelopment of the Olympic Park. Working with two engineering companies (Arup and Atkins), temporary venues were removed and new infrastructure erected to ensure a lasting legacy in this part of London. The site itself was reopened as the Queen Elizabeth Olympic Park in April 2014. Many of the original sporting complexes (the Olympic Stadium, Aquatic Centre and Velodrome) are still there, and the Olympic Village has been converted into 2800 flats and apartments. Some of the infrastructure, retail developments such as Stratford's Westfield Shopping Centre, Stratford International and regional train stations, and tourist attractions such the ArcelorMittal Orbit (a large piece of public art) are still in place and are key features of the park.

Construction is continuing to develop the park. The International Quarter will be a mixed-use development with 7000 additional homes on the edge of the park. The design of the development in the heart of the park encourages workers and residents to make use of the extensive cycle and walking trails and to use the leisure and sporting facilities to promote a

Yellow = Players, Orange = Attitudes and actions, Purple = Futures and uncertainties

healthier and more creative place to live and work. Transport for London and the Financial Conduct Authority are just two organisations intending to have offices in the park, which should eventually create 25,000 new jobs.

Newham Borough Council, in partnership with the London Legacy Development Corporation (LLDC), is responsible for continuing to develop the park for the rest of this century. This brings benefits to local residents, including tickets to events in the stadium and use of facilities for school sports. Six new Olympic boroughs have been designated, and already the park regeneration has lowered unemployment from 13 per cent in 2012 to 9 per cent in 2015. Newham Borough Council is responsible for running the London Aquatics Centre, making this iconic venue as affordable as any other local leisure centre. There is continuing work to convert the Olympic Media Centre into the 'East London Tech City' and 'Here East', both a technology and start-up hub creating flexible space for art and design companies to develop new products and share ideas. The Smithsonian Institute plans to open a museum, and two universities (UCL and London College of Fashion) intend to have campuses here.

ACTIVITY

What is the role and importance of different players in the successful redevelopment of the Olympic Park in east London?

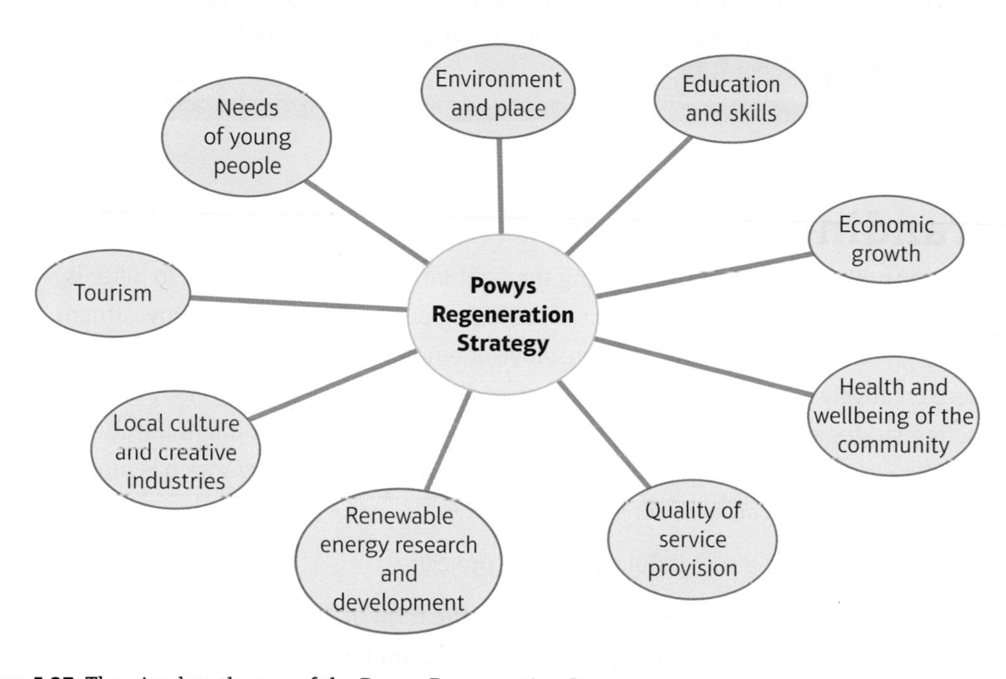

Figure 5.27: The nine key themes of the Powys Regeneration Strategy

CASE STUDY: The Powys Regeneration Partnership

The Powys Regeneration Partnership focuses on nine key themes, as shown in Figure 5.27.

Powys straddles a large portion of North and South Wales. Powys Council wants to exploit this region of beautiful landscape in order to make the most of opportunities for green tourism, the agricultural industry (through food) and the creative industries. There is also a specific aim to build on expertise in renewable energy and develop a low-carbon green economy around wind and water.

Powys suffers from a skills shortage. Lack of university places and qualifications, together with a long-standing emphasis on agriculture and tourism, means many jobs are part of a low-wage economy. EU structural funds have been used to encourage more competitive businesses to grow, partly by rejuvenating a local chamber of commerce. Other successful schemes include establishing Transition Towns in Machynlleth, Rhayader and Llandrindod Wells, where locally based loyalty cards and local currency encourage people to spend money supporting locally

ACTIVITY

Use a spider diagram or mind map to describe the interacting roles and importance of different players in rural regeneration strategies like that in Powys.

produced goods and services and protecting the unique characteristics that bring vitality to local areas.

Another aspect of the plans involves harnessing local talent and skills, as well as volunteers. Powys Council estimates that the area has close to 26,000 volunteers and that their work contributes £173.6 million to the local economy. Since the plans revolve around better use of blogs and forums to network people together, many rural communities need better high-speed broadband. Hopefully this will reverse out-migration of young people and business owners, make homeworking more realistic and encourage small business formation and growth as a means of **diversification** in the rural economy. Better investment in ICT has already started to come through the Carmarthenshire Community Broadband Partnership, installing two broadband masts to address black spots, as well as 'Powys Connections' programmes to provide advice and grant funding to micro-businesses wishing to specialise in ICT services.

The county is susceptible to changes that directly affect agriculture, including extreme weather. The Cambrian Mountains Initiative and EcoDyfi are both initiatives encouraging carbon storage and water regulation. The Green Investment Bank is making £1 billion of finance available for research and development, including £200 million for low-carbon technology and £860 million incentives to encourage renewable heat. There are also plans to spend £30 million improving the energy-efficiency of housing stock.

Rebranding strategies

The two case studies above show how having the right infrastructure and fabric in place is essential, in both urban and rural areas, for encouraging personal and business investment. In an increasingly globalised, competitive and consumer-orientated world, cities have been forced to think more creatively in order to continue attracting visitors and investment. Many have turned to marketing strategies to rebrand their image and reputation. Often this might include slogans – such as 'the Big Apple' for New York – to change the image of a city; New York has more recently been rebranded as 'the city that never sleeps'.

There are two elements to a successful rebranding project:

- **Regeneration** – investment in an area, perhaps in the form of infrastructure or other elements of physical fabric. Hopefully this triggers a process of **cumulative causation** (the multiplier effect, summarised in Figure 5.10) via employment opportunities in sports arenas, cultural centres, heritage sites, and shopping and leisure facilities. The overall change is sometimes described as **urban renewal.**
- **Re-imaging** – the area is 'sold' or 'advertised' with new packaging to change the impression investors have of the area and attracting people to it. Sometimes **rebranding** might simply mean changing the name of the city: for example, Saigon was renamed Ho Chi Minh City (1976), and Madras became Chennai (1996).

Ultimately, the intention is to 'sell' a place to potential customers, and they could belong to many types of groups, so the nature of the rebranding might vary according to the target group, some of whom are summarised in Table 5.4.

Yellow = Players, Orange = Attitudes and actions, Purple = Futures and uncertainties

Table 5.4: The players benefiting from regeneration

Visitors	Residents and employees	Business and industry	E[illegible]
• Business visitors who may be attending a conference or sales trip • Non-business visitors who may be tourists or travelling to visit relatives and friends	• Professionals (doctors, university teachers, etc.) • Skilled employees • Local wealthy people • Young people • Investors and entrepreneurs • Semi-skilled and unskilled workers	• Older heavy industry • New hi-tech industry • Quaternary sector employees	• Internat[illegible] businesses • International travel companies

Urban rebranding and place identity

As already noted, the British nostalgic emotional attachment to the past can be a powerful way to make sure regeneration strategies are successful. In 2008 a publication by Historic England noted 20 schemes in the UK that had used **place identity** to successfully integrate conservation of iconic and historic sites into commercially successful and distinctive redeveloped urban spaces. In urban areas, this probably means taking advantage of Victorian **industrial heritage**. In rural areas, it might mean historical and literary references to how the place was portrayed in the past; for example writers or artists who lived in or were inspired by the area.

The following two case studies illustrate how both a rural and an urban area have used links to the past to successfully attract tourists nationally and from abroad.

CASE STUDY: The Titanic Quarter, Belfast

Belfast was once famous for its shipbuilding, and particularly for the construction of the ill-fated *Titanic*, at the Harland and Wolff shipyards at the neck of the River Lagan and Belfast Lough. The Titanic Quarter in Belfast, named after the city's most famous product, has become one of Europe's largest waterfront developments. Shaped around a marina, it comprises historic maritime landmarks, including the huge yellow lifting cranes that are an iconic symbol of Belfast's industrial heritage, luxury shops, postmodern workspace architecture and a university campus. Belfast Harbour continues to operate as a private company providing the logistical framework for successful trade, albeit on a significantly smaller scale than in the past.

Right at the head of old slipway where the *Titanic* was built is the 'Titanic Belfast' visitor centre. The spectacular architecture is shaped like the bow of a ship and the materials reflect water and ice crystals. The old White Star Line logo has been incorporated into the design. The centre welcomed a million visitors in its first year of opening, 70 per cent of them from outside Northern Ireland, and who principally came to visit the Titanic Quarter. The project won the top prize in the UK Regeneration and Renewal awards in 2012.

In 2015, Deloitte's were commissioned to evaluate the success of the project, and concluded that £105 million of additional tourism had resulted, sustaining 893 additional jobs in the Belfast economy. Demand for flats in the area now outstrips supply. Land of the Giants, an outdoor arts venue, is now based in the area, as well as the annual BBC Proms in the Park. New water-based leisure activities have also begun.

...egener...

...T/ICT

ACT...

...ad Historic ...gland's *Guide to Constructive Conservation* to investigate how industrial heritage has been used to shape urban regeneration strategies.

Figure 5.28: The Titanic Quarter, Belfast

Belfast itself has suffered from socio-political unrest since the 1960s. Although this ended formally in 1998, the 'Troubles' have continued to limit investment from outside. In a desire for sustainability, the city is looking to the post-industrial world with a range of rebranding strategies:

- A £4.9 million project to transform the former Harland and Wolff headquarters into a 4-star hotel shows a continuing demand for regeneration in the Titanic Quarter.
- The two Harland and Wolff gantry cranes are still in use, as the old dry docks are now used to construct offshore wind turbines.
- The older 'Paint Halls' have found a new use as film studios; the most high profile of which has been the filming of *Game of Thrones*.
- The Short Brothers factory, now owned by Bombardier, continue to build wings for aircraft on the site of Belfast City Airport, itself redeveloped and linked more closely with Belfast City Centre.

On the back of this development and infrastructure, a number of companies have started to invest in the area, notably Audi, Premier Inn and the Public Records Office for Northern Ireland. The Northern Ireland Science Park nearby has begun to grow again with Citi and Intel investing in the IT and software industry, and this is helping raise aspirations for young people in East Belfast enrolling in Belfast Metropolitan College.

Extension

Decide whether any other categories might have been more useful for rebranding Belfast, or whether the city could have taken advantage of more than one:

- Sport
- Cultural facilities
- Technology and science
- Retail development
- Education
- Leisure and entertainment.

Rural rebranding strategies

Rural areas, and particularly the agricultural economy, have changed considerably since the 1990s. With concern over inflated production and the environmental impact of intensification, there has been a shift in focus towards wider rural development, including support ecosystem services and preservation of cultural landscapes.

These changes to the rural economy are an example of diversification, described by some geographers as a shift towards a **post-production** rural economy – a controversial term, not least because agricultural production is still the most important use of rural landscapes. Because tourism is increasingly a vital mechanism of growth in rural areas (driven perhaps by continued investment in regional airports, such as Leeds Bradford), some geographers prefer to describe these changes as the creation of 'multifunctional' areas.

Yellow = Players, Orange = Attitudes and actions, Purple = Futures and uncertainties

CASE STUDY: Brontë Country

'Brontë Country' is the name given to a region in the West Yorkshire and East Lancashire Pennines. Many visitors come here every year to experience the bleakness and desolation that inspired books such as *Wuthering Heights* and *Jane Eyre*, authored by the Brontë sisters, who lived in Haworth (Figure 5.29). The Fair Trade Way links Fairtrade villages and towns in the Bradford area, taking visitors through the countryside and historical sites that appear in the Brontë sisters' books.

Figure 5.29: The village of Haworth

Alongside the literary and cultural associations, there are many are other tourist attractions in the area:

- The Keighley Bus Museum and Keighley and Worth Valley Railways.
- The Pennine Bridleway and outdoor activities including horse riding and mountain biking routes.
- The industrial village of Saltaire (near Bradford), designated a UNESCO World Heritage Site.

Some farms have taken advantage of EU grants and subsidies to **diversify** their use of land and create specialised products. For example, Skipbridge Farm (between Harrogate and York) is now a successful wedding venue, offering creative and bespoke weddings incorporating its small animal holdings, a bed & breakfast, cottages and 'glamping' tents.

ACTIVITY

TECHNOLOGY/ICT SKILLS

Good photographic analysis requires the observer to be empathetic and interpretative – to imagine being in a particular place and able to draw conclusions from the range of visible information.

Use the five prompt questions in Figure 5.30 to evaluate the geography of your local area.

[Source: Jessica King]

Take an image of any chosen area and fill in the following...

1. **Camera App.** What do you see in the picture? What are your first impressions? Who or what is the main focus?

2. **Maps.** Where are you? What is significant or insignificant in the surrounding area?

3. **Zoom In.** When you look closer, can you see more than meets the eye?

4. **Panorama App.** What is the bigger picture? Are there any wider implications that you may miss at first sight?

5. **Instagram App.** What filter (or perspective) are you viewing your image through? Would other people see what you can see? What changes have occurred over time?

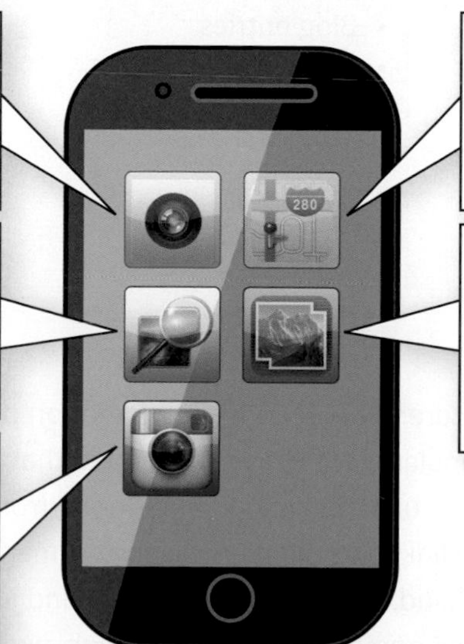

Figure 5.30: Looking through the 'geographical lens'

CHAPTER 5

How successful is regeneration?

Learning objectives

4A.10 To understand how to assess the success of regeneration using economic, demographic, social and environmental measures

4A.11 To understand the criteria that different urban stakeholders use to judge the success of urban regeneration

4A.12 To understand the criteria that different rural stakeholders use to judge the success of rural regeneration

Measuring the success of regeneration

Regeneration often involves spending large sums of public money on re-invigorating the local economy, to help business start-ups, create jobs, reskill the workforce and make once-prosperous areas successful and liveable again. In 2006 the Audit Commission (now the National Audit Commission) produced a series of indicators of successful economic regeneration. These act as a measure of success of the schemes and whether the expense was justified.

Table 5.5: Different ways to assess the success of regeneration

Aspect of regeneration	Economic regeneration	Social progress	Living environment
Measurement	• Income • Poverty • Employment	• Reduction in inequality • Reduction in deprivation • Demographic changes – increased life expectancy, reduced health deprivation	• Reduced pollution levels • Reduction in abandoned and derelict land
Potential fieldwork and research	• Census data • Neighbourhood statistics • Historical and current data	• Interviews • Newspaper accounts • Acorn scores • Blog entries • Social media	• Photography and comparison with historical photos • Fieldwork – environmental quality surveys and pollution indicators • Drosscape is an urban design framework which guides surveys

Measures of success

It is important to use multiple measurements to judge regeneration projects appropriately. If a town or city is a prosperous and popular place to live, it is hoped that its high levels of employment and skills can help those in long-term unemployment or suffering from associated mental health issues. Economic growth is strongly linked to the physical environment as well as to the social and cultural infrastructure. On the one hand, growth provides funds and justifies projects to improve the social, economic and natural environment. On the other, high-quality liveable spaces improve

Yellow = Players, Orange = Attitudes and actions, Purple = Futures and uncertainties

the image and reputation of an area. This, in turn, makes the area more attractive and appealing for businesses, families and tourists to come to and invest in (a positive multiplier effect).

The measurements summarised in Table 5.5 list the quantitative and qualitative sources used to judge how much everyone in a town or rural area has benefited from change. Qualitative sources can be helpful to understand the viewpoints held by different stakeholders, particularly as lived experience through places – both pre- and post-regeneration – provides a valuable perspective about the real impact of projects.

Judging the success of urban regeneration projects

Urban redevelopment strategies are not always successful. Some are made with particular political motives. The need to rebuild bombed urban centres after the Second World War created an opportunity for urban planners like Patrick Abercrombie to put their plans into practice. Their ideas influenced the 'comprehensive redevelopment planning' that was eventually characterised by a 'brutalist' style of architecture, with concrete-heavy, quick-to-build and functionally efficient urban designs. In many cities these designs – probably not built with universal consent – quickly showed their age and are now often described as dull and 'harsh'.

Modern regeneration strategies take into account a much broader range of public opinion, and this has resulted in many urban spaces being contested; disagreement remains as to whether the changes have resulted in positive or negative impacts for all local stakeholders.

CASE STUDY: Regeneration in Plymouth

Plymouth is a city long associated with the UK's naval history. With direct access to the English Channel, the city's Devonport Dockyard was the largest naval base in Western Europe. Its strategic importance meant that it was heavily bombed during the Second World War and many of its buildings had to be rebuilt. Although the Navy continues to provide many jobs in Plymouth, defence-budget cuts mean that the economy now relies on innovative services and the cultural sector, particularly through marine and maritime businesses, for example those found at Tamar Science Park.

A number of key changes and investments have occurred in Plymouth:

- In 2014 the government signed a 'city deal' to bring millions of pounds of investment and transfer dockyard land from the MoD to the council, creating 1200 new jobs.
- The city was awarded £670,000 of funding from the Crown Estates to improve coastal areas.
- Through Plymouth University, which has 27,000 students, the city has placed a strong focus on education, with many new university buildings, including the £20 million Marine Building housing one of the largest and most advanced wave tanks in the world.
- Princess Yachts and Wrigley are two companies that have maintained their presence in the city.
- Culture has featured heavily in the plans, for example in the redevelopment of Grade I listed Royal William Yard – previously a supply depot and now home to many apartments, restaurants and Ocean Studios, a creative space for aspiring artists.
- Plymouth Hoe, a huge green space overlooking the English Channel, is a popular space for MTV and other music events.
- Refurbishment of the centrally located Guildhall has included bakery and café markets. The Guildhall now hosts many community events.

Despite these developments, there is still huge inequality in the city, where 90 per cent of houses are in council tax bands C, B and A, which often reflect low wages and aspiration. To address this imbalance, government funding from the New Deal for Communities, together with the city's Social Investment Fund, has been used to rebuild the community of Devonport, a residential area near the dockyards. Lessons learned from previous developments have ensured that working-class residents have not been forced out of their homes. A range of housing types have been constructed, to rent, buy or part-buy. Since house prices have remained affordable (averaging £165,000), the decision to regenerate is a much less contested decision, and a successfully revived

ACTIVITY

One of the earliest and most controversial projects in Plymouth was to demolish the 1950s Drake Circus shopping centre and rebuild it as a modern covered shopping venue. Reaction from local residents at the time was mixed.

1. Comments on local news websites are easy to find, choose four representing a range of views.

2. What do they tell you about whether people think Plymouth made the right decisions about regeneration in the city centre?

urban village has resulted. Due to its affordability, many new graduates decide to stay in the city, to live in this idyllic waterfront location.

Figure 5.31: Drake Circus, Plymouth, before and after regeneration

Table 5.6: Key changes in Plymouth between 2001 and 2011 (%)

	2001 census	2011 census
All households owning their accommodation outright	25.5	27.2
All households owning their accommodation with a mortgage or loan	37.3	31.5
Good health	66.7	33.5
Day-to-day activities limited	20.5	10.0
Full-time employee	39.1	36.1
Part-time employee	12.9	14.8
Self-employed	5.6	6.8
Unemployed	3.2	4.2
People aged 16 and over with five or more GCSEs grade A–C, or equivalent	21.2	16.8
People aged 16 and over with no formal qualifications	28.9	22.3

Maths tip

Standard deviation is explained in several places in this book, especially in Chapter 2, page 97 and Chapter 3, page 132. Two-axis scattergraphs can be found in Chapter 1, page 50 and Chapter 6, page 306. With a single-axis scattergraph, values are plotted at the same distance away from one vertical axis. Quartiles can be worked out from a single-axis scattergraph by finding the mid-point in the scatter of values, and then finding the mid-point of the upper half and the mid-point of the lower half: In this way, the range of values is divided into four or quarters, hence the term quartiles.

ACTIVITY

CARTOGRAPHIC SKILLS

1. a. Pick one census statistic, and download ward-level 2011 census data for a town or city (for example Plymouth) into an Excel spreadsheet.

b. Find a census ward outline map of your chosen place, or satellite image with the ward boundaries shown. You could draw the boundaries on a satellite image if one is not available from the planning department of your chosen place.

c. Find the same ward-level data for the census in 2001. Add this data to your spreadsheet.

d. Find the positive or negative change between the two sets of data (2001 and 2011) by using Excel functions (or a calculator), and record this on the spreadsheet.

2. Consider the range of the different change values and calculate categories, perhaps using quartiles or standard deviation or a single-axis scattergraph. Allocate a colour to each category of values; this should be a logical sequence.

3. Produce a choropleth map on your outline ward map or satellite image (GIS).

4. Describe and explain the pattern of change shown by your completed map. Investigate if the areas that have improved had any regeneration strategies between 2001 and 2011.

Yellow = Players, Orange = Attitudes and actions, Purple = Futures and uncertainties

Whether these projects will be enough to revive the fortunes of Plymouth is yet to be seen; they can benefit some individuals more than others. The small extract of census data in Table 5.6 suggests that some changes have been positive (limited activities) and others negative (health and educational attainment). This makes it important to understand the perspective of different stakeholders in their experience of living in the city and what the reality of life is. Table 5.7 shows a number of descriptions of the changes in Plymouth, made by the local MP (recorded in Hansard) and others on the *Guardian* website, reflecting on the successes. They show the importance of urban regeneration schemes being matched by investment in the surrounding infrastructure both across the city and regionally.

Table 5.7: The view of different stakeholders about changes in Plymouth

'Plymouth has been fairly successful with inward investment. Some 25 major foreign-owned manufacturing companies in Plymouth employ more than 12,000 people. Tourism employs a similar number, and call centres are a recent, fast-developing sector, which now employs over 4,000.' (Local MP in Hansard)	'Our local Evening Herald supports the new deal programmes … the paper has featured success stories of what young people, lone parents, and, more recently, older people have achieved … a key part in helping Plymouth people to see their city as a can-do place.' (Local MP in Hansard, talking about local paper)
'I live 20 miles outside the city, aside from the new Drake Circus (filled with the usual high street chains), the rest is depressingly run-down, and there is little to make me want to travel in.' (Rural resident)	'I've just graduated, but there are no suitable jobs in Plymouth for me to come back to. I will have to look to the larger cities like Bristol or London, perhaps those in the Midlands.' (University graduate)
'The Royal William Yard is not welcoming, too imposing. Purpose-built places like that filled with middle-class shops and shoppers offer little for the rest of us.' (Plymouth resident)	'Getting to and fro is the real problem. The M5 stops at Exeter and then the miles grind by. The train is little better, it can be 5.5 hours from London.' (Plymouth resident)

AS level exam-style question

Explain two ways in which social measures can be used to assess the success of regeneration schemes. (4 marks)

Guidance

Take care to develop each point – you need to be clear why a change in each measurement means an improvement or a benefit resulting from the regeneration scheme.

A level exam-style question

Explain why different stakeholders assess the success of regeneration using contrasting criteria. (6 marks)

Guidance

A successfully developed point here relates the viewpoint of the stakeholder about whether a scheme was successful to the criteria (economic, social or other) they used.

Evaluating rural regeneration

The Causeway Coast, Northern Ireland

Rural areas such as the North Antrim coast have used a variety of strategies to encourage restructuring of the local economy. The coastline is world famous for the basalt columns of the stunning Giant's Causeway (Figure 5.32). Through the recent sporting success of local residents like Rory McIlroy, the coast is increasingly being recognised, with the return of the Open Golf Championship to Portstewart and the decision to film on location there for the TV series *Game of Thrones*.

There are been some successful 'bottom-up' local projects to encourage the agricultural economy:

- Causeway Coast and Glens Borough Council manage a successful Causeway specialist market at the Diamond shopping centre in Coleraine (the county town) once a fortnight, helping local businesses market local foods and crafts, gourmet food and drink, and seasonal fruits and vegetables.
- North Antrim Community Network has been an important voice and influence in helping rural communities retain their local identity.

However, the North Coast needs some 'top-down' regeneration strategies to focus the North East Region Rural Development Programme. Table 5.8 (taken from the Moyle Council Rural Development Plan (2012–15)) summarises some of the key aims and objectives of this plan, as well as the changes it is hoped will occur, which could be the ways of judging success.

Table 5.8: Summary of the rural development targets in the North East Regional Rural Development Programme for Northern Ireland

Focus	Aim	Success criteria
Rural diversification	• £8.5 million to support rural businesses • £55,000 to support 100 farming families as they diversify into activity-based tourism • Creation of 20 new business projects and 40 new businesses	• The number of grants awarded: capital funding, marketing, technical support, representing • 400 jobs created • 3 per cent economic growth • 300 new businesses
New rural workspaces	• Establish 40 new enterprises, as well as 40 micro-enterprises focused on the social economy (networking) • Capital funding of up to £50,000	• 20 per cent new business • 700 new rural economy jobs • 300,000 kw of new renewable energy
Tourism	• New activity-based tourism projects: walking, cycling, water-based activities and indoor wet-weather activities • Enhance local tourism amenities: signage, picnic areas, parking areas, better marketing • Regional food tourism initiative, e.g. the 'Menu of Moyle'	• 10 per cent increase in number of tourists • 50 new jobs • 60 new businesses • 200,000 more kw/h of renewable energy from tourist infrastructure
Basic services and transport	• Village renewal: provision for childcare, youth, elderly population • Tackle low educational attainment • Tackle dereliction, vacant properties and associated poor environmental conditions	• Better quality of life for 5000 villages • 10 jobs created • 10 projects for children and young people • Appointment of Village Renewal Officer • 50 rural development projects
Conservation and rural heritage	• Replace the ferry to Rathlin Island and fund projects to restore and conserve historical buildings	• Support for 20 rural projects • 5000 more visitors to rural heritage sites

ACTIVITY

TECHNOLOGY/ICT SKILLS

Use the NI census website to collect data to see whether the targets in Table 5.8 have been, or are being, met.

The management issues on the North Coast are significant, partly because of the large size of the area. However, the number of stakeholders is a particular challenge; the list includes:

- the National Trust
- Moyle District Council
- local farmers
- Translink (Northern Ireland's public transport network)
- renewable energy companies
- local hoteliers and restaurateurs.

Making changes to rural places affects the meaning other stakeholders attach to that space, so creating a 'contested space'. The change alters both the reality and their image of the place.

Yellow = Players, Orange = Attitudes and actions, Purple = Futures and uncertainties

CASE STUDY: Contested space at the Giant's Causeway

The Giant's Causeway on the Causeway Coast has been a World Heritage Site since 1986, in recognition of its outstanding natural beauty. The Causeway (Figure 5.32) is Northern Ireland's most important heritage resource and one of its most visited tourist attractions. Its dynamic landscape is also a contested space, meaning that the needs of visitors and other stakeholders may conflict with one another and with the needs of conservation.

Figure 5.32: The Giant's Causeway, Northern Ireland

In 2013 a new £18 million Giant's Causeway Visitor Centre opened at the top of the Causeway. The access road was upgraded to turn it into a bus route, complete with bus stop and turning circle right at the outcrop that forms the Causeway. Not only does this make it the only UNESCO site in the world with a bus stop, but – as researchers at Queen's University, Belfast have suggested – this kind of transport provision inevitably changes the impression people have of the environment. Its accessibility means that tourists perceive the area to be safe, whereas in reality, the high-energy waves and popular desire to scramble over the rocks expose visitors to great danger. Installing safety equipment subtly degrades the quality of the scenery. Building the Bushmills Tramway and the Park & Ride bus link from nearby Bushmills increases the number of visitors, but perhaps beyond the carrying capacity of the site.

All these changes are important for the economy of the Causeway Coast. However, the concern of locals is that tourists dominate and are ruining neighbourliness because they only stay seasonally. They say that the increase in traffic pollutes the air and the new buildings degrade the scenery (Figure 5.33). However, some geographers argue that the idea of 'rural' is itself a social construction, and means something different to every stakeholder. Certainly, people often have a romantic sense of 'rural', perceiving it to be the opposite of 'urban'.

The portrayal of a beautiful rugged coastline on postcards and tourist posters suggests a sense of harmony, but what is not shown in a picture postcard is a community where, as in urban areas, the elderly can be marginalised, young people disadvantaged, small business owners like farmers struggle, and community services might be in decline. In the context of Northern Ireland's recent past, there is also a patchwork of multiple and overlapping traditions and sectarian divides. When it comes to the Causeway Coast as a whole, change and development, while bringing economic growth, also bring contests between protecting a fragile environment and making the most of the economic potential created by tourist access to Northern Ireland's natural heritage.

ACTIVITY

TECHNOLOGY/ICT SKILLS

1. Use mapping software to look at the different photos taken by visitors to the Giant's Causeway.

2. Look on the internet to read comments left by visitors about the Giant's Causeway. Make a Table of positive and negative impressions.

3. Assess the impacts of tourist development.

Figure 5.33: The North Antrim Coast, Northern Ireland

Extension

To what extent does the socio-political context of places like Northern Ireland make it difficult to conduct genuine geographical research into the success of regeneration projects?

Fieldwork exemplar: Regenerating places

Enquiry question

How successful are the regeneration and rebranding of the Titanic Quarter in Belfast?

Fieldwork hypotheses

Null hypothesis (H_0): The Titanic Quarter has not helped to regenerate the Belfast Harbour area.

Alternative hypothesis (H_1): The Titanic Quarter is a successful regeneration strategy for the Belfast Harbour area.

Locating the study

Belfast is the capital of Northern Ireland, located at the head of Belfast Lough, a long deep inlet that connects Belfast with the Irish Sea. The area is sheltered, and Belfast has for a long time been a gateway maritime trading hub. Heavy industry characterised some of the surrounding areas, including shipbuilding (notably of the *Titanic*). However, post-war industrial change resulted in serious decline for East Belfast (100,000 jobs lost), exacerbated by civil unrest in the 1970s and 80s. The Titanic Quarter is part of the ongoing regeneration of the harbour area. Belfast Harbour is still a centre for freight and passenger trade, including cruise ships, to the north and west of the UK. George Best City Airport is to the north-east, and there are industrial estates and the SSE (Odyssey) Arena for large-scale cultural events.

Methodology

Three zones were examined around Belfast Harbour: the still heavily industrialised north-west dockyard zone (ferry and freight terminals); the industrial park to the north-east of the harbour zone, near Holywood Exchange and Belfast City Airport; and the most recent, post-industrial, south-east zone around the Titanic Quarter and SSE Arena. In each zone, five different sites were profiled. At each site, data was collected to do the following:

1. Judge the potential for redevelopment (using a bipolar scale measuring, for example, variety of design, amount of green space, pavement quality and noise pollution).

2. Assess the environmental quality (also using a bipolar scale measuring signage, building materials, cleanliness, colour and light and access).

3. Map land use using the RICEPOTS models, where different land-use types were scored according to whether they were more likely to be found in an industrial or a post-industrial setting (Residential = 8, Industrial = 1, Commercial = 5, Entertainment = 7, Public buildings = 6, Open space = 5, Transport = 2, Services = 3).

Groups of students standardised their judgements for environmental quality and potential for redevelopment. They used apps on digital devices to make judgements about sound levels. The land use was mapped onto satellite images of the area taken in 2007.

It was predicted that the lowest potential for redevelopment would be in the Titanic Quarter, which would also see the highest score for the RICEPOTS survey, which in turn would show more use of land for residential, open space and entertainment. The highest potential for redevelopment and lowest environmental quality score would be at the dockyard zone, which would also score the lowest for the RICEPOTS survey, with land use mainly being for industry, services and transport.

The largest risks for observers were from traffic or unfriendly approaches from strangers. To manage these, they stayed together in small groups all the time, taking great care when crossing all roads, using proper crossing points where available. Clothing appropriate to the cold weather was also worn to reduce the risk of hypothermia.

Data presentation

The group data was inputted into a spreadsheet. XY co-ordinates were established for each site at each zone. This allowed the data to be plotted using mapping software to produce located bar graphs superimposed on maps of the area to show spatial patterns. The land-use map was finalised by plotting polygons using mapping software and making quantitative and qualitative judgements about the changes to land use since the satellite photo was taken in 2007. Using online mapping, it was also possible to track changes to the area back to 2001, particularly in surrounding neighbourhoods, and judge the addition of new services and other changes over a period of 15 to 20 years.

Zone	Development potential score
Titanic Qu NW (Visitor Centre)	95
Belfast Harbour - Dufferin Road	78
Belfast Harbour - Sinclair Road	74
Belfast Harbour - Ferry Terminals	68
Airport Road West (W)	68
Airport Road (SW)	66
Belfast Harbour - Duncrue Street	59
Titanic Qu SE (Harland & Wolff)	56
Titanic Qu NW (Dock & Pump House)	55
Airport Road West (E)	53
Airport Road West (NE)	49
Titanic Qu SW (Sydenham SSE Arena)	0
Titanic Qu (NE) - Queens Road NE	0
Belfast Harbour - Corporation Street	0
Airport Road West (S)	0

Analysis and conclusion

The greatest potential for redevelopment was found in the industrial zones to the north-west (scores of between 59 and 78) (see Table 5.9), compared to relatively lower scores around the industrial park (49–53). There were some unusually high scores in the Titanic Quarter (95 just to the north of the Titanic Visitor Centre). Looking at the RICEPOTS survey, the conclusion was that a significant number of commercial and services-related infrastructure remained, with some vacant land about to be redeveloped. This suggested that there is potential for the Titanic Quarter to be redeveloped further, which in the short term might make the residential areas less desirable. The environmental quality scores were highest in the Titanic Quarter for areas that have been successfully redeveloped (15), compared to Belfast Harbour, which had the lowest (8). The industrial estate always scores middle-ish values, but the recent move to the site by the Department of Finance significantly raised the land-use score around the Airport Road to the north-east of Belfast City Airport. This would be regarded as a positive sign that the area is regarded as a successful regeneration, and investment by the public sector is likely to trickle down into the surrounding area, particularly to the nearby Holywood Exchange retail park.

The industrial park has been a successful regeneration scheme, and the Titanic Quarter, although it is still early days, shows signs that it will be successful too.

Evaluation

The fieldwork yielded clear differences between three different zones in the area around Belfast Harbour. It demonstrated that regeneration has been successful in recent years (high scores for the industrial park) and was likely to continue being successful, as the Titanic Quarter continues to develop to make the best use of the available brownfield sites.

There was systematic coverage of the zones and different sites. However, the Google Earth graphs showed clearly anomalous data for the Titanic Quarter, where the potential for redevelopment for one site was much higher. The group responsible for that site re-examined their data and concluded that they had accidentally reversed the scoring system.

One weakness of the methodology was the timing, since fieldwork was conducted in the late morning and early afternoon, when there were few people in the area. The methodology did not include any assessment of the types of people who work or live in the area, or any judgement about their lived experiences – and even if it had, there were few people around during the day. It might have been better to judge the success of the Titanic Quarter during the weekend. In addition, further studies could have been conducted into the residential neighbourhoods surrounding the harbour (e.g. Knocknagoney and North Belfast) to determine whether the communities immediately adjacent to the harbour had felt the benefits of change.

Yellow = Players, Orange = Attitudes and actions, Purple = Futures and uncertainties

Summary: Knowledge check

Through reading this chapter and by completing the tasks and activities, as well as your wider reading, you should have learned the following and be able to demonstrate your knowledge and understanding of 'Regenerating places' (Topic 4A).

a. How is economic activity classified?
b. How does economic activity affect social indicators?
c. What are four place functions?
d. What are the reasons places might change?
e. How can we measure economic change over time?
f. What are the global, national and regional influences on change?
g. How does a spiral of growth create a successful place?
h. How does a spiral of decline create deprivation?
i. How can people engage in their local communities?
j. What affects levels of engagement?
k. Why is there conflict about regeneration?
l. What decisions does the UK government make?
m. How do local governments try to attract investment?
n. Who plays a role in making decisions about regeneration?
o. What strategies are used to regeneration urban and rural areas?
p. How is industrial and literary history used to define place identity?
q. How does rebranding help improve the image of areas?
r. How do we measure the success of economic regeneration?
s. Why is regeneration sometimes contested?
t. How does lived experience alter meaning and reality of places?

As well as checking your knowledge and understanding through these questions, you also need to be able to make links and explain ideas, such as:

- the ways in which type of employment is linked to health and life expectancy
- how working in different economic sectors affects quality of life
- how global, international, regional and national influences change the function of places, and types of jobs for people in those places
- how land use reflects employment sectors and quality of life
- why low quality of life might sometimes be an indicator of the need for regeneration
- why some consequences of urban decay can also be causes (a spiral of decline)
- how global change triggers a spiral of decline in some places but makes other regions more successful
- why poor quality of life might cause greater political engagement
- whether migration to an area might cause positive or negative economic and social change
- how national infrastructure investment affects growth or decline in different places
- how investment by the government can support local decision-making
- how the industrial past can create problems but also stimulate successful regeneration based on place identity
- how rural rebranding strategies like diversification are linked to national and global changes.

Preparing for your AS level exams

Sample answer with comments

Using named examples, assess the extent to which rebranding is the most effective way of regenerating urban areas. (12 marks)

Some places use sport as a catalyst, like Stratford, which held the 2012 London Olympics. Affordable homes have made it easier for local residents to have money to spend in the area, and Westfield Shopping Centre was built so that residents have a wider variety of shops to choose from, and the inclination to spend more. Therefore the residents need more employment opportunities to be able to afford shopping at the new shops. Businesses and banks will invest in the area, creating more jobs and stimulating the economy. As this puts the multiplier effect into motion, it allows the area to be more sustainable through economic growth – encouraging more services to be set up such as travel infrastructure, educational facilities and medical facilities. These services will create more job opportunities. Transport links were set up to get people from all parts of the country and the world to Stratford. The Eurostar went straight from Paris to Stratford in a matter of hours. Sport has left a legacy and still attracts many people today.

A clear focus on an urban area, showing good understanding of the multiplier effect, a vital part of a regeneration strategy. However, there is little reference to the idea of rebranding.

Birmingham is marketing itself as an 'easy-access, heart of England' place to visit. Leisure is being promoted to deal with the impacts of post-industrial decline. The Sea-Life Aquarium is appealing for children. Local companies like Chiltern Railways want to provide easy-access transport to attractions like Cadbury World and the Think Tank, which help to boost the economy and remind people of Birmingham's industrial heritage. The intention is to make a visit to Birmingham different from any other place, which is something changing image alone cannot achieve. Regeneration is needed too. The government needs to put money into expanding infrastructure (e.g. Chiltern Railways) and clearing former factories destroyed in World War 2. Often cities need a masterplan to co-ordinate the interests of local people, many of whom might need retraining and have a nostalgic attachment to Birmingham's history, which is why the Black Country Museum is also mentioned.

A much better focus on rebranding and image here. This addresses the question more directly, because it shows how changing people's image of Birmingham ultimately leads to regeneration through new investment. The paragraph starts to show judgement about the importance of image, noting that other factors, e.g. investment in infrastructure, local business and industrial heritage, are also important.

Changing the image of a place means marketing it to attract visitors. The Titanic Quarter in Belfast is the new name for part of Belfast Harbour that was in urgent need of regeneration after deindustrialisation led to the decline of shipbuilding. The name is a reminder of the city's industrial heritage. The Titanic Visitor Centre claims to be the only centre located on the site of its historical theme, and has been very successful since its opening in 2012, bringing many tourists from outside Northern Ireland. The multiplier effect can be seen in the hotels and conference centre facilities in the Titanic Quarter, the filming of *Game of Thrones* and the expansion of university accommodation into the area. Rebranding has played a crucial role, initiating change in this brownfield site.

This final paragraph better addresses the question by starting to make a judgement about the role of rebranding alongside the consequent rejuvenation of an area.

Verdict

This is a decent student answer. The majority of marks in this type of question are for analysis and evaluation of geographical information from different contexts, with only a few marks for knowledge and concept recall. Most of the answer describes the features of change in different places, so to get better marks the candidate would need to spend more time judging how important rebranding is for regenerating an area.

Exam tip

It would be easy to muddle questions on this topic with those on the Diverse places topic because the themes overlap. However, there are subtle differences in emphasis, which means that you must stick to the topic that you have studied.

Preparing for your A level exams

Sample answer with comments

Evaluate the importance of attachment to place for a community when considering the need for regeneration. (20 marks)

The longer people live in a place, the greater their attachment to it. Some migrant groups who arrived in the UK post-WW2 clustered together in different places around the UK, particularly in London. For example, the Bangladesh community settled in Tower Hamlets, particularly around the East London Mosque in Whitechapel. Because the mosque became the first to be allowed to broadcast the call to prayer, this area has become a focal point for the Muslim community in East London and many families have lived there since the 1970s. Whitechapel Road Street Market stretches out along Whitechapel Road (A11) and every kind of Asian speciality food and product can be found there.

This geographical knowledge is relevant but there are some inaccuracies in place detail. This knowledge could be used more fully to explain why the community is so attached to the area rather than just describing where people are.

Because Whitechapel is a more deprived part of London, the need for regeneration is higher. Its overall Index of Multiple Deprivation position is 5,700. Income deprivation is particularly high, so when a new Tesco development opened in 2010, there was anger about the threat to local shops catering to local groups of people. On the other hand, some parts of the community in Whitechapel try to ban 'suggestive' advertisements and signs advertising alcohol in some shops. This an important way to protect against changes to local culture, but there have also been reports of homophobic abuse and clashes with the EDL.

This geographical knowledge provides some evidence, but it would be more accurate to note that the IMD is a composite index and that in some aspects of it (e.g. education) Whitechapel did better than average. Although the interpretation is partially correct, it could link in better with other information presented. The answer may also be losing sight of what the question is about.

Whitechapel Vision is a plan to transform Whitechapel when Crossrail arrives in 2018. The Royal London Hospital has already been rebuilt. The plans include 3,500 new homes, and a civic hub that will create 5,000 new jobs overall. House prices are expected to increase significantly. There has been criticism by locals that local traders were not consulted, and that some parts of the plans specifically aimed to bring new people into the area. There are concerns that this would change or dilute the local culture. Because Whitechapel is located next to one of London's wealthiest districts (the financial hub around Liverpool Street), there are very different views in the area about priorities, and election outcomes are diverse. The area has cheaper housing and young professionals looking for affordable housing are starting to move into Whitechapel. Because they work in the financial district, they want to vote for the party that will lower taxes the most or the one that strengthens the economy. However, students in Whitechapel's' Somali community want the government to help disadvantaged people from different backgrounds.

This continues to use ideas about people's views and lived experience. It's good to identify contrasting attitudes of different players towards changes involving regeneration. Additional evidence could be added to prove that there is such a difference between Liverpool Street and Whitechapel, especially in terms of regeneration schemes.

In conclusion, attachment to place is very important because the ethnic background, and time someone has spent living in a place affects the way in which they vote, whether they want to see improvements to the quality of life and lower levels of deprivation, and whether they approve of schemes to bring investment and regeneration to the area.

In longer answers it is important to come to a conclusion that reflects the previous paragraphs and the question. It would help to have a more balanced answer overall, perhaps addressing the spiral of decline in Whitechapel, and how conflict between different players hinders attempts to agree on priorities for regeneration. Comparing contrasting locations would have allowed breadth and enabled contrasting points to be made.

Verdict

This is an average answer, and perhaps on the short side for 20 marks (a timing issue?); some key ideas are missing. The command word 'evaluate' asks for a judgement about the role of people's view of their place in the types of regeneration that have occurred. There are some suggestions of why certain groups may like something and others not like it, but these are not really attached to a regeneration scheme or community project. It is good to include the different players and their views but this is as far as this answer goes. A possible answer is of course that 'attachment to place' is not as important as other factors, such as the need for infrastructure or housing stock.

Dynamic places

Diverse places

Introduction

The traditional view of the UK is as a 'green and pleasant land'. With satellite technology, it is now possible to add some facts to this perception. For example, the amount of woodland covering the UK (12.7 per cent) is nearly double the land area used for urban landscapes (6.8 per cent) – which may be a surprise to those living and working in large towns and cities.

But the UK has a growing population, which increases pressure on green spaces with the demand for new housing, road improvements and intensive farming. Pressures also exist from climate change. There is great diversity in physical geography across the UK, with variations in rock type, topography and climate, and this helps to determine human landscapes and living spaces. Enclosed farmland covers 40 per cent of the UK's land area, dominating much of lowland England such as East Anglia, while mountains, moorlands and heath (18 per cent of the UK's land area) dominate highland regions.

While physical geography may underpin some population patterns, socio-cultural factors are also relevant. There are political differences too, as seen in voting patterns and devolved areas within the UK. Economic changes involving a shift from secondary to tertiary employment have brought variations in quality of life in both urban and rural areas. Population factors (quantitative, e.g. age structure, and qualitative, e.g. perception) also vary considerably between rural and urban areas and increase the diversity of places within the UK.

In this topic

After studying this chapter, you will be able to discuss and explain the ideas and concepts contained within the following enquiry questions, and provide information on relevant located examples:

- How do the population characteristics of places in the UK vary?
- How do different groups of people view urban and rural living spaces in the UK?
- Why are there tensions within some UK communities?
- How successfully are the culturally and demographic issues managed?
- How do the geographical characteristics of your local living space compare with contrasting UK locations?

Figure 6.1: Superimposed images showing contrasting urban and rural places in the UK. Which type of place do you live in? Which type of place would you like to live in, and why?

CHAPTER 6

Synoptic links

Many physical and human factors influence urban and rural places in the UK. The physical factors, such as a long coastline, and resources such as coal have had a historical influence on shaping places. Secondary industries developed near energy resources, ports enabled trade with the world, and modern transport has improved links and to Europe and the world. Waves of immigration from the Caribbean, Commonwealth countries and Eastern Europe, have added cultural diversity to places. Political factors include community groups, pressure groups, local government (e.g. planning), big businesses (including TNCs), special authorities (e.g. National Parks), national government (e.g. policies on house building, coastal defences, regeneration), and European Union government (e.g. the Common Agricultural Policy, regional development schemes).

Useful knowledge and understanding

During your previous studies of Geography (KS3 and GCSE), you may have learned about some of the ideas and concepts covered in this chapter, such as:

- Urban areas in the UK (e.g. distribution, structure, functions, change)
- the processes shaping urban areas in the UK
- quality of life variations in UK urban areas
- strategies for improving quality of life in UK urban areas
- the distribution pattern of the UK population
- UK population change and challenges
- population migration (both to and within the UK)
- rural areas in the UK (e.g. agriculture, natural environment, settlements)
- human activity shaping upland and lowland UK landscapes
- differences between the urban core and the rural periphery of the UK
- links between urban areas and rural areas of the UK
- challenges for rural and urban areas of the UK
- the role of globalisation in shaping places in the UK
- studies of your local area (fieldwork research).

This chapter will consolidate and extend this knowledge and understanding.

Skills covered within this topic

- Analysis of choropleth maps (e.g. demographic variations).
- Interpretation of population pyramids and other graphical methods of representing population data.
- Use of location quotients.
- Use of indices to show contrasts between places in terms of socio-cultural factors.
- Construction and interpretation of flow-line maps to show patterns of migration.
- Interpretation of oral/written accounts (social media articles, newspapers) of communities and residents of places that portray their lived experiences.
- Interpretation of historical and contemporary Ordnance Survey evidence of change in rural and urban places.
- Interpretation and use of scattergraphs.
- Statistical testing of measures of central tendency and distribution (e.g. standard deviation), including correlations, of data on UK urban and rural areas.
- Interpretation and evaluation of qualitative information sources (advertising, tourist information, art, music, photographs, film, literature).

Fieldwork

This topic includes a fieldwork element, examining one aspect of an urban or rural living space. Basic techniques such as land-use surveys, to confirm what the place is like, can be combined with questionnaires, to find out what people think about the place, and the use of secondary data (e.g. the census and Index of Multiple Deprivation) to find out the area's socio-cultural and socio-economic characteristics. This will be a local study like the example at the end of this chapter; guidance on fieldwork skills is covered in a separate section of this book.

CHAPTER 6

How do population structures vary?

Learning objectives

4B.1 To understand why population structures vary from place to place and over time

4B.2 To understand why population characteristics vary from place to place and over time

4B.3 To understand how past and present connections have shaped the demographic and cultural characteristics of places

Population structures

Population density patterns and processes

In 2015 the UK had a total population of just over 65 million people. Not surprisingly, the large majority (84 per cent) of this population lived in England, because this is where most habitable lowland exists, along with a warmer climate in the south and the major employment areas of Greater London, the West Midlands and Greater Manchester. The average population density in England was 413 people per km^2, while Scotland had the lowest overall density of 68 people per km^2, due to highlands, a less hospitable climate, fewer employment opportunities and isolation in peripheral areas (Figure 6.2). The **built environment,** or urban landscape, accounted for 10.6 per cent of England, 4.1 per cent of Wales, 3.6 per cent of Northern Ireland and only 1.9 per cent of Scotland.

Population dynamics

Since 1960 the UK population has grown by about 12 million, but not at a constant rate:

- The 1960s had an annual **population growth** rate of 0.61 per cent, as the 'baby boom' that had started with the peace after 1945 continued. The economy was growing strongly, which encouraged families to have children because they could afford to bring them up and the future looked good. Immigration, from the Caribbean in particular, added to the population – encouraged by the UK government, because a larger workforce was needed to fuel economic growth.
- In the 1970s and 1980s the annual population growth rate slowed to an average of 0.14 per cent, as the economy weakened and recession, deindustrialisation and higher unemployment occurred. **Fertility rates** decreased as more females entered employment and followed careers, and there were fewer job opportunities to attract migrants.
- The 1990s were a period of social and economic adjustment, and the growth rate increased a little due to late 'baby boomers' having their children.
- Not until the 2000s did the population growth rate increase significantly again (to an annual rate of 0.64 per cent), as the UK economy successfully adjusted away from the secondary to the tertiary sector. Waves of **international** immigrants entered the country, often from new EU countries (such as Poland). The birth rate increased, partly because most immigrants were young adults (in 2013 a quarter of all live births were to mothers who had been born outside the UK), while the death rate slowly declined as a result of better health care (see Figure 6.3).

Yellow = Players, Orange = Attitudes and actions, Purple = Futures and uncertainties

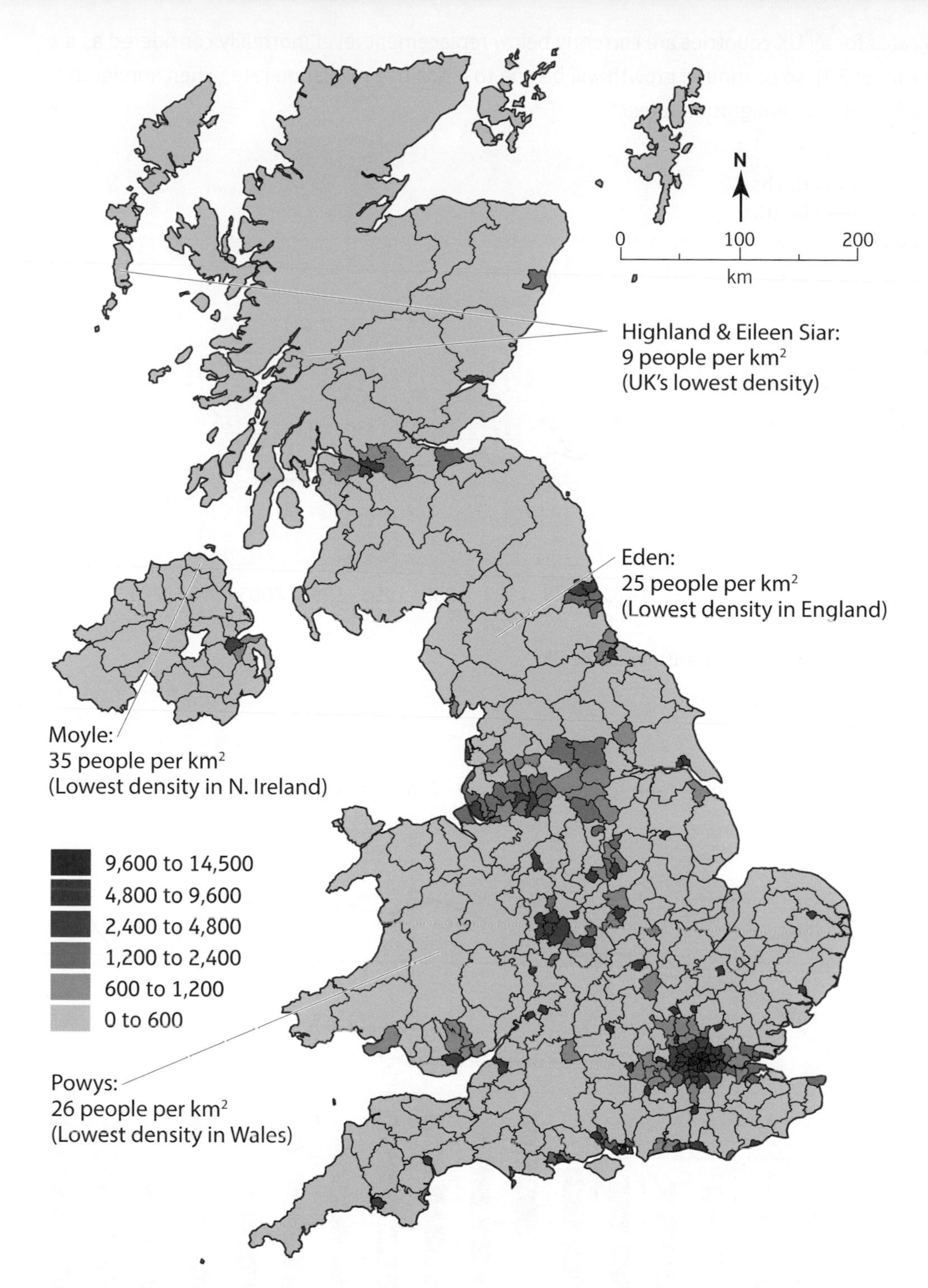

Figure 6.2: The population density of the UK in 2015, people per km^2

ACTIVITY

CARTOGRAPHIC SKILLS

Study Figure 6.2 and atlas maps. Describe and explain the pattern of population density in the UK.

- Since 2010 the annual population growth rate has increased again (to 0.71 per cent), despite the recession that started in 2008 causing wages to stagnate or decrease. This growth has been largely due to immigration, linked to a perception that the UK was coping better with recession than many other EU countries. Net positive migration was higher than natural increase, contributing 17 per cent more to the UK population growth rate (see Figure 6.4).

The Office for National Statistics (ONS) predicts further population increases in the 2020s, but at a slightly lower growth rate, with the UK population reaching over 67 million by 2020. However, fertility rates for all UK countries are currently below replacement level (normally considered as a fertility rate of 2.1), so continued growth will be due to a positive migration rate (when immigration rates are higher than emigration rates).

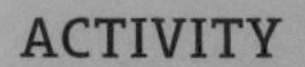

ACTIVITY

Compare the influence of natural change and net migration on population change in the UK from the 1990s to the 2010s.

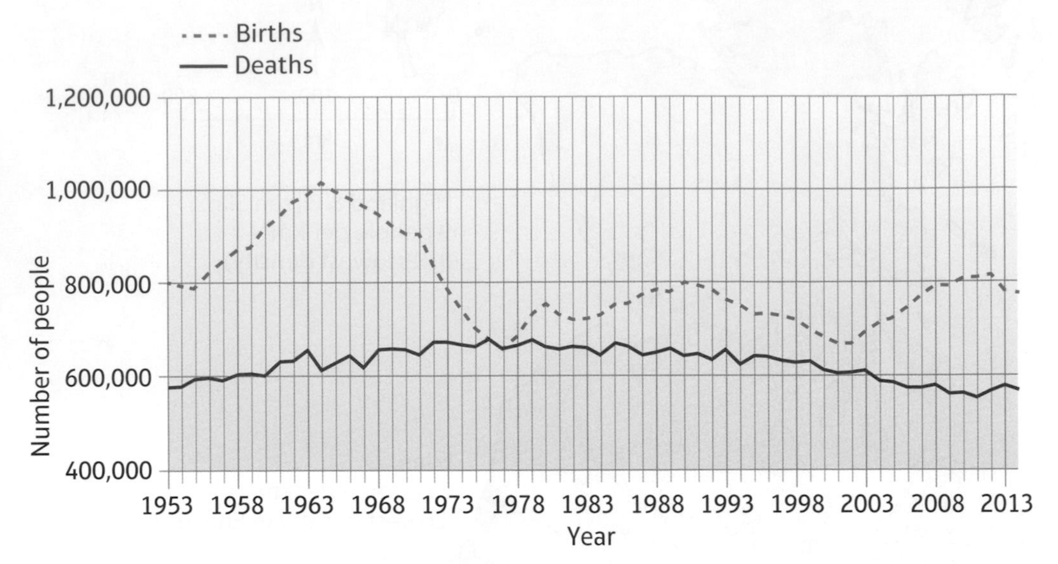

Figure 6.3: Recent demographic transition in the UK

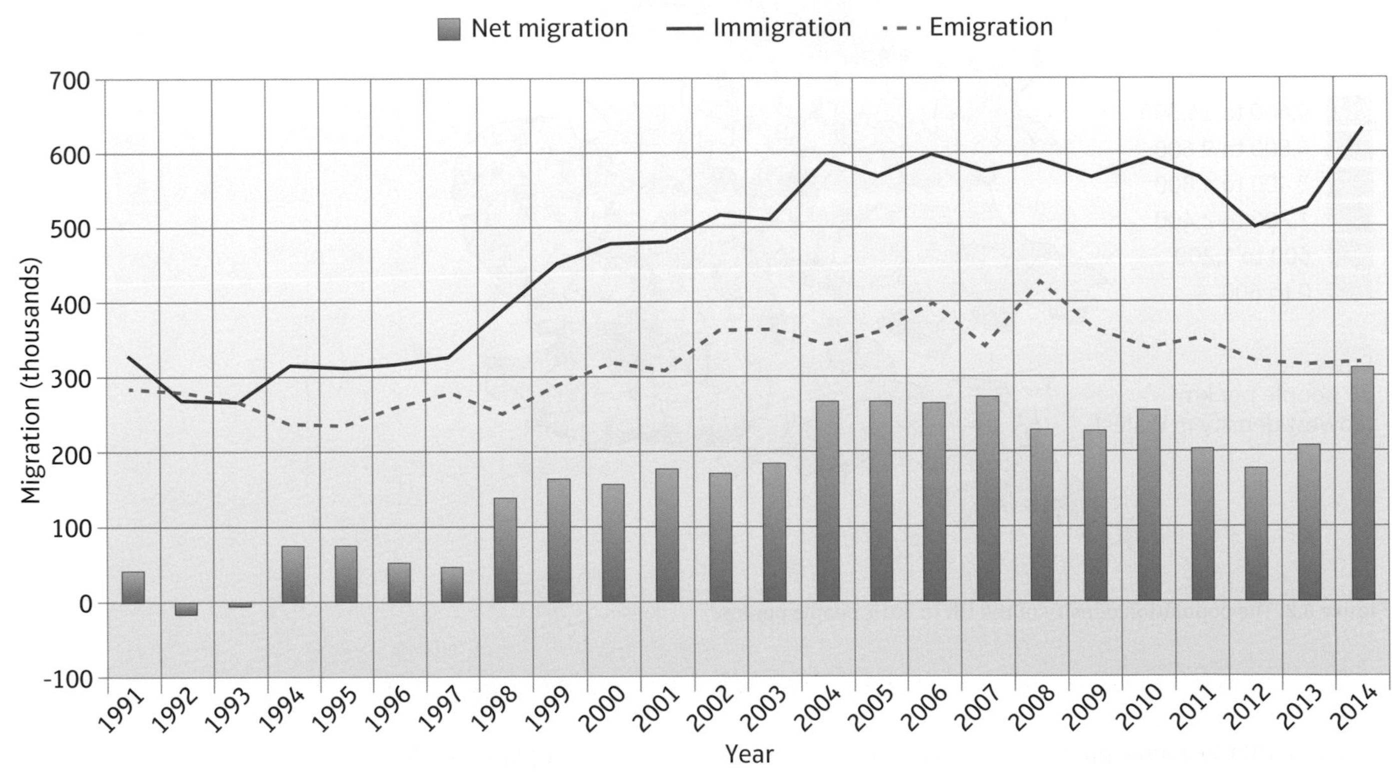

Figure 6.4: Immigration and emigration rates for the UK

The rural–urban continuum

UK averages (or means) hide the great variation that exists within the four countries, and within the regions of these countries. For example, 80 per cent of the UK population live in urban areas (defined as settlements of over 10,000 people), which means that large parts of the UK are sparsely populated. In England 11 per cent of the land area is currently urban, but these densely populated areas depend on links to the services and goods provided by rural hinterlands. A **rural–urban continuum** exists, ranging from the high densities of inner urban areas to the low densities of the remotest rural areas. Figure 6.5 shows the rural-urban continuum in East Anglia based on road journey times from north London. The pattern is distorted where there are motorways, but it is clear that there is an increase in rurality away from London and there are no other large cities influencing the region. Also, note the zones in which Holbrook and Happisburgh, featured case studies, are found.

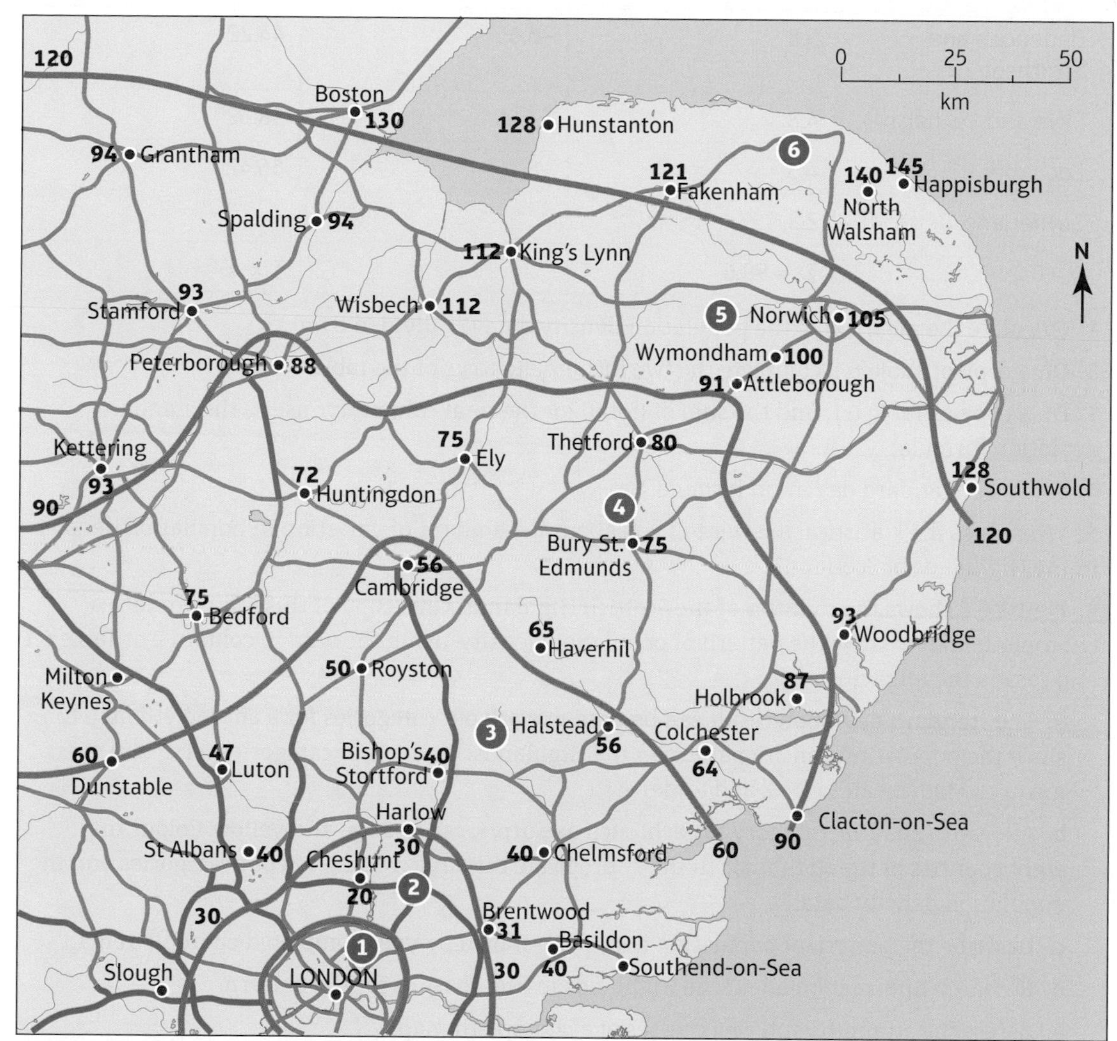

Key: Isolines (grey) and numbers indicate road journey times in minutes from London's North Circular Road

1. Urban core
2. Suburbs, urban sprawl, green belt, ring road
3. Accessible rural, dormitory settlements, expanded towns, commuter settlements
4. Intermediate rural, suburbanised villages, market towns
5. Rural, villages and towns away from main roads change little, farming dominant
6. Remote rural, tourism, National Parks, some declining villages, retirement coastal settlements, some villages dominated by 'second home' owners

A roads

motorways

Figure 6.5: The rural-to-urban continuum in East Anglia

ACTIVITY

CARTOGRAPHIC SKILLS

On a copy of Figure 6.5 or similar map, produce a similar isoline (isochrone) map based on train journey times from London.

ACTIVITY

NUMERICAL AND STATISTICAL SKILLS

Table 6.1: Selected data for the Highlands of Scotland districts (2013)

District	Population density (people per km^2) (x)	$(x-\bar{x})$	$(x-\bar{x})^2$
Nairn	30.6	18.15	329.42
Inverness	26.8		
Caithness	14.4	1.95	3.80
Ross and Cromarty	10.6	−1.85	3.42
Badenoch and Strathspey	5.8	−6.65	44.22
Skye and Lochalsh	4.8	−7.65	58.52
Lochaber	4.3	−8.15	66.42
Sutherland	2.3		
$n = 8$	$\sum x = 99.6$		$\sum(x-\bar{x})^2 =$

Synoptic link

To understand population distribution patterns, you need a broad knowledge and understanding of physical geographical factors – such as climate, topography and resources – and human geographical factors – such as accessibility to jobs and services. Urban population densities are linked to urban processes such as urbanisation and reurbanisation, counter-urbanisation and decentralisation. Rural population densities are linked to the success of primary industries such as farming, forestry and fishing and tertiary industries such as tourism, as well as to influences from urban areas. (See page 284.)

1. Calculate the mean ($\bar{x}$) of the population density data ($\sum x$ divided by n)

2. On a copy of Table 6.1, complete the two incomplete rows of the table.

3. On a copy of Table 6.1, find the sum of (total) for the final column for use in the standard deviation formula.

4. Solve the standard deviation formula. $s = \sqrt{\frac{\sum(x-\bar{x})^2}{n}}$

5. What does the statistical answer tell you about the amount of variation in population density in the Highlands of Scotland?

6. Figure 6.6 shows the counties of the Scottish Highlands. On a copy of this map, produce a choropleth map to show the pattern of population density, using the data in column 2 of Table 6.1 and taking the following steps:

a. The standard deviation result can be used to work out categories for a choropleth map to show the population density pattern in the Highlands. Create four categories: +1sd and over, mean to +1sd, mean to -1sd, and under -1sd.

b. Allocate a shading category to each, such as purple, red, orange and yellow. Colour the eight counties of the Highlands in the appropriate colour according to your categories and the population density data.

c. Describe the pattern of population density as shown by your completed choropleth map.

d. Research information about the Highlands to find reasons for the pattern.

e. Assess the strengths and weaknesses of a choropleth map.

One factor which shows evidence of the rural-urban continuum is population density. Highest densities are in the urban areas, but even in rural zones there will be variations based on employment opportunities and accessibility. The statistical and mapping skills exercises show the zones and diversity in the Highlands of Scotland.

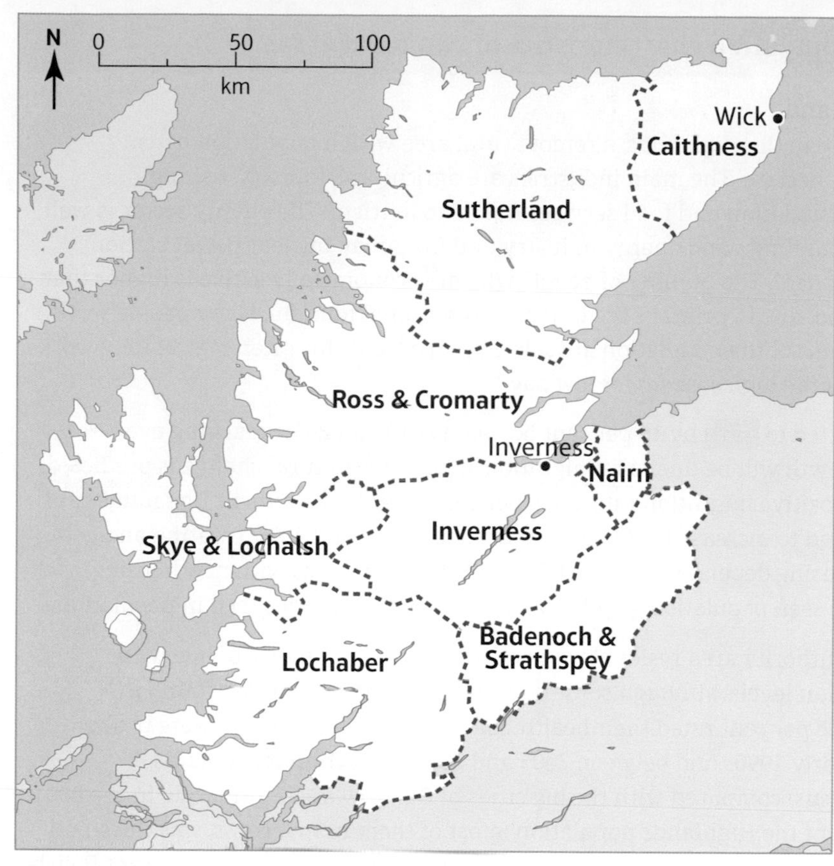

Figure 6.6: Counties of the Highlands of Scotland

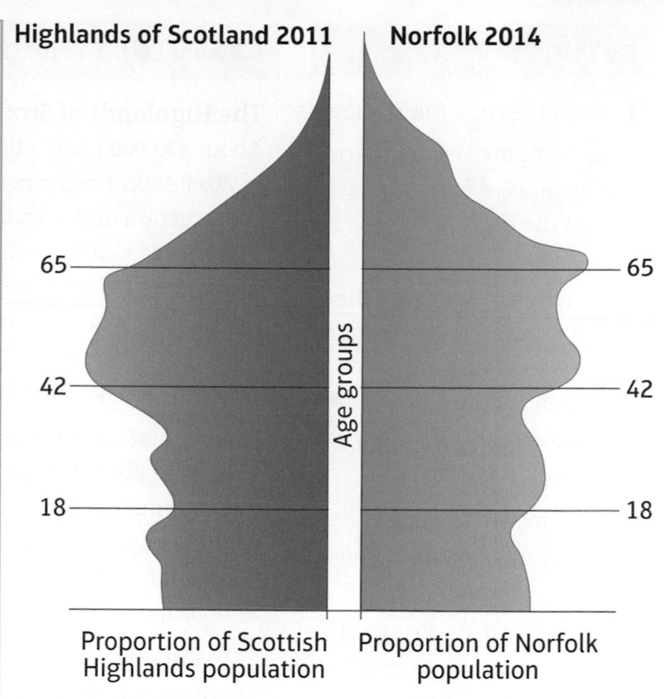

Figure 6.7: Sketch population pyramid showing the age structure of Norfolk and the Highlands of Scotland, 2014

Population characteristics

Diversity, connections and influences

Not only does population density vary within regions of the UK, but other population characteristics also vary. The **population structure** refers to the number of males and females in different age groups at a given time and place, and may be shown graphically in a population pyramid (Figure 6.7). Some general trends are clear, such as urban areas having younger populations and rural areas older. In developed countries such as the UK, people of all ages and backgrounds have the opportunity to live and work in any rural or urban place. This possibility has arisen due to modern communications networks, such as broadband, and the flexible location of many tertiary jobs. However, as yet, there has not been a significant change to established population structures as shown by Figures 6.7 and 6.8. Figure 6.8 shows the population structure of Liverpool and the inner London borough of Newham, and these contrast a lot with rural population structures in terms of age groups.

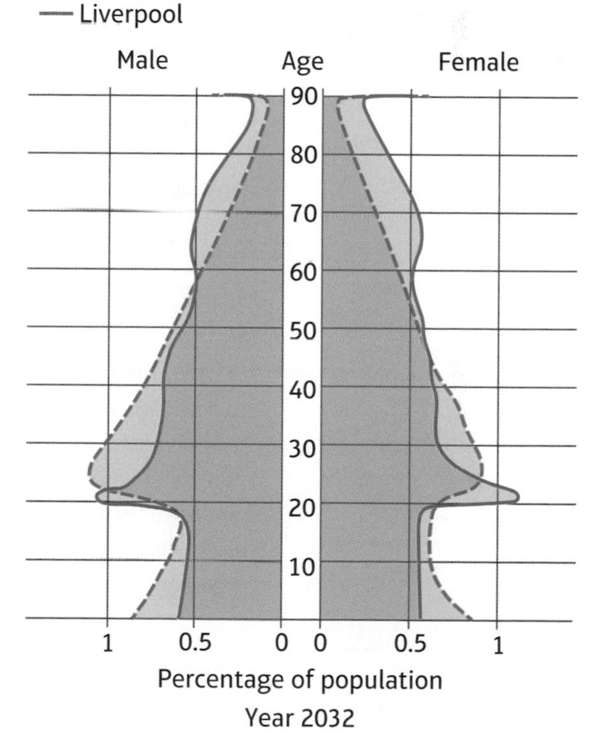

Figure 6.8: The projected future population structure of Newham and Liverpool

CASE STUDY: The population characteristics of two rural areas

The Highlands of Scotland

About 233,000 people live in the Highlands, a remote rural area with a population density in 2011 of 0.1 people per hectare. The main industries are agriculture, forestry, fishing, construction and accommodation and food services linked to tourism. The energy sector is well established, with an expanding wind energy industry and Dounreay nuclear power station (now being decommissioned). The number of people who are economically active is higher than in other areas of Scotland, due to primary sector jobs and seasonal tourism. Many people work from home and 15 per cent of the population are self-employed, but this often means they work long hours (especially in the high season) for low pay.

The population is predicted to grow by 15 per cent between 2010 and 2035, reaching over 250,000 in 2031. This growth will be due to immigration: a negative natural change is predicted, but accompanied by a positive migration rate of 0.5 per cent (Highland Council). The number of retired people is predicted to increase by 79 per cent between 2010 and 2035, creating an ageing population and an increasing dependency ratio (1.74 in 2010). Urban areas with new housing, such as Inverness South, have seen population growth, while other areas have declined in population.

In the Highland Local Authority area residents generally have a high quality of life, with relatively low **deprivation** levels, although some deprived neighbourhoods are found in towns (e.g. Wick). Only 16 per cent rated their health below 'good' in 2011. There were peaks in immigration in the early 1990s and between 2003 and 2010, but **ethnicity** remains homogeneous (2011 census) compared with the big cities of Glasgow and Edinburgh. In the 2001 census only 0.8 per cent of the Highlands population, most of them in Inverness, was classified as belonging to an ethnic minority. The 2011 census showed an increase, with 1.5 per cent Polish and 0.8 per cent Asian, and 5.4 per cent of the population born outside the UK.

Norfolk, East Anglia

About 880,000 people live in Norfolk, a more accessible rural area than the Highlands, with a population density of 1.6 people per hectare (2011). Agriculture dominates the landscape, but due to the significant regional city of Norwich and the proximity of London by train, there is a strong tertiary sector and an expanding quaternary sector. The population is growing (at a rate of 7.6 per cent from 2004 to 2014), but at a slower rate than England or the East of England.

Most of the population increase is among people over the age of 65, who have chosen the county for retirement, moving from London and the South East and then living longer (low **mortality rate** and long **life expectancy**). In 2014 more than 23 per cent of the population in the county was aged over 65, which was 5.8 per cent above the England average (Figure 6.7). The population is therefore ageing, despite a small increase in the number of children. There is a small but growing non-white British (i.e. ethnic) population (7.6 per cent of the total population in 2011). In 2012–13 there was a net influx of just 2700 international migrants into Norfolk, mainly to work on farms.

ACTIVITY

1. Read the case study above.
   a. Compare the population densities of East Anglia and the Highlands of Scotland.
   b. Why do you think they are different even though they are both rural regions of the UK?
2. Study Figures 6.7 and 6.8 carefully.
   a. Compare the age structure of Norfolk and the Highlands.
   b. Compare and suggest reasons for the projected future population structures of Liverpool and Newham.

AS level exam-style question

Explain why large numbers of retired people are found in many UK rural areas. (6 marks)

Guidance

Be careful to think about several possible reasons. The migration of retired people is one, but there is also the fact that older rural people may be unable to move. Refer to examples covered in this chapter.

A level exam-style question

Explain why population density varies within UK rural regions. (6 marks)

Guidance

It is best to consider this through examples (e.g. East Anglia and the Highlands of Scotland). You will need to think about the influence of both physical and human factors.

Yellow = Players, Orange = Attitudes and actions, Purple = Futures and uncertainties

Causes of change in UK rural areas

Rural settlements (fewer than 10,000 people) consist mostly of villages and hamlets found dispersed in areas of countryside. Their main function is residential (housing) but there may be some services such as a pub, a post office, a church and a village shop, although these have been in decline. Some types of village may have a clear land use pattern because they have been 'suburbanised' by commuters and exurbanites moving in. These are people who used to live in urban areas and who remain urban-orientated. Once, rural settlements used to consist of farms (isolated dwellings), villages for farmworkers and services linked to farming, and market towns where farmers could sell their products and access tertiary services such as banks. However, many traditional rural communities have become dormitory settlements or **commuter villages**. This is as a result of:

- the increased mechanisation of farming, reducing the need for farmworkers
- a wider variety of job opportunities for young rural adults
- influences from and recreational opportunities in urban areas
- the development of faster transport networks and widespread car ownership.

Suburbanisation has converted rural land into urban settlements, a process that is usually irreversible. This happens mostly in accessible rural zones – such as East Anglia (see Figure 6.5) – within an hour's travel time of a nearby city. In remote rural areas, traditional farming may still be found, such as crofts in the Highlands of Scotland and sheep farms in Wales, but many of them rely on government support to enable the farmers to stay and make a living. Consequently, in these remote areas there is an ageing farm population, with fewer young adults to take over: villages have become depopulated as young adults move out to towns and cities where there is a wide selection of jobs that are better paid than farm work. In addition, farm mechanisation has reduced the need for labourers.

Within agriculture itself there have been changes as food prices have fallen or become uncertain, and many farmers have diversified to have a wider economic base. Farmers are now generating additional income through tourism (for example offering B & B or farm stays), recreational activities (such as paintballing, off-road driving or pony trekking), renewable energy (for example wind turbines), converting barns into workshops, or offices adding value to products (such as making ice cream) or selling land for housing.

In accessible rural areas, counter-urbanisation is leading to population growth as more people move in from cities. This may keep services open, but often they become more expensive farm shops and food halls. Changing land ownership may alter the socio-economic mix and increase property prices beyond what rural young adults can afford. With the internet and computerisation of tertiary jobs, a small but increasing trend is to work more from home and not commute. Smaller-scale migrations are also affecting rural areas: for example, some rural settlements in scenic or coastal locations have become centres of retirement, adding to the ageing population structure (as in North Norfolk), and some have received significant numbers of international migrants to work in agriculture or food processing (as in Boston, in Lincolnshire).

Extension

Use the internet for the latest information on what is happening to the population of Norfolk.

ACTIVITY

TECHNOLOGY/ICT SKILLS

1. Choose *one* of the tourist trails listed below and then investigate relevant websites before writing answers to the tasks:
   a. *Game of Thrones* filming locations (Northern Ireland)
   b. Dylan Thomas Country (Wales)
   c. Robert the Bruce Trail (Scotland)
   d. Constable Country (East Anglia)
   e. Thomas Hardy Country (Dorset)
   f. Downton Country (Oxfordshire, Berkshire)
   g. *Pride and Prejudice* tour (Derbyshire).
2. Identify the natural landscape and traditional culture portrayed by the tourist trail.
3. Explain the image of UK rural areas that is shown or created.

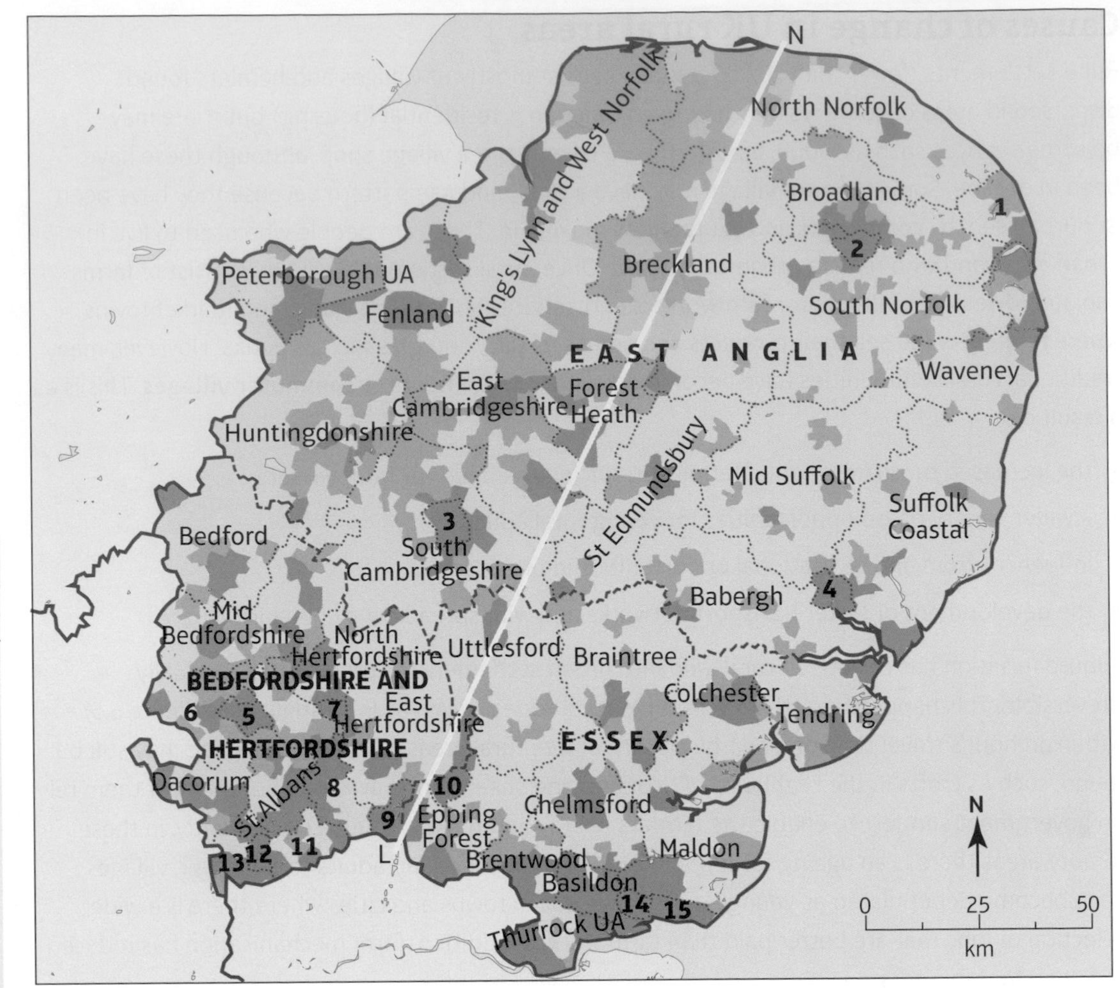

— Regional boundary
---- NUTS 2 boundary
.... Local or unitary authority boundary
Urban population over 10,000 – less Sparse
Town and fringe – less sparse
Town and fringe – sparse
Village, hamlet and isolated dwellings – less sparse
Village, hamlet and isolated dwellings – sparse

1 Great Yarmouth
2 Norwich
3 Cambridge
4 Ipswich
5 Luton UA
6 South Bedfordshire
7 Stevenage
8 Welwyn Hatfield
9 Broxbourne
10 Harlow
11 Hertsmere
12 Watford
13 Three Rivers
14 Castle Point
15 Southend-on-Sea UA

Figure 6.9: Population and spatial zones of East Anglia

ACTIVITY

CARTOGRAPHIC SKILLS

1. Study Figure 6.9 and atlas maps.
   a. Describe and explain the changes in population and spatial zones found along the line L–N (from the edge of London to the North Norfolk coast).
   b. Summarise the regional, national, international and global influences that have shaped UK rural areas. Perhaps use a table format for your notes. Influences include transport routes (regional), government policy (national), foreign businesses (international), and the internet (global).

Causes of change in UK urban areas

Urban settlements (over 10,000 people) are built-up areas with many functions and services, including residential areas, shops, hospitals, schools, industries, transport hubs and offices. Urban areas usually have a clear structure based on land-use zones, which can be represented by urban models. These models have become more complex over time, reflecting the **demographic changes** that have taken place relative to population size. Figure 6.10 is a composite model based on earlier models such as by Burgess and Hoyt.

Yellow = Players, Orange = Attitudes and actions, Purple = Futures and uncertainties

The 1960s

During the economic boom and **industrialisation** of the 1960s, many UK cities had plenty of secondary and tertiary jobs available, and their populations grew. Immigrant workers arrived from the Caribbean, India and Pakistan. The national government encouraged immigration and 472,000 people arrived between 1955 and 1962, to fill the many jobs available. The immigrants moved mainly into inner urban areas, which was where they could afford housing or found people of a similar background already settled (Figure 6.10).

The spatial area of cities also grew, with the development of suburban industrial parks, as shown in the outer suburbs on Figure 6.10. These parks became economic hubs – central points for business and industry. London, however, did not grow during this time, because of a process of decentralisation:

- Purpose-built 'new towns' such as Milton Keynes, established in 1967, were developed to relieve the problems of over-population in London.
- Richer people started to live in towns and rural areas outside the city and commute to work, using the cheap rail travel available.

Between the 1930s and 1950s there had been little investment in infrastructure or housing stock due to the war and post-war recovery, so the quality of living conditions in the cities was declining. Social (council) housing was built in the 1950s for the many people who were earning low wages after the war, but there was still a great demand in the 1960s for housing. A peak was reached in 1968 when 425,000 houses were built and half of UK families owned their own homes. However, the tower blocks built in the 1960s and early 1970s proved to be poorly designed and constructed.

The 1970s

Economic conditions in some UK cities worsened and the number of secondary jobs decreased, causing people to move away from declining cities. Liverpool was one of these: its population was 750,000 in 1961, but only 500,000 in 1981. A significant number of people lived in poor conditions – dilapidated housing with no modern amenities such as an inside toilet – and, due to unemployment, they were unable to improve their quality of life.

Redevelopment began in many urban areas to try and improve the living spaces. This consisted mostly of demolishing older, often terraced, housing and factories, then building high-rise flats on the same land in inner urban areas, or new council estates in the suburbs. London's population also shrank but this was partly because of some government offices moving out of London to areas of economic decline (such as South Wales or NE England) and reduced rates of immigration.

The 1980s

A global economic recession affected the whole country and many factories closed (deindustrialisation), leaving behind derelict, unsightly urban land and buildings. Often these areas were in the CBD frame zone, and became areas of discard with very little use (Figure 6.10). Many external factors, especially competition from countries such as Japan and South Korea, **transnational corporations** (TNCs) and the 'container' revolution in transporting cargo, reduced the number of jobs in factories, ports and warehouses.

The national government set up urban development corporations (UDCs) to undertake large-scale redevelopment of inner-city areas. One example was London Docklands, which was started in 1981 and became the largest redevelopment scheme in the world. This is an example of assimilation, where an inner urban place becomes similar to the central business district (Figure 6.10). Enterprise zones offering tax benefits were also established and financial markets were deregulated to attract businesses into urban areas, including foreign direct investment (FDI).

Counter-urbanisation – migration from urban areas to the rural–urban fringe and **accessible** rural areas, and commuting to work by rail or motorway – started during this period. This migration was largely voluntary in the London area, but in some places such as Birmingham it was part of urban planning. The push and pull factors linked to the desire for better living space (Figure 6.11) were very strong. This started to spread cities outwards (urban sprawl), a process that continued into the twenty-first century and can be seen in the structure of modern cities (see Figure 6.10).

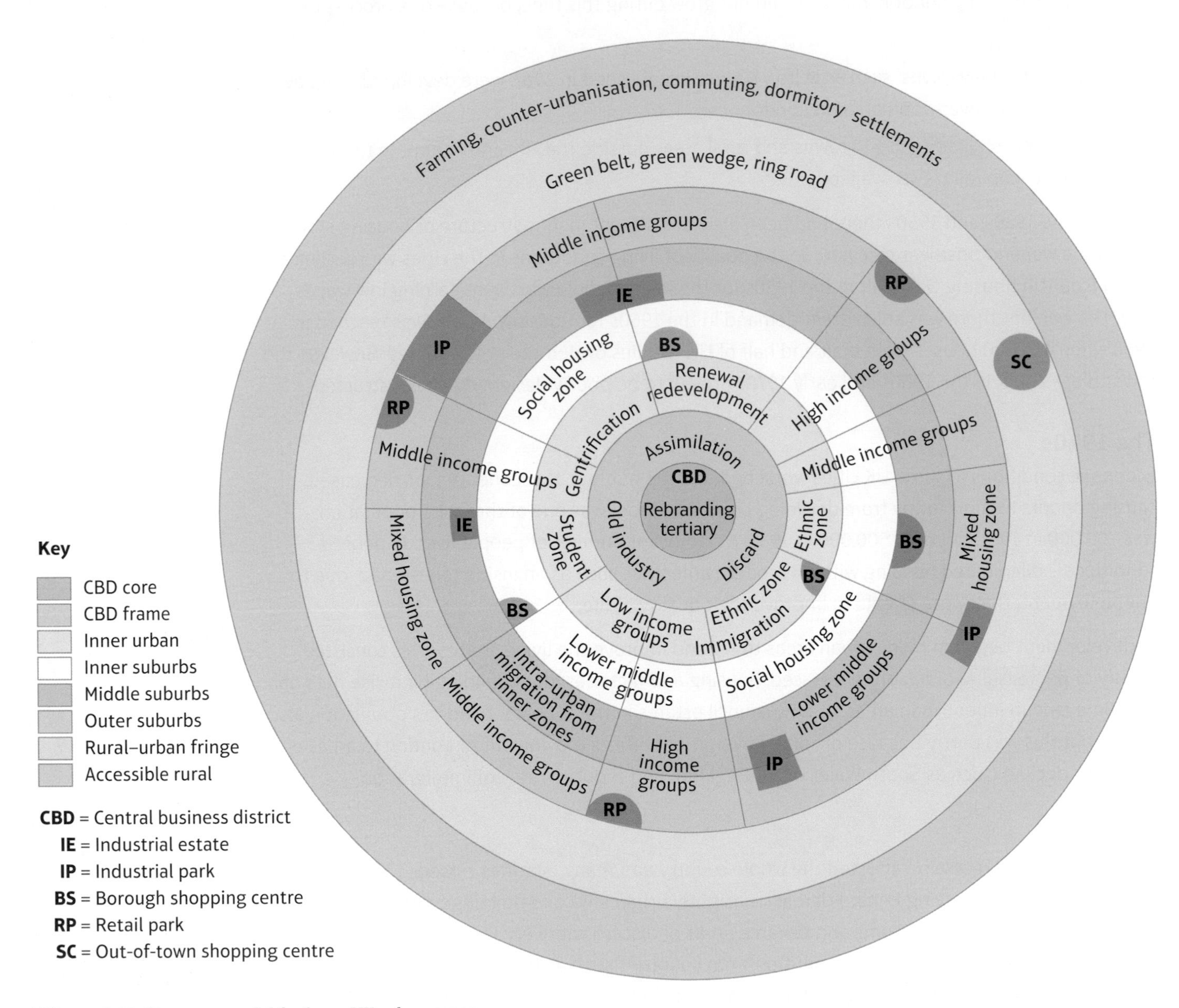

Figure 6.10: Structure model for large UK urban areas

Yellow = Players, Orange = Attitudes and actions, Purple = Futures and uncertainties

Push out of large urban areas		Pull into accessible rural areas
High crime rates Poor-quality housing stock Overcrowding Air pollution Deprivation Derelict land and buildings	Cheap rail fares Improved motorway network Increasing car ownership Socio-economic status	Lower crime rates Green open spaces Lower air pollution Less crowding Higher quality of life

Figure 6.11: Flow diagram showing push and pull factors in counter-urbanisation

The 1990s

The high-rise flats built during the 1960s and 1970s were later recognised as a failure. Many were antisocial, a focal point for crime, infested with cockroaches, ants, mice and rats, and poorly built. In some cases they were demolished (Figure 6.12) and replaced by new housing in neighbourhood street layouts. One example is the City Challenge scheme for the Holly Street Estate in Hackney, London, built between 1992 and 1997. High streets had been declining, as consumer patterns changed to a preference for more **accessible** suburban or out-of-town centres. The discarded parts of the inner urban areas and central business districts (CBDs) decayed (Figure 6.10). To counteract this suburbanisation of retailing, the regeneration of city centres began, with pedestrianisation and the renewal of historic buildings.

London, however, as the capital and the largest city in the UK, drew in light industries and service industries for its large number of consumers. London attracted young adult workers, not only for the job opportunities but also for entertainment and other services.

The growth of the UK population since the 1990s has been due to net migration, rather than natural increase (see Figure 6.4). The significant increase in the migration rate since 1998 has partly been because of political and economic instability in Africa, Eastern Europe and the Middle East.

ACTIVITY

1. Carry out an in-depth study of your local place. Make notes on:
   - a. the characteristics of the area
   - b. processes that have caused change over time
   - c. the positive and negative results of the changes
   - d. the impact of the changes on people's identity.

Figure 6.12: The demolition of an old housing estate in Liverpool

Synoptic link

The population of a place may change as a result of two processes: (a) natural change, which is the difference between the birth rate and death rate, natural increase being where births are higher than deaths, or (b) the migration rate, which is the difference between immigration (people moving into a place) and emigration (people moving out of a place), a positive migration rate being where immigration is higher than emigration. (See page 276.)

The 2000s

By the new century, suburbanisation had established functions around the edges of UK cities, such as retail parks, industrial parks and out-of-town shopping centres (see Figure 6.10). Rules about freedom of movement within the EU allowed immigration from countries joining the EU after 2004 and 2007; these people sought better-paid jobs in UK cities and rural areas, adding further to the cultural diversity of UK towns and cities.

The 2007/08 financial collapse prompted a worldwide recession, which proved devastating for those secondary and tertiary businesses that were only marginally profitable, and increased unemployment and the rate of high-street shop closures. Continued foreign competition, especially from China, affected jobs in the UK and caused **internal migration** of those searching for work. All UK cities, and some rural towns, developed complex ethnic mixes due to successive waves of immigration from the Caribbean, other Commonwealth countries and Eastern Europe. Net migration increased the UK population by over 240,000 people per year on average between 2004 and 2014. The number of temporary student immigrants also increased during the late 2000s, peaking between 2009 and 2011.

The 2010s

The recession continued to affect certain sectors of the economy (for example, construction) and consumer spending patterns. Consequently, redevelopment slowed, and high streets found it difficult to attract retail investment. Charity shops and discount stores often moved into the vacant shops. Suburban growth continued, and some infilling of inner urban areas that had not previously been redeveloped took place. Initiatives, especially rebranding and regeneration, aimed to attract fresh private investment, new businesses and people back into inner urban areas. Urban councils had little money of their own, because of the economic recession and reductions in central funding from the national government. Immigration continued to increase population numbers in many urban places, with arrivals from eastern Europe and others travelling through Europe, as dramatically shown by refugees travelling in small boats to the Canary Islands, Lampedusa and the Greek Islands.

ACTIVITY

1. Compare the urban places shown in Figures 6.12 and 6.13, consider social and economic opportunities in your answer.
2. Summarise the regional, national, international and global influences that have shaped UK urban areas.

London, however, as a 'world city' with a top international reputation – further enhanced by hosting the Olympic Games in 2012 – did attract new businesses and young adults. There has been an increasing number of people living in single-person households, as well as 'super-rich' foreigners, who often invested in city centre property (Figure 6.13).

Figure 6.13: South Kensington, London, a wealthy urban area

Yellow = Players, Orange = Attitudes and actions, Purple = Futures and uncertainties

CHAPTER 6

How do different people view diverse living spaces?

Learning objectives

4B.4 To understand that urban places are seen differently by different groups because of their lived experience and perception of those places

4B.5 To understand that rural places are seen differently by different groups because of their lived experience and perception of those places

4B.6 To understand the range of ways of evaluating how people view their living spaces

Evaluating living spaces

Factors affecting perception

How do people feel about their living spaces in the UK? A 2014/15 Office for National Statistics (ONS) personal well-being survey showed that Northern Ireland had the highest rating by people in terms of 'life satisfaction', including the highest ratings in 'happiness' and 'worthwhile'. The other UK countries had similar overall ratings. Most regions in England had improved their personal well-being ratings since 2011/12 (Figure 6.14). The South East having the fewest dissatisfied respondents for 'life satisfaction', 'worthwhile' and 'happiness' scores, and the West Midlands region had the fewest anxious respondents. The North East had consistently more respondents who were dissatisfied or anxious according to each of the criteria and in all but 'happiness' had not seen an improvement since 2011/12.

How people perceive their living space depends on a combination of lived experience, their stage in the **life cycle**, environmental factors, and their socio-economic and socio-cultural background (see Table 6.2).

Table 6.2: Factors influencing perceptions of urban and rural areas

Human factors	Accessibility factors	Environmental factors
Age (life cycle)	Access to employment	Levels of pollution
Family composition	Employment characteristics	Levels of crime
Family size	Access to services (e.g. health, education, shops)	Amount of open space
Level of education	Access to public transport	Housing conditions
Ethnic background	Car ownership	Type of open space
Health status	Access to broadband	Physical geography (e.g. climate, scenery)
Gender	Access to recreation	Traffic congestion
Level of income	Affordable housing	
Socio-economic status	Access to community activities (including children and teenagers)	
Cultural beliefs (religion)		

ACTIVITY

1. Study Table 6.2.
   a. Discuss this list with your friends and family and then suggest which other factors should have been included.
   b. Which factor do you think has the largest influence on people's perceptions? Justify your choice.

ACTIVITY

Study Figure 6.14. Describe and suggest reasons for the patterns of satisfaction shown.

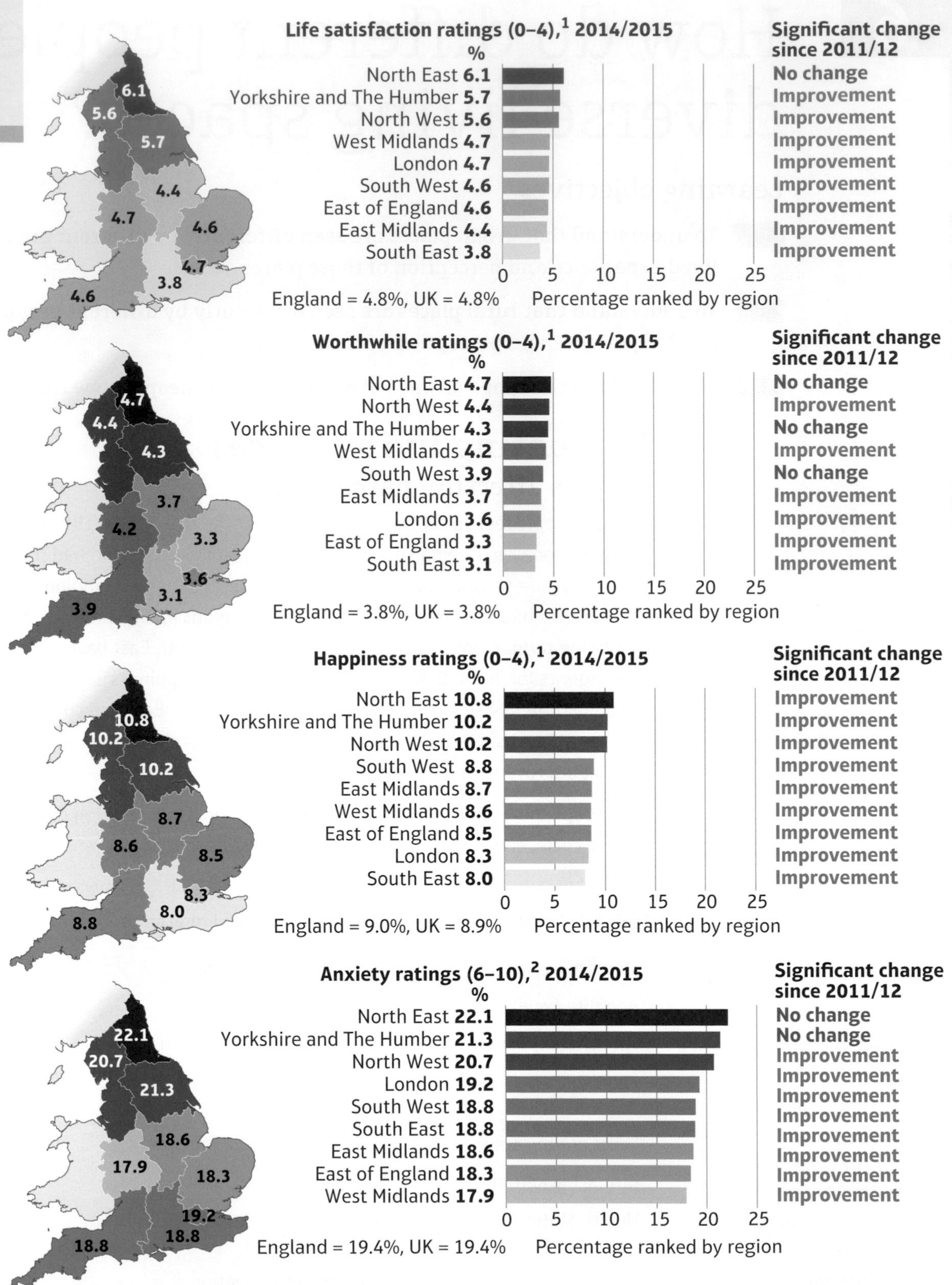

Ratings on 11-point scale (0 'not at all' to 10 'completely'). Adults aged 16 and over were asked 'Overall, how satisfied are you with your life nowadays?', 'Overall, to what extent do you feel the things you do in your life are worthwhile?', 'Overall, how anxious did you feel yesterday?'

1. Percentage of respondents answering 0–4, those least satisfied.
2. Percentage of respondents answering 6–10, those most anxious.

Figure 6.14: Maps showing patterns of personal well-being at lowest levels in England

Yellow = Players, Orange = Attitudes and actions, Purple = Futures and uncertainties

Perceptions of urban places

CASE STUDY: Local perceptions of Liverpool, an urban area

Impact 08, a study investigating the impacts on Liverpool of being the European Capital of Culture, revealed a range of **perceptions** of the city from diverse urban communities. In 2009 86 per cent of respondents believed that Liverpool had a positive future; they felt that redevelopment and regeneration were improving the city and that a major strength was its people. There were some concerns, however: 35 per cent of people were most concerned about crime and antisocial behaviour, although 49 per cent believed that crime was falling across Liverpool as a whole. The positives and negatives, and changes in perception, are summarised in Figure 6.15.

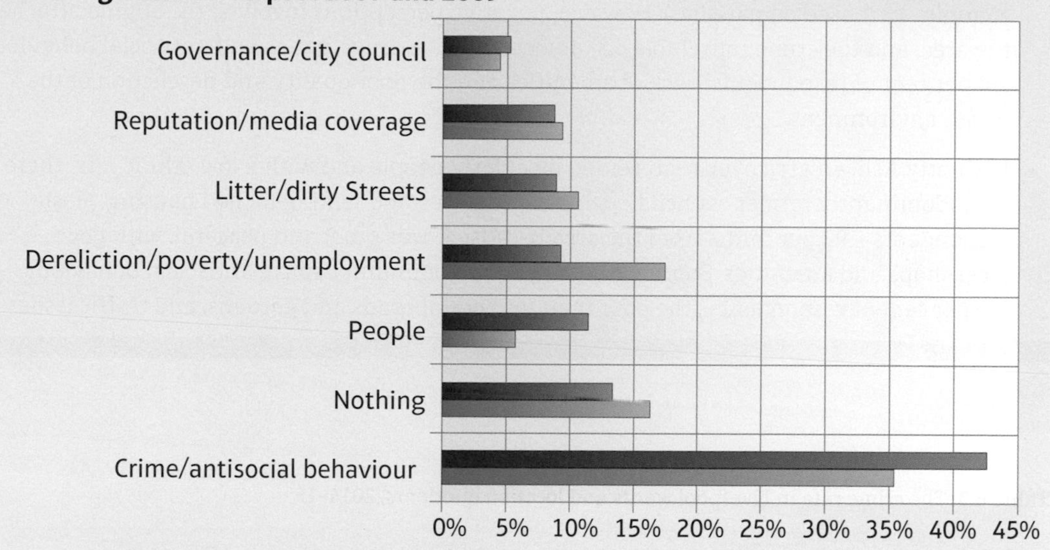

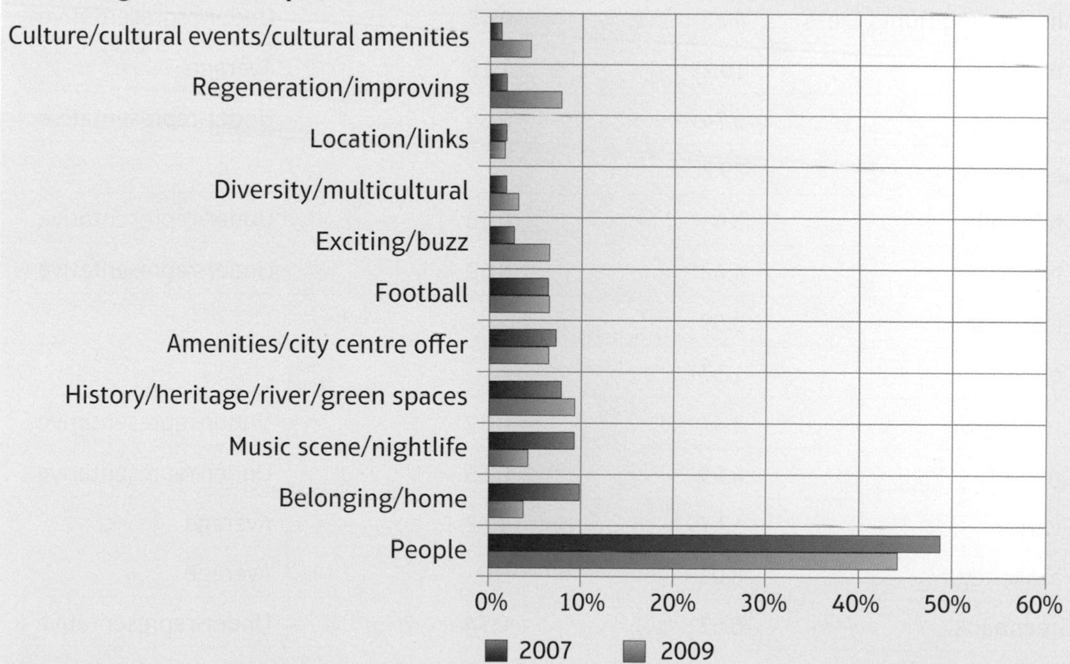

Figure 6.15: The best (bottom) and worst (top) things about Liverpool

There were some differences in perception between different neighbourhoods in Liverpool:

- In the city centre, an area with many 20- to 29-year-olds, there was above city average deprivation (see Table 6.4) and a large ethnic mix. Some 86.7 per cent of respondents said that they liked living there because they were close to amenities, their place of work, nightlife, cafes and restaurants, and they thought that the diversity of people was positive and that regeneration was improving the area. The main negative for this area was the antisocial behaviour associated with some nightlife.

- In Aigburth, a prosperous suburb with mature families, high employment and a low ethnic mix, 97 per cent of respondents liked living in the area because it was a peaceful, quiet community with good transport links, access to services and green open space. The main issues concerned the lack of local shops, large traffic flows and the lack of parking spaces. Only 23 per cent of respondents mentioned crime and antisocial behaviour.
- In Kirkdale, an area of high economic deprivation and with many older white British people, 85.2 per cent of the respondents liked the area, with 69 per cent citing the strong sense of community (based on parishes). However, negative perceptions involved the stigma attached to the area, and the crime rate (Table 6.3) associated with drugs, gangs and antisocial behaviour (50 per cent of respondents), lack of amenities and the poor quality and dereliction of the urban environment.
- In Knotty Ash, an area over-represented by elderly people and with a low ethnic mix, there is predominantly former council housing stock with some semi-detached housing. Most respondents – 92 per cent – liked the area because it was quiet and peaceful, with good local shops and amenities. Concerns included low community spirit, antisocial behaviour (50 per cent of respondents), the poor maintenance of roads and gardens, and traffic issues.

ACTIVITY

Read the case study of perceptions in Liverpool. Consider the characteristics of the people and the urban neighbourhood they live in. Explain to what extent these characteristics, when combined, influence the contrasting perceptions of Liverpool. Information on page 285 and in Figures 6.12 and 6.16 and Tables 6.3 and 6.4 may also be useful.

Table 6.3: The crime rate in Liverpool wards and location quotients, 2014–15

Census ward	Crime rate %	Location quotient	LQ category
Allerton and Hunts Cross	4.28	0.48	Under-representative
Anfield	10.29	1.15	Average
Belle Vale	5.74	0.64	Under-representative
Central	29.2		
Childwall	3.4	0.38	Under-representative
Church	4.42	0.49	Under-representative
Clubmoor	8.59		
County	12.77		
Cressington	3.47	0.39	Under-representative
Croxteth	4.95	0.55	Under-representative
Everton	12.79	1.42	Average
Fazakerley	8.08	0.9	Average
Greenbank	6.63	0.74	Under-representative
Kensington and Fairfield	11.26	1.25	Average
Kirkdale	10.94	1.22	Average
Knotty Ash	7.63	0.85	Average
Mossley Hill	2.93		

Yellow = Players, Orange = Attitudes and actions, Purple = Futures and uncertainties

Census ward	Crime rate %	Location quotient	LQ category
Norris Green	6.25	0.7	Under-representative
Old Swan	8.29	0.92	Average
Picton	9.37	1.04	Average
Princes park	9.54	1.06	Average
Riverside	21.09		
Speke-Garston	9.75	1.09	Average
St Michael's	6.43	0.72	Under-representative
Tuebrook and Stoneycroft	10.27	1.14	Average
Warbreck	6.9	0.77	Average
Wavertree	5.42	0.6	Under-representative
West Derby	6.17	0.69	Under-representative
Woolton	4.37	0.49	Under-representative
Yew Tree	4.42	0.49	Under-representative
	Liverpool av.= 8.98		

ACTIVITY

NUMERICAL AND STATISTICAL SKILLS

Study Table 6.3. To understand the significance of a spatial pattern, a simple division calculation called a location quotient (LQ) can be used to give a ratio – in this example *% crime rate in a ward* divided by *average % crime rate in Liverpool*. Most of the table has been completed with these calculations.

1. On a copy of Table 6.3:

  a. Complete the missing five LQ calculations. (LQ categories are: Under 0.75 = under-representative; 0.75 to 1.5 = average; 1.5 to 3 = moderately over-represented; and over 3 = highly over-represented.)

  b. Complete the final column for the five calculations.

  c. Using these categories, produce a choropleth map on Figure 6.16.

  d. With reference to your completed chloropleth map and Table 6.3, describe the pattern of crime in Liverpool in 2014–15.

ACTIVITY

Compare the concerns of people in Liverpool with people in Newham. In what ways are they similar or different?

Figure 6.16: Map of Liverpool Census wards

Yellow = Players, Orange = Attitudes and actions, Purple = Futures and uncertainties

CASE STUDY: Local perceptions of Newham, an urban area

Newham Council's aim is that the East London borough will become a major business location, where people choose to live, work and stay (Table 6.4). The Newham Household Panel Survey of 2013 suggested that progress was being made towards this aim, with 87 per cent of respondents seeing a cohesive community with people from different backgrounds getting on well. For example, the 16-to-24 age group had most friends from different ethnic backgrounds (62 per cent), and racially motivated attacks were declining.

There has been stabilisation of ethnic diversity in the borough because of reduced 'in and out' movements, and a slowing of the 'white flight' to outer urban or accessible rural areas. Results of the survey showed that:

- 77 per cent of respondents were satisfied with their life overall (ratings above London and UK averages)
- 76 per cent were satisfied with their health
- 75 per cent agreed that their quality of life was good
- 87 per cent said that they felt safe during the day
- 82 per cent were at least fairly satisfied with their accommodation
- over 80 per cent, especially families, rated schools and refuse collections highly
- shops and transport were perceived to have improved (the Olympic effect), with over 80 per cent of respondents rating them highly
- the over-65 age group rated housing and health services highly.

However, there were several concerns, linked to poverty, with 41 per cent of Newham's population in relative poverty and 55 per cent of children in poor households; 63 per cent of respondents received some form of benefit payment, with 25 per cent receiving housing benefit and 69 per cent living in social housing (affordable, sometimes provided by the council). Housing was perceived as an issue, with 11 per cent of respondents dissatisfied (twice the national average) especially among the 35-to-44 age group and those renting from a private landlord. Housing services had the lowest rating of the council services. Older people, the black ethnic group, the unemployed and long-term residents were less satisfied with their social housing than other groups.

The second perceived issue was crime and antisocial behaviour: 51 per cent of respondents said that they were worried about crime, even though statistics showed that crime was decreasing in the borough (down 8 per cent between 2012 and 2013). Of the 45 per cent of respondents who said that they did not feel safe to go out after dark, females and the Asian ethnic group were the most concerned groups. Specific concerns were car crime (including break-in), burglary and teenagers hanging around (66 per cent of respondents viewed the latter as a problem), with the Indian ethnic group and older long-term residents most concerned about crime.

AS level exam-style question

Explain why some urban places are perceived as undesirable. (6 marks)

Guidance

Remember that the negatives may be real or imagined, and that perceptions will vary according to the background characteristics of different groups of people.

A level exam-style question

Explain why levels of deprivation vary in an urban place that you have studied. (8 marks)

Guidance

Consider urban structure and the changes within your chosen place, as well as the characteristics of the population groups living in the various urban zones.

ACTIVITY

Using Table 6.4, compare the characteristics of UK urban and rural places. How diverse are these characteristics?

Table 6.4: Comparison of UK rural and urban places

	Liverpool	Newham	Happisburgh, Norfolk	Holbrook, Suffolk
	Urban local authority	*Urban borough local authority*	*Village in remote rural ward*	*Suburbanised village ward*
Population size, 2011	466,415	307,984	2386	2467
% of adults over 65	17%	8.5%	29.8%	24.8%
% born outside UK	9%	53%	2.64%	4.9%
Largest minority ethnic groups (% of population)	Black African (1.8%) Chinese (1.7%) Arab (1.2%) Indian (1.1%)	Indian (13.8%) Black African (12.3%) Bangladeshi (12.1%) Pakistani (9.8%)	Mixed race (0.4%)	Indian (0.3%) Bangladeshi (0.2%)
% with health below good	22.8%	16.9%	23.3%	14.6%
% single-person 65+ households	11.9%	6.9%	14.6%	13.4%
% single-person (other) households	27.3%	19.2%	11.8%	13.1%
% households not deprived	32.9%	25%	39.8%	56%
% households with deprivation in three or more dimensions	11%	10.9%	3.9%	1.2%
% travelling to work by car	54.7%	21.9%	78.2%	62.4%
% travelling to work by train	5.1%	12.5%	1%	6%
% cycling or walking to work	14.9%	8.8%	6.7%	17.9%
% working from home	2.7%	2.4%	10.9%	10.3%
% owned tenure	46.9%	33.3%	76.8%	68.7%
% social rented tenure	27.8%	29.6%	7.1%	9.2%
2015 IMD rank position*	7458 (av. for city)	7510 (av. for borough)	8484	26342
2015 IMD crime rank*	11051 (av. for city)	4713 (av. for borough)	29209	29813
2015 IMD barriers to housing rank*	23463 (av. for city)	1908 (av. for borough)	280	11,161
2015 IMD living environment rank*	8058 (av. for city)	7766 (av. for borough)	1367	8866
Average house price (Oct 2015)	£202,475	£270,000	£191,000**	£221,250**

Source: Calculations from neighbourhood statistics (ONS) and IMD 2015 (*Note: for the Index of Multiple Deprivation (IMD) the lower the rank position number, the worse the deprivation i.e. 1 = worst out of 32,844 areas). (Liverpool improved between 2010 and 2015.) (**Note: rural district average)

Perceptions of rural places

CASE STUDY: Local perceptions of Holbrook, Suffolk, a rural area

The parish of Holbrook, a village of about 2500 people, is 13 km south of the town of Ipswich and 133 km from London. It is in an intermediate rural district (see Figures 6.5 and 6.9) with easy access to the intercity railway network at Ipswich and Manningtree for commuters, and dominating the Shotley Peninsula rural area. Consequently it has developed into a suburbanised village with a greater range of services than might be expected for its size: it has a primary school, a secondary school, a small Co-op supermarket, two pubs, a church, a chapel, a village hall and a doctors' surgery.

In 2002 the Holbrook Action Project Team undertook a survey of local perceptions of the area. It revealed that most people worked within 16 km of the village, so it was *not* a typical 'dormitory' village with large numbers travelling to London or Norwich and just returning to their homes to sleep. The major travel issues were getting to the hospital in Ipswich (30 to 40 minutes away), but this was partly due to lack of information distribution. There was a desire for better public transport – including cheaper fares, more frequent service, more relevant routes and consideration for disabled people, and more cycle paths (Table 6.4). In addition, 36 per cent found pavements poor, and 56 per cent wanted to see safer roads. There was a wish for more children's out-of-school organised playgroups, but Reade Field, with its play area and playing fields, and the village hall, were regarded as successes for the community.

Community information was found to flow well through 'word of mouth' and noticeboards: 57 per cent of respondents felt that the parish council was aware of local concerns and feelings. To improve the village environment, it was felt that maintaining woodlands, common land and hedges was a top priority, as was cleaning up the shore of the nearby Stour estuary. Over half of respondents were in favour of environmentally sensitive streetlights. While it was felt that the number of litterbins was adequate, 60 per cent thought litter was an issue, and 70 per cent were concerned about dog fouling.

The survey found uncertainty about local business development in the village. Respondents were against any large-scale house building and most were only in favour of converting redundant buildings or allowing small-scale developments (infilling by single houses or small groups of no more than ten houses). In terms of crime, people were most concerned about poor coverage of the area by the police, antisocial behaviour (especially dog fouling and litter) and vandalism. People shopped outside the village because of wider choice, lower cost and easier parking. Respondents experienced mobile phone reception difficulties, but only 17 per cent of respondents had internet reception difficulties.

CASE STUDY: Local perceptions of Happisburgh, Norfolk, a rural area

Happisburgh is a coastal village in North Norfolk, with a population of about 2400 people. It is 32 km north of Norwich, in what is a sparsely populated rural area with quite high deprivation levels (see Table 6.4). It is an isolated settlement with few village services other than a church, a primary school and a cafe. The few local businesses are mostly linked to tourist accommodation, and they are not busy out of the summer season.

A significant geographical feature of the village is the rapid coastal erosion, with 30 properties along Beach Road having disappeared or been demolished since 2000 due to the retreat of the soft rock cliffs. In 1998 the Coastal Concern Action Group was formed to campaign for social justice for those affected by coastal erosion in the village (Figure 6.17). Surveys were conducted as part of the Pathfinder Pilot Project submission for managing the North Norfolk coast in 2009.

ACTIVITY

Compare the perceptions and concerns of the three featured rural areas: Holbrook, Happisburgh and Caithness and Sutherland. Why are the concerns and issues similar or different in these rural places?

The survey found that residents were proud of the traditional village, with its mix of long-term residents and new villagers, and no domination by second homes. Respondents were also proud of the traditional Norfolk heritage, buildings and countryside, and the fact that it is surviving coastal erosion that has led to widespread recognition.

Respondents recognised the strong community spirit in the face of adversity, with residents contributing to village life. (Happisburgh won the *Eastern Daily Press* Pride in Community Award for small villages in 2007.) However, they were concerned about the external image of the village conveyed by negative **media** reports that the whole village was going to disappear. This was causing a reduction in property values and a decline in the number of people wanting to move into the village. There was concern that the government was not listening to locals, and therefore that social justice was not being achieved. There was also anxiety about loss of tourism owing to the cliff recession, which resulted in poor access to the beach as well as debris on the beach and land that was not cleared away (Figure 6.17).

Respondents hoped for increased use of local services such as the post office to prevent them closing and for investment in the village to protect villagers from economic decline and falling property values. People wanted better pavements, cleaner areas (including less dog fouling), play equipment for children, access to the beach, and drainage to cope with heavy rainfall. A need for affordable housing to keep young families in the community and give them a future was recognised, but any developments should be discreet and appropriate, avoiding adverse impacts on the quality of life of existing residents. Residents hoped for cliff drainage and coastal defences so that Happisburgh could be protected.

Residents' fears included the loss of tourist facilities (e.g. the caravan site and tea shop) and bus routes, the reduced economic viability of the village and its services, and the increasing size of agricultural vehicles moving on country lanes. There were fears that a push by councils for more house-building would lead to negatives such as affordable housing not going to those it was designed for, the loss of open countryside views, and the creation of a dormitory village of commuters to Norwich, which would destroy the community. A significant fear was the 'no active intervention' Shoreline Management Policy, which meant no government funds for coastal protection and no assistance for those who lost their homes, and that a coastal buffer zone would become a wasteland. This situation could affect the economic vitality of the village, as people would not buy properties and would visit only to see the negative impacts of coastal erosion.

Figure 6.17: Houses in Happisburgh nearing the cliff edge, 2010 (now demolished as part of the Pathfinder Pilot Project)

Synoptic link 

UK settlements are found in many diverse geographical situations. For some, their situation has changed due to physical processes, and what were once locations fit for purpose are no longer so. This is true for some coastal communities that are now vulnerable to coastal erosion linked to rising sea levels, higher storm surges and stronger winds producing larger destructive waves, all of which are attributable to climate change (global warming). (See pages 147–148.)

Yellow = Players, Orange = Attitudes and actions, Purple = Futures and uncertainties

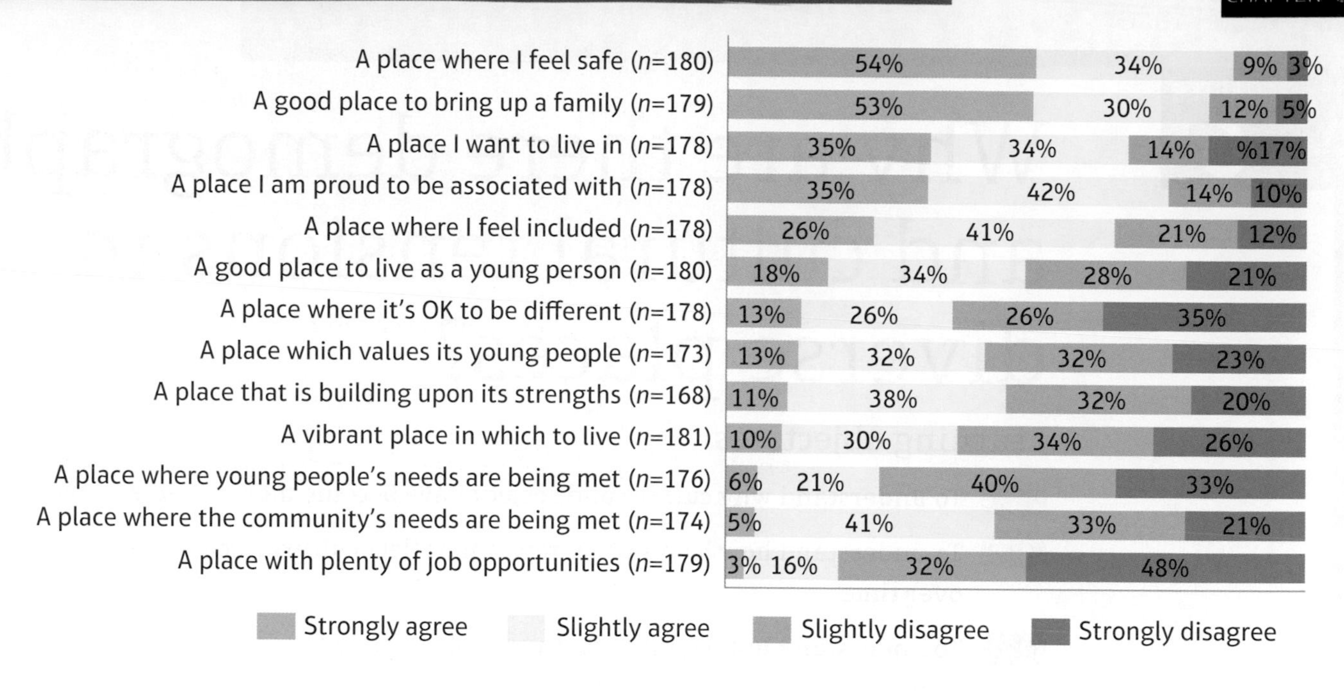

Figure 6.18: Opinions of young people about living in Caithness and Sutherland (2015)

ACTIVITY

GRAPHICAL SKILLS

Study Figure 6.18. Describe and explain the issues facing young people in parts of the Highlands of Scotland.

CASE STUDY: Local perceptions of Caithness and Sutherland, a rural area

Census data from 2011 and other recent surveys show that young adults are likely to leave the Highlands of Scotland, and that only 16 per cent of the population is between 15 and 29 years of age. Caithness and Sutherland (C&S) is part of this area in the north of Scotland. Despite a small increase in young adults between 2001 and 2011, a further decrease is expected by 2037. This is a great concern for the sustainability of this very remote rural area.

Communications are poor, despite a train service on the east coast and a small airport at Wick. In C&S, jobs are mostly in public services, primary industries, or linked to the decommissioning of the Dounreay nuclear power station. A 2015 survey (Highlands and Islands Enterprise) into the attitudes and aspirations of young people indicated that those in C&S were strongly connected to, and proud of, their local community. Of those surveyed, 51 per cent were 'committed stayers', planning to live and work in the Highlands; this was above the Highlands average of 43 per cent. By contrast, 22 per cent were 'committed leavers', compared to 40 per cent for the Highlands (Figure 6.18).

There are still many obstacles for young people, but the majority were involved in the community, felt safe, and were optimistic about life in the Highlands in the future, due to access to higher education (at Inverness), modern communications and career opportunities in creative industries. In Caithness and in additional areas in Sutherland (e.g. Golspie) 10,000 homes have been built, with more planned, and access to 'next generation' broadband has been provided. This connectivity was a boost for retaining young people who desire digital services and infrastructure, and also for the growth of businesses and new business start-ups, which could diversify the economic base.

ACTIVITY

1. Investigate what people think about the place and community in which you live. Use **media** sources (secondary data) such as research blogs, newspaper articles and carefully selected social media. Use fieldwork methods (primary data) such as interviews.
    a. Explain why the views discovered are similar to, or different from, the examples given in this section of the chapter.
    b. How might the contrasting perceptions and opinions of people in your local place cause misconceptions and potential conflicts?

CHAPTER 6

Why are there demographic and cultural tensions in diverse places?

Learning objectives

- **4B.7** To understand why culture and society have become more diverse in the UK
- **4B.8** To understand how levels of segregation reflect cultural, economic and social changes over time
- **4B.9** To understand how changes to diverse places may lead to tension and conflict

Diversity in UK culture and society

Internal migration flows

Internal migration is common in the UK: 2.9 million people moved from one place to another in 2014. This internal movement involves complicated transfers of people between the regions, with flows in both directions (see Table 6.5), and cities with high turnovers. For many decades there was a pronounced north-to-south drift of people in England, particularly to London and the South East, and then a second movement (counter-urbanisation) into the neighbouring regions, such as the East of England, causing **regional disparities**.

However, it now appears that the drift to the South East from northern regions is slowing, with many people instead moving into the East Midlands from elsewhere. This perhaps reflects the economic readjustment that has taken place. The largest negative flow in 2014, outside London, was a movement out of Yorkshire and Humber (Table 6.5). Interestingly, nearly 45,000 people left England for Scotland and, while Edinburgh was a popular destination, so too were the Highlands, which had a net gain of 3230 people between 2009 and 2013 (see the case study on page 280).

From 2012 to 2014 more people left London than migrated to it. Newham was one of the boroughs with the highest outflows: a net loss of 24,920 between 2009 and 2013 (see the case study on page 293.) Many of those choosing to leave the capital were over 30 years of age, white British, and families with children. Reasons for this movement include:

- the high cost of housing in London and the relatively cheaper housing in the accessible rural regions around the capital
- social and **environmental quality** factors such as the desire to be close to peaceful, green open spaces
- the desire to avoid ethnic tensions and deprivation factors
- better social and educational opportunities for children in the rural areas.

International migrants who may have first moved to London because of the job opportunities are also now choosing to move elsewhere (for example, Leicester). Older people from London are retiring to countryside areas – for example, North Norfolk gained 1920 people between 2009 and 2014 – or moving to be closer to their families. London has gained people in the 20-to-30 age

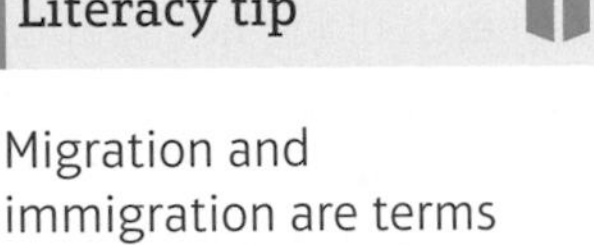

Literacy tip

Migration and immigration are terms used in this chapter in their geographical definitions. Migration takes place when a person changes their permanent residence and lives in the new location for at least a year. International immigration occurs when someone crosses the border of a country to live in it as a citizen for at least a year (so temporary workers and students are not counted).

Yellow = Players, Orange = Attitudes and actions, Purple = Futures and uncertainties

group, many of them new graduates seeking work and others arriving from abroad. Higher salaries and social and recreational opportunities in the capital also attract this age group. People who live in the South East often enjoy a high quality of life, as well as proximity to the world city of London.

Other big UK cities are also losing people to neighbouring areas (counter-urbanisation), again because of the shortage of affordable housing, the stressful living environment and inequalities brought about by deprivation factors, including crime (see Table 6.3). For example, Liverpool gained 65,590 people but lost 70,333 from 2009 to 2013.

ACTIVITY

Using Table 6.5 and ONS maps, describe and suggest reasons for the internal migration movements in the UK in 2013–14.

Table 6.5: Internal migration data for UK regions, June 2013 to June 2014

	FROM											
TO	North East	North West	Yorks & Humber	East Midlands	West Midlands	East	London	South East	South West	Wales	Scotland	Northern Ireland
North East	–	7010	11,180	3590	2510	3240	4320	4710	2350	990	3450	700
North West	*6640*	–	**23,070**	11,070	14,800	7460	13,840	11,700	7400	*10,120*	*5730*	*2080*
Yorks & Humber	**10,720**	**22,980**	–	**19,350**	8650	9350	10,960	10,710	5750	2900	3630	660
East Midlands	3110	10,900	*19,840*	–	**19,090**	20,130	15,160	18,530	7460	3060	2650	570
West Midlands	2250	14,090	8570	*17,840*	–	9110	16,600	15,570	13,490	8640	2360	640
East	2890	7290	8120	15,930	7900	–	*70,140*	29,330	9970	3310	3100	750
London	5890	*16,100*	14,250	14,990	15,460	**37,120**	–	**65,570**	*20,910*	6270	**6440**	1430
South East	4130	11,550	10,340	15,500	14,260	*30,860*	**105,180**	–	**36,380**	7720	5320	1090
South West	2360	8800	6720	9100	*17,100*	13,260	23,150	*48,920*	–	**12,220**	3570	750
Wales	950	10,450	2760	3170	8840	3500	5450	8810	11,860	–	1310	400
Scotland	4160	7290	4810	3170	2930	4390	6920	7230	3980	1730	–	**2610**
Northern Ireland	690	1860	660	640	570	830	1350	1260	700	360	2090	–

Note: The largest (**bold**) and second largest (*italic*) movements are picked out in the table

Internal migrations are often selective, and segregation may occur on an age basis, with urban centres being places for younger adults, suburbs or accessible rural areas for families, and the remoter rural areas for older people. These are generalisations, of course, but there is a pattern. Different social and economic patterns become established in these locations, for example city centre nightlife and rural real ale pubs. Different age groups also prefer different living spaces for the services and amenities they offer:

- Older people want good access to medical care, bungalows and sheltered apartments to live in, access to shops nearby and access to public transport.

Extension

On an outline map of UK regions produce a flow line map to show the pattern of internal migration in the UK in 2013–14. Use data from Table 6.5 or find more recent data.

1. For this practical exercise, London will be excluded because of the large numbers involved, so there are 11 regions. Divide these up between a group of students (for example take two regions each).

2. For your allocated regions, draw proportional arrows on your outline map to show the movement in and out of each region. The width of the arrow represents the number of people moving. All students must use the same width scale: Data ranges from 360 to 36,380, this is a very large range. Some discussion may be useful but one possibility is 1 mm = 1000 migrants. Colour the in-migration arrows in blue and the out-migration arrows in red.

3. Compare completed maps to confirm your understanding of the internal migration patterns in the UK and check this with your answer to the Activity above when working out a suitable scale.

- Young families wish for access to good schools, play areas and housing with sufficient bedrooms.
- Retired people may move to seaside towns and villages for a quieter life and lower pollution levels, but issues relating to an ageing population may develop, such as a lack of provision of the services they need.

Segregation

Tension and conflict

Divisions can arise in rural areas when there is an influx of exurbanites, bringing with them an idea of the **rural idyll**. This rural idyll may include a happy family life, peace and quiet, a safe environment, plenty of green open space, good health and low pollution levels, and a rustic work ethic. However, agriculture is now very commercialised and intensive, with large machinery moving between fields and operating long hours at harvest time, and during a farming year smelly manure may be added to the soil, which might bring complaints from those whose idealised image of a rural idyll is spoiled. There is also a high level of deprivation in some rural areas that is not directly visible – as shown by the Index of Multiple Deprivation (IMD) – for example:

- many workers on low pay or seasonal pay
- single-parent families
- unemployment
- elderly people living alone
- manual workers nearing retirement
- unaffordable housing
- young adults and school leavers with few job prospects unless they leave the area, which can result in **social exclusion**.

International migration flows

International migration has been a major feature of the UK's changing demography since the mid-twentieth century. It has happened in waves and is linked to economic factors. Between 2014 and 2015 more than 636,000 immigrants entered the UK (10.4 per cent of them from the Indian subcontinent), while 300,000 emigrated; this was the highest ever annual positive migration rate. Immigration from other EU countries amounted to 265,000 people, with work their main reason for entering the UK: 64 per cent of these had a definite job to go to; 31 per cent of them were from Romania or Bulgaria (EU2 countries) and 84 per cent of these said that their move was job-related. There were also 29,000 asylum applications, with 12.8 per cent from Eritrea, 9.8 per cent from Sudan and 8.3 per cent from Iran (ONS data 2015).

There is much debate in the media and within communities about the positive and negative impacts of large-scale immigration and **cultural diversity**. The most popular destinations for immigrants are London, the South East and the East – and 11 per cent of the population of eastern England was born abroad. UK urban areas from Glasgow to Birmingham to Luton also attract migrants because of the job opportunities they provide and the presence of ethnic communities already there.

Yellow = Players, Orange = Attitudes and actions, Purple = Futures and uncertainties

CASE STUDY: UK international migration, 1948–68

After 1945 the UK needed a large number of unskilled and low-skilled workers in order to rebuild the economy after the war. The small inflow of European migrants was insufficient and so the government turned to the countries of its former Empire that were gaining independence from UK rule (such as India, which gained independence in 1947).

In 1948 the British Nationality Act was passed to give Commonwealth citizens' free access to the UK, and large numbers moved from the West Indies in the 1950s and from the Indian subcontinent during the 1950s and 1960s. By 1970 there were about 1.4 million non-white residents in the UK, a massive increase from the few thousand in 1945. The 1962 and 1968 Commonwealth Immigration Acts imposed restrictions and slowed down this movement, and since then further laws have introduced more restrictions.

The UK has become a multiracial society and, to outlaw racial discrimination and help ensure equality, various Race Relations Acts (the last in 2000) have been passed. They were superseded by the Equality Act in 2010.

Ethnic tension and segregation

For international migrants, **segregation** or **social clustering** may occur, either by choice or by a combination of forces beyond the control of individuals and families (including house prices). As well as **ethnicity** influencing groups of people to live in the same area, other factors such as socio-economic status, wealth, types of employment and community factors such as health, crime and levels of education also influence where ethnic groups live. For instance, it has been found that many Polish migrants head for Ealing in London or Northern Ireland, many Iraqis to Hull, many Bulgarians to Herefordshire, many Zimbabweans to Leicester, many Lithuanians and Latvians to Peterborough, and many Slovaks to Warrington.

The living spaces in these urban areas develop the services and characteristics to match the ethnic mix, with shops, restaurants and places of worship and entertainment serving the ethnic community. For example, in Newham a 2009 study for the council showed that the majority of the Indian and Pakistani ethnic groups used the Green Street, East Ham, and Forest Gate local shopping areas, the first two for both ethnic non-food and food, and the latter for ethnic food (Figure 6.19), while the black African and Caribbean communities used Stratford local shopping area the most for ethnic food and non-food (Figure 6.20).

Segregation can be both positive and negative:

- Positives include the improved social networking within a community of people with similar backgrounds and the sharing of common services that cater to their needs, such as shops, restaurants and community centres.
- Negatives include the division of a place into disparate parts, which may then lead to misunderstanding and mistrust between different ethnic groups.

Perhaps, over time, later generations from different ethnic groups will become assimilated, enhancing the multicultural population of the UK, but this does not appear to have happened as yet (Social Integration Commission, 2014).

Extension

Investigate the changes made to Newham by developments for the Olympic Games in 2012. Consider the roles of local community groups, Newham Borough Council, the national government, the Olympic Delivery Authority, and the Westfield Group in a) changing land use patterns, and b) creating challenges and opportunities for local people.

ACTIVITY

CARTOGRAPHIC SKILLS

1. Study the 1:25,000 Ordnance Survey map extract showing Newham Borough in London.
   - **a.** Using map evidence, describe the character of the urban area.
   - **b.** What map evidence suggests that this may be a popular area for immigrants?
   - **c.** Identify any evidence of services for ethnic groups.
2. Suggest what other sources of information may reveal more evidence than the OS map.

Figure 6.19: Shops in Newham

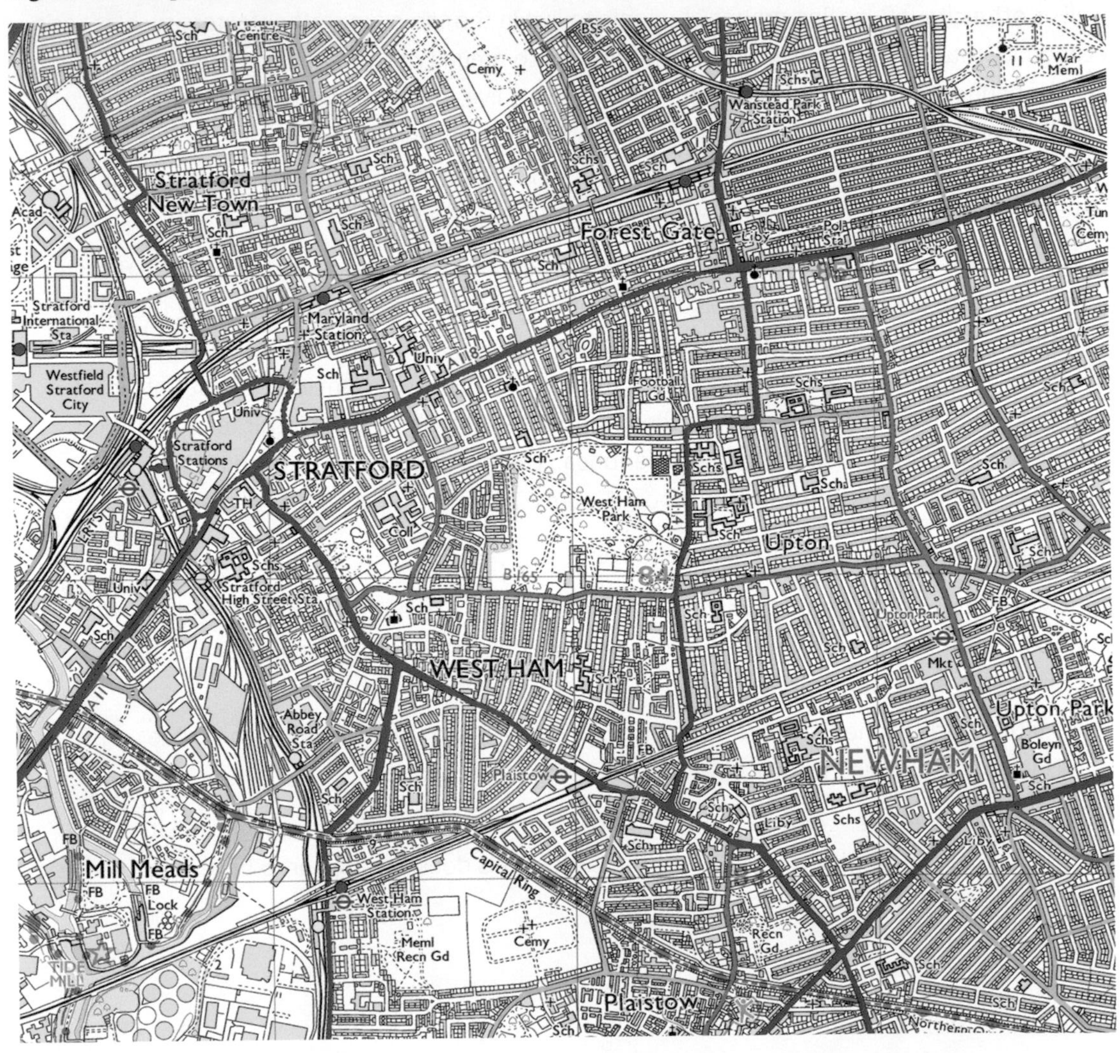

Figure 6.20: Ordnance Survey map extract for Newham Borough, London (1:25,000 scale) © Crown copyright 2016 OS 100030901

Yellow = Players, Orange = Attitudes and actions, Purple = Futures and uncertainties

CASE STUDY: Ethnic tension in Boston, Lincolnshire, a rural area

In 2001 the largest immigrant group in Boston was German, but they only made up 0.4 per cent of the 55,753 population. By 2011 the population of this remote, rural town (see Figure 6.5) had grown to 64,600, largely because of international immigration, with 10.6 per cent of the population from the new EU countries of Poland, Lithuania, Latvia and Romania. A 2014 estimate suggested that the proportion had increased to 17 per cent.

There are perceived positives regarding the immigrants:

- They are providing a much-needed workforce for the area, especially for manual work on farms (vegetable and flower-picking). There were only 1500 native jobseekers available in the area.
- The larger population helps to keep services going and creates a wider variety of jobs in the community (e.g. in maternity services and primary schools).
- The immigrants have opened shops and services for local people and are keeping retailing going in Boston despite the economic recession, leaving no empty shop spaces.
- Immigrants have brought social vitality to a remote rural town, which might otherwise have had an ageing population.

There are also some perceived negatives:

- Pressures on some services have increased because of the larger number of people (for example, primary schools and increased waiting times at GP surgeries).
- There is increased competition for jobs and reduced pay levels, as immigrants are prepared to work for less and do longer hours of work.
- Some immigrants do not speak fluent English, and some children start school without knowing any English, which hinders their integration into the community.
- A minority of immigrants abuse the benefits system.
- Some send money (remittances) to their families back in the home country, rather than spending it in the local area.

Extension

Investigate the latest immigration situation in Boston, Lincolnshire, by using media articles from the internet (such as BBC News, MailOnline, *The Independent*, *The Guardian*, *The Mirror* and *The Telegraph*).

Places reflect people

As well as creating a distinctive place, some migrants may adopt an existing place because of its character. This is true for some Russians in the upper socio-economic group, including billionaires: they have been attracted to Chelsea and South Kensington in London because the Russian Embassy is nearby, it is an expensive and exclusive area with shops such as Harrods as well as specialist shops and cafes catering for Russian customers in their ethnic enclave. Wealthy Russians come to London because it is an important world city with theatres and other entertainment, has private schools for their children, and because the flight time to Moscow is under four hours. They also trust in British 'bricks and mortar' for investing their money, which they perceive is more secure than keeping their money in Russia, because London is a stable world financial market.

Literacy tip

It is sometimes difficult to find the correct terminology to refer to groups of people. For example, the use of the word 'class' by geographers has largely been abandoned. Know the meaning of the alternatives, such as 'socio-economic group', 'socio-cultural group' and 'ethnic group', and when and how to use them.

AS level exam-style question

Explain how internal migration within the UK has affected recent demographic patterns. (6 marks)

Guidance

Remember that demographic means population, so consider a range of characteristic patterns such as age, ethnicity and socio-economic status and link these to how and why they have been created by specific movements of people inside the UK. Be careful not to write about international migration.

A level exam-style question

Explain how international immigration has diversified the culture of the UK. (8 marks)

Guidance

Recall the waves of migration since 1945 and identify the cultural characteristics that each of these ethnic groups has brought, and their contribution to UK society. Mention specific places that illustrate the points you make.

CASE STUDY: Ethnic tension in Luton, Bedfordshire, an urban area

In 2009 Luton became an example of a place where a small number of people caused tension within a **culturally diverse** community. Divisions had became apparent in 2001 when many people in Luton failed to complete the national census, showing that they were not **politically engaged**. A few of the town's 20,000-strong Muslim community were linked to terrorism between 2004 and 2009. Between 2001 and 2011 the Asian population grew by 49 per cent and the black population by 69 per cent, and other ethnic minorities such as Turkish also grew.

In 2009 The *Independent* ran the headline: 'Luton: the enemy within?' after an anti-war protest by about 20 Muslims, with what most regarded as extreme views, attempted to disrupt a homecoming parade by British troops returning from Afghanistan. The majority of Luton residents of all backgrounds supported the parade, but this small protest increased ethnic tension. Having opposing pressure groups in Luton has not helped those who are trying to reduce tensions within the community.

ACTIVITY

Glasgow is the most ethnically diverse Scottish city but, as the city has been redeveloped and regenerated, some people perceive some migrant groups as a threat to further improvements in the quality of the urban area. Carry out an investigation into how the fears of some Glaswegians have led to the socio-economic exclusion of some ethnic minority groups.

Yellow = Players, Orange = Attitudes and actions, Purple = Futures and uncertainties

CHAPTER 6

How successfully are cultural and demographic issues managed?

Learning objectives

4B.10 To understand that the management of cultural and demographic issues in diverse places can be assessed in several ways

4B.11 To understand that different urban stakeholders have different criteria for assessing the success of managing change in urban communities

4B.12 To understand that different rural stakeholders have different criteria for assessing the success of managing change in rural communities

Measuring management

Criteria for assessing management

A March 2015 survey commissioned by the Royal Mail produced a list of the best places to live in the UK (Table 6.6), but the result will have depended on the criteria used for measuring this. And did these places arise by accident, or was there some planning behind their success? Even experts are not sure – there is a lack of a comparable, time-related database. The Royal Mail survey of post code areas used the criteria of schools, open space, commuting, employment, and housing. Figure 6.22 shows that there is no correlation between employment change and people's satisfaction with the urban place. There are two significant anomalies, with Cardiff experiencing a drop in employment but people remaining happier than expected, while London has the largest increase in jobs but has a lower satisfaction rating. This supports the idea that management of places is perhaps judged mainly by non-economic criteria (Table 6.7).

Table 6.6: The best two places to live in each UK country (Royal Mail survey, March 2015)

UK country	Top place (postcode and name)	Second place (postcode and name)
England	CH63 – Bebington, Wirral	IP5 – Kesgrave, Ipswich (Figure 6.21)
Northern Ireland	BT64 – Craigavon, County Armagh	BT10 – Finaghy, Belfast
Scotland	G78 – Neilston, East Renfrewshire	G64 – Torrance, East Dunbartonshire
Wales	SA6 – Morriston, Swansea	SA7 – Birchgrove, Swansea

ACTIVITY

1. Find photographs of the eight places shown in Table 6.6.

- **a.** Using this photographic information, identify what all these places have in common.
- **b.** Does this fully explain their high position in the survey?

Figure 6.21: Kesgrave (IP5), a suburb of Ipswich

ACTIVITY

Study Figure 6.22. To what extent does the data suggest that the management of these UK cities was successful? Explain your answer.

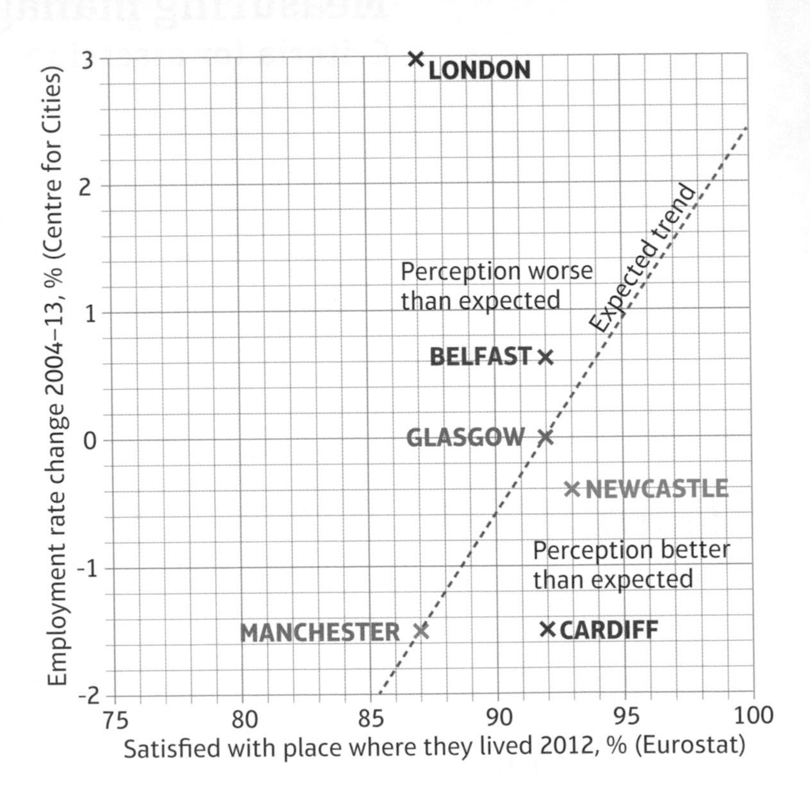

Figure 6.22: Scattergraph showing levels of satisfaction with the management of selected UK cities

Managing change in urban communities

Urban places have experienced a variety of government schemes and initiatives over time, aimed at managing, developing and improving urban areas. For example:

- between 1945 and 1965, **planning** aimed to decentralise London, with the building of new towns such as Milton Keynes and Crawley, and Development Areas were established
- between 1965 and 1980 there were Improvement Areas and Housing Action Areas

Yellow = Players, Orange = Attitudes and actions, Purple = Futures and uncertainties

Extension 

Investigate the Royal Geographical Society series '21st Century Challenges: The Thames Gateway' for the positives and negatives of this regeneration project.

- between 1980 and 1990 there were enterprise zones such as the Isle of Dogs, urban development corporations (UDCs) for areas such as London Docklands and city action task forces, for example in Toxteth, Liverpool
- between 1990 and 2000 there were the Single **Regeneration** Budget, New Deals for Communities and City Challenge
- from 2000 there were Sustainable Communities, such as Merseyside and the Thames Gateway, and Estate Regeneration.

The European Union also introduced schemes that have influenced UK urban areas, such as URBAN II (with 11 places in the UK in 2015, including the Thames Gateway) and City of Culture (Glasgow in 1990, Liverpool in 2008). There are a mixture of local small-scale schemes (bottom-up) and larger schemes (top-down) that try to improve urban areas. An example of a top-down scheme is the improvements made to Stratford as part of the London 2012 Olympics, which provided a major park, swimming pool, velodrome, and stadium, as well as housing and places of employment for local people. An example of bottom-up schemes are those linked to the 'transition network' initiative, amongst its aims are to support local economies and the natural environment (for example, Totnes in England and Kinsale in Ireland).

Progress in urban living spaces since 2011

A report on English cities by the Department for Communities and Local Government (DCLG) in 2011 showed that cities are important focal points of economic growth and provide essential infrastructure and key assets in a globalised world. They have a wide range of professional services, intellectual resources in higher educational establishments, cultural resources, shops and businesses that make them inviting places to live and work, and modern transport and connectivity through broadband. Improvements since 2000 have included employment growth, lower unemployment and lower crime rates, but the 2007/08 recession reduced the funds available for regeneration from both private and public sources, which slowed the progress of improvement strategies.

The performance of UK cities has been mixed, with strong performances in the South and East, and also in some northern cities such as Leeds, Manchester and Sunderland. Others have had falling employment: Stoke –12 per cent, Blackburn –7 per cent and Luton –4 per cent. Further regeneration initiatives (such as, in retail and construction) will be important for the future, because there are still economic, social and regeneration challenges, such as:

- increased economic competition from emerging countries
- making the transition to a low-carbon economy
- preparing for climate change
- investing for the ageing population
- the effect of international migration on jobs and socio-cultural factors.

Local Enterprise Partnerships

Urban cores are linked with their hinterlands (see Figure 6.23), so solving the problems of urban areas may spread benefits to rural areas. One strategy that involved decentralised decision-making for this reason was the Local Enterprise Partnership (LEP) initiative; in 2015 there were 38 of these, covering geographical regions rather than urban governments (for example Liverpool City region), so that money could be combined from different sources. The LEPs, run by local authorities and the private sector, aimed to provide jobs and reduce barriers to employment such as low education levels, with a focus on deprived areas.

Extension

Decision-making: is it better to build houses on brownfield or greenfield sites? Explain your answer.

Liverpool City Region LEP used EU Structural and Investment Funds (2014 to 2020) to work with local partners on improving economic areas (such as the Blue/Green Economy which focuses on only using the resources you have (blue) and being environmentally friendly (green), sense of place and connectivity. National government departments involved with LEPs include the Department for Communities and Local Government, with EU regional development funds, and the Department for Work and Pensions, with EU social funds. Funding was available for strategies that would help provide equal access to lifelong learning or improve education and training for workers.

A regional growth fund (RGF) was aimed at diversifying the economy, especially in areas dependent on public sector (government) employment. The emphasis was on helping those places with a comparative disadvantage, such as the North – as measured by research and development expenditure or foreign direct investment. The 'Big Society' agenda promoted a 'bottom-up' approach, giving local people the means to be involved in making improvements to their neighbourhoods. Local concerns were about the lack of affordable housing and also about house-building nearby or on rural land (greenfield sites) – so-called nimbyism. Other concerns related to image, crime and lack of connectivity, including fast broadband (see Figure 6.24).

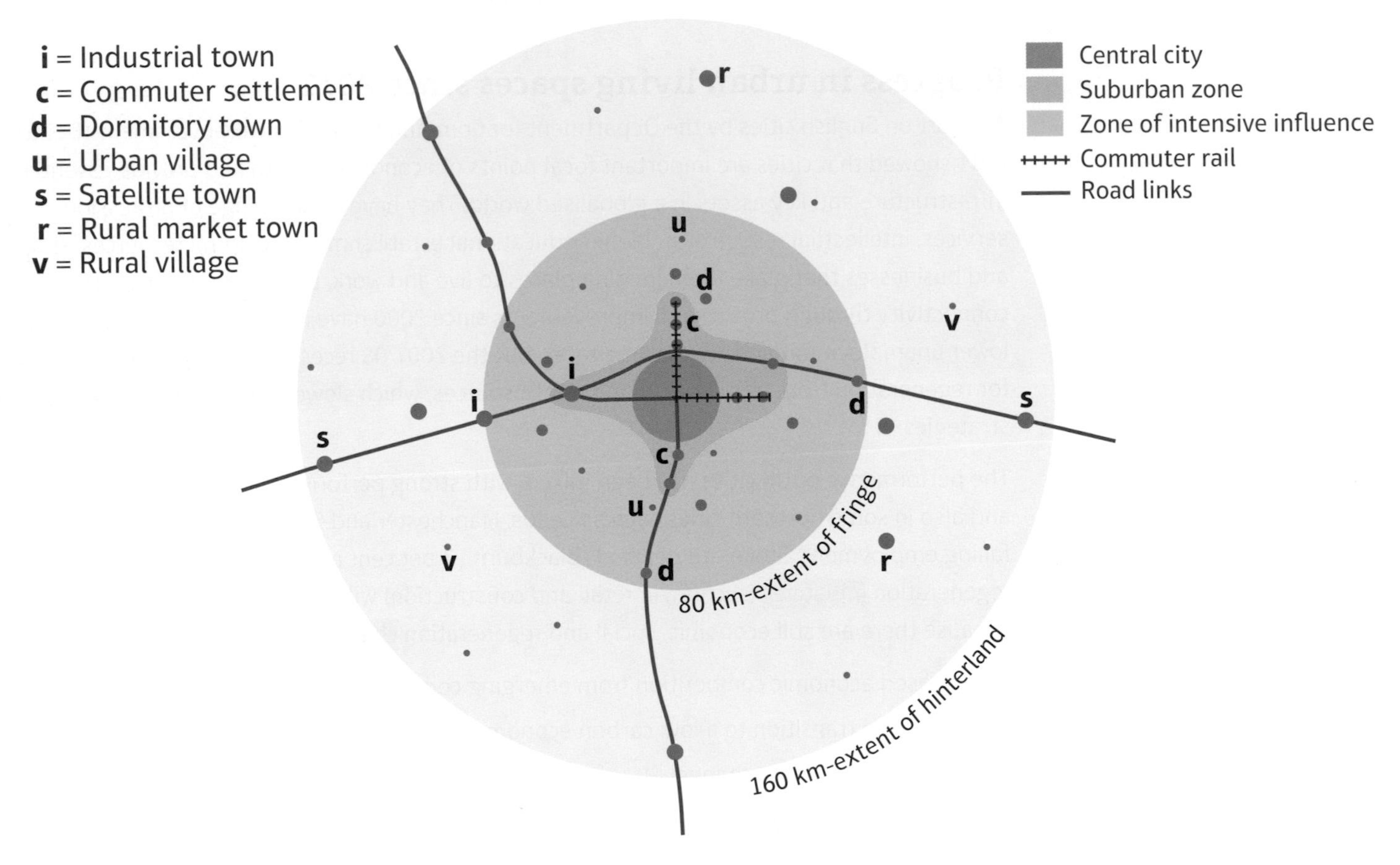

Figure 6.23: The urban core and its hinterland

Yellow = Players, Orange = Attitudes and actions, Purple = Futures and uncertainties

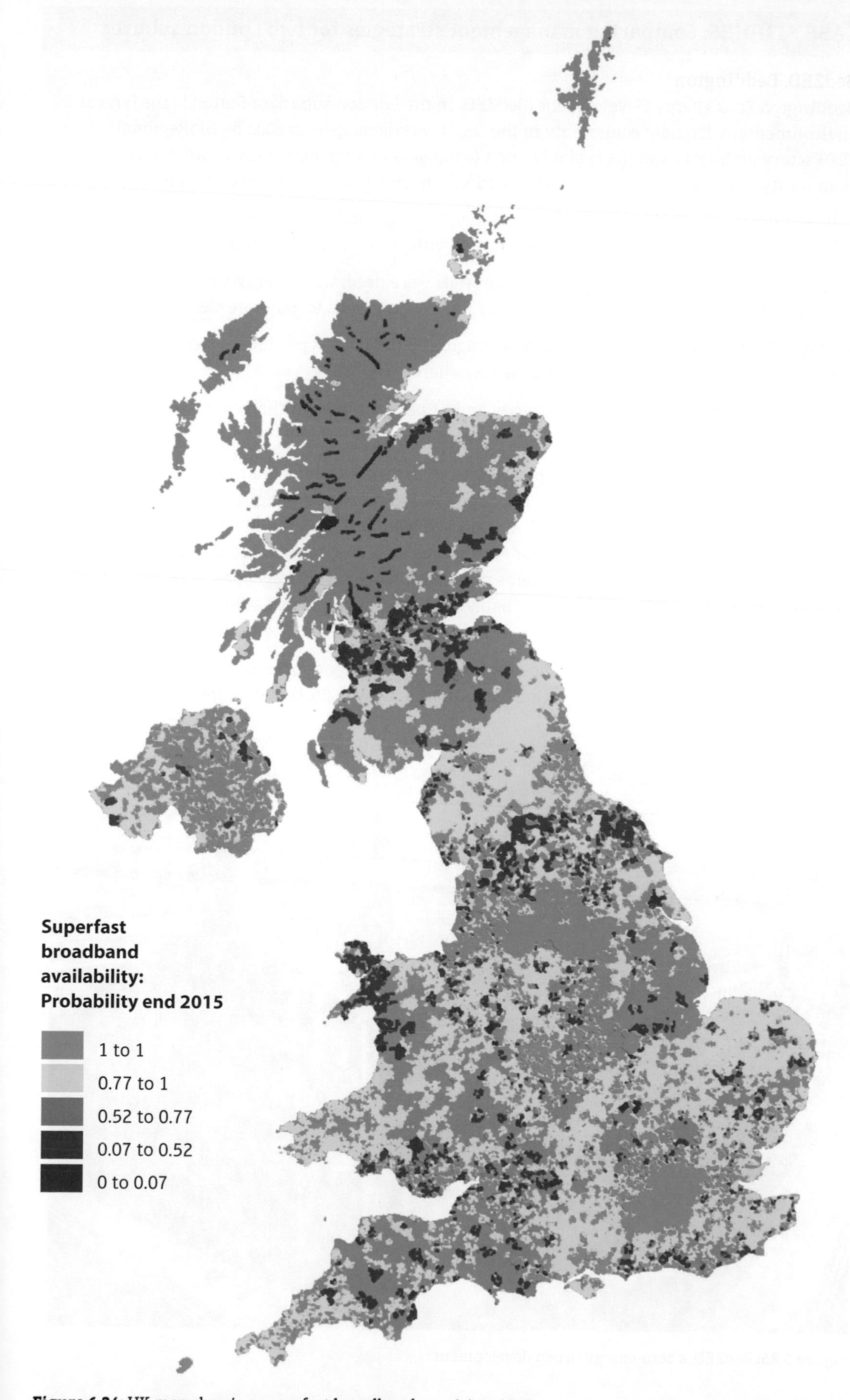

Figure 6.24: UK map showing superfast broadband provision, 2015

ACTIVITY

CARTOGRAPHIC SKILLS

1. Study Figure 6.24.
   a. Describe the pattern of superfast broadband provision in the UK.
   b. How may the lack of this connectivity affect people and businesses in places that do not have it?

ACTIVITY

1. Compare the improvements made in Bedzed and Brixton. Consider economic, social, demographic and natural environmental characteristics in your answer.

2. To what extent are any improvements due to local initiatives (from the community), area initiatives (from local council or organisations), and national initiatives (from national government departments).

CASE STUDIES: Comparing management strategies for two London suburbs

BedZED, Beddington

Beddington Zero Energy Development (BedZED) in the London suburb of Sutton is the largest environmentally friendly community in the UK. It was developed in 2002 by BioRegional, ZEDFactory architects and the Peabody Trust (a housing association) as an affordable eco-community with 100 homes. It used a brownfield site and attempted to be carbon neutral:

- Its buildings were well insulated and designed to retain and recycle heat, with south-facing windows capturing heat and solar exchangers with wind cowls to circulate air (Figure 6.25).
- Mostly recycled or renewable building materials were used, with energy-saving and water-saving devices and water meters, as well as photovoltaic cells to generate electricity.
- Water consumption is half the London average, and the houses produce 37 per cent fewer CO_2 emissions than the same size of housing elsewhere.
- The houses use 45 per cent less electricity and 81 per cent less gas than comparable dwellings; and 60 per cent of waste is recycled.

The development is near public transport hubs and has a car club for sharing. With the emphasis on affordability, a quarter of the homes are social housing, a quarter shared-ownership and the remainder freehold. A combined heat and power unit was planned but did not work, so a gas-fired communal boiler system was installed instead. BedZED residents on average know 20 of their neighbours well – more than twice the average for the local area – and there is a strong sense of community, probably because residents have a similar life philosophy (i.e. eco-friendly) and 84 per cent felt that community facilities were better than elsewhere.

BioRegional say that government policies are needed to influence others to go 'green'; however, Sutton Council does plan to create a 'One Planet Borough' by 2025, based on the BedZED experience.

Figure 6.25: BedZED, a zero-energy urban development

Yellow = Players, Orange = Attitudes and actions, Purple = Futures and uncertainties

Angell Town, Brixton

In October 2015 the London *Evening Standard* launched a campaign ('The Estate We're In') focused on regenerating the Angell Town estate in Brixton, an inner London suburb. This housing estate was built in the 1970s but the housing deteriorated very quickly, and even in the twenty-first century its community of 4000 people was struggling to overcome socio-economic problems within a large ethnic population. Crime data for the area, while showing considerable variation from year to year, listed an average of about 350 crimes per month. This matched the pattern reported in a 2015 crime report for London, which showed teenagers involved with gangs and offences with knives.

Some regeneration has helped, for example closing off underpasses, converting garages into shops, building social and private housing, fitting CCTV, improving policing (including 'hot spot' policing) and organising neighbourhood meetings. The local people were seen as the key resource to bring about improvements in the community.

Locals wanted to make even more changes for the better but they needed help. The Angell Town pilot project was an alliance of community residents and donors, attempting to bring positive change to this disadvantaged housing estate that had experienced fatal stabbings and shootings. Funds of £150,000 were initially raised to transform the estate for all age groups, males and females, but with an emphasis on the young. Residents identified their needs and made suggestions and applied for small grants (£1000 and £5000). Charities and community groups could apply for more, between £2000 and £20,000.

The Angell Town pilot used donations of £100,000 from the Citi banking group, £100,000 from Lambeth Council and £50,000 from the London *Evening Standard's* Dispossessed Fund. Schemes included a new all-weather football pitch with training and mentoring by Football Beyond Borders and Lambeth Tigers (£30,296); boxing training and job readiness delivered by Dwaynamics (£17,341); Block Workout gym and fitness (£15,000); business start-up training by Tree Shepherd (£12,940); My London offering trips to learn more about the city (£5000); and a food and craft market by It's Your Local Market (£5000).

Residents, some of whom have lost relatives to violence, continue to be viewed as the biggest asset, and they lead all the schemes. Police also operate a Community Payback scheme, under which offenders complete high-visibility community service.

The funds were managed by the London Community Foundation, using Angell Town as a pilot to discover whether the strategy could be expanded to other London estates. The expansion to other estates had funding of £1.2 million (£400,000 from Citi banking group, £400,000 from the government, £200,000 from the *Evening Standard's* Dispossessed Fund, £100,000 from Mount Anvil, £100,000 from Linklaters and £50,000 from Citygrove).

Extension

Investigate the Prince of Wales' Foundation for Building Community, which also aims to listen to local people and use their knowledge of their living environment to bring about positive changes.

ACTIVITY

1. Use the police.uk website to investigate neighbourhood crime data over the time period of your studies.

- **a.** Monitor changes in the number and types of crime found in Angell Town, Brixton, using the postcode SW9 7JP.
- **b.** Does the data suggest that the initiatives started in 2015 are making any difference?

2. Compare the crime data for Angell Town with your own neighbourhood or settlement.

ACTIVITY

Study Figure 6.26. To what extent may house prices reflect the success of managing problems in UK regions and the places within them?

AS level exam-style question

For one strategy that you have studied for improving UK urban places, explain the reasons for its success or failure. (6 marks)

Guidance

Make sure you clearly identify the strategy. Include evidence of success or failure in your answer, but do not just be descriptive; remember to explore the reasons for success or failure (or both). Don't put more than one strategy in your answer – you will only get marks for one!

A level exam-style question

For one strategy that you have studied for improving UK rural areas, explain its success or failure. (6 marks)

Guidance

Make sure you clearly identify the strategy. Include evidence of success or failure in your answer, refer to a range of factors and assess the degree of success across the range. Don't put more than one strategy in your answer – you will only get marks for one!

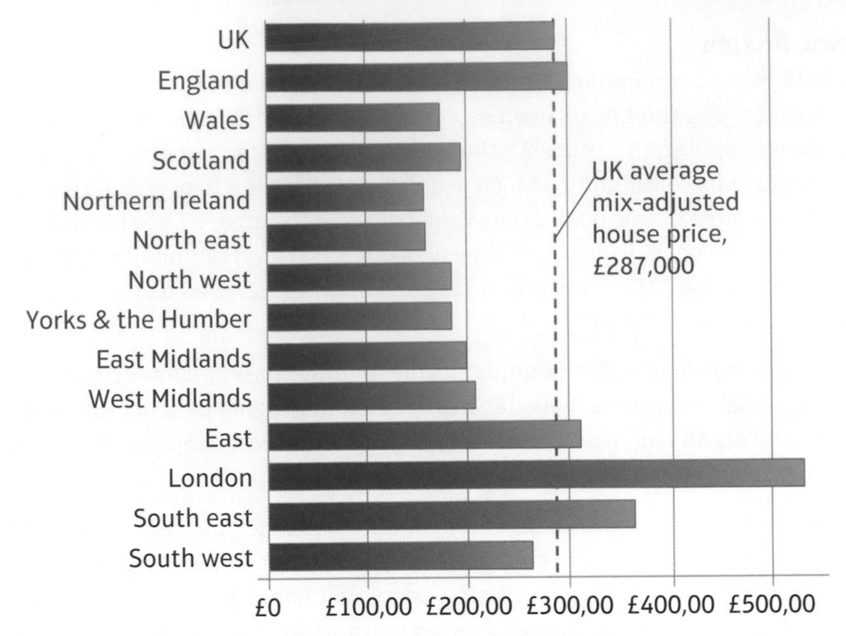

Figure 6.26: Average mix-adjusted UK house prices by region, October 2015

CASE STUDY: Building trust between different groups

Ethnic segregation and tensions have been identified as a big issue for UK urban places. It is known that if people from different backgrounds (such as different ethnicities and ages) interact, there is less friction. Distrust can develop as a result of **social exclusion** – when groups of people live in isolated or segregated communities.

Some communities have successfully promoted greater integration. For example, in Plaistow in the London borough of Newham, local community leaders have promoted interaction:

- The Barking Road Community Centre, once known as the Newham African-Caribbean Resource Centre, was renamed after suggestions from Newham Borough Council to make it more inclusive.
- Community facilities such as youth clubs in Plaistow are places where different groups can mix.

Research shows that young people – and about half of Plaistow residents are under 30 years old – are likely to integrate better than other age groups, partly because they have better access to jobs and opportunities elsewhere in London. This access, according to a 2012 survey for the Migration Advisory Committee, should help decrease ethnic tensions, since deprivation is the main cause of tension in places of diversity. It can also be concluded that economic strategies creating jobs, wealth and opportunities, as well as social strategies reducing segregation and providing affordable housing, should also reduce ethnic tension and division.

Managing change in rural communities

In 2012 the national government established 14 Rural and Farming Networks in England (e.g. Farming Food and Rural Network East) to identify and give opinions on local issues and concerns to the Department for Food and Rural Affairs (DEFRA). The networks brought together people from rural communities, rural businesses, and food and farming industries, in a 'bottom-up' approach.

Yellow = Players, Orange = Attitudes and actions, Purple = Futures and uncertainties

Progress in rural living spaces

A bottom-up approach has long been at the core of the LEADER European Union initiative, which started in the 1990s to help rural development through partnerships. The latest LEADER phase (2014–20) supports 21 schemes in Scotland, for example, and in England £165 million has been set aside to help rural communities with economic growth: £100m for rural business, £20 million for superfast broadband extension to remote areas (see Figure 6.24), £25 million to promote rural tourism, and £20 million for community-based renewable energy schemes.

In 2015 the UK government announced 15 new rural enterprise zones (REZs) to encourage private investment. It aimed to do this by streamlining planning permission, providing superfast broadband, giving discounts on business taxes and reinvesting taxes in the local area. In north Norfolk, two REZ sites have been established:

- Coltishall (a 26-hectare site) will focus on digital and low-carbon businesses
- Wells-next-the-Sea (a 7.4-hectare site) will focus on offshore wind energy.

ACTIVITY

Consider the information about strategies available in north Norfolk, and the characteristics and issues found in the area (identified earlier in this chapter). Evaluate the likely success of these strategies from the perspectives of national government, local government and local people.

LEPs in rural areas

Some Local Enterprise Partnerships cover rural regions. The New Anglia LEP aims to create 95,000 new jobs, 10,000 new businesses and 117,000 more homes by 2026. This is a part of the larger Rural Development Programme for England (RDPE), which aims to improve agriculture, the environment and rural life, by:

- increasing farm productivity
- supporting micro- and small businesses
- boosting rural tourism
- providing rural services
- providing cultural and heritage activities (linked to tourism)
- increasing forestry productivity.

CASE STUDY: The Eden Project, Cornwall

One approach to improving rural living spaces is the development of large-scale projects such as the Eden Project, an educational charity that opened officially near St Austell in March 2001. It was funded by Lottery money (£56 million), EU regional development funding and South West Regional Development Agency funding (together equalling £50 million) and £20 million in loans, as well as support from other local and national organisations and businesses. A former china clay quarry was converted into two huge enclosures consisting of adjoining domes housing thousands of plant species and replicating natural biomes. The project now attracts over a million visitors a year.

The Eden Project has brought several benefits:

- It provides many jobs for local people in an area of high unemployment and little economic diversity. The project employs 650 people directly.
- It has greatly benefited local hotels, restaurants and shops that gain more customers. The biggest increase in customers is in travel and accommodation (2011 survey), creating extra jobs in catering and accommodation. St Austell, the closest town, received the most positive effects, followed by Penzance and Plymouth (Figure 6.28).

- It has been estimated that the Eden Project has contributed over £1 billion to the local economy.
- It tries to buy from local suppliers (£7 million spent a year) and so creates 'knock-on' benefits for the area; for example, it is calculated that a further 200 local jobs were secured in addition to the people it employs directly.
- It has diversified into music events (Eden Sessions), a youth hostel within its grounds (opened in 2014), and it offers apprenticeships and degree courses.
- Future plans include developing geothermal energy, a learning village and a hotel.

However, as with many large projects, there are some disadvantages:

- Traffic in the area has increased as a result of the large number of visitors, causing congestion on narrow rural lanes.
- Some local firms find recruiting staff more difficult because of competition with the Eden Project for workers.
- The St Austell area has benefited a lot, but areas less close to the project have lost investment and tourists.
- Some other local firms and attractions have experienced a fall in their visitor numbers and turnover as result of the attraction of the Eden Project.
- One project on its own is not enough to bring more rural people out of poverty and there is still significant deprivation in the area. Four out of 20 neighbourhoods in the St Austell and Mevagissey district were in the lowest 20 per cent nationally in the 2015 IMD.

Figure 6.27: The Eden Project, a visitor attraction in Cornwall, built in a disused quarry

ACTIVITY

1. Would a large single strategy like the Eden Project help your nearest rural area? Explain your answer in as much detail as possible.

2. Investigate the Campaign to Protect Rural England (CPRE) website to discover the latest concerns and strategies in rural places.

a. Are these concerns changing, decreasing or increasing?

b. Are the strategies helping to improve rural places in England?

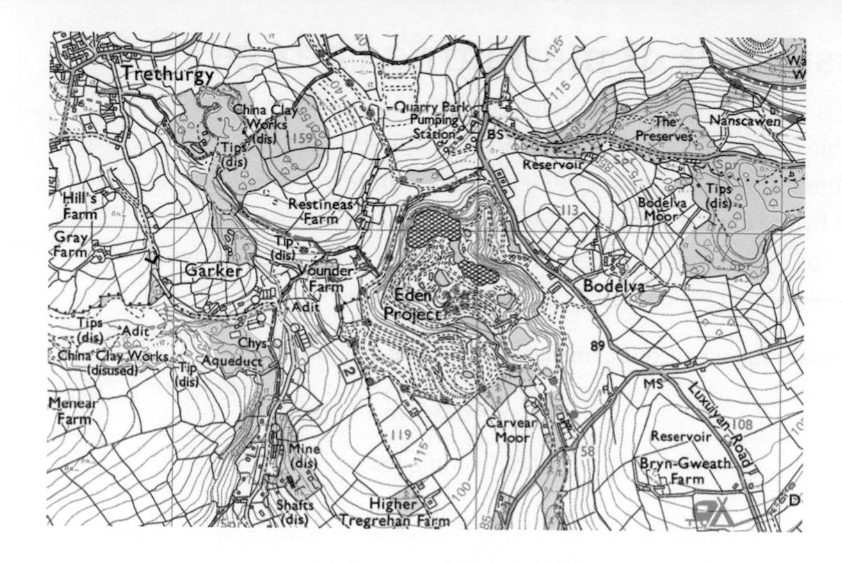

Figure 6.28: Ordnance Survey map extract showing the Eden Project and surrounding rural area (1:25,000 scale) © Crown copyright 2016 OS 100030901

ACTIVITY

CARTOGRAPHIC SKILLS

Study the Ordnance Survey map extract showing the Eden Project in Cornwall (Figure 6.28). Using map evidence, assess the positive and negative impacts of the Eden Project on this rural place – for example the expansion of tourism-related functions, change to landscapes and road systems.

Strategies in Scotland

In Scotland, a Less Favoured Area Support Scheme (LFASS) operates up to 2018, when it is due to be replaced by another EU-funded scheme called Areas Facing Natural Constraints (AFNC). Scotland receives £65.5 million per year to help farmers and crofters operate their businesses, to reduce the amount of abandoned land, to maintain the farming landscape in the countryside and to support sustainable farming systems. In 2015 there were also grant schemes available, such as:

- the Crofting Agricultural Grant Scheme, which provided money for capital projects and for managing common grazing lands (up to £25,000 per individual or £125,000 for a crofting group)
- Croft House Grant Scheme, which was designed to attract and retain people within crofting communities by allocating money for repair, improvement or rebuilding of houses
- a broadband scheme to support communities wishing to create and improve broadband connectivity.

Literacy tip

Make sure that you understand a range of financial terms, such as investment, budget, credit, funding, grant, loan, tax incentives, reinvesting, discounts and recession. Most strategies for improving UK places involve money in some way.

Assessing the success of managing change

It is difficult to measure the success of strategies to solve cultural and demographic issues in UK urban or rural places. Table 6.7 suggests some indicators, but different groups of people (stakeholders) will have different priorities and so the weighting of these indicators will vary, even in the same place. As shown by the case studies in this chapter, rural and urban places in the UK are diverse, and while there may be some similar concerns (such as, crime in urban areas, few young adults in rural areas), each stakeholder group desires different improvements.

Table 6.7: Some indicators for measuring the success of managing UK urban and rural areas

Population growth	Level of private investment
Increase in job opportunities	Declining crime levels
Increase in higher-paid jobs	More cultural events
Increase in educational attainment	Better Happiness indices
Lower deprivation (IMD)	Better Quality of Life indices
Increased retail sales	Better appearance of built environment
Increase in number of services	More public investment
Increase in property values	Increased community involvement
Increase in affordable housing supply	Improved connectivity (transport and internet)
	Declining pollution levels

Another reason why it may be difficult to assess the success of strategies is that databases are not always available or comparable. For example, Scotland completed an IMD in 2012, while England and Wales completed theirs in 2015. And while international immigrant numbers are known for entry to the UK, it is not known exactly where these people move to in the UK. In addition, there are several ways of measuring 'happiness' and quality of life. The Office for National Statistics (ONS) recognises that there is not enough information to make complete judgements, and it is also difficult to *quantify* local involvement success or the influence of planning restrictions. Different stakeholders will use different criteria: national and regional governments will be interested in economic growth and job creation, while local communities may be more concerned about perceived ethnic tension and crime rates.

ACTIVITY

1. Study the strategies available to the Highlands of Scotland as outlined in the text. Considering the characteristics and features of the Highlands region (covered earlier in this chapter), evaluate the relevance and likely success of these strategies.

2. Investigate the website of Highlands and Islands Enterprise (HIE) to discover the latest schemes and strategies. Assess how useful you find this website in determining whether or not the strategies and schemes have been successful.

Extension

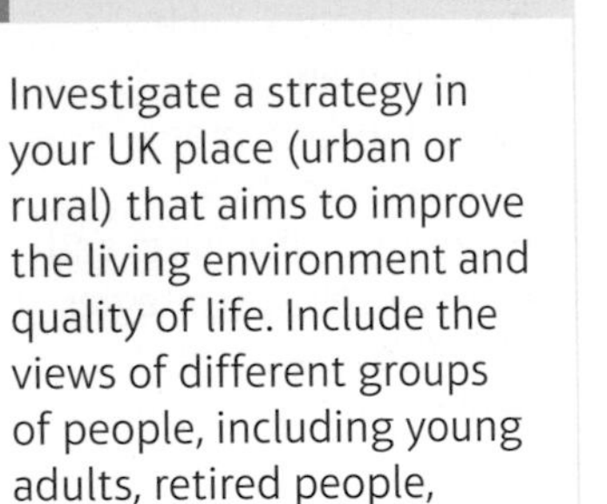

Investigate a strategy in your UK place (urban or rural) that aims to improve the living environment and quality of life. Include the views of different groups of people, including young adults, retired people, local government, local businesses and ethnic minority groups. Assess whether or not this strategy has been successful.

Extension

Investigate a range of different measures used to judge the success of diverse UK places (see Table 6.7). To what extent do the media use these measures to create an image of different places?

Yellow = Players, Orange = Attitudes and actions, Purple = Futures and uncertainties

Fieldwork exemplar: Diverse places

Enquiry question

What is the extent of deprivation in the city of Norwich?

Fieldwork hypotheses

Null hypothesis (H_0): The pattern of urban deprivation within the city of Norwich is random.
Alternative hypothesis (H_1): Urban deprivation decreases with distance from the CBD to the outer suburbs of Norwich.

Locating the study

The city of Norwich (population over 400,000) is East Anglia's regional centre. It has a large agricultural hinterland and a strong tertiary employment sector. Norwich is centrally located in Norfolk at the confluence of the Rivers Yare and Wensum. It is the focal point of a radial road network, including the A11 and A47. As with other UK cities, it has experienced many urban processes during its development, including deindustrialisation, suburbanisation and regeneration. It is large enough to have all urban zones, as predicted by urban land-use models, all services and functions, and a diverse population living in the residential zones.

Methodology

An east-to-west stratified transect 16 km (10 miles) long through the CBD was selected to cover all urban zones twice; along this route, data was collected at 26 systematic points. In this way, biased data collection was avoided and a comprehensive range of primary data, including qualitative externality surveys, and secondary data, including quantitative Index of Multiple Deprivation (IMD 2010), was available for all urban zones. The externality surveys examined a wide range of deprivation indicators, such as conditions of buildings, accessibility, proximity to services, and air and noise pollution, which were individually scored by a group from direct observation. Standardised tables were used to ensure that all observers recorded in the same way.

The mean group score was then calculated for each transect point to negate anomalous judgements. To assist judgement, digital noise and air pollution meters were used to provide accurate read-outs from three random readings. The IMD was available from Norwich City Council Planning Department via the internet. This is professionally collected data, which ensures accuracy. The largest risks for observers were from traffic and unfriendly approaches from strangers. We managed these by staying together in small groups all the time, taking great care and paying attention when crossing all roads, and using proper crossing points where available, with close supervision by a teacher. Clothing appropriate to the wet weather conditions was also worn to reduce the risk of hypothermia.

Data presentation

The quantitative data could be analysed statistically through standard deviation and Spearman's correlation analysis. The latter allowed a correlation to be established between distance from CBD and the deprivation assessment. Visual presentation included a linear choropleth map superimposed on a street map of the city (Figure 6.29); the coloured zones clearly showed the pattern of deprivation across the city and the map enabled a comparison with neighbourhoods and Super Output Areas.

Analysis and conclusion

The correlation analysis showed that there was no correlation (+0.03) between deprivation and distance from the CBD, instead suggesting that there is a random pattern of deprivation in Norwich. Rather than a gradual change, the data presentation clearly shows fluctuations from east to west across the city. The range in scores was 30.8 with a mean of +3.15.

1. The eastern suburbs (Thorpe St Andrew) had low deprivation due to modern housing, good road links, access to services such as supermarket and sports centre, and employment opportunities in small businesses. Pollution was also low compared with other points.

2. In the western suburbs (Bowthorpe), the group observations suggested low deprivation, with access to many green open spaces, modern housing, a factory estate and local shops. However, the IMD indicated higher deprivation here (IMD rank 7952) than in some inner urban zones, in terms of socio-economic variables such as employment.

3. The central areas had the largest contrast, with Mancroft ward the most deprived area (IMD rank 5441) in Norwich, but within this area is the major shopping district and tertiary employment with well-maintained modern and historic buildings. However, 27 per cent of Mancroft's population is classified as income-deprived and it has the highest crime rates along the transect studied. Clearly the poor people in this area are unable to access the well-paid jobs of the CBD.

The data disproves the hypothesis and the primary and secondary data are sometimes contradictory. This shows that there are a multitude of variables affecting deprivation in Norwich, such as regeneration not benefiting poorer people, which led to the random pattern overall.

Evaluation

All urban zones in Norwich were covered and systematic points allowed statistical techniques to be used to analyse primary data. This enabled comparison with the professionally collected Index of Multiple Deprivation. However, large areas to the north and south of the city were not surveyed and the 26 points were observed at different times of the day, which may have affected qualitative judgements due to the flows within an urban system. A major difficulty was that the primary data was for points and the secondary data was for areas, so data means had to be used during comparisons. The results may be useful to urban planners and neighbourhood groups in identifying where to make improvements. Further studies would involve studying the whole city of Norwich, or extending the survey to contrasting UK cities.

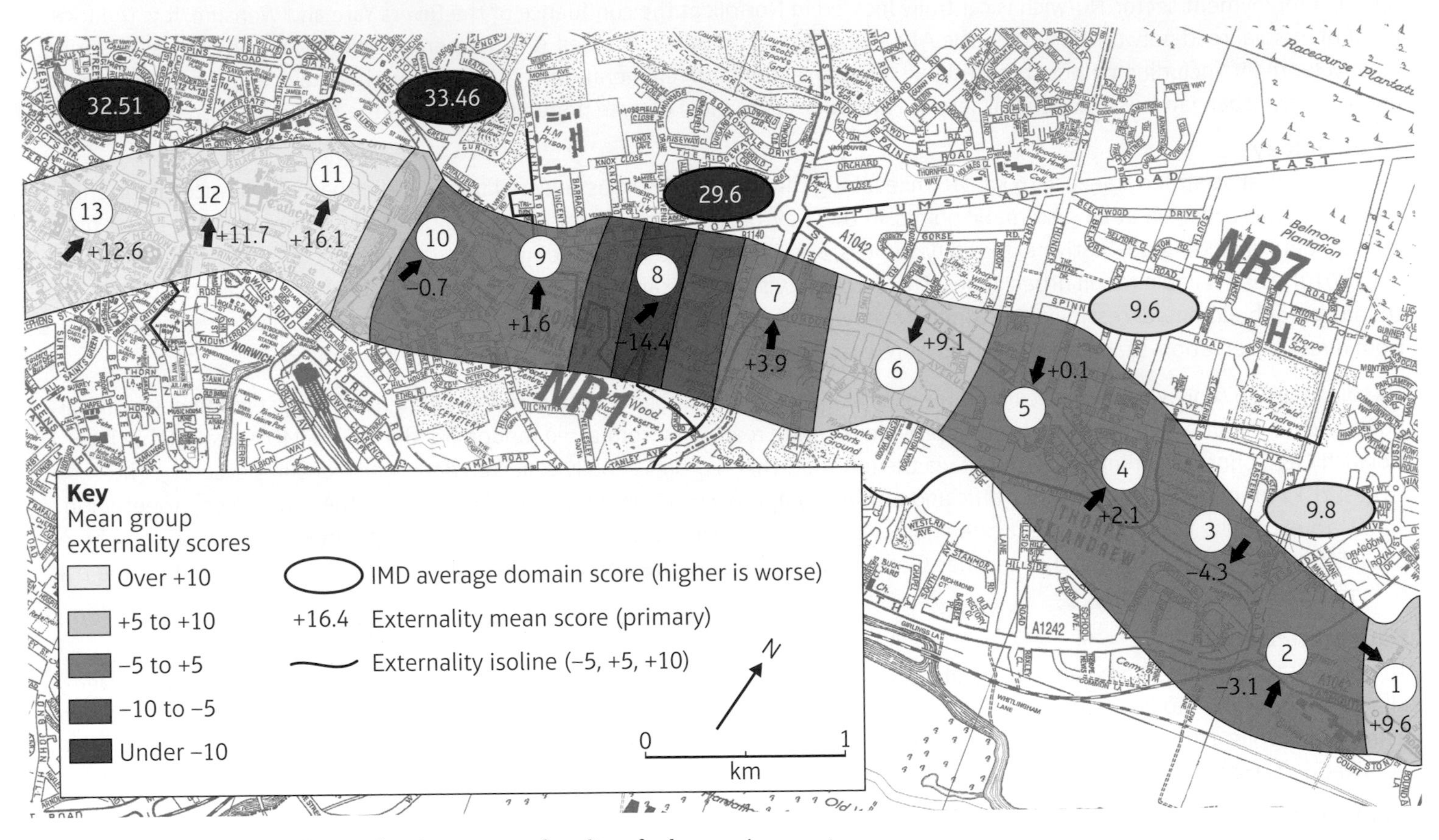

Figure 6.29: Linear choropleth map showing pattern of quality of urban environment

Summary: Knowledge check

By reading this chapter and completing the tasks and activities, as well as your wider reading, you should have learned the following, and be able to demonstrate your knowledge and understanding of diverse places (Topic 6B).

- **a.** Why have UK places changed at different rates and developed different population characteristics?
- **b.** What is the rural–urban continuum?
- **c.** Why are there cultural and economic differences between and within places?
- **d.** What is the role of natural change and migration (internal and international) in establishing patterns in places?
- **e.** What are the characteristics of, and influences on, your local place?
- **f.** Which processes have influenced places over time?
- **g.** Why do some places become more desirable, and others less desirable, as living spaces?
- **h.** How do different groups of people perceive their living spaces?
- **i.** How do different perceptions develop and how can they be measured?
- **j.** Why are places becoming more culturally diverse?
- **k.** How are tensions caused within communities?
- **l.** What different patterns exist according to factors such as crime, happiness, deprivation and housing costs?
- **m.** How can the management of issues in places be judged accurately?
- **n.** Who are the different stakeholders in the decision-making process for places?
- **o.** Has the management of rural and urban places been successful?

As well as checking your knowledge and understanding through these questions, you also need to be able to make links and explain the following ideas:

- There is a considerable variation in cultural and demographic patterns throughout the UK.
- The main processes are clear, but there are significant differences at the local level, and structural parts of urban and rural areas still have contrasting patterns.
- There is a lack of reliable databases to enable analysis of a place at different time periods, or to enable comparison of different places.
- Qualitative data based on opinions is difficult to quantify, and may change quickly over time.
- Government strategies have changed over time, partly because of changing governments and political philosophies, but also due to global influences.
- A north-south divide may still exist in the UK, and in England, but some evidence suggests that it may be diminishing. This may be due to gradual economic and demographic restructuring rather than the result of any direct policy or strategy.
- A wide range of players (stakeholders) is involved in decision-making that will affect rural and urban places, from EU government to individual residents.
- National government policy and laws may have indirect effects on places, such as immigration policy and laws, or the support of specific regions or initiatives.

Preparing for your AS level exams

Sample answer with comments

With reference to UK rural areas that you have studied, assess the reasons why perceptions of these areas vary. (12 marks)

People may see places in the UK in different ways depending on their background. For example, people of different ages may want different things – the elderly may wish for peace and quiet, while the young want to party and enjoy lots of activities. Ethnic groups may want jobs so that they can improve their lives while White British people may want areas with plenty of services.

The candidate identifies some groups of people to consider, but has not yet made it clear that they have noticed that the question does say specifically 'rural' areas. In such questions it is important not to write about the wrong area (i.e. urban in this case).

In places like the Highlands of Scotland there is an ageing population and crofting is a difficult form of farming and the crofters perceive their life as hard and are grateful for financial assistance from the government. Young adults living in the area tend to move out as there are few job opportunities and services are often far away, so they perceive the area as undesirable. Tourists and second-home owners may view the area as a peaceful, quiet, unpolluted environment and so are keen to visit, to get away from their busy lives in cities. Young adults like to live in cities such as Liverpool, especially the central area, because they have access to services and entertainment when they want them, but older people prefer to be in the suburbs because there are lower crime rates and less deprivation.

The candidate has a suitable range of groups and a suitable area in the Highlands of Scotland, but the situation is more complex than stated. For example, many young adults in Caithness and Sutherland would like to work and live in the Highlands. The mention of Liverpool gains no marks because it is an urban and not a rural area.

In East Anglia there are contrasting settlements; commuters, especially those with children, have chosen to live in suburbanised villages because they offer a peaceful environment and bigger houses but are still within reach of their place of work and primary school. In North Norfolk there are retirement villages in quite remote locations and people here are worried about losing their homes to cliff erosion, but are proud of their community spirit. So, perceptions vary according to various reasons such as age, family circumstances, and life cycle stage, as these determine the aims of people.

The answer continues to show understanding and offers a competent discussion with sound reasons. However, there is a lack of detail (e.g. no named villages in East Anglia) and perhaps Boston could be mentioned to broaden the perceptions to include immigrant workers.

Exam tip

It would be quite easy to muddle questions on the Regenerating places topic with those on the Diverse places topic because there is an overlap in themes. However, there are subtle differences in emphasis, which means that you must stick to the topic that you have studied.

Verdict

This is an average answer to the question (Level 2 – 5 to 8 marks). It demonstrates mostly relevant knowledge and understanding and makes some relevant connections. It applies knowledge and understanding to produce a partial answer, and overall the answer is fairly balanced and coherent but could have been improved by including more specific examples and more detailed explanations.

Preparing for your A level exams

Sample answer with comments

With reference to examples, evaluate the different methods available to judge the success of UK rural strategies in bringing sustainable change. (20 marks)

Measuring the success of strategies to make improvements to rural areas is not easy because there may not be any data from before the strategy to allow a comparison with afterwards, some of the improvements may not be quantifiable, and some successes may only become clear after a very long period of time. One rural strategy in the UK is the Eden Project in Cornwall, which is based on tourism but has had a multiplier effect on the surrounding area, and there are other schemes such as those in National Parks and Rural Enterprise Zones.

It is clear that the candidate recognises the complexity and difficulty of making judgements, and they correctly identify rural strategies. However, they appear to have not recognised sustainability in the question, and the meaning of this could have been explored from the social, economic and natural environment perspectives. Also, perhaps it is not wise to start with a negative about data and information, although this is a type of evaluation; it would be better mentioned in a conclusion.

The Eden Project used a derelict brownfield site and so improved its appearance immediately. It also created 6,500 jobs and so is a significant employer within the St Austell area of Cornwall, which helps to keep young adults within the rural communities instead of moving away to cities such as Exeter or Bristol. It has also helped local businesses by buying products from them, like foods for the café, spending £7 m a year and increasing the number of jobs supported. However, several local districts were still very deprived (lowest 20% nationally) according to the 2015 IMD. There is a Rural Development Programme for England (RDPE), which is trying to improve agriculture, the environment and rural life, by increasing farm and forestry productivity and supporting micro and small businesses. It is too soon to make a judgement on whether this has been successful or not. In Scotland there is a Less Favoured Area Support Scheme (LFASS) and another EU funded scheme called Areas Facing Natural Constraints (AFNC). Scotland got £65.5m a year to help farmers or crofters by maintaining the farming landscape and supporting sustainable farming. In 2015 there were grant schemes available, such as Crofting Agricultural Grant Scheme, which provided money for big projects, there was also a Croft House Grant Scheme to keep people within crofting communities by allocating money for repair, improvement or rebuilding of houses.

This paragraph lists a suitable range of rural strategies with some facts (one inaccurate: 650 jobs for Eden Project, not 6,500). There are sustainable themes but they appear to be more accidental than deliberate. The main problem is that the question is not really being answered.

Rural areas need many things: more services such as post offices, affordable houses, younger people, and more jobs rather than just those in farming to be sustainable in the 21st century. These can be measured in a variety of ways such as through use of the Census (e.g. 2011) or the Index of Multiple Deprivation (e.g. 2015), or through questionnaires with rural people. In this way quantitative and qualitative data can be combined to show the overall success of rural strategies.

Here the candidate mentions some general ways of measuring change from which to judge the success of strategies but does not give detail; also there is little evaluation, i.e. which is 'best' and from which point of view. However, mentioning that this data needs to be combined is an appropriate point to make.

Exam tip

For the longest essays, even in an examination, it helps to plan. People do this in different ways, but a small outline plan, based on the question set, before starting to write usually ensures that the question is answered.

Verdict

This is an average answer to the question. (Level 3 – 11 to 15 marks). It contains mostly relevant knowledge and understanding, a few connections, a partial but coherent interpretation of the question, and an incomplete conclusion. The answer would have been improved if closer attention had been paid to what the question is asking: 'Evaluate the methods of judging success…' and giving more detail of the specific methods and the ways they are evaluated (i.e. from the economic, social and environmental points of view).

Appendix: Answers to statistical exercises and other calculations

Chapter 1: Tectonics

Activity on page 28

Table showing selected data for the Tohoku tsunami wave in the Pacific Ocean

Location	Distance from epicentre (km)	Arrival time (minutes after earthquake)	Velocity of tsunami wave (km/hr)	Tsunami height at the location (m)
Midway Island	4100	302	814.57	1.27
Kahului, Hawaii	6345	486	783.33	1.74
Vanuatu	6345	596	638.76	0.69
Hilo, Hawaii	6535	503	779.52	1.41
Crescent City, California	7835	671	700.60	2.02
Port San Luis, California	8500	668	763.47	1.88

For example: 302 minutes divided by 60 = 5.03333 hours; 4100 km divided by 5.03333 hours = 814.57 km/hr

Activity on page 48

Hypotheses for statistics: There are two types of hypotheses – Null (H_0) and Alternative (H_1). The traditional approach is to establish a **Null Hypothesis** before a statistical analysis, which is expected, using geographical knowledge and understanding, to be wrong. The **Alternative Hypothesis** is the one expected to be correct. For example, a Null Hypothesis may be that there is no significant correlation between the magnitude of an earthquake and the number of deaths and injuries, whereas an Alternative Hypothesis may be that there is a significant positive correlation between the magnitude of an earthquake and the number of deaths and injuries.

Table showing correlation between magnitude and deaths

Earthquake	Magnitude	Rank	Deaths and injuries	Rank	d	d^2
Chile 1960	9.5	1	7,223	4	3	9
Chile 1960	8.2	19	0	17.5	1.5	2.25
Chile 2010	8.8	5	12,558	3	2	4
Chile 2015	8.3	14.5	28	12	2.5	6.25
Indonesia, Banda Sea 1963	8.3	14.5	0	17.5	3	9
Indonesia, Sumatra 2004	9.1	3	228,900	1	2	4
Indonesia, Sumatra 2005	8.6	7.5	1,653	6	1.5	2.25
Indonesia, Sumatra 2007	8.4	10.5	186	10	0.5	0.25
Indonesia, Sumatra 2012	8.6	7.5	22	13	5.5	30.25
Japan, Hokkaido 1994	8.3	14.5	393	9	5.5	30.25
Japan, Hokkaido 2003	8.3	14.5	755	7	7.5	56.25

Earthquake	Magnitude	Rank	Deaths and injuries	Rank	d	d^2
Japan, Honshu 1968	8.2	19	434	8	11	121
Japan, Honshu 2011	9.0	4	26,113	2	2	4
Peru 2001	8.4	10.5	2,816	5	5.5	30.25
Russia, Kuril Islands 1963	8.5	9	0	17.5	8.5	72.25
Russia, Kuril Islands 2006	8.3	14.5	0	17.5	3	9
Russia, Kurilskiye 1969	8.2	19	0	17.5	1.5	2.25
Russia, Severo Kurilskiye 2013	8.3	14.5	0	17.5	3	9
USA, Alaska 1964	9.2	2	139	11	9	81
USA, Alaska 1965	8.7	6	17	14	8	64
n = 20						$\sum d^2 = 546.5$

Spearman's rank correlation analysis: Formula and calculation:

$$r_s = 1 - \frac{6\sum d^2}{n^3 - n}$$

$$1 - \frac{6 \times 546.5}{8000-20} = 1 - \frac{3279}{7980} = 1 - 0.411 = +0.589$$

(Significant at the 95% confidence level, reject null hypothesis, therefore there is a positive correlation between earthquake magnitude and number of deaths and injuries i.e. as one goes up so does the other).

Chapter 2: Glaciation

Activity on page 98

Student t-test

Table showing clast size data for 2 sites on an outwash plain in Iceland

Column 1 **Site 1** – clast long-axis in mm (x_1)	Column 2 ($x_1 - \bar{x}_1$)	Column 3 $(x_1 - \bar{x}_1)^2$	Column 4 **Site 2** – clast long-axis in mm (x_2)	Column 5 ($x_2 - \bar{x}_2$)	Column 6 $(x_2 - \bar{x}_2)^2$
30	–10	100	12	–2	4
51	11	121	10	–4	16
35	–5	25	13	–1	1
48	8	64	14	0	0
35	–5	25	16	2	4
29	–11	121	18	4	16
33	–7	49	15	1	1
50	10	100	16	2	4
49	9	81	12	-2	4
$\sum x_1 = 360$		$\sum (x_1 - \bar{x}_1)^2 = 686$	$\sum x_2 = 126$		$\sum (x_2 - \bar{x}_2)^2 = 50$
$N_1 = 9$		$s_1 = 8.73$	$N_2 = 9$		$s_2 = 2.36$
$\bar{x}_1 = 40$			$\bar{x}_2 = 14$		

$$t = \frac{\bar{x}_1 - \bar{x}_2}{\sqrt{\frac{s_1^2}{n_1} + \frac{s_2^2}{n_2}}}$$

So Student's t-test result = 8.63

Activity on page 99

Table showing clast shape data for 2 sites on an outwash plain in Iceland

Site 1				Site 2			
a, b, c axes in mm	a + b	2c	Cailleux Index values	a, b, c axes in mm	a + b	2c	Cailleux Index values
30, 14, 2	44	4	11.00	12, 10, 9	22	18	1.22
51, 21, 4	72	8	9.00	10, 10, 8	20	16	1.25
35, 21, 5	56	10	5.60	13, 12, 11	25	22	1.14
48, 27, 7	75	14	5.36	14, 14, 13	28	26	1.08
35, 27, 4	62	8	7.75	16, 15, 13	31	26	1.19
29, 14, 3	43	6	7.17	18, 17, 14	35	28	1.25
33, 9, 4	42	8	5.25	15, 13, 11	28	22	1.27
50, 25, 6	75	12	6.25	16, 15, 12	31	24	1.29
49, 23. 14	72	14	5.14	12, 12, 9	24	18	1.33
			$\sum x_1 = 62.52$				$\sum x_2 = 11.02$
			$n_1 = 9$				$n_2 = 9$
			$\bar{x}_1 = 6.95$				$\bar{x}_2 = 1.23$

Chapter 3: Coasts

Activity on page 132

Student t-test

Table showing pebble size data for two sites on a beach in the UK

Column 1 **Site 1** – pebble long-axis in mm (x_1)	Column 2 ($x_1 - \bar{x}_1$)	Column 3 ($x_1 - \bar{x}_1)^2$	Column 4 **Site 2** – pebble long-axis in mm (x_2)	Column 5 ($x_2 - \bar{x}_2$)	Column 6 ($x_2 - \bar{x}_2)^2$
83	17.67	312.23	28	−4.89	23.91
67	1.67	2.79	39	6.11	37.33
71	5.67	32.15	22	−10.89	118.59
46	−19.33	373.65	27	−5.89	34.69
62	−3.33	11.09	45	12.11	146.65
49	−16.33	266.67	51	18.11	327.97
76	10.67	113.85	31	−1.89	3.57
65	−0.33	0.11	29	−3.89	15.13
69	3.67	13.47	24	−8.89	79.03
$\sum x_1 = 588$		$\sum(x_1 - \bar{x}_1)^2 = 1126.01$	$\sum x_2 = 296$		$\sum(x_2 - \bar{x}_2)^2 = 786.87$
$n_1 = 9$		$s_1 = 11.19$	$n_2 = 9$		$s_2 = 9.35$
$\bar{x}_1 = 65.33$			$\bar{x}_2 = 32.89$		

$$t = \frac{\bar{x}_1 - \bar{x}_2}{\sqrt{\frac{s_1^2}{n_1} + \frac{s_2^2}{n_2}}}$$

$$t = \frac{65.33 - 32.89}{\sqrt{\frac{11.185^2}{9} + \frac{9.35^2}{9}}}$$

$$t = \frac{32.44}{\sqrt{\frac{125.22}{9} + \frac{87.42}{9}}}$$

$$t = \frac{32.44}{\sqrt{23.62}}$$

$t = \frac{32.44}{4.86}$ $t = 6.67$ (note that rounding may give a slightly different value to a computer)

Fieldwork calculation on page 165

Chi-squared (Chi²) test

$$x^2 = \sum \frac{(O - E)^2}{E}$$

O = the frequencies observed

E = the frequencies expected

$\sum$ = the 'sum of'

Chapter 4: Globalisation

Activity on page 203

a) 0.9% b) 46.9% c) plot curve d) 0.8%

e) 50.8% g) 0.66

Chapter 5: Regenerating Places

None

Chapter 6: Diverse Places

Activity on page 278

(a) 99.6 ÷ 8 = 12.45

(b) and (c)

Table of population density data, Highland areas of Scotland and Standard Deviation calculation

District	Population density (people per km²) (x)	$(x-\bar{x})$	$(x-\bar{x})^2$
Nairn	30.6	18.15	329.42
Inverness	26.8	14.35	205.92
Caithness	14.4	1.95	3.80
Ross and Cromarty	10.6	−1.85	3.42
Badenoch and Strathspey	5.8	−6.65	44.22
Skye and Lochalsh	4.8	−7.65	58.52
Lochaber	4.3	−8.15	66.42
Sutherland	2.3	−10.15	103.02
n = 8	$\sum x = 99.6$		$\sum(x-x)^2 = 814.74$

Standard deviation formula and calculation:

$$s = \sqrt{\frac{\sum(x-\bar{x})^2}{n}}$$

(d) $\sqrt{814.74 \div 8} = \sqrt{101.8425} = 10.092$

(e) Even in this peripheral rural region of the UK there is a lot of variation in population density, with areas having some significant towns (e.g. Inverness), while others only have isolated dwellings (e.g. Sutherland).

Activity on page 278

(a) and (b) The suggested categories will be: Over 22.54 (purple), 22.54 to 12.45 (red), 12.45 to 2.36 (orange), under 2.36 (yellow). Thus Inverness will be purple, and Sutherland will be yellow for example.

Table showing 2014-2015 Crime Rate in Liverpool wards and location quotients

Census ward	Crime rate %	Location Quotient	LQ category
Allerton and Hunts Cross	4.28	0.48	under representative
Anfield	10.29	1.15	average
Belle vale	5.74	0.64	under representative
Central	29.2	3.25	highly over
Childwall	3.4	0.38	under representative
Church	4.42	0.49	under representative
Clubmoor	8.59	0.96	average
County	12.77	1.42	average
Cressington	3.47	0.39	under representative
Croxteth	4.95	0.55	under representative
Everton	12.79	1.42	average
Fazakerley	8.08	0.9	average
Greenbank	6.63	0.74	under representative
Kensington and Fairfield	11.26	1.25	average
Kirkdale	10.94	1.22	average
Knotty Ash	7.63	0.85	average
Mossley Hill	2.93	0.33	under representative
Norris Green	6.25	0.7	under representative
Old Swan	8.29	0.92	average
Picton	9.37	1.04	average
Princes park	9.54	1.06	average
Riverside	21.09	2.35	moderately over
Speke-Garston	9.75	1.09	average
St Michael's	6.43	0.72	under representative
Tuebrook and Stoneycroft	10.27	1.14	average
Warbreck	6.9	0.77	average
Wavertree	5.42	0.6	under representative
West Derby	6.17	0.69	under representative
Woolton	4.37	0.49	under representative
Yew Tree	4.42	0.49	under representative
	Liverpool av. = 8.98		

In this example: *% crime rate in a ward* divided by *average % crime rate in Liverpool*. LQ categories are: Under 0.75 = under representative, 0.75 to 1.5 = average, 1.5 to 3 = moderately over represented, and over 3 = highly over represented.

Glossary

Abandoned land: Spaces in urban areas that were previously used for industry, now derelict and awaiting regeneration. Could also be applied to rural areas where farming is no longer taking place.

Ablation: The output of snow and ice from a glacier by melting, calving of icebergs, sublimation or evaporation.

Ablation till: Till deposited by melting ice from stationary or retreating glaciers forming landforms, such as terminal and recessional moraines.

Abrasion: Material plucked from the bedrock is frozen into the glacier ice, and as the glacier moves downslope this material will rub against the valley sides and floor, wearing them away by a process similar to sandpapering. Also applies to the action of waves and rivers.

Accessibility: How easy it is to travel to or communicate with a place. Distance and transport routes to other places are usually the most important factors.

Accumulation: The input of snow and ice to a glacier by precipitation, avalanches and wind deposition.

Active landscape: A landscape currently experiencing glaciation, with active glacial processes and landform development.

Active layer: The upper part of the ground surface in an area of permafrost, which regularly thaws during the summer months. Unlike the permafrost below, it is highly mobile, due to frequent freeze-thaw cycles and meltwater saturation, caused by the impermeable nature of the permafrost underneath.

Administrative centres: Places, usually larger county towns or cities, that make decisions about how to organise infrastructure and economic activity in the surrounding area, e.g. Aylesbury, Buckinghamshire.

Amenity value: The benefit to people of a resource that they can use to improve their lives. An example is a beach, which has a high amenity value as it can be used for recreation, or a natural area such as a forest.

Arête: A narrow, knife-edged rocky ridge between two adjacent corries.

Ash: The very fine particles of rock ejected during a volcanic eruption. These articles form part of the tephra, which is a term for all sizes of ejected volcanic material.

Asymmetric economic shock: An economic event that has a disproportionate impact (one large and one small) on economies, due to the contrasting structures of those economies.

Avalanche: A sudden downslope movement of snow and ice (can be of different sizes and dry, wet or slab).

Backshore: The cliff or sand dunes, plus the upper beach closest to the land.

Basal melting: Melting at the base of a glacier produces meltwater which lubricates and increases the rate of movement. Melting may be caused by pressure melting, geothermal or frictional heat.

Basal slip: Occurs when the base of the glacier is at the pressure melting point, which means that meltwater is present and acts as a lubricant, enabling the glacier to slide more rapidly over the bedrock.

Beach morphology: The surface shape of a beach, with features such as storm ridges, berms, beach cusps, runnels. A pebble beach will have more prominent features than a sandy beach, due to the steeper angle of rest of the particles (greater friction between them).

Benioff zone (Wadati-Benioff zone): The linear zone where a descending oceanic plate is in contact with a continental plate as it is subducted. This is a zone of earthquake activity created by the friction between the two plates, which creates stresses and strains which are suddenly released at earthquake foci down to a depth of about 700 km.

Built environment: An urban area with buildings and infrastructure, such as roads and railways.

Capital: Productive assets/goods or a financial stake in productive activity.

Capital markets: A system of intensive financial investment (e.g. into banks, stock exchange, bonds) often by foreign companies or individuals.

Carbon cycle: The process by which carbon moves from the atmosphere into the Earth (geology) and its biosphere, and then back again.

Cirque/corrie/cwm: An amphitheatre or armchair-shaped large hollow high on a mountainside with a steep back wall and sides and a rock lip (or threshold).

Cirque glacier: Snow and ice accumulate in hollows on a mountainside and the ice becomes thicker to form a circular mass, which eventually flows out over a threshold to become a valley glacier.

Climate change: Long-term significant change in the average climate that a given region of the world experiences. It may be natural, as shown by Milankovitch cycles, or human-enhanced, as shown by the emission of greenhouse gases over the last 200 years. The term is often used to refer to the current warming of the planet.

Coastal recession: The retreat of a coastline due to coastal erosion or sea-level rise or submergence.

Collision plate boundary: Where two plates move towards each causing a very slow collision which is likely to cause folding and faulting of crustal rocks and the uplift of continental crust to form fold mountains (a process known as an orogeny). Also see **convergent plate boundary**.

Commercial: Places, usually towns or cities, where the major economic activity is trade, shopping or financial services, e.g. City of London.

Community adaptation: People within communities, either whole or parts of settlements, work together to change their way of life so that the impact of a tectonic hazard event is not as hazardous.

Community preparedness: People within communities, either whole or parts of settlements, work together to get ready for a tectonic hazard by having stronger, aseismic homes and workplaces, emergency procedures and well-trained rescue and aid services.

Commuter village: A rural settlement close to a large city that has become home for many people working in the urban area. They journey to and from the city on a daily basis (i.e. commute), usually by road or train.

Composite indicator: A development indicator which measures more than one variable, such as the Human Development Index.

Concordant coast: A coastline where the geology is arranged parallel to the shore (this may be in terms of the resistance of the rock strata (layers) or the folding of the rock strata).

Connectedness: A measure of how connected different people are through different communication links (fibre-broadband, road, rail).

Conservative plate movement: Where two plates meet and move alongside each other in a similar direction or opposite direction, usually at different speeds. Friction between the two plates is great and stresses and strains build up to create shallow earthquake foci. Also see **transform fault plate boundary**.

Constructive plate boundary: Where two plates move in opposite directions, leaving a zone of faulting and a gap into which magma from the asthenosphere rises. Also see **divergent plate boundary**.

Consumer society: A society that places great value on the purchase of goods and services.

Consumption: The purchase of goods and services by consumers.

Continuous permafrost: Found at the highest latitudes where virtually all the ground is frozen as there is very little, if any, surface melting. It may extend to depths of several hundred metres.

Convergent plate boundary: Where two plates move towards each other and at the boundary the denser oceanic plate (basaltic) is subducted beneath the less dense continental plate (granitic), creating surface features such as a trench, and deep features such as the Benioff zone. Also see **destructive plate boundary**.

Core regions: Regions with globally significant amounts of economic activity compared to other regions.

Cost-benefit analysis: A comparison of the costs of doing something, compared with the benefits gained from doing it. Commonly used in coastal management to decide on coastal defences; for example, by comparing the financial costs of building a 3-km-long sea wall with the financial value of the land uses along that 3 km that would be protected.

Crag and tail: A very large mass of hard rock with a steep stoss and a gently sloping tail of deposited material.

Crime rate: The amount or level of crime in a place. Sometimes this is given as the number of crimes over a set period, or the number of crimes per number of people. There are several sub-categories of crime, and rates may be given separately for these.

Cryosphere: The cold environments of our planet where water is in its solid form as ice; this includes sea ice, lake ice, river ice, snow cover, glaciers, ice caps, ice sheets and permafrost.

Cultural diffusion: The spread of cultural ideas and way of life between individuals and cultures.

Cultural diversity: The number or range of different population groups with different lifestyles in a place, usually linked to ethnicity but also to socio-economic groups or age groups.

Cultural enrichment: The addition of ideas, actions and meanings that are the result of the arrival of new people to an area, e.g. EU migration to Boston, Lincolnshire.

Cultural erosion: The loss or dilution of a specific culture due to cultural diffusion. A change in ideas and traditions.

Culture: The way of life, especially the general customs and beliefs, of a particular group of people at a particular time.

Dalmatian coast: Named after the coastal area on the eastern side of the Adriatic Sea, where the geology is parallel to the shore, giving parallel long islands and long inlets, probably due to submergence.

Decolonisation: The process by countries which were colonies of empires gain their independence.

Deindustrialisation: The mass closure of industries in regions traditionally associated with secondary industrial production, also features high unemployment levels. Partly due to a global shift in production from developed to developing countries.

Demographic changes: Changes in the population characteristics of a place. This could be numbers of people or types of people (i.e. cultural change).

Demographic characteristics: Data about a group of people, such as their age, gender or income.

Demographics: The study of population and population change.

Depopulation: A significant and sustained decline in the population size of a region or country.

Deprivation: When people lack the things they would expect to have in the 21st century, such as jobs, a certain level of income, affordable housing, access to services such as schools and health care.

Deregulation: The process of reducing or removing rules governing economic activity in a country, with the aim of encouraging investment.

Derelict land: Previously used land in cities that has fallen into disrepair because buildings have closed and no one is there to maintain them. Also see **abandoned land**.

Destructive plate boundary: Where two plates of contrasting densities meet, the denser oceanic crust is forced under the less dense oceanic crust into the soft asthenosphere, where it slowly melts to become part of the upper mantle material, or return to the crust as features like batholiths or volcanoes. Also see **convergent plate boundary**.

Development: Usually considered as economic growth leading to an improvement in the standard of living. Development can be measured in other ways, such as freedom, equality or the well-being of the natural environment (i.e. sustainable development).

Diaspora: The movement of a population away from their homeland.

Disaster: An event, such as an earthquake, that disrupts normal conditions to a point where a place or community cannot adjust and there is significant loss of life and injuries, and/or financial costs.

Disaster trends: What has been happening to the number of severe tectonic hazards over time. Care must be taken when comparing historic trends with recent trends, due to improvements in monitoring and reporting.

Discontinuous permafrost: The permanently frozen ground is relatively shallow and fragmented by patches of unfrozen ground or talik. The surface layer of the ground melts during the summer months.

Discordant coast: A coastline where the geology is arranged perpendicular (at right-angles) to the shore, this may be in terms of the resistance of the rock strata (layers) or the folding of the rock strata.

Divergent plate boundary: Where two plates move apart from each other so that there is a gap in the crust. This may be under an ocean, such as the Mid-Atlantic Ridge, or on land, such as the East Africa Rift Valley. Molten material from the asthenosphere (upper mantle) can rise up and add new rock to the trailing edges of the plates.

Diversification (rural diversification): In times of low farm income farmers seek alternative incomes, such as bed & breakfast, wind turbines, off-roading trails, and conversion of farm buildings to offices or workshops. It may also link to the political decision by the EU to encourage farmers to find alternative business uses for their land other than growing surplus food.

Diversity: The degree of variation within a population e.g. age, ethnicity, culture.

Drumlin: Smooth elongated mounds of till, with a long-axis parallel to the direction of ice movement and with a steep stoss and gentle lee.

Dynamic equilibrium: Where a natural system tries to achieve a balance by making constant changes in response to influences within the system. For example, within a sediment cell the coast is constantly rebalancing itself through the processes linked to sources, transfers and sinks.

Earthquake (seismic) waves: The shockwaves created by the release of tension at a focal point. There are several types of wave movement which travel through or around the Earth. The strongest effects will be at the epicentre which is immediately above the focus.

Economic sector: A way to group jobs that have a similar purpose, e.g. primary sector jobs involve extraction of the Earth's resources (mining/fishing/farming). Other sectors are secondary, tertiary and quaternary.

Economies: The amount of money being transferred between different players leading to wealth creation.

Elite migrants: Migrants able to move between countries due to their wealth, fame or valuable skills.

Emergent coast: A coastline that is advancing relative to the sea level of the time, this may be due to large amounts of deposition or due to isostatic rebound, for example.

Engagement (levels of engagement and political engagement): The decision by individual members of the public to get involved in tackling political issues, e.g. through volunteering, running for election.

Englacial: Within the ice e.g. glacier.

Entrainment: Small rock fragments entrained by basal ice freezing around them so that they are dragged along.

Environmental Impact Assessment: As part of the planning stage, all large projects must carry out a study of what the likely impacts will be of the project. A focus is often on the natural environment to ensure that it remains sustainable, but can also include the impacts on people (social and economic environments).

Environmental quality: The condition and attractiveness of the surroundings within which people live. This includes both the natural and the built environment – open space and levels of air, water and land pollution.

Environmental refugees: People who have to leave their homes or face a very high risk of being killed by a natural event. There are an increasing number of people who live by the coast who may have to move very soon due to higher sea levels resulting from climate change. Officially a refugee is someone who crosses an international border, which can cause difficulties for those who are displaced within their own countries when they lose their homes and land due to sea level rise or storm surges with tropical cyclones but cannot get international help.

Erratic: A rock that has been transported and deposited by ice for some distance from its source. It has a different lithology to the bedrock on which it is deposited.

Esker: Long, narrow, sinuous (winding or meandering) ridges of fluvioglacial sand and gravel that have been deposited from an ice tunnel.

Ethical consumption: The purchase of goods and services determined by an ethical or moral criteria, such as eco-friendly, free trade, or welfare of animals and workers.

Ethnic composition: Information about the ethnic characteristics of a group of people,

Ethnicity: The cultural background of a group of people, often based on religion or country of origin.

Eustatic: Change in sea level, which can be up or down: up when ice melts after an ice age, or up due to thermal expansion due to the present global warming; down during an ice age when more water is stored as ice.

Export processing zones: A geographic area where goods may be landed, handled, manufactured or reconfigured, and re-exported without the payment of customs duties or taxes.

Fertility rate: The number of children born to a woman during her lifetime.

Fluvioglacial: Refers to erosion or deposition caused by meltwater from glacier ice.

Focal depth: The depth at which an earthquake starts (focus). It is divided into shallow, intermediate and deep. Shallow earthquakes have the greatest impacts, as the seismic waves have not lost as much of their energy by the time they reach the surface.

Foreign direct investment (FDI): A controlling interest or investment in a business by a foreign entity. Investment within a country originating from outside that country.

Foreign investment: When a company from overseas spends money on a factory or shop (e.g. opening one) in the UK, e.g. Microsoft in Reading.

Foreshore: The lower part of the beach which is covered twice a day at high tide; the part of a beach that receives most regular wave action.

Freeze-thaw weathering: When water freezes in the cracks and joints of rock, it expands by up to 10 per cent of its volume, weakening the rock and causing disintegration through repeated freeze-thaw cycles.

Frost heave: The upward dislocation of the soil and rocks by the freezing and expansion of soil water. As the ground freezes, large stones become chilled more rapidly than the soil. Water below such stones freezes and expands, pushing up the stones and forming small domes on the ground surface, sometimes creating regular patterns.

Gated communities: Neighbourhoods of towns or cities where houses are designed with gates and fences to improve privacy or safety.

Gentrification: Renewal, renovation or rebuilding of older and deteriorating buildings in order to create more up-market places for middle-class residents to live, often displacing poorer residents.

Geological structure: The arrangement of rocks in layers, or folds, and the joints and bedding planes within them.

Glacial: A cooler interval of time during the thousands of years of an ice age, when glacier ice advances.

Glacial crushing: The direct fracturing of weak bedrock because of the weight of ice above it.

Glacial erosion: The removal of loose weathered material by glacier ice, including plucking, abrasion, crushing and basal meltwater.

Glacial trough: A U-shaped valley with steep sides and a wide, flat floor carved from a former river valley by a glacier.

Glacier mass balance: The difference between accumulation of snow and ice and the ablation of snow and ice.

Glacier outburst flood: A large and abrupt release of water from a subglacial or proglacial lake. They are known as *jökulhlaup* in Iceland.

Global hub: A location that is well connected within many networks to other key locations in the world.

Global shift: The relocation of industries from OECD countries to NICs, predominately from North America to Latin America and South and South East Asia.

Globalisation: The growing economic interdependence of countries worldwide through increasing volume and variety of cross-border transactions in goods and services, freer international capital flows, and more rapid and widespread diffusion of technology. Also shown through culture, lifestyles, and global processes such as climate change and natural disaster impacts.

Glocalisation: Adapting global goods or services to meet the needs of local people. Alternatively, using local suppliers.

Governance: How a place or area is managed by different levels of government. The policies regarding hazards can affect how prepared and resilient a place or area is.

Grading: Describes sediments that are layered according to clast size, with coarse sediments at the base, grading upward into progressively finer ones.

Greenhouse Earth: A time during which the overall temperature of the Earth is higher than average due to higher levels of greenhouse gases in the atmosphere. This may be caused by increased levels of volcanic activity or changes in the carbon dioxide and methane balance between the oceans and the atmosphere. Also linked to Milankovitch cycles.

Haff coast: A lowland coast where the long bars and lagoons are parallel to the shore.

Hanging valley: A tributary U-shaped valley high above the main glacial trough floor, often with a waterfall.

Hazard: Something that has a negative impact on people in any way. Also see **natural hazard**.

Hazard management cycle: A sequence of actions and decisions related to a place or area being prepared for or responding to the threat of a natural hazard.

Hazard profile: An analysis of different types of hazard, or actual hazard events, based on a range of criteria. This allows a useful comparison to be made.

Hazard resistant design: Buildings of any type that have been made to withstand and reduce the impact of a hazard event. For example, in an earthquake most people are killed when roofs fall on them, so preventing this from happening saves many lives. These designs do not have to be expensive or involve high-technology.

High-energy environment: Part of a coast where the wave action is mostly in the form of large destructive waves. The potential for erosion is much greater, although the rate of change may depend on the rock type. These coasts tend be rocky.

Holocene: The current part (epoch) of the Quaternary geological time period, which started approximately 10,000 years ago and extends up to the present day. It is an interglacial and is characterised by the development of human civilisations, and increasing anthropogenic influences on Earth's systems.

Hot spot: An intra-plate location, oceanic or continental, where magma from the mantle has broken through a weak point in the crust. There appear to be two types: one associated with individual upwelling from near the core-mantle boundary, and the second from the top of two large mantle plumes.

Hydro-meteorological hazards: Hazards associated with flowing water (e.g. river floods) or the weather (e.g. lightning storms or tropical cyclones). These may combine with tectonic hazards, for example in the creation of lahars.

Ice cap: A dome-shaped ice mass (less than 50,000 km^2), which covers mountain peaks and plateau areas. Such as in Iceland.

Ice field: A large mass of ice on sea (at least 10km diameter) or on land where several glaciers have connected together.

Ice margin: Environments at the edge of the glacial ice where a combination of glacial and fluvioglacial processes occur.

Ice sheet: The largest masses of ice on land, which extend over areas greater than 50,000 km^2. The two major ice sheets on Earth today are those covering much of Greenland and Antarctica.

Ice wedges: When dry areas of the active layer refreeze, the ground contracts and cracks. Ice wedges will form when meltwater enters the crack during the summer and freezes at the start of winter. Repeated thawing and refreezing of the ice widens and deepens the crack, enlarging the ice wedge

Iceberg calving: When chunks of ice break off the front of an ice sheet, ice shelf or glacier snout and fall into the sea. They form large floating masses called icebergs.

Icehouse Earth: A global ice age occurs when ice sheets cover a larger proportion of land due to the overall temperature of the Earth being lower than average. During this time, the climate fluctuates between very cold glacials, when ice advances, and warmer interglacials, when ice retreats. Linked to Milankovitch cycles and the low levels of greenhouse gases in the atmosphere.

Ice-sheet scouring: Glacial erosion over extensive areas beneath ice sheets. This process creates landforms such as roche moutonnée, knock and lochan, and crag and tail.

Idyll: Used to describe a place that maximises the positives of a living space. It is often used in relation to rural areas with low pollution levels and plenty of green open space, unspoilt natural areas, traditional (old-fashioned) ways of life and jobs.

Imbrication: Refers to sediment in which the clasts have a preferred orientation and dip caused by a strong current.

Impermeable: A rock that does not allow rainwater to pass through it because there are only a limited number of cracks or pores within it.

Industrialisation: The rapid growth of secondary industries such as factories, including the creation of secondary sector jobs. In the UK this was mostly in the 19th century and is known as the industrial revolution, but there was also industrial growth in the 1960s.

Inequality: Differences in income and wealth, and well-being, between individuals, groups within a community or communities within a society.

Infrastructure investment: When money is spent by the government on projects to connect major towns and cities, e.g. High Speed 2.

Innovation: A new idea, more effective invention or process within an industry, service or community.

Intensity: The amount of damage created by a tectonic event or other hazard. For earthquakes, the Mercalli Scale is used to express the level of damage.

Interglacial: A warmer interval of time during the long time span of an ice age, when glacier ice retreats.

Internal deformation: Occurs when the weight of glacier ice and gravity causes the ice crystals to deform, so that the glacier moves downslope very slowly. The angle of slope will also determine the amount of deformation.

Internal migration: The movement of people within a country, where the move is a change of permanent residence for at least one year.

International migration: The movement of people from one country to another, where the move is a change in permanent residence for at least one year.

Intra-plate: Tectonic activity that is found away from the plate boundaries and closer to the middle of a tectonic plate. Locations are less frequent and require separate explanations, such as the influence of mantle plumes or ancient fault lines.

Inward investment: The addition of money into a local economy, perhaps by a foreign company.

Isostatic: The change in land level, which can be up or down. During an ice age the weight of ice pushes the crust down into the soft upper mantle where it is thickest; at the same time where the ice is thin or doesn't exist, the land may be tilted upwards. After an ice age the crust that was pushed down rebounds slowly back upwards, while at the same time causing other areas to sink.

Jokulhlaups: A tectonic hazard common in Iceland, where volcanic activity underneath ice caps creates large volumes of meltwater which suddenly burst out as floodwaters under the ice, when the water pressure reaches a certain level.

Kame: Undulating mounds of fluvioglacial sand and gravel deposited on the valley floor at the glacier snout.

Kame terrace: Flat, linear deposits of fluvioglacial sand and gravel deposited along the valley sides.

Kettle hole: Circular depressions, often forming lakes in outwash plains, when ice lobes are separated from an ice front, covered with debris and a hollow forms when the ice melts.

Knock and lochan: A lowland area with alternating small rock hills (knock) and hollows often containing small lakes (lochan).

Lahar: A mixture of meltwater from snow and ice on top of an active volcano and tephra (volcanic material such as ash) from eruptions that travels very quickly down existing river valleys, reaching some distance away from the volcano. Heavy rain, perhaps created by cloud seeding by ash eruptions, will also create the same hazard.

Land use zoning: Regulations that restrict human activities and construction in certain zones where there is a hazard risk. For example, in regions where there is a tsunami risk, making sure that all residential property and valuable services are on higher land above the likely level of a tsunami.

Landslide: A mass movement of rock and soil down a steep slope under the influence of gravity, perhaps triggered by an earthquake loosening material.

Lava flow: Molten magma that reaches the Earth's surface is known as lava. It will flow down the sides of a volcano until it cools and solidifies. Basaltic lavas flow faster than andesitic lavas, for example, because of different viscosities.

Liberalisation: To reduce and remove rules governing economic activity.

Life cycle stage: Each person experiences various stages during their lifetime, such as a child living with parents, an independent single young adult, a young married couple, a married couple with children, an older married couple, an old person on their own. At each stage the opinions and needs of the person change.

Life expectancy: The average number of years an individual is likely to live from birth.

Liquefaction: A hazard common in lowland areas during an earthquake. Groundwater and loose soil and sediments are shaken so that the ground loses cohesion and acts like a fluid, therefore any buildings or infrastructure lose their solid base and subside and crack.

Lithology: The rock types and their general characteristics.

Little Ice Age: A time of cooler Holocene temperatures between 1300 and 1870, when glaciers advanced. The causes are still debated, but there is evidence to suggest that volcanic emissions and variations in solar output may have contributed. It caused hardship for people in temperate areas such as Europe.

Littoral zone: The coastal zone in which sediments are moved around between the land, beach and sea.

Lived experience: During a person's lifetime they will have different experiences, depending on their family situation, family culture, educational experience, life cycle, living spaces and personal interests. These affect their judgements about places and situations and lead to their perceptions, views and opinions.

Loch Lomond Stadial: The last glacial advance in the UK, which occurred 12,000 to 10,000 years ago, marking the end of the Pleistocene.

Lodgement till: Till deposited by actively moving ice, forming landforms such as drumlins. It is lodged or pressed into the valley floor beneath the ice.

Loess: Accumulations of windblown (aeolian) deposits.

Low-energy environment: Part of a coast where the wave action may be absent due to being sheltered. The potential for deposition is very high, either by the sea or river and sea currents. These coasts tend to be lowland areas such as estuaries or sand dunes.

Macro-scale: Large-scale landforms, e.g. glacial troughs, cirques, pyramidal peaks.

Magnitude: The amount of energy released by a tectonic event. For earthquakes this is best measured on the Moment Magnitude Scale, and for volcanoes the Volcanic Explosivity Index.

Mantle convection: The movement of mantle material in cells when heated by radiation from the Earth's core. It is thought that there may be several layers of convection.

Mantle plumes: There are two huge mantle plumes (Pacific and African); these are hotter areas of the mantle that move upwards underneath the crust and push it up. They can cause weak points in the crust which can become hot spots. These mantle plumes are irregular in shape. Attempts are being made to create mathematical models of them.

Mass movement: The downslope movement of material such as soil, rock, mud or snow, under the influence of gravity.

Media: The various ways of presenting information to people, including printed material such as journals and magazines; broadcasts through radio and television; electronic communication such as online newspapers and blogs.

Megacities: Cities with a population of 10 million or more.

Mega-disaster: When a major hazardous event becomes catastrophic and more than a disaster. For example, the scale of the impacts are unusually great or very severe with huge numbers of deaths, loss of buildings and infrastructure, or long-lasting impacts on normal social and economic systems.

Meltwater channel: A V-shaped channel cut into bedrock and deposits under, along and in front of an ice margin,

Mercalli Scale: An earthquake intensity scale based on 12 levels of damage to areas.

Meso-scale: Medium-scale landforms, e.g. roche moutonnée, ribbon lakes and drumlins.

Micro-scale: Small-scale landforms up to 1m long, e.g. striations.

Migration: The movement of people from one place to another for at least one year with the intentions of settling permanently in the new location.

Milankovitch cycles: Three cyclical changes in the orbit, axis and tilt of the Earth, which changes the amount of solar radiation received by the Earth causing the glacial-interglacial cycles, the longest of which is 100,000 years.

Modify loss: Reduce the impact of losses experienced from a tectonic hazard, for example by insuring belongings and property.

Modify the event: Alter the natural hazard itself in order to change its likely impacts. Earthquakes cannot be changed, but some volcanic activity can be modified, such as by diverting lava flows.

Modify vulnerability: Vulnerability is a key factor in determining the impact of a hazard, so making people less vulnerable will reduce the scale of a disaster.

Moment Magnitude Scale: The most accurate earthquake magnitude scale, it measures the total energy released by an earthquake.

Moraine: The collective term for depositional landforms composed of till.

Mortality rate: The number of deaths per thousand people due to a specific cause of death or age group, for example the cancer mortality rate or the child mortality (deaths under the age of five years).

Multicultural: The existence, acceptance, or promotion of multiple cultural traditions within a single geographic area, such as a country or city.

Multiple-hazard zone: A country or region that experiences several natural hazards, such as earthquakes, volcanoes, river floods, landslides and tropical cyclones. Some analyses may include man-made hazards as well.

Natural hazard: A physical geographical event, tectonic, hydrological or meteorological, which has a negative impact of people through causing injury or death, loss of property, or disruption to the normal way of life.

Nearshore: The section of the littoral zone between the low-tide level and the deeper offshore water.

Negative feedback: Within a natural system a process may cause a change that is reduced or dampened in the future, for example a rockfall due to wave erosion may protect the base of a cliff and reduce the amount of erosion.

Nivation: A combination of processes weathering and eroding the ground beneath a snow patch. They include freeze-thaw weathering, solifluction and meltwater erosion.

Offshore: The section of the littoral zone that consists of deeper water in which waves maintain their shape and speed, furthest from the land.

Offshoring: Relocation of a business process from one country to another.

Outsourcing: The contracting-out of a business process to another company.

Palaeomagnetism: When magmas and lavas solidify, the iron minerals in the rock align with the Earth's magnetic field, permanently recording the direction. When oceanic crust rocks were studied it was found that rocks had recorded magnetic fields in opposite directions.

Patterned ground: The collective term for stone and ice-wedge polygons and stone stripes caused by frost heave and other periglacial processes.

Perception: The view of a place or issue based on feelings and experience; a qualitative judgement.

Periglacial: Environments near to glaciers but not experiencing moving ice. They are dominated by frozen ground and seasonal freeze-thaw processes.

Permafrost: Permanently frozen ground where the subsoil temperatures are below 0°C for at least two consecutive years

Permeable: A rock that has cracks, joints and bedding planes within it so that rainwater can pass through.

Pingo: Where water is able to seep into the upper layers of the ground in large quantities and then freeze. The expansion of an ice lens causes the overlying sediments to heave upwards into a dome, or pingo, which may rise as high as 50m. During a warmer period the ice melts and the dome collapses, leaving a hollow.

Planning: The decision-making process of a national, regional or local body (government, county council, local authority or planning department), which decides where to locate things like new houses and where to regenerate, redevelop or renew.

Plant succession: The change to a plant community because the growing conditions (moisture, humus, shelter, soil stability) have changed enough to allow new plants to establish themselves and take over. In ecosystems such as salt marshes and sand dunes, succession takes place with distance from the sea and finishes with a climax community (usually woodland).

Pleistocene: The first part (epoch) of the Quaternary geological time period, which started approximately 2.6 million years ago and ended 10,000 years ago when the Holocene epoch began. It is often known as the Ice Age, as it was characterised by over 50 glacial-interglacial cycles.

Plucking: Occurs where rocks are weakened or well-jointed and where there is meltwater present at the base of the glacier. The meltwater penetrates into joints and around obstacles and then freezes onto the rock. As the ice moves, it exerts an immense pulling force onto the attached rock, which is plucked from its position. A process also known as quarrying.

Polar glacier: The glaciers in cold polar glacial environments, such as Greenland and Antarctica. The glaciers are frozen onto the bedrock beneath and melting only occurs at the surface during the short summer season. They move more slowly than temperate glaciers.

Policy decisions: Significant decisions made by a government in the national interest, perhaps when local interests are conflicting, e.g. fracking in Lancashire.

Political engagement: The willingness and ability of people to get involved in voting for governments (local and national) or joining pressure groups that are trying to bring about change.

Population density: The number of people living within an area, usually measured per square kilometre. Core and urban areas have higher population densities, while peripheral and rural areas have lower population densities.

Population growth: The increase in the numbers of people living within a defined area or place, usually given as a percentage per year.

Population structure: The numbers of people within each age group and by gender. It is usually shown in a bar graph known as a population pyramid, with age groups in five-year bands (0–4, 5–9, etc.) with males on the left and females on the right.

Positive feedback: Within a natural system a process may cause a change that is increased or exaggerated in the future, for example during the formation of a spit, the deposition shelters areas behind it, encouraging further deposition.

Post-production: Developing a place after decline, often through a marketing strategy to re-image the place so that people see it differently and more positively, e.g. bohemian creative communities in cities.

Pressure and Release model: A simple model of cause and effect, summarising the pressures created by natural hazards and indicating how these pressures could be reduced.

Privatisation: Transferring ownership of a business, agency, service or property from the public sector (government-owned) to the private sector.

Productivity: The quantity of goods or services produced per worker per hour.

Proglacial: Environments located at the front of a glacier, ice cap or ice sheet and dominated by fluvioglacial processes.

Proglacial lake: A lake formed in front of an ice margin, often trapped between ice and higher land.

Protectionism: The economic policy of controlling trade between countries by means of tariffs on imported goods, restrictive quotas, and regulations that disadvantage foreign companies and protect domestic ones.

Pyramidal peak: A pointed mountain peak with at least three sides and corries.

Pyroclastic flows: A very fast-moving cloud of superheated gases and ash travelling at ground level.

Quality of life index: An attempt to objectively quantify the life-satisfaction of people living in a particular place.

Rapid onset: A hazard that happens very quickly, with no or little warning, such as an earthquake.

Rebranding: Creating a new look or reputation for an area, often relying on an area's industrial past or literary fame (e.g. Black County/Beatrix Potter Country); altering the feel and attitude people have towards it.

Recycling: The conversion of waste materials into reusable resources.

Regelation creep: A large bedrock obstacle (>1m wide) causes an increase in pressure, which causes the ice to plastically deform around the feature (creep). Smaller obstacles (<1m wide) will cause pressure-melting, increasing ice movement by basal slip. The ice refreezes on the down glacier (lee) side of the obstacle. The process of melting under pressure and refreezing is known as regelation.

Regeneration: The process of improving a rural or urban place by making positive changes. These include knocking down derelict buildings and building new ones (redevelopment), improving the existing buildings and area (renewal) or changing the image of a place through redesign and publicity. See **rebranding** and **re-imaging.**

Regional disparity: The economic, and perhaps cultural, gap between different parts of a country, with a wealthy core region and a poorer peripheral region.

Re-imaging: Part of a regeneration strategy by changing the image or name of a place and therefore how people view it, e.g. renaming Saigon as Ho Chi Minh City.

Relict landscape: A landscape that is not currently experiencing glacial activity, but features fossilised glacial landforms due to past glaciations.

Remittance: A transfer of money by a foreign worker back to their home country.

Resilience: The ability of a community to resist the impacts of a hazard by adapting and recovering.

Response curve: A graph designed to show the time taken for a place to recover from a natural hazard and make changes to be more prepared for the next event.

Ribbon lake: A long, narrow lake along the floor of a glacial trough.

Risk: The potential that a community or place will face losses due to a hazard, such as a volcanic eruption.

Roche moutonnée: A mass of bare rock on the valley floor with a smooth stoss (up-valley side) and a steep, jagged lee (down-valley side).

Rural–urban continuum: The whole range of area and place types, from the remotest peripheral rural area to the central business district of a large city. It includes accessible rural areas, the rural–urban fringe and suburbs.

Sandur: A flat expanse of fluvioglacial debris in front of a glacier snout, ice sheet or ice caps.

Sea floor spreading: The movement of the oceanic crust away from a constructive plate boundary, as recorded by the magnetic stripes in the basaltic rock (palaeomagnetism).

Sediment cell: Also sometimes known as a littoral cell, is part of a coast that is linked together by all of the processes, such as cliff erosion (sources), longshore drift (transfer), and depositional landforms (sinks). It also involves sediment movements by the wind, tidal currents, storm activity and onshore/offshore movements.

Segregation: The separation of a group from other groups, either by force or through the choice of that group.

Sink estates: Council housing estates in Britain that score badly on the Index of Multiple Deprivation.

Slab pull: At a subduction zone the descending part of the oceanic tectonic plate pulls the rest of the plate with it.

Slow onset: A hazard that happens very slowly with plenty of evidence and warning, such as a drought.

Social clustering: Groups of people frequently living close to people of a similar background to feel more comfortable in their daily lives. This may be voluntary, or partly forced by economic factors such as poverty and affordability of housing. Sometimes known as polarisation.

Social exclusion: The inability of a group of people to become involved in the cultural activities of a place. This may be because the group are very poor or an ethnic minority.

Socio-economic impacts: The effects on people (social) and businesses and employment (economic) of an event or process.

Solifluction: The downslope movement of the saturated active layer under the influence of gravity.

Sorting: Refers to the range of clast sizes in a sediment. A deposit comprising a mixture of clast sizes is poorly sorted, while one comprising mostly clasts of the same size is well sorted.

Spatial predictability: The area or place where a hazard or event may take place with some certainty.

Speed of onset: How fast a natural hazard may occur. Also see rapid onset and slow onset.

Spiral of decline: An ongoing series of problems in an area, where one problem can lead to others which in turn reinforce the original problem, e.g. Middlesbrough.

Sporadic permafrost: Occurs where the mean annual temperature is only just below freezing and permafrost patches cover less than 50 per cent of the landscape.

Stakeholders: An individual or group of people who have an interest in the outcome of decisions made to change urban or rural areas, e.g. the Aylesbury Society.

Stratification: The formation of layers, e.g. in fluvioglacial deposits.

Subaerial processes: The combination of weathering processes and mass movements that are affecting coastal land above the level of the sea (literally 'under the air'). However, weathering can also be found on the foreshore and backshore, such as on a wave-cut (shore) platform.

Subduction: The process involving the descent of an oceanic plate into the upper mantle beneath a continental plate.

Sub-glacial: The environment beneath the glacier ice, subject to immense pressure from the overlying weight of ice. Beneath temperate glaciers there may be large volumes of meltwater.

Sublimation: Occurs when a substance changes state from a solid to a gas without passing through the liquid phase.

Submergent coast: A coast that is sinking relative to the sea level of the time. This may be due to sea-level rise, or isostatic and tectonic sinking.

Supra-glacial: On the surface of the glacier.

Tariffs: Taxes applied to goods when they are traded across international borders.

Tectonic resistant design: The construction and architecture of buildings to make them stronger in the event of an earthquake, tsunami or volcanic eruption, so that they are less likely to collapse. Those designed to withstand earthquake waves are known as aseismic.

Temperate glacier: Temperate or warm-based glaciers occur in alpine glacial environments, such as the European Alps, Norway and New Zealand. The temperature of the ice is often close to zero and mild summer temperatures cause much melting. They move faster than polar glaciers.

Terrestrial processes: The natural physical processes that operate on the land, mainly rivers, but also ice and the wind.

Till: Deposits of angular rock fragments (clasts) in a finer matrix (rock flour and clay), it is unstratified (not layered), unsorted (mixture of clast sizes) and contains erratics (rocks from several sources).

Till plain: A large, relatively flat plain of till in a lowland area (also called ground moraine).

Trade bloc: A type of intergovernmental agreement, often as part of a regional intergovernmental organisation, such as the EU, where regional barriers to trade (tariffs and non-tariff barriers) are reduced or eliminated among the participating states.

Transform fault plate boundary: Where two plates meet at a major fault in the Earth's crust, usually with movement in contrasting directions, creating friction, strain/stress and earthquakes. Usually the fault system is complicated and not just one fault.

Transnational corporation (TNC): A very large business with factories and other operations in more than one country around the world.

Truncated spur: Steep valley sides formed when the gently sloping interlocking spurs of a river valley have been cut off by glacial erosion.

Tsunami: A sequence of huge waves created by a large displacement of seawater, usually by an undersea earthquake.

Tundra: The treeless vegetation of dwarf shrubs, sedges and grasses, mosses and lichens found in periglacial environments.

Urbanisation: An increase in the proportion of a population that lives in urban areas, a result of rural to urban migration which causes growth of urban settlements.

Valley glacier: Moving body of ice within a valley, sometimes described as a tongue. Glaciers are powerful erosion agents, shaping the landscape in both polar and alpine environments

VEI (Volcanic Explosivity Index): The scale used to measure the magnitude (energy release) of a volcanic eruption.

Vulnerability: The weaknesses of people in a situation where they are exposed to risk. This is a major factor when considering the impacts of a tectonic hazard on people.

Wadati-Benioff zone: An alternative name for the Benioff zone, recognising the contribution of the two main scientists who initially investigated this part of a subducting tectonic plate. See **Benioff zone**.

Water column displacement: The movement of a volume of seawater above the point at which the seabed was moved up or down by an earthquake event, such as a thrust.

Water cycle: The continuous, cyclical processes which circulate water between the Earth's oceans, atmosphere and land.

Index

Acknowledgements

Picture credits

The authors and publisher would like to thank the following individuals and organisations for permission to reproduce material in this product.

Photographs

(Key: b-bottom; c-centre; l-left; r-right; t-top)

Alamy Images: Adrian Wojcik 31, Archimage 253, Construction Photography 230b, David Morgan 144, Steven Dusk 286, EPA 116-117, David Gartland 262r, Ian Dagnall Commercial Collection 259, JLBvdWOLF 258, Bjanka Kadic 237, Justin Kase zsixz 17br, Loop Images Ltd 157, 254, Wayne Perry 262l, Stan Pritchard 230t, Sebastian Remme 249t, Jack Sullivan 222–223, 228, Homer W Sykes 249b; **Tom Chance:** 310; **©Crown copyright, 2013:** UKTI inward Investment Report 2014 / 15 – Published June 2015 by UK Trade & Investment URN: UKTI / 15 / 43 17bc (left); **EUMETSAT:** copyright 2010 EUMETSAT 36b; **Flickr:** kloniwotski / Flickr (CC BY-SA 2.0) 172-173; **Fotolia.com:** Dmitry Pichugin 17cr, Surangaw 266; **Getty Images:** Bloomberg 17bc (right), ChinaFotoPress 206, Ann E. Yow-Dyson 211, Jonas Gratzer 151, LOCOG 242, Loop Images 162, Aurelien Meunier 208, Arlan Naeg 17tl, Dana Stephenson 17cl, VisitBritain / Jason Hawkes 314; **GSHAP:** (UN / IDNDR) 43; **Mamdouh M. El-Hattab:** 145t, 145b; **John Frost Historical Newspapers:** 229; **Rick Lawrence/Samskara Design:** 231; **Lindsay Frost:** 17bl, 18–19, 37, 143, 272–273, 272–273/b, 296, 302, 306, 17bl, 18–19, 37, 143, 272–273, 272–273/b, 296, 302, 306; **Mukta Dinwiddie Maclaren Architects:** 163; **NASA:** 17tr; **National Geophysical Data Center:** K. Jackson. U.S. Air Force 72; **NOAA; :** 33; **North News & Pictures:** Paul Kingston 239t, 239b; **The Oxford Science Park:** 252; **Pearson Education Ltd:** John Pallister 107; **Press Association Images:** Martin Rickett 285; **Reuters:** Erik de Castro 149, Kyodo 32, Sigit Pamungkas 36t; **Science Photo Library Ltd:** Sputnik 105; **Shutterstock.com:** 129, alice-photo 251, Konstantin Kopachinsky 76, Mikadun 68-69, Norbert A 109, Khoroshunova Olga 75, 80, Pecold 265, Grigorii Pisotsckii 104; **Worldmapper.org:** Copyright Benjamin D. Hennig (Worldmapper Project) 47; **www.imagesource.com:** 101

Cover images: *Front:* **Getty Images:** Poorfish

All other images © Pearson Education

Figures

Figure on page 67 from Unpublished map by Kathleen Compton: Iceland showing uplift rate; data from: Compton, K., Bennett, R. A. and Hreinsdóttir, S. (2015), Climate-driven vertical acceleration of Icelandic crust measured by continuous GPS geodesy. Geophys. Res. Lett., 42: 743–750. doi: 10.1002/2014GL062446; Figure 1.25 from Map of Megacities, http://www.megacities.uni-koeln.de/documentation/, By kind permission of Prof. Dr. Franke Kraas; Figure 1.26 from Henning, B.D. Map: Global earthquake intensity and population distribution, http://www.viewsoftheworld.net/wp-content/uploads/2011/03/EarthquakeDensityMap.png, © 2011 Benjamin D. Hennig; Figure 1.27 after *Geophysical loss events worldwide 1980 – 2014*, Munich RE (NatCatSERVICE 2015) p.5, Number of events, © 2016 Münchener Rückversicherungs-Gesellschaft, NatCatSERVICE ; Figures 1.27, 1.28 after EM-DAT: The OFDA/CRED International Disaster Database – www.emdat.be – Universite Catholique de Louvain, Brussels – Belgium; Figure 1.30 adapted from *Environmental Hazards*, Macmillan (Park, C.C. 1991), by kind permission of Dr Chris Park; Figure 1.31 after The Disaster Risk Management Cycle, TorqAid. Please note that the latest versions of the Disaster Risk Management Cycle (DRMC), together with other related illustrations, can be found on the TorqAid – www.torqaid.com website.; Figure 1.32 after Lava flow hazard at Mount Etna (Italy): New data from a GIS-based study, *Special Paper of The Geological Society of America*, 396, Figure 6A (Behncke, B., Neri, M. and Nagay, A. 2005), with permission from the Geological Society of America; Figure 2.2 from *What's After the Day After Tomorrow? A science perspective on the science fiction movie*, Woods Hole Oceanographic Institution (Cook, J. 2004) https://www.whoi.edu/, © Woods Hole Oceanographic Institution https://www.whoi.edu/; Figure 2.3 adapted from *Edexcel A2 Geography: Student Book*, Pearson Education (Byrne, P. et al 2009) p.186, © Peter Byrne, Viv Pointon, Steph Warren and Nigel Yates,2008, Reproduced by permission of Pearson Education Ltd; Figure 2.6 from Yearly Averaged Sunspot Numbers 1610 – 2010, http://www.nasa.gov/mission_pages/sunearth/news/solarcycle-primer.html#.VnWxyvmLTIU NASA/MSFC; Figure 2.11 from Observed Distribution of Permafrost Types, http://www.grida.no/graphicslib/detail/observed-distribution-of-permafrost-types_1637#, Riccardo Pravettoni, UNEP/GRID-Arendal; Figure 2.12 adapted from *Fundamentals of Physical Geography, 2nd Edition* (Pidwirny, M. 2006) Figure 10ag-2, http://www.physicalgeography.net/fundamentals/10ag.html, by kind permission for Professor M. Pidwirny; Figure 2.16 from *Climate Change 2007: The Physical Science Basis. Working Group I Contribution to the Fourth Assessment Report of the Intergovernmental Panel on Climate Change*, Cambridge University Press (Solomon, S., D. Qin, M. Manning, Z. Chen, M. Marquis, K.B. Averyt, M. Tignor and H.L. Miller (eds.) 2007) Figure 4.15 p.359; Figure 2.23 adapted from *Landscapes Resource Pack*, Learning and Teaching Scotland © Crown copyright 2012, Contains public sector information licensed under the Open Government Licence (OGL) v3.0.http://www.nationalarchives.gov.uk/doc/open-government-licence.; Figures 2.24, 2.37 from Landranger Map 115 Snowdon/Yr Wyddfa, Ordnance Survey, © copyright 2016 OS 100030901; Figure 2.28 from Explorer Map 342 Glasgow, Ordnance Survey, © copyright 2016 OS 100030901; Figure 2.33 from Greenland Ice Sheet, *Arctic Report Card*, 2015, Fig. 3.4 (Tedesco, M. et al.), http://www.arctic.noaa.gov/reportcard/greenland_ice_sheet.html; Figure 2.40 from The Alps, as defined for application of the Alpine Convention (2010), http://www.eea.europa.eu/data-and-maps/figures/the-alps-as-defined-for, © European Environment Agency, Alpine Convention 2010; Figure 3.5 from *Geomorphology and Hydrology*, Longman (Small, R.J. 1989) 0582355893, Reproduced by permission of Pearson Education Ltd; Figure 3.6 adapted from *Geology Explained in South and East Devon*, David & Charles (Perkins, J.W, 1971) Fig.71, p.152, by kind permission of F+W International; Figure 3.19 from Landranger map 113 Grimsby, Ordnance Survey, © copyright 2016 OS 100030901; Figure 3.20 after A 150-year record of coastal dynamics within a sediment cell: Eastern England, *Geomorphology*, 179, pp.168–185 (Montreuil A and Bullard J. 2012); Figure 3.22 from Holocene Sea Level, http://www.globalwarmingart.com/wiki/File:Holocene_Sea_Level_png, by kind permission of Robert Rohde; Figure 3.23 from Holocene land- and sea-level changes in Great Britain, *Journal of Quaternary Science*, Vol.17 (5-6), pp.511–526 (Shennan, I. and Horton, B. 2002), Copyright © 2002 John Wiley & Sons, Ltd.; Figure 3.26a from Improving Coastal Vulnerability Index of the Nile Delta Coastal Zone, Egypt, *Earth Science and Climate Change*, Vol.6(293), p.7 (El-Hattab M.M. 2015), doi:10.4172/2157-7617.1000293, Copyright © 2015 El-Hattab M.M.; Figure 3.26b from Improving Coastal Vulnerability Index of the Nile Delta Coastal Zone, Egypt, *Earth Science and Climate Change*, Vol.6(293), p.4 (El-Hattab M.M. 2015), doi:10.4172/2157-7617.1000293, Copyright © 2015 El-Hattab MM; Figure 3.28 from Kiribati composite Seal Level 1949 to March 2016, https://eyesonbrowne.wordpress.com/tag/sea-level-rise/, EyesOnBrowne – Kiribati Sea levels, R. Browne, 16 Oct 2013; Figure 3.31 from *Counting the Costs: Climate Change and Coastal Flooding* The Climate Council of Australia (Steffen W., Hunter J. and Hughes L. 2014) Figure 10, p.22, with permission from Climate Council of

Australia; Figure 4.4 from Topic 5: Bridging the Development Gap, *6GEO3 Unit 3 Contested Planet*, Slide 9 (Edexcel); Figure 4.5 from OpenData product 1:250,000 Scale Colour Raster Map extract centred on NO 02954 96044 or between that grid reference and NO 14579 87836 or NH 90257 13577, Contains OS data © Crown copyright and database right 2016; Figure 4.6 from *The World Economy: Historical Statistics*, Development Centre Studies, OECD Publishing, Paris (Maddison, A. 2003) p.73, DOI: http://dx.doi.org/10.1787/9789264104143-en, Reproduced with permission of the OECD; Figure 4.9 from Topic 5: Bridging the Development Gap, *6GEO3 Unit 3 Contested Planet*, Slide 2 (Edexcel); Figure 4.14 after ONS, Contains public sector information licensed under the Open Government Licence (OGL) v3.0.http://www.nationalarchives.gov.uk/doc/open-government-licence.; Figure 5.3 adapted from Wages throughout the country: how does your area compare?, *The Guardian*, 24 November 2011 (Ball, J.), http://www.theguardian.com/news/datablog/2011/nov/24/wages-britain-ashe-mapped, Copyright Guardian News & Media Ltd 2016; Figure 5.4 from *Identity in Britain*, Policy Press (Thomas, B. and Dorling, D. 2007) Figure 7.22, p.249, ISBN 9781861348203, © Bethan Thomas and Daniel Dorling 2007. Reproduced by permission of Policy Press; Figure 5.10 adapted from *Socio-economic Models in Geography*, Methuen (Chorley, R.H & Haggett, P. 1967) p.258, reprinted by permission of the publisher Taylor & Francis Ltd Books UK; Figure 5.13 after ONS, Office for National Statistics licensed under the Open Government Licence v.3.0.; Figure 5.14 after *Official Statistics: English Indices of Deprivation 2015* (Department for Communities and Local Government 2015), Contains public sector information licensed under the Open Government Licence (OGL) v3.0.http://www.nationalarchives.gov.uk/doc/open-government-licence.; Figure 5.15 from *Edexcel AS Geography*, Pearson Education Ltd (Byrne, P. et al 2008) p.204, 1846903211, © Peter Byrne, Viv Pointon, Steph Warren and Nigel Yates,2008, Reproduced by permission of Pearson Education Ltd; Figure 5.18 from *Audit of Political Engagement 12, The 2015 Report* (The Hansard Society 2015) p.19 © Hansard Society 2015; Figure 5.23 from Fracking push gets go-ahead across UK as ministers tighten safeguards, *The Guardian*, 28/07/2014 (Mason, R.), Copyright Guardian News & Media Ltd 2016; Figure 6.2 adapted from Compendium of UK Statistics, The Neighbourhood Statistics Service http://www.neighbourhood.statistics.gov.uk/HTMLDocs/dvc134_c/index.html, Office for National Statistics licensed under the Open Government Licence v.3.0.; Figure 6.3 from *Overview of the UK Population*, 5 November (ONS 2015) Fig.1, p.5, Office for National Statistics licensed under the Open Government Licence v.3.0.; Figure 6.4 from *Overview of the UK Population*, 5 November (ONS 2015) Fig.2, p.6, Office for National Statistics licensed under the Open Government Licence v.3.0.; Figure 6.8 adapted from Office for National Statistics, www.neighbourhood.statsitics.gov.uk/HTMLDocs/dvc183/#20/180/202/null/null/true/false/na/1, Office for National Statistics licensed under the Open Government Licence v.3.0.; Figure 6.9 from Portrait of East Anglia, *Regional Trends*, 41, Map 5.2 (Corke, S. and Wood, J. 2009), Office for National Statistics licensed under the Open Government Licence v.3.0.; Figure 6.14 from *Statistical Bullletin: Personal Well-being in the UK, 2014/15*, 23 September 2015, ONS Map 2, Office for National Statistics licensed under the Open Government Licence v.3.0.; Figure 6.15 from *Neighbourhood Impacts: A longitudinal research study into the impact of the Liverpool European Capital of Culture on local residents*, Impacts 08 (Melville, R., Rodenhurst, K., Campbell, P. and Morgan, B. 2010) Fig.22 and Fig.24; Figure 6.18 from *Our Next Generation: Young people in Caithness and Sutherland: Attitudes and Aspirations: Research Report: September 2015*, Highlands and Islands Enterprise (2015) p.31; Figure 6.20 from Explorer map 162: Greenwich and Gravesend, Ordnance Survey, © copyright 2016 OS 100030901; Figure 6.24 after Superfast broadband availability forecast 2015, http://point-topic.com/wp-content/uploads/2013/08/Point-Topic-UK-superfast-broadband-availability-forecast-2015.png, Copyright © 2013 by Point Topic, reproduced by permission.; Figure 6.26 from House Price Index: October 2015, http://www.ons.gov.uk/economy/inflationandpriceindices/bulletins/housepriceindex/october2015, Office for National Statistics licensed under the Open Government Licence v.3.0.; Figure 6.28 from Landranger map 200 Newquay & Bodmin, Ordnance Survey, © copyright 2016 OS 100030901; Figure 6.29 from *Street map of Norwich*, G. I. Barnett & Son Limited © 2005 G. I. Barnett & Son Limited

Tables

Table 1.5 based on data from Munich Re and USGS; Table 1.8 based on information in Degg M. and Homan J. (2005) Earthquake Vulnerability in the Middle East pp 54 to 66 in Geography (Spring 2005) Volume 90 (1); Table 1.9 based on information from WorldRiskReport 2015, pub. Bündnis Entwicklung Hilft (Alliance Development Works), and United Nations University – Institute for Environment and Human Security (UNU-EHS); Table 1.10 based on data from NOAA; Table 1.11 based on data from UN and Lloyd's City Risk Index and Cambridge University: Centre for Risk Studies; Table 1.12 on data from Niederlaender, E. (2006) Statistics in focus – Population and Social Conditions 10/2006 Health, Causes of death in the EU, pub. Eurostat; Table 1.13 from D. Guha-Sapir, R. Below, Ph. Hoyois – EM-DAT: The CRED/OFDA International Disaster Database – www.emdat.be – Université Catholique de Louvain – Brussels – Belgium.; Table on page 171 adapted from New Analysis Shows Global Exposure to Sea Level Rise, *Climate Central* (Strauss, B. & Kulp, S.), http://sealevel.climatecentral.org; Table 4.4 based on data from Statistical bulletin: UK Trade: September 2015 (ONS 2015); Table 5.7 contains Parliamentary information licensed under the Open Parliament Licence v3.0; Table 6.1 adapted from Highland profile – key facts and figures, http://www.highland.gov.uk/info/695/council_information_performance_and_statistics/165/highland_profile_-_key_facts_and_figures/2; Table 6.3 adapted from Merseyside Police prepared by Liverpool City Council, Citysafe Team, 2015 http://liverpool.gov.uk/council/key-statistics-and-data/data/crime/, Contains public sector information licensed under the Open Government Licence v3.0; Table 6.5 adapted from Office for National Statistics, www.neighbourhood.statistics.gov.uk/HTMLDocs/dvc25/migrationmatrices.2014.xls, Office for National Statistics licensed under the Open Government Licence v.3.0.

Text

Extract on page 66 from Melting Ice Spells Volcanic Trouble, *New Scientist* (Pearce, F.), https://www.newscientist.com/article/dn26923-melting-ice-spells-volcanic-trouble/, © 2015 Reed Business Information – UK. All rights reserved. Distributed by Tribune Content Agency; Extract on page 66 from Melting Ice Spells Volcanic Trouble, *New Scientist*, https://www.newscientist.com/article/mg22530081-300-melting-ice-spells-volcanic-trouble/, © 2015 Reed Business Information – UK. All rights reserved. Distributed by Tribune Content Agency; Extract on page 155 adapted from Has the great climate change migration already begun?, *The Guardian*, 15/09/2014 (Harman, G.), Copyright Guardian News & Media Ltd 2016; Extract on page 170 adapted from China tops new list of countries most at risk from coastal flooding, *CarbonBrief* (McSweeney, R. 2014), http://www.carbonbrief.org/; Extract on page 220 adapted from Are creative people the key to city regeneration? by Edwin Heathcote, FT.com, 19 May 2014, © The Financial Times Limited. All Rights Reserved.; Newspaper Headline on page 245 from Middlesbrough really is the worst place to live: Damning verdict of Location, Location, Location, *The Daily Mail*, 04/08/2009 (Revoir, P.); Newspaper Headline on page 245 from Is Middlesbrough the worst place to live in the UK?, *The Guardian*, 05/08/2009 (Wainwright, M.), Copyright Guardian News & Media Ltd 2016; Newspaper Headline on page 245 from What can Middlesbrough teach Britain?, *The Telegraph*, 22/09/2010 (Corrigan, T.); Extract on pages 252–253 from *The Role of Historic Buildings in Urban Regeneration, Eleventh Report of Session 2003-04*, House of Commons (ODPM: Housing, Planning, Local Government and the Regions Committee 2004), Contains public sector information licensed under the Open Government Licence (OGL) v3.0.http://www.nationalarchives.gov.uk/doc/open-government-licence.